Propositional Logics

The Semantic Foundations of Logic

Propositional Logics

2nd Edition

Richard L. Epstein

with the assistance and collaboration of

Walter A. Carnielli
Itala M. L. D'Ottaviano
Stanisław Krajewski
Roger D. Maddux

New York Oxford
Oxford University Press
1995

Oxford University Press

Oxford New York Toronto
Delhi Bombay Calcutta Madras Karachi
Kuala Lumpur Singapore Hong Kong Tokyo
Nairobi Dar es Salaam Cape Town
Melbourne Auckland Madrid

and associated companies in
Berlin Ibadan

Published by Oxford University Press, Inc.,
198 Madison Avenue, New York, New York 10016

Library of Congress Cataloging-in-Publication Data
Epstein, Richard L., 1947–
The Semantic foundations of logic /
Richard L. Epstein.
p. cm. Includes bibliographical references and index.
Contents: v. 1. Propositional Logics, 2/e;
v.2. Predicate Logic
ISBN 0-19-508761-5
1. Logic. 2. Logic, Symbolic and mathematical.
3. Semantics. I. Title.
BC71.E57 1995 160—dc20 94-13917

9 8 7 6 5 4 3 2 1

Printed in the United States of America
on acid-free paper

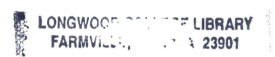

Dedicated to

Peter Eggenberger
Harold Mann
and
Benson Mates

with gratitude for their encouragement and guidance

Preface

This book grew out of my confusion. If logic is objective, how can there be so many logics? Is there one right logic, or many right ones? Is there some underlying unity that connects them? What is the significance of the mathematical theorems about logic that I've learned if they have no connection to our everyday reasoning?

The answers I propose revolve around the idea that what one pays attention to in reasoning determines which logic is appropriate. The act of abstracting from our reasoning in our usual language is the stepping-stone from reasoned argument to logic. We cannot take this step alone, for we reason together: logic is reasoning that has some objective value.

For you to understand my answers—or perhaps better, conjectures—I have retraced my steps: from the concrete to the abstract, from examples, to general theory, to further confirming examples, to reflections on the significance of the work. In doing so I have had to begin at the beginning: What is logic? What is a proposition? What is a connective? If much seems too well known to be of interest, then plunge ahead: the chapters can be read more or less independently (which explains the occasional repetitions); the introduction is a tour guide for the more experienced traveller. But the path I have chosen is not always the most familiar. At places I think I have found an easier way, though because it is new, or perhaps simply because I may not know it well, the way may seem more difficult.

I want to see where this path leads, whether the path, in the end, seems smoother and takes us to higher ground with a better view. So I may not always stop to argue each principle at length. It is the trip as a whole I hope you will find refreshing.

> In the discussions of the wise there is found unrolling and rolling up, convincing and conceding; agreements and disagreements are reached. And in all that the wise suffer no disturbance.
>
> Nagasena

Come, let us reason together.

Preface to the Second Edition

When I was asked to publish the second volume of my series *The Semantic Foundations of Logic* with Oxford University Press, I suggested we do a new edition of *Propositional Logics,* too. There were a few corrections that colleagues had pointed out to me, and I thought I could clean up the text a bit.

I was wrong. There were a lot of corrections that were needed, both to the technical work and the exposition. I have revised the entire text, with more changes than I could easily list here. Among the most significant changes are the correction or simplification of many axiomatizations, the addition of examples of formalization of ordinary reasoning, and the addition of exercises to make the text more suitable for individual or classroom use. Still, the numbering of the chapters and the bulk of the material remain the same.

I want to thank P. H. Rodenburg and Piergiorgio Odifreddi for pointing out errors and suggesting improvements. And I am grateful to Arnon Avron for his close reading of the text, along with suggestions for corrections, and the correspondence we developed about the nature of the presentation. The late George Hughes was, as always, a great source of comment and encouragement. David Isles used revised versions of Chapters I–III and V in his class, and his comments have been useful. Walter Carnielli helped me with the research on valid deductions in Chapter IV.G.8, which appeared previously in *Reports on Mathematical Logic, 26,* and is reprinted here with permission.

And again, Benson Mates and Peter Eggenberger, with their comments and encouragement, made me feel part of the long tradition and community of those who study of logic.

Again, to all these and any others I may have inadvertently forgotten, I am most grateful.

There is no end, but only a continual beginning . . .

Acknowledgements

The story of this book began in Wellington, New Zealand. Working with the logicians there in 1977, Douglas Walton and I developed relatedness logic.

In 1978 I met Niels Egmont Christensen, who led me to see that a slight variation in the work on relatedness logic could model his ideas on analytic implication. Later that year at Iowa State University I began to question what was the "right" logic. From then until I left I.S.U. Roger Maddux challenged me and helped me to clarify my intuitions technically. He and Donald Pigozzi introduced me to nonclassical logics and the algebras of them. In 1980, Roger Maddux, Douglas Walton, and I wrote a monograph that contained the basis of much of the technical work of Chapters V–VIII. In 1980 and 1982 I gave lectures on propositional and predicate logics at I.S.U. where Howard Blair, William Robinson, and later Gary Iseminger challenged me to explain my philosophical assumptions. I am grateful that while at Iowa State University I was given ample time for research, due particularly to Dan Zaffarano, Dean of Research.

In 1981 I visited the University of Warsaw on an exchange sponsored by the U.S. National Academy of Sciences and the Polish Academy of Sciences. There I met and began collaborating with Stanisław Krajewski, whose insights led me to clarify the relationship between formal languages and the languages we speak. Part of our joint work is the chapter on translations between logics, which was influenced by discussions with L. Szczerba.

In 1982 I moved to Berkeley where it was my good fortune to meet Peter Eggenberger and Benson Mates. They are fine teachers: listening to my early inchoate ideas, reading my confused analyses, guiding my reading in philosophy, they have helped me understand what I was trying to say. The shape of many of the discussions in this volume comes from conversations I had with them.

In 1984 I lectured to a group of Brazilians at Berkeley on what was still a series of separate papers. It was through the urgings of Walter Carnielli that I then made the decision to turn those papers into a book and, finally, to publish the work on propositional logics as a separate volume. Much of the form and outline of this volume was developed in discussions with him. He, Newton da Costa, and Itala D'Ottaviano read versions of several of the chapters, and later in 1986 Karl Henderscheid read a draft of the entire volume. Their questions and criticisms substantially improved the exposition.

In 1985 the Fundação de Amparo à Pesquisa do Estado do São Paulo provided me with a grant (number 84/1963–2) to visit Brazil and lecture at the VII Latin American Symposium on Mathematical Logic. That lecture was published in the proceedings of the conference as *Epstein, 1988,* and parts of it are reprinted in Chapter IV with permission of the American Mathematical Society.

In 1987 the Fulbright Foundation awarded me a fellowship to lecture and do research at the Center for Logic, Epistemology and History of Science at the University of Campinas and at the Universidade Federal da Paraíba in Brazil. Those visits gave me an opportunity to collaborate with Walter Carnielli and Itala D'Ottaviano and resulted in Chapters IV.G and IX. Part of Chapter IX appeared in *Reports on Mathematical Logic, 22* and is reprinted here with permission.

In 1987 I also visited the University of Auckland and met Stanisław Surma who gave me many useful suggestions for the book. Later David Gross helped me with the presentation of Chapter I.

To all these people, and any others I have inadvertently forgotten, I am most grateful. Much that is good in this book is due to them; the mistakes and confusions are mine alone. It is with great pleasure I thank them here.

> In spite of everything, a man was given a chance to get a little peace. He allotted himself a task, and, while performing it, realized that it was meaningless, that it was lost among a mass of human endeavors and strivings. But when a pen hung in air and there was a problem of interpretation or syntax to solve, all those who once, long ago had applied thought and used language were near us. You touched the delicate tracings warmed by their breath, and communion with them brought peace. Who could be so conceited as to be quite sure that he knew which actions were linked up and complementary; and which would recede into futility and be forgotten, forming no part of the common heritage? But was it not better, instead, to ponder the only important question: how a man could preserve himself from the taint of sadness and indifference.
>
> Czeslaw Milosz

And I am grateful to Harold Mann, who helped me ponder that question.

Contents

III Relatedness Logic: The Subject Matter of a Proposition – S and R –

VI Modal Logics
– S4, S5, S4Grz, T, B, K, QT, MSI, ML, G, G* –

VIII Many-Valued Logics
– $\mathbf{L_3}$, $\mathbf{L_n}$, $\mathbf{L_{\aleph}}$, $\mathbf{K_3}$, $\mathbf{G_3}$, $\mathbf{G_n}$, $\mathbf{G_{\aleph}}$, **S5** –

The Semantic Foundations of Logic

Contents of other Volumes

Predicate Logic (Oxford University Press, 1994)

Classical Logic (projected)

A General Framework For Semantics for Predicate Logics (projected)

Introduction

Why are there so many logics? Is there no one right way to reason, no one notion of necessity, of objectivity?

There seems to be a unity to the ways we reason, a structural and conceptual unity, based on common semantic assumptions. In this introduction I will describe that unity in general terms and outline the contents of this book. In doing so I will use a few technical terms of logic, so someone new to the study of logic may prefer to proceed directly to the text.

The proper place to begin is with classical propositional logic. There are good reasons why it is so widely accepted. Something is right about it. It's not the whole story, but it is a fundamental part, a standard of reference for all other logics.

Classical logic is fundamental because it is the simplest symbolic model of reasoning we can devise, once certain assumptions are made about the nature of logic. In classical logic a proposition is abstracted to its truth-value and its form only, relative to the propositional connectives. I set out those assumptions in Chapter I, and develop classical logic in Chapter II.

Something like the division of propositions into the true and the false, which is basic to classical logic, seems basic to all reasoning. Every logician, in the end, divides propositions into those that are acceptable and those that are not. I argue throughout that it is correct in each case to understand these divisions as the division into the true propositions and the false. We seem to have no direct access to the world, but only our uncertain perceptions. Therefore, to call a sentence true seems at best a hypothesis we hope to share with others, and a realistic humility demands that we see the distinction between logical and pragmatic grounds for rejecting a proposition as a matter of degree, not of kind.

Moreover, even were we to agree that there is something in the world that objectively determines whether 'Ralph is a dog' is true or false, it seems to me that there is nothing in the world external to us that can determine in the same way the truth-value of 'If Ralph is a dog, then George is a duck.' The truth-value of this sentence depends not only on whether Ralph is a dog and whether George is a duck, but on how we are interpret 'if . . . then . . .'. For most of us, for most logics, the truth-value of an 'if . . . then . . .' sentence depends on more than just the truth-values of the antecedent and consequent. Modalities, tenses, accessibility to understanding, constructive mathematical content, or subject matters may enter in, and logics, including logics that seem to reject the true-false dichotomy of propositions, have been based on all these aspects, as I show in Chapters III and V–IX.

A general form of semantics arises from the view that all these logics comprise a spectrum. Each, except for classical logic, incorporates into the semantics some aspect of propositions other than truth-value and form. Each logic analyzes an 'if . . . then . . .' proposition classically if the aspects of antecedent and consequent are appropriately connected, while rejecting the proposition otherwise. As we vary the aspect, we vary the logic. This overview is presented in Chapter IV. Mathematicians may prefer to begin with Chapter IV.G if they wish to confine their studies to technical and structural questions.

To show that this general form of semantics unifies many logics, both structurally and conceptually, I present chapters on analytic implication, modal logics, intuitionistic logic, many-valued logics, and a paraconsistent logic. In each, I first present an introduction to the assumptions of the logic, along with the standard semantics in terms that I hope are a reasonably accurate reflection of how the logic is commonly understood. Then I show how the logic can be understood as part of the overview, giving semantics within the general framework that I believe conform to and reflect the intuitions of the original practitioners of the logic. But I do not feel I have to show that my reading of intuitionistic logic, for example, is in absolute accord with the intuitionists' understanding in order to justify it, only that it is a way to grasp their logical analyses strongly enough to give a projective knowledge of their work, based on well-motivated semantic assumptions.

This general form of semantics is not intended to replace other semantics. For example, under certain assumptions possible-world semantics are a good explanation of the ideas of modal logics. But providing uniform semantics that are in reasonable conformity with the ideas on which various logics are based allows for comparisons and gives us a uniform way in which to approach the sometimes overwhelming multiplicity of logics.

It is equally important that this general form of semantics provides a simple tool for incorporating into logic various aspects of propositions that previously have been treated informally. We now have a framework in which to formalize, discuss, and compare different notions as they affect our reasoning. For example, in Chapter V I present a case study of how to use the general framework to develop a logic that incorporates a notion of referential content of propositions.

The referential content of a proposition is not, however, a primitive notion. It depends on the referential content of the predicates and names of which the proposition is composed. The internal structure of propositions matters, and predicate logic is a further test of the aptness of these ideas. A general form of semantics for predicate logics is based on the idea that when predicates and names are primitive, it is the aspects of these in addition to their extensions that determine the truth-value of a proposition. Many more assumptions about reasoning and language must be made in order to establish a logic that deals with the internal structure of propositions, and this I have attempted in *Predicate Logics* (see the Table of Contents).

The semantic framework I set out in Chapter IV is a very weak general form of logic, a general form that becomes usable only upon the choice of which aspect of propositions we deem to be significant. But then is logic relative to the logician? Or does a notion of necessary truth lie in this very general framework? It seems to me that it is our agreements about how we will reason together that determine our notion of objectivity. I do not mean by this only active, explict agreements, but also implicit ones, what may be said to be our common background. Most of our agreements are implicit and not necessarily freely made: lack of disagreement I understand as agreement. I discuss this in Chapters I and XI.

Throughout I have tried to find and make explict fundamental assumptions or agreements on which our reasoning and logic are based. I have repeated the statement of certain of these assumptions in different places, partly because I want the chapters to be as self-contained as possible, but also because it is important to see these assumptions and agreements in different contexts and applied differently to be able to grasp their plausibility and pervasiveness.

What I am doing here could be seen as founding logic in natural language and reasoning, a basis as fundamental to mathematical logic as physics is to the study of differential equations. When nonconstructive assumptions are used to do either the mathematics of logic, that is, prove theorems about our formalizations, or to apply logic to a particular subject matter, such as arithmetic or geometry, we can see precisely where they are needed. Those assumptions I treat as abstractions from experience, for that is how I understand abstract things. However, they need not be viewed that way, and I have attempted to provide alternate readings of the technical work based on the assumption that intangible, nonsensible abstract things are as real or more real than the objects we daily encounter. Most of the discussion of these matters is in Chapter I and in the development of classical logic in Chapter II, particularly Section II.G. In Chapter IV I point out specific nonconstructive, infinitistic abstractions of the semantics that we usually make in pursuing metalogical investigations.

In Chapter II I also present a Hilbert-style formalization of the notion of proof and syntactic deduction, which I use throughout the book. The metalogical investigations that I concentrate on generally concern the relation between the semantic and syntactic notions of consequence, and whether or how those can be represented in terms of theorems or valid formulas by means of a deduction theorem.

The many examples of logics presented in this book allow us to consider the extent to which one logic or way of seeing the world can be reduced to another by a translation. In Chapter X I present a general theory of translations, developed in collaboration with Stanisław Krajewski.

There are many important subjects in the study of propositional logics that I do not deal with here. I have not discussed the algebraic analyses of propositional logics; for that you can consult *Rasiowa, 1974* and *Blok and Pigozzi, 1989.* I have made no attempt to connect this work with the categorial interpretation of logic, for which you can consult *Goldblatt, 1979.* Nor have I dealt with other approaches to the notion of proof in propositional logics. And there are many propositional logics I have not discussed, quite a few of which are surveyed in *Haack, 1974,* and *Gabbay and Guenthner, 1989,* which also discuss philosophical issues, as well as in *Marciszewski, 1981.*

This is not the story of all propositional logics. But I hope to have done enough to convice you that it is a good story of many logics that brings a kind of unity to them.

Propositional Logics

I The Basic Assumptions of Propositional Logic

A. What is Logic?

Logic is the study of how to reason, how to deduce from hypotheses, how to demonstrate. As presented here, logic is concerned with providing symbolic models of acceptable reasoning.

What do we mean by 'acceptable'? Is logic concerned only with the psychology of how people reason, setting out pragmatic standards? I, or you and I together, can reflect on our rules for reasoning but those cover only very simple cases. We are led, therefore, to formal systems, devised to reflect, model, guide, and/or abstract from our native ability to reason. These formal systems are based on our understanding of certain notions such as truth and reference, and those in turn seem to be dependent on (*i*) how we understand the world and (*ii*) how the world really is.

But is there any difference between (*i*) and (*ii*)? And if so, is it a difference we perceive and can take into account? We must ask these questions in doing logic, for they concern how we will account for objectivity in our work and to what extent we shall see our systems as prescriptive, not just a model of what is done, but what should be done.

B. Propositions

Let us begin by asking what objects, what things we are going to study in logic.

1. Sentences, propositions, and truth

When we argue, when we prove, we do so in a language. And we seem to be able to confine ourselves to *declarative sentences* in our reasoning.

For our purposes here I will assume that what a sentence is and what a declarative sentence is are well enough understood by us to be taken as *primitive,* that is, undefined in terms of any other fundamental notions or concepts. Disagreements about some particular examples may arise and need to be resolved, but our common understanding of what a declarative sentence is will generally suffice.

So we begin with sentences, written (or uttered) concatenations of inscriptions (or sounds). To study these we may ignore certain aspects, such as what color ink they are written in, leaving ourselves only certain features of sentences to consider in reasoning. The most important of these aspects for logic are called *truth* and *falsity.*

I will not try to explain truth and falsity here. In general we understand well enough what it means for a simple sentence such as 'Ralph is a dog' to be taken as true or to be taken as false. For such sentences we can regard truth as a primitive notion, one we understand how to use in most applications, while falsity we can understand as the opposite of truth, the not-true. Our goal, then, is to formalize truth and falsity in more complex and controversial situations, leading us, according to various conceptions of truth, to various formal logics.

Which declarative sentences are true or false, that is, have a *truth-value*? Some, it would seem, are too ambiguous, such as 'I am half-seated', or nonsensical, such as '7 is divisible by lightbulbs'. But if only sentences that are completely objective, precise, and unambiguous are true or false, then 'Strawberries are red' can be neither true nor false: Which strawberries? What hue of red? Measured by what instrument or person? And then we couldn't analyze the following:

(1) If strawberries are red, then some colorblind people cannot see strawberries among their leaves
Strawberries are red
Therefore:
Some colorblind people cannot see strawberries among their leaves

Surely this is an example of acceptable reasoning, reasoning that is important for us to formalize, for this is how we actually reason. And yet, I suspect, any attempt to make the sentences in (1) fully precise will fail. At best we can redefine terms, using others that may be less vague or ambiguous. But always we have to rely on our common understanding. What we need in order to justify our example as acceptable reasoning is that we may treat 'Strawberries are red' and the other two sentences in the example as if they have truth-values, not that they are precise and unambiguous. All declarative sentences, except perhaps those in highly technical work such as mathematics, are in some way imprecise. This imprecision is an essential component of communication, I believe, for no two persons can have exactly the same thoughts or perceptions and hence must understand every linguistic act somewhat differently.

It is sufficient for our purposes in logic to ask whether we can agree that a particular sentence, or class of sentences as in a formal language, is declarative and whether it is appropriate for us to hypothesize a truth-value for it. If we cannot agree that a certain sentence such as 'The King of France is bald' has a truth-value, then we cannot reason together using it. That does not mean that we adopt different logics or that logic is psychological; it only means that we differ on certain cases. The assumption that we agree that a sentence has a truth-value, that the imprecision of the sentence is inessential, is always there, even if not explicit.

But then is truth agreement? The word 'agreement' may be too strong, and 'convention' even worse. Almost all our conventions, agreements, assumptions are implicit, tacit. They needn't be conscious or voluntary. Many of them may be due to physiological, psychological, or perhaps metaphysical reasons: for the most part we shall never know. Agreements are manifested in lack of disagreement and in that we communicate. To be able to see we have made, or been forced into, or simply have an agreement is to be challenged on it. In Chapter XI I will discuss further the notion of agreement and how it relates to an explanation of the objectivity of logic.

My goal in this series of books is to find, or perhaps devise agreements upon which to found logic, agreements sufficiently fundamental and universal to account for not just one logic, but many, perhaps all logics. The agreement with which I begin summarizes our discussion to this point.

Propositions A *proposition* is a written or uttered sentence that is declarative and that we agree to view as being either true or false, but not both.

Again, our agreements need not be explicit. For example, if I say 'Cats are nasty' and you disagree with me, then I know that you consider that sentence to be a proposition, even if we haven't explicitly said that.

From now on I will often say a proposition has a truth-value, since we've agreed to view it as if it does, though we need not agree on which truth-value it has.

But how can I say that this definition is fundamental when many logics have been based on very different conceptions of propositions?

2. Other views of propositions

Consider one such view: what is true or false is not the sentence, but the "meaning" or "thought" expressed by the sentence. Thus 'Ralph is a dog' is not a proposition;

it expresses one, the very same one expressed by 'Ralph is a domestic canine'.

Platonists take this one step further. A *platonist,* as I shall use the term, is someone who believes that there are abstract objects not perceptible to our senses that exist independently of us. Such objects can be perceived by us only through our intellect. The independence and timeless existence of such objects account for objectivity in logic and mathematics. In particular, propositions are abstract objects, and a proposition *is* true or *is* false, though not both, independently of our even knowing of its existence. Thus the following, if uttered at the same time and place, all express or stand for the same abstract proposition:

(2) It is raining
Pada deszcz
Il pleut

It is argued that the word 'true' can only be properly applied to things that cannot be seen, heard, or touched. Sentences are understood to "express" or "represent" or "participate in" such propositions.

Those who take abstract propositions as the basis of logic argue that we cannot answer precisely the questions: What is a sentence? What constitutes a use of a sentence? When has one been used assertively or even put forward for discussion? These questions, they say, can and should be avoided by taking things inflexible, rigid, timeless as propositions. But then we have the no less difficult questions: How do we use logic? What is the relation of these formal theories of mathematical symbols to our arguments, discussions, and search for truth? How can we tell if this utterance is an instance of that abstract proposition? It's not that taking utterances of sentences as propositions raises questions that can be avoided. For example, were we to confine logic to the study of abstract propositions, argument (1) would be defective: the sentences there could not be taken to express propositions because of their lack of precision.

There are still other views of what kind of thing a proposition is; *Williamson, 1968* compares several from a viewpoint similar to mine. Most notably, Frege has taken the thought of a sentence to be what is true or false. I find it difficult to understand how two people can have the same thought, which is in any case not a material thing, so I will direct you to *Frege, 1918,* for his explanation. In the chapters below I will consider arguments that there are not two truth-values, but many, or that it makes no sense to classify a proposition as true or false, but only as assertible or not assertible.

But in the end the platonist, as well as the person who thinks a proposition is the meaning of a sentence or a thought, reason in language, using declarative sentences that they call 'representatives' or 'expressions' of propositions. Can we not reason together by concentrating on these sentences?

For me to reason with one who understands propositions differently it is not necessary that I believe in abstract propositions or thoughts or meanings. It is enough that we agree that certain sentences are—or from his viewpoint represent—propositions. Whether such a sentence expresses a true proposition or a false proposition is as doubtful to him as whether, from my view, it is true or is false. From my perspective, the platonist conception of logic is an idealization and abstraction from experience; from his perspective I mistake the effect for the cause, the world of becoming for the reality of abstract objects. But we can and do reason together using sentences, and to that extent my definition of 'proposition' can serve him, though he might prefer another word for it. Then in constructing a particular logic we can take other views of propositions into account as added weight to the significance of the word 'proposition'.

C. Words and Propositions as Types

Suppose now that we are having a discussion. An implicit assumption that underlies our talk is that words will continue to be used in the same way, or, if you prefer, that the meanings and references of the words we use won't vary. This assumption is so embedded in our use of language that it's hard to think of a word except as a *type,* that is, as a representative of inscriptions that look the same and utterances that sound the same. I do not know how to make precise what we mean by 'look the same' or 'sound the same'. But we know well enough in writing and conversation what it means for two inscriptions or utterances to be *equiform.* And so we can make the following agreement.

Words are Types We will assume that throughout any particular discussion equiform words will have the same properties of interest to logic. We therefore identify them and treat them as the same word. Briefly, *a word is a type.*

This assumption, while useful, rules out many sentences we can and do normally reason with quite well. For example:

Rose rose and picked a rose

If we subscribe to the assumption that words are types, we shall have to distinguish the three equiform inscriptions in this sentence. We can use some device such as 'Rose_1 rose_2 and picked a rose_3' or '$\text{Rose}_{\text{name}}$ $\text{rose}_{\text{verb}}$ and picked a $\text{rose}_{\text{noun}}$'.

Further, if we accept this agreement, we must also avoid words such as 'I', 'my', 'now', or 'this', whose meaning or reference depends on the circumstances of their use. Such words, called *indexicals,* play an important role in reasoning, yet our demand that words be types requires that they be replaced by words we can treat as uniform in meaning or reference throughout a discussion, such as

'Richard L. Epstein', 'Richard L. Epstein's', 'March 9th, 1991', and so on.

Now suppose I write down a sentence that we take to be a proposition:

Socrates was Athenian

Later I want to use that sentence in an argument, say:

If Socrates was Athenian, then Socrates was Greek
Socrates was Athenian
Therefore:

But we have two distinct sentences, since sentences are inscriptions. How are we to proceed?

Since words are types, we can argue that these two equiform sentences should both be true or both false. It doesn't matter to us where they're placed on the paper, or who said them, or when they were uttered. Their properties for logic depend only on what words (and punctuation) appear in them in what order. Any property that differentiates them isn't of concern to reasoning.

We couldn't make this argument were we to allow indexicals in our reasoning. If first I say 'I am over 6 feet tall', and then you say 'I am over 6 feet tall', we would not be justified in assuming that these two utterances have the same properties of concern to logic. Yet formalized versions of self-referential sentences such as '**a** is false', where the letter '**a**' names the last quoted sentence, can introduce a form of indexicality that leads us to classify equiform sentences differently (see, for example, *Epstein, 1992*). Avoiding such problem sentences for now, let us make the following assumption to simplify our work:

Propositions are Types In the course of any discussion in which we use logic we will consider a sentence to be a proposition only if any other sentence or phrase that is composed of the same words in the same order can be assumed to have the same properties of concern to logic during that discussion. We therefore identify equiform sentences or phrases and treat them as the same sentence. Briefly, *a proposition is a type*.

It is important to identify both sentences and phrases, for in argument (1) above we want to identify the phrase 'strawberries are red' in the first sentence with the second sentence.

The device I just used of putting *single quotation marks* around a word or phrase is a way of naming that word or phrase, or any linguistic unit. We need some such convention because confusion can arise if it's not clear whether a word or phrase is being used *as* a word or phrase, as when I say 'The Taj Mahal has eleven letters', where I don't mean the building has eleven letters, but that the phrase does. When we use this device we'll say that we have *mentioned* the word or phrase that is in quotation marks, and that the entire inscription including the quotes is a *quotation name* of the word or phrase. Otherwise, we simply *use* the word or phrase, as we normally do. We are justified in using quotation names because we have agreed to view words and propositions as types. Mentioning a linguistic unit can also be done by italicizing or putting the phrase in display format.

I use these devices for mentioning linguistic units with some reluctance, for there is not always a clear distinction between using a word and mentioning it. Moreover, when we write 'and' do we mean a string of symbols or the word with all its aspects? If we mean the word, then when we write 'Ralph is a dog' do we mean those words in that order, or do we mean the proposition? The linguistic unit intended must be inferred from the context, and sometimes it's not even clear to the user of the convention.

I will also use single quotation marks for quoting direct speech.

The device of enclosing a word or phrase in *double quotation marks* is equivalent to a wink or a nod in conversation, a nudge in the ribs indicating that I'm not to be taken literally, or that I don't really subscribe to what I'm saying. Double quotes are called *scare quotes,* and they allow me to get away with "murder".

D. Propositions in English

I want to impose an additional restriction on propositions. It will be hard for us to agree that a particular sentence is a proposition if we are speaking different languages. Therefore, throughout this book I will deal only with propositions in English or some formalized version of English.

Some argue that since modern logic is done by people speaking many different languages, it should not be considered so closely connected to one language, English, as I draw it in this volume. Abstract or mathematical notions such as function and object suffice. But if logic does not grow out of reasoning as we do it in our daily lives, how are we to use it? And how are we to justify the methods of reasoning our logic endorses? I start with what we have—reasoning in English —and look for abstractions and idealizations that I hope can serve speakers of different languages.

Exercises for Sections A–D

1. Give an example of formal modeling that is prescriptive in a discipline other than logic. Give another example that is descriptive.

2. a. Which of the following are declarative sentences?
   Ralph is a dog
   I am 2 meters tall
   Is any politician not corrupt?
   Power corrupts

Feed Ralph
Did you feed Ralph?
Why can't the English teach their children how to speak?
Strike three!
Love is not love that alters when it alteration meets
No se puede vivir sin amar
Ralph believes that George is a goose
Ralph didn't see George
Whenever Juney barks, Ralph gets mad
If anyone should say that cats are nice, then he is confused
If Ralph should say that cats are nice, then he is confused
If Ralph should say that cats are nice, then Ralph is confused
I now pronounce you man and wife
Would that I were rich
I wish I could get a job
$2 + 2 = 4$
$\int_1^x \frac{1}{t}\, dt = \ln x$
$\frac{d\, e^x}{dx} = e^x$
Herman Melville wrote 'Moby-Dick'
Herman Melville wrote Moby-Dick
There are an odd number of stars in the universe
Pada deszcz
The sentence to the right is true $\quad 2 + 2 = 4$
The sentence to the right is true $\quad 2 + 2 = 5$

b. What is a declarative sentence?
c. List all uses of indexicals in the phrases in (a).
d. Explain why we cannot take sentence types as propositions if we allow the use of indexicals.
e. Which of the phrases in (a) are propositions? Which are true?
f. Are sentence types abstract objects?

3. If a proposition must be completely precise and unequivocal, how could we reason with 'Strawberries are red'?

4. Describe and distinguish the following as candidates for objects having truth-values, considering especially the extent to which one can be viewed as an abstraction or idealization of another:
   a. Sentences as inscriptions or utterances

b. Sentence types
c. Thoughts
d. Abstract propositions
e. Propositions as defined in the text (Section B.1)

5. Explain and comment on the use of quotation marks in the following.
   a. Murder is something English
   b. 'Murder' is something English
   c. 'Murder,' she said
   d. These exercises are "murder"
   e. 'Dog' means 'any of a large and varied group of domesticated animals related to the fox, wolf, and jackal'
   f. 'Dog' means the same as 'any of a large and varied group of domesticated animals related to the fox, wolf, and jackal'
   g. 'Dog' means any of a large and varied group of domesticated animals related to the fox, wolf, and jackal
   g. 'Dog' means dog
   h. 'triangle' $\equiv_{\text{Def}}$ 'a three-sided polygon'

6. Distinguish the conception of agreements as a basis for logic, as described in Section B, from conventionalism and from pragmatism.

E. Form and Content

We begin with propositions that are sentences. There are two features of such sentences which contribute to our reasoning and proofs: their *syntax*, by which we mean the analysis of their form or grammar, and their *semantics*, by which we mean an analysis of their truth-values and meaning or content. These are inextricably linked: the choice of what forms of propositions we'll study will lead to what and how we can mean, and the meaning of the forms will lead to which of those forms are acceptable.

Often forms are chosen as primary, as when a logic is presented solely as a collection of forms of acceptable sentences and ways to syntactically manipulate those. It often seems easier to gain agreement on some few acceptable forms than on questions of content. That is because forms can be exhibited and we can, each of us, invest these with our own meanings. That is, forms are (comparatively) objective. Consider:

All men are mortal
Socrates is a man
Therefore:
    Socrates is mortal

Is this an example of valid reasoning acceptable on the basis of its form only? It's often said so. But why that form? How can we distinguish as valid the form of that argument from the form of:

All men are mortal
Socrates is mortal
Therefore:
Socrates is a man

Only by reference to the notions of truth and meaning.

I'll emphasize these notions in creating models of reasoning. It's in the semantics above all, I believe, that agreement must be reached. Without making explicit the assumptions that lead to our choice of acceptable forms, the objectivity of those forms can be based on implicit misunderstandings between us. My goal is to make explicit the disagreements as well as the agreements. Moreover, by studying what the forms mean before asking which are acceptable I hope to make the formal systems easier to understand.

F. Propositional Logic and the Basic Connectives

There are so many properties of propositions that could affect logic that we must begin by restricting our attention to only some of them. In this volume we will consider only properties of propositions as wholes and ways to connect propositions to form new ones. *We will ignore the internal structure of propositions except insofar as they are built from other propositions in specified ways.* This is what is called the study of *propositional logic.*

There are many, many ways to connect propositions to form a new proposition. Some are easy to recognize and use. For example, 'Ralph is a dog and dogs bark' can be viewed as two sentences joined by the connective 'and'. Note that to view 'and' as a connective of sentences we need to assume that, for example, 'Dogs bark.' and 'dogs bark' are equiform.

Some other common connectives are: 'but', 'or', 'although', 'while', 'if . . . then . . .', 'only if', 'neither . . . nor . . .', and so on. We want to strike a balance between choosing as few as possible to concentrate on, in order to simplify our semantic analyses, and as many as possible, so that our analyses will be broadly applicable.

Our starting point will be the four traditional basic connectives of logic: 'and', 'or', 'if . . . then . . .', and 'not'. We must be a bit careful with 'not', as it is to operate on propositions. In English it can occur in many different ways in a sentence, so let's agree that we'll study it as the connective that precedes a sentence as in *It's not the case that . . .* , or other uses that can be assimilated to that understanding of it. These four connectives will give us a rich enough grammatical basis to begin our logical investigation. But whether they will be enough, or be the most suitable connectives, are questions we will have to raise in relation to each particular type of semantic analysis.

These English connectives have many connotations and properties, some of which may be of no concern to us in logic or in a particular type of semantic analysis; for example, in American English 'not' is usually said more loudly than the surrounding words in the sentence. Therefore, let us replace these connectives with formal symbols to which we will give fairly explicit and precise meaning in each semantic analysis, based on our understanding of the English ones.

symbol	*what it will be an abstraction of*
$\wedge$	'and'
$\vee$	'or'
$\neg$	'it's not the case that'
$\rightarrow$	'if . . . then . . .'

Thus a complex sentence we might study is 'Ralph is a dog $\wedge$ dogs bark', corresponding to the earlier example.

A further formal device is important in reducing ambiguity: *parentheses.* In ordinary speech we might say, 'If George is a duck then Ralph is a dog and Dusty is a horse'. But it is not clear which of the following is meant:

If George is a duck, then: Ralph is a dog and Dusty is a horse
If George is a duck, then Ralph is a dog; and Dusty is a horse

Such ambiguity should have no place in our reasoning. Let us require the use of parentheses to enforce one of these two readings:

George is a duck $\rightarrow$ (Ralph is a dog $\wedge$ Dusty is a horse)
(George is a duck $\rightarrow$ Ralph is a dog) $\wedge$ Dusty is a horse

To lessen the proliferation of quotation marks in naming linguistic items, I will assume that *a formal symbol names itself* when confusion seems unlikely. Thus I might say that $\wedge$ is a formal connective.

Here is some terminology that goes with these symbols:

The sentence formed by joining two sentences by $\wedge$ is called a *conjunction* of the two; each of the original propositions is a *conjunct* that we *conjoin* with the other. When $\vee$ joins two sentences we obtain a *disjunction* by *disjoining* the *disjuncts.*

The sentence formed by putting $\neg$ in front of another is called the *negation of* it; it is a *negation.* Thus the negation of 'Ralph is a dog' is '$\neg$(Ralph is a dog)', where the parentheses are used for clarity.

The symbol $\rightarrow$ is called variously *the arrow, the conditional,* or *the implication sign.* The result of joining two sentences with it is called a *conditional* or *implication*; the proposition on the left is called the *antecedent* and the one on the right the *consequent.*

The connectives $\wedge$, $\vee$, $\rightarrow$ are *binary,* joining two propositions, while $\neg$ is *unary.*

Before we begin our first semantic analysis of these connectives in Chapter II, I want once more to stress that though the views I present here are one way to motivate and understand the technical work that follows, they are not the only way. The assumptions I have laid out in this chapter will seem clearer and more reasonable as they are used and argued for in many different contexts in the following chapters.

Exercises for Sections E and F

1. Classify each of the following as a conjunction, disjunction, negation, conditional, or as belonging to none of those categories.
   a. Ralph is a dog $\wedge$ dogs bark
   b. Ralph is a dog $\rightarrow$ dogs bark
   c. $\neg$ cats bark
   d. Cats bark $\vee$ dogs bark
   e. Cats are mammals and dogs are mammals
   f. Ralph is a dog
   g. $\neg$ cats bark $\rightarrow$ ($\neg$ (cats are dogs))
   h. Cats aren't nice
   j. Dogs bark $\vee$ ($\neg$ dogs bark)
   k. It is possible that Ralph is a dog
   l. Either Ralph is a dog or Ralph isn't a dog

2. For each sentence in Exercise 1 that is a conjunction, specify its conjuncts; if a disjnunction, specify its disjuncts; if a conditional, specify its antecedent and consequent; if a negation, specify what it is a negation of.

3. a. Write a sentence that is a negation of a conditional whose antecedent is a conjunction.
   b. Write a sentence that is a conjunction of disjunctions each of whose disjuncts is either a negation or has no formal symbols in it.

4. Write a sentence that might occur in daily speech that is ambiguous but which can be made precise by the use of parentheses. Indicate at least two ways to parse it using parentheses.

5. List at least four words or phrases in English not discussed in the text that are used to form a proposition from one or more propositions and which you believe are important to a study of reasoning.

6. Give an example of a use of 'not' that cannot be assimilated to a use of 'It is not the case that'.

G. A Formal Language for Propositional Logic

1. Defining the formal language

We have no dictionary, no list of all propositions in English, nor do we have a method for generating all propositions: English is not a fixed, formal, static language. But by using variables we can introduce a rigid formal language to make precise the syntax of the propositions we will study.

Let $p_0, p_1, \ldots$ be *propositional variables.* These can stand for any propositions, but the intention is that they'll be the ones whose internal structure won't be under consideration. Our formal language is built from these using the connectives $\neg, \rightarrow, \wedge, \vee$ and parentheses. This will be the *formal object language.* To be able to talk about that language, indeed even to give a precise definition of it, we need further variables, the *metavariables* $A, B, C, A_0, A_1, A_2, \ldots$ to stand for any of $p_0, p_1, p_2, \ldots$ or complex expressions formed from those. The analogue of a sentence in English is a *well-formed formula,* or *wff.* Here's how we generate wffs.

The Formal Language $L(\neg, \rightarrow, \wedge, \vee, p_0, p_1, \ldots)$

a. (p_i) is a wff for each $i = 0, 1, 2, \ldots$
b. If A, B are wffs, then so are $(\neg A)$, $(A \rightarrow B)$, $(A \wedge B)$, and $(A \vee B)$.
c. Only such concatenations of symbols as arise from repeated applications of (a) and (b) are wffs.

This definition of the formal language of propositional logic could suffice, but clause (c) is not entirely clear. There are two ways to specify wffs that are offered as more precise.

The platonist conceives of the formal language as a complete infinite collection of formulas. And infinite collections are real, though abstract, entities, he says, so he defines:

> $L(\neg, \rightarrow, \wedge, \vee, p_0, p_1, \ldots)$ is the smallest collection containing (p_i) for each $i = 0, 1, 2, \ldots$ and closed under the formation of wffs, that is, if A, B are in the collection, so are $(\neg A)$, $(A \rightarrow B)$, $(A \wedge B)$, and $(A \vee B)$.

One collection is said to be smaller than another if it is contained in or equal to the other collection.

The platonist argues that his definition is clearer and more precise because it replaces the somewhat vague and ambiguous condition (c) with strict criteria on collections. Whatever clarity is gained, however, depends on our accepting completed infinite totalities, an assumption at odds with some of the logics we study.

An alternative way to clarify (c) is to convert it to an inductive definition, one that uses the structure of the natural numbers.

An Inductive Definition of wff

We assign the number:

1 to (p_i) for each $i = 0, 1, \ldots$

$n+1$ to $(\neg A)$, $(A \rightarrow B)$, $(A \wedge B)$, $(A \vee B)$ in case the maximum of the numbers assigned to A and to B is n

A concatenation of symbols is then a *wff* if it is assigned some $n \geq 0$.

This definition has the advantage of being acceptable for most, perhaps all logics. As well, it stratifies wffs so that we can give proofs about the language using induction (see Chapter II.H).

But, still, we need to know that there is only one way to parse each wff. It may seem obvious that parentheses ensure that the reading of each wff is unambiguous, but this requires a demonstration, which I give in Chapter II.J.1, after we have considered the role of mathematics in such analyses.

Excessive parentheses can make it difficult to read a formal wff, for example,

$$(((p_0) \rightarrow ((p_1) \wedge (p_{32}))) \rightarrow ((((p_{13}) \wedge (p_6)) \vee (p_{317})) \rightarrow (p_{26})))$$

So I sometimes use an informal convention to eliminate parentheses: $\neg$ binds more strongly than $\wedge$ and $\vee$, which bind more strongly than $\rightarrow$. Thus $\neg A \wedge B \rightarrow C$ is to be read as $(((\neg A) \wedge B) \rightarrow C)$. I'll also dispense with the outermost parentheses and those surrounding a variable. Informally, I sometimes use [ ] or { } in place of the usual parentheses. So the example I gave above could be written informally as:

$$(p_0 \rightarrow p_1 \wedge p_{32}) \rightarrow ([(p_{13} \wedge p_6) \vee p_{317}] \rightarrow p_{26})$$

As well, whenever I write an extended conjunction or disjunction without parentheses, I will assume the conjuncts or disjuncts are associated to the *left*; for example, $p_1 \wedge p_2 \wedge p_3 \wedge p_4$ will abbreviate $((p_1 \wedge p_2) \wedge p_3) \wedge p_4$.

Formal wffs exemplify the forms of propositions. To talk about the forms of wffs of the formal language, we use *schema*: formal wffs with the propositional variables replaced by metavariables. These are the skeletons of wffs; for example, $p_{136} \vee \neg p_{136}$ is an *instance* of $A \vee \neg A$.

2. Realizations: semi-formal English

The formal language is meant to give us the form of the propositions we will study. No wff such as $p_0 \wedge \neg p_1$ is true or false; it is the skeleton of a proposition. It is only when we first fix on a particular interpretation of the formal connectives and then assign propositions to the variables, such as 'p_0' stands for 'Ralph is a dog' and 'p_1' stands for 'Four cats are sitting in a tree', that we have a semi-formal proposition, 'Ralph is a dog $\wedge$ $\neg$(four cats are sitting in a tree)', one which we agree can be viewed as having a truth-value. We may read this as 'Ralph is a dog and it's not the case that four cats are sitting in a tree' so long as we remember that we've agreed that all that 'and' and 'it's not the case that' mean will be captured by the interpretations we will give for $\wedge$ and $\neg$.

A *realization* is an assignment of propositions to some or all of the propositional variables, relative to some particular interpretation of the formal connectives. The *realization of a formal wff* is the formula we get when we replace the propositional variables appearing in the formal wff with the propositions assigned to them; it is a *semi-formal wff.* The *semi-formal language* for that realization is the collection of realizations of formal wffs all of whose propositional variables are realized. I call the semi-formal language of a realization *semi-formal English* or *formalized English.* I will use the same metavariables A, B, C, A_0, A_1, A_2, . . . for semi-formal wffs, too.

For example, we could take as a realization:

(1)	p_0	is assigned	'Ralph is a dog'
	p_1	is assigned	'Four cats are sitting in a tree'
	p_2	is assigned	'Four is a lucky number'
	p_3	is assigned	'Dogs bark'
	p_4	is assigned	'Juney is barking loudly'
	p_5	is assigned	'Juney is barking'
	p_6	is assigned	'Dogs bark'
	p_7	is assigned	'Ralph is barking'
	p_8	is assigned	'Cats are nasty'
	p_9	is assigned	'Ralph barks'
	p_{47}	is assigned	'Howie is a cat'
	p_{312}	is assigned	'Bill is afraid of dogs'
	p_{317}	is assigned	'Bill is walking quickly'
	p_{4318}	is assigned	'If Ralph is barking, then he will catch a cat'
	p_{4319}	is assigned	'Ralph is barking'

Then some of the semi-formal wffs of the realization would be:

Ralph is a dog $\wedge$ $\neg$(four cats are sitting in a tree)

Bill is afraid of dogs $\wedge$ Ralph barks $\rightarrow$ Bill is walking quickly

Juney is barking loudly $\rightarrow$ Juney is barking

Ralph is a dog $\wedge$ dogs bark $\rightarrow$ Ralph barks

Actually, these are only abbreviations of semi-formal wffs, using the conventions for informally deleting parentheses given above. The second, for example, is an abbreviation of the realization of $(((p_{312})\wedge(p_9))\rightarrow(p_{317}))$.

A realization is linguistic, a formalized fragment of English. Because it is linguistic, I have used quotation marks to indicate that I am mentioning pieces of language and not using them. The quotation marks are signals for you to understand what I mean; they are not part of the realization. That is why the realization $p_0 \to p_3$ is 'Ralph is a dog → dogs bark', and not ''Ralph is a dog' → 'Dogs bark''.

Note that not all variables need be realized. Nor need we assign distinct propositions to distinct variables: the realization of p_7 is the same as the realization of p_{4319}.

The propositions we assign to the variables are *atomic*. They are the simplest propositions of the realization, ones whose internal structure will not be under consideration in this particular analysis. So the assignment of 'If Ralph is barking, then he will catch a cat' to p_{4318} is a bad choice, for it will not allow us to make a semantic analysis based on the form of that proposition in relation to 'Ralph is barking'. Generally speaking, the atomic propositions of a realization shouldn't include any English expressions that we've agreed to formalize with our formal connectives, though applying that rule usually depends on the particular semantic analysis of the connectives we choose.

Wffs that contain formal connectives are called *compound* or *complex*. Before we can speak of their realizations as propositions we must decide upon interpretations of the formal connectives.

Exercises for Section G

1. Why do we introduce a formal language?
2. Identify which of the following are formal (unabbreviated) wffs:
   a. $(p_1) \vee \neg(p_2)$
   b. $((p_1) \to (p_2))$
   c. $((p_1 \vee p_2) \to p_2)$
   d. $(\neg(p_1)(p_2) \wedge (p_1))$
   e. $(\neg(\neg(p_1)) \vee \neg(p_1))$
   f. $((\neg(\neg(p_1))) \vee \neg(p_1))$
   g. $((\neg(\neg(p_1))) \vee (\neg(p_1)))$
3. Abbreviate the following wffs according to our conventions on abbreviations:
   a. $((((p_1) \to (p_2)) \wedge (\neg(p_2))) \to (\neg(p_1)))$
   b. $((((p_4) \wedge (p_2)) \vee (\neg(p_6))) \to ((p_7) \to (p_8)))$
4. Give an example of a formula that is:
   a. A conjunction, the conjuncts of which are disjunctions of either atomic propositions or negated atomic propositions.

b. A conditional whose antecedent is a disjunction of negations and whose consequent is a conditional whose consequent is a conditional.

5. Explain how the inductive definition of 'wff' differs from the other two definitions in the text. What advantages can you see in using one of these definitions rather than the others?

6. Distinguish the following: ordinary language, the formal language, a semi-formal language.

7. a. In realization (1), give the realization of:

   i. $((p_8 \wedge p_{4318}) \wedge p_7) \rightarrow p_1$

   ii. $(p_0 \wedge p_1) \rightarrow p_2$

   iii. $\neg(p_4 \wedge \neg p_5)$

   iv. $p_3 \rightarrow \neg\neg p_6$

   v. $\neg(p_{312} \wedge p_7) \wedge \neg p_{317}$

   vi. $p_{312} \wedge p_7 \rightarrow \neg p_{317}$

   b. The following wff is the realization of what formal wff in realization (1):

   Four cats are sitting in a tree $\wedge$ four is a lucky number $\rightarrow$
   $\neg$(If Ralph is barking then he will catch a cat $\rightarrow$ Howie is a cat)

   c. Exhibit formal wffs of which the following could be taken to be realizations:

   i. All cats are nasty $\rightarrow$ Howie is nasty

   ii. $\neg$((Ralph is a dog $\vee$ $\neg\neg$Ralph barks) $\vee$ Ralph is a puppet) $\rightarrow$
   no number greater than 4 billion is a perfect square

8. What assumptions do you need to make to show that the three definitions of wffs all result in the same collection?

II Classical Propositional Logic – PC –

In this chapter we'll look at the simplest logic that can be developed from the assumptions of Chapter I. In doing so, we will consider for the first time the idea of a model, the logical form of a proposition, formalization of the idea of proof, and the notion of a logic. In Section J I recapitulate the definitions in mathematical format, after first discussing the role of mathematics and infinitistic assumptions in the study of logic.

A. The *Classical Abstraction* and the *Fregean Assumption*

What are the simplest interpretations we can prcvide for the formal connectives that will be consonant with the assumptions we made in Chapter I? We abstract away every property of a proposition other than its form and that semantic property that makes it a proposition: truth-value. That is, in this our first semantic analysis, we will make the following assumption.

The Classical Abstraction The only properties of a proposition that matter to logic are its form and its truth-value.

If the only aspects of a proposition that matter are its truth-value and form, then the truth-value of a complex proposition can depend only on the connectives appearing in it and the truth-values of its parts. If this were not the case then the truth-values of 'Ralph is a dog' and of 'cats are nasty' wouldn't determine the truth-value of 'Ralph is a dog $\wedge$ cats are nasty'. But if the truth-values don't, and we've agreed that no other property of these propositions matters, what could determine the truth-value of the complex proposition? If there is something nonfunctional, something transcendent, that occurs when, say, $\wedge$ connects two sentences, then how are we to reason? From the truth of one proposition how could we deduce the truth of another? Without some regularity reasoning cannot take place. This is a fundamental assumption that we will make throughout this volume; I name it after Frege, who clairified and emphasized it in his work.

The Fregean Assumption The truth-value of a proposition is determined by its form and the semantic properties of its constituents.

If propositions using 'and', 'or', 'not', and 'if . . . then. . .' can only be understood within the context of all of language, then we'll just have to make do with our formal abstractions if we want to give a model of reasoning.

B. Truth-Functions and the *Division of Form and Content*

The simplest propositions are those that contain no formal connectives, for example, 'Ralph is a dog' or 'Every bird sings'. They are *atomic*. Since they have no form of significance to propositional logic, the *Classical Abstraction* limits us to considering only their truth-values. The *Fregean Assumption* then tells us that if we join two atomic propositions with $\wedge$, $\vee$, or $\rightarrow$, or place $\neg$ in front of one, the truth-value of the resulting proposition must depend on only the truth-value of those atomic propositions. That is, the connectives must operate semantically as functions of the truth-values of the constituent atomic propositions; they are *truth-functions*.

Which truth-functions correspond to our connectives? I'll let you convince yourself that the only reasonable choices for $\neg$ and $\wedge$ are given by the following tables, where I use p and q to stand for atomic propositions, and 'T' to stand for 'true', 'F' for 'false':

p	¬p
T	F
F	T

p	q	p∧q
T	T	T
T	F	F
F	T	F
F	F	F

That is, if p is T, then $\neg$p is F; if p is F, then $\neg$p is T. And p$\wedge$q is T if both p and q are T; otherwise it is F.

For $\vee$ there are two choices, corresponding to an inclusive or exclusive reading of 'or' in English. The choice is arbitrary; it's customary now to use the inclusive version, as in 'p or q or both'.

p	q	p∨q
T	T	T
T	F	T
F	T	T
F	F	F

The table we use for $\rightarrow$ is:

p	q	p→q
T	T	T
T	F	F
F	T	T
F	F	T

The second row is the essence of 'if . . . then . . .': a true antecedent does not have a false consequence. The first row is also fundamental, at least so long as we ignore all aspects of atomic propositions except their truth-values.

For the last two rows, consider that if we were to take both to result in F, then p → q would be evaluated the same as p∧q. If we were to take the third row as T and the fourth as F, then p→q would be the same as q; were the third row F and the last T, then p → q would be the same as q → p. Rejecting these counter-intuitive choices, we are left with the table that is most generous in assigning T to conditional propositions: p → q is true so long as it's not the case that p is T and q is F.

An example from mathematics will illustrate why we want to classify conditionals with false antecedent as true. I hope you'll agree that the following is true of the counting numbers:

If x and y are odd, then the sum of x and y is even

If we were to take either of the last two rows of the table for '→' to be F, the formalization of this proposition would be false, for we could provide false instances of it: $4 + 8 = 12$, which is even, and $4 + 7 = 11$, which is not even. The formalization of 'if . . . then . . .' we have chosen allows us to deal with cases where the antecedent "does not apply" by treating them as vacuously true.

Given any atomic propositions p and q, we now have a semantic analysis of ¬p, p→q, p∧q, and p∨q. But what about:

(1) (Ralph is a dog ∧ dogs bark) → Ralph barks

Here the antecedent is not a formless entity that is simply true or false: it contains a formal connective that must be accounted for. How shall we analyze this conditional compared to:

(2) (Ralph is a dog → dogs bark) → Ralph barks

Certainly the antecedents p∧q and p→q are to be evaluated by different methods. But if both are evaluated as true or both as false, is there anything that can distinguish them semantically? We have agreed that the only semantic value of a proposition that we shall consider is its truth-value. So if two propositions have the same truth-value, they are semantically indistinguishable and should play the same role in any further semantic analysis. We make the following assumption.

The Division of Form and Content If two propositions have the same semantic properties, then they are indistinguishable in any semantic analysis, regardless of their form.

Thus to evaluate '(Ralph is a dog ∧ dogs bark) → Ralph barks', we first determine the truth-value of 'Ralph is a dog ∧ dogs bark', and then only the truth-value of that proposition and the truth-value of 'Ralph barks' matter in determining the truth-value of the whole. We proceed in the same manner in determining the truth-value of (2) so that if 'Ralph is a dog $\wedge$ dogs bark' and 'Ralph is a dog $\rightarrow$ dogs bark' have the same truth-value, then the two conditionals (1) and (2) will have the same truth-value.

The *Division of Form and Content* imposes a sharp distinction between syntax and semantics. Suppose we were to take the view that any use of more than three negations at the beginning of a wff is only for emphasis, and then interpret $\neg$ by:

$$\neg A \text{ is true if and only if } \begin{cases} A \text{ is false and } A \text{ is not of the form } \neg\neg\neg B \\ \quad \textit{or} \\ A \text{ is true and } A \text{ is of the form } \neg\neg\neg B \end{cases}$$

If we accept the *Division of Form and Content,* then we have to classify the number of appearances of $\neg$ at the beginning of a proposition as a *semantic* value of the proposition.

Adopting the *Division of Form and Content,* we have that the tables for $\neg$, $\rightarrow$, $\wedge$, and $\vee$ above apply to compound as well as atomic propositions:

(3)

A	¬A
T	F
F	T

(4)

A	B	A ∧ B
T	T	T
T	F	F
F	T	F
F	F	F

(5)

A	B	A ∨ B
T	T	T
T	F	T
F	T	T
F	F	F

(6)

A	B	A → B
T	T	T
T	F	F
F	T	T
F	F	T

These are the *classical truth-tables* for $\neg$, $\rightarrow$, $\wedge$, $\vee$, also called the *classical evaluations* or *classical readings* of $\neg$, $\rightarrow$, $\wedge$, $\vee$.

Note that I've used 'and', 'or', 'not', and 'if . . . then . . .' to explain the tables for the formal connectives. This isn't circular: we are not defining or giving meaning to 'and', 'or', 'not', 'if . . . then . . .' but to $\wedge, \vee, \neg, \rightarrow$. I must assume that you understand the ordinary English connectives.

Have we really restricted ourselves by looking at only these four connectives? What about others that may be important to reasoning? Our assumptions tell us that they must be treated as truth-functions. In Section J.3 we'll see that it's possible to define every truth-functional connective of propositions from just these four formal connectives.

C. Models

$L(\neg, \rightarrow, \wedge, \vee, p_0, p_1, \ldots)$ is the formal language. No wff such as $p_0 \wedge \neg p_1$ is true or false: it is only when we make an assignment of propositions to the variables, such as p_0 stands for 'Ralph is a dog' and p_1 stands for 'Four otters are sitting on a log', that we have a semi-formal proposition 'Ralph is a dog $\wedge$ $\neg$(four otters are sitting on a log)', one which we agree can be viewed as having a truth-value.

We may not know the truth-value of, say, 'Ralph is a dog' and wish to test the hypothesis that it's true. After we specify an assignment of propositions to the variables we then have to specify which of the propositions are to be taken as true and which as false. Compound propositions can then be evaluated using the truth tables. These assignments and evaluation are what we call a *model*. Writing real(p_0), real(p_1), ... for the realizations of the variables $p_0, p_1, \ldots$, respectively, a model can be presented schematically:

I

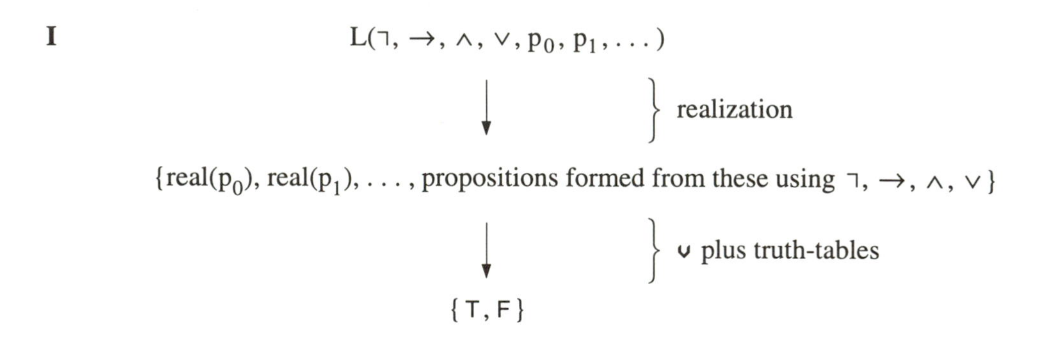

Note that in doing this we implicitly agree to view any semi-formal sentence as a proposition. We do this because we have agreed on how to understand the formal connectives. From now on I will use the *metavariables* $A, B, \ldots A_0, B_0, \ldots$ to range over wffs in the formal language, sentences in the semi-formal language, or English sentences taken as propositions, trusting to the context to make it clear which is meant in those cases where the distinction matters.

In the diagram v is a *valuation*, a way we've agreed to assign T or F to those propositions assigned to the propositional variables (v is the lowercase letter, to be distinguished from the connective $\vee$). Then v plus the truth-tables uniquely determine the truth-value of any complex proposition of the semi-formal language because there is only one way to parse each of those, as we'll see in Section J.1 below. We view the truth-tables as a method to extend v to all propositions of the semi-formal language and write $v(A) = T$ or $v(A) = F$ to indicate that the proposition A is true or is false in this model.

When, in later chapters, there may be a question of which semantic analysis is under discussion, I will call these *classical models*.

Exercises for Sections A–C

1. Why do we take $A \to B$ to be false if A is false?

2. a. What is a truth-functional connective?
   b. Give at least two examples in which 'not' in English is not used truth-functionally? Why then do we formalize 'not' truth-functionally?

3. Which of the following italicized phrases could reasonably be construed as a truth-functional connective in the given sentence?
   a. *If* sodium burns, *then* it is not a metal
   b. You can't make an omelette *without* breaking eggs
   c. *If* Ralph passed that test, *then* I'm a monkey's uncle
   d. The sky is blue *because* the sun is yellow
   e. Earl climbed the ladder *and then* painted the eaves
   f. Ralph went to the movie *and* bought popcorn
   g. *If* the moon is made of green cheese, *then* $2 + 2 = 4$
   h. $2 + 2 = 4$ *or* the Implicit Function Theorem is false
   j. *It's not the case that* Napoleon lost at Waterloo
   k. *Either* $2 + 2 = 4$ *or* $4 + 4 = 7$
   l. *If* Richard Nixon were a dog, *then* he would bark too much
   m. *Ralph believes* that Cedar City is a seaport
   n. There is *no* one who can lift 500 kgs
   o. *Neither* Ralph *nor* Dusty is a dog

4. Assign truth-values to the atomic propositions in realization (1), p. 15, and evaluate the truth-value of the realizations of each of the wffs in Exercise 2.a, 3.a, 3.b, and 7, on pp. 16–17.

5. Assign truth-values to the atomic propositions in realization (1) that will make the following have truth-value F, if that is possible.
   a. Ralph is barking $\to$ cats are nasty
   b. (Ralph is a dog $\vee$ dogs bark) $\wedge$ $\neg$(Juney is barking)
   c. Ralph is barking $\wedge$ cats are nasty $\to$ Ralph is barking
   d. Dogs bark $\vee$ $\neg$ (dogs bark) [as a realization of $p_6 \vee \neg(p_6)$]
   e. Dogs bark $\vee$ $\neg$ (dogs bark) [as a realization of $p_6 \vee \neg(p_3)$]
   f. $\neg$(cats are nasty $\to$ $\neg\neg$ (cats are nasty))
   g. ((Ralph is a dog $\vee$ dogs barks) $\wedge$ $\neg$(Ralph is dog)) $\to$ dogs bark

6. Distinguish between a realization and a model.

7. Give an example of reasoning that takes into account some aspect of propositions besides truth-values.

8. Show that each of the following is independent of the others in the sense that we can give an assignment of truth-values to propositions of a semi-formal language that respects that one but not the other two:
   a. The *Classical Abstraction*
   b. The *Fregean Assumption*
   c. The *Division of Form and Content*

D. Validity and Semantic Consequence

1. Tautologies

Some propositions we take to be true from our experience, such as 'Dogs bark'. The truth of others, however, follows solely from the agreements we have made about the meaning of the logical connectives. For instance, we might argue that relative to the classical interpretation of the connectives 'Ralph is a dog or Ralph is not a dog' is true due solely to its form because we would formalize it as '(Ralph is a dog) $\vee$ $\neg$(Ralph is a dog)', which can be taken as a realization of, for example, $p_1 \vee \neg p_1$. In any model the realization of that wff will be true. Hence it is the form of the original proposition that ensures its truth. Using formal wffs for generality and precision, we make the following definition.

Classical Tautologies A formal wff is a *classical tautology* or *classically valid* if in every classical model its realization is evaluated as true; in that case we write $\models A$. A semi-formal proposition is a classical tautology if it is the realization of a wff that is a classical tautology.

We can speak of a proposition in ordinary English being a tautology if we understand by this that there is a straightforward formalization of it into semi-formal English that is a tautology. We also say that a scheme of wffs is a tautology if every instance of it is a tautology.

So relative to classical propositional logic, the following example is a tautology:

(7) $(p_1 \vee \neg p_1)$

And so, then, is:

(8) (Ralph is a dog) $\vee$ $\neg$(Ralph is a dog)

But in intuitionistic logic, as we shall see in Chapter VII, though (8) is true in any model in which the words are interpreted as we normally understand them, the proposition is not valid because there are realizations of (7) that are not true.

Our goal is to extend these notions to propositions of ordinary English. We can say that a proposition in ordinary English is a tautology if there is a formalization of it on which we feel certain we'll all agree and which is valid. For example,

Ralph is a dog or he isn't a dog

This is a tautology for classical logic because (8) is an obvious formalization of it.
But what is obvious may not always be so obvious. Consider:

(9) Ralph is a dog $\wedge$ George is a duck $\rightarrow$ Ralph is a dog

This is a classical tautology because we can take it to be a realization of the following classically valid wff:

(10) $(p_1 \wedge p_2) \rightarrow p_1$

Of course it also could be taken as a realization of $(p_{32} \wedge p_{47}) \rightarrow p_{32}$. But that doesn't matter, since that, too, will be valid if (10) is valid. The propositional form of (9) is fixed because it is given in a semi-formal language. But now consider:

(11) If Ralph is a dog but George is a duck, then Ralph is a dog

We try a few replacements of the atomic propositions in it and we always seem to get a true proposition, suggesting that it is the form of (11) that makes it true:

If Dusty is a horse but Marilyn Monroe is a woman, then Dusty is a horse

If Marilyn Monroe eats grass but George eats grass,
then Marilyn Monroe eats grass

These replacements reflect an implicit classification of 'but' as a connective. Arguing for the formalization of 'but' as $\wedge$, we do indeed get (11) as a tautology. But whereas pointing to one formal wff that is a tautology which (9) could realize is enough to justify (9) as a tautology, for (11) much stronger arguments for the appropriateness of the formalization must be made, since the propositional form of (11) is not fixed as the form of (9) was. For example, the following replacement makes us doubt the formalization of 'but' as $\wedge$:

If Richard Nixon is honorable but Richard Nixon lied,
then Richard Nixon is honorable

And if 'but' is not treated as a connective, we could have this replacement:

(12) If Richard L. Epstein eats grass as often as Marilyn Monroe eats grass,
then Richard L. Epstein is a midget

This assumes that the form of (11) is $p_{16} \rightarrow p_{12}$, which isn't a valid wff. Indeed, (12) isn't even true on a reading of 'if . . . then . . .' as the classical conditional, since the antecedent is true but consequent false. The classification of informal propositions as tautologies depends on what constitutes a good formalization of an informal proposition, which we'll consider in Section E.2 below.

Form can determine falsity, too: 'Ralph is a dog ∧ ¬(Ralph is a dog)' will be evaluated as false no matter what value is assigned to 'Ralph is a dog'. We define:

> A proposition or formal wff that is evaluated as false regardless of the truth-values of its atomic constituents is called a (*classical*) *anti-tautology* or *contradiction*

Anti-tautologies are false due solely to their form, here relative to the classical interpretation of the connectives.

The notion of tautology we've defined formalizes the following informal characterization.

Tautologies A tautology is a proposition that is not only true, but remains true on (uniform) replacement of the atomic propositions appearing in it by any other atomic propositions.

The replacement must be uniform: the same words must be replaced in the same way throughout the proposition.

This informal notion of validity, current at least since *Bolzano, 1837* (vol. II, p. 198), does not speak of meaning or possible models, but relies only on the notion of a proposition being true or false and recognition of syntactic categories. As such it seems simpler than our semantic analyses.

Yet the informal notion does rely on a notion of possibility: for a proposition to be valid all possible substitution instances of it must be true. And there is the difficulty. How can we survey all possible substitutions? Using semantic notions we can make classifications of validity or invalidity based on general observations about models.

Note that this is a notion of tautology or validity relative to the propositional logic parsing of propositions. We do not consider the internal structure of atomic propositions.

2. Semantic consequence

Form can determine relations among propositions, too. If the proposition 'Ralph is a dog ∧ Howie is a cat' is true in a model, then each of its constituents must be true, too. Equally, if both 'Ralph is a dog' and 'Howie is a cat' are assigned T in a model, then 'Ralph is a dog ∧ Howie is a cat' must be assigned T. From the truth of one proposition or a collection of propositions, we can conclude the truth of another, based solely on their form. Using Greek letters for collections of propositions or wffs, we make the following definition for formal wffs.

Classical Semantic Consequence A wff A is a *classical semantic consequence* of a collection of wffs Γ, written '$\Gamma \models A$', if in every model in which every wff in Γ is evaluated as true, so, too, is A; we also say that the pair Γ, A is a classical semantic consequence.

The same terminology can be applied to semi-formal propositions viewed as realizations of formal wffs, and then in turn to ordinary English propositions, as described for validity. Writing '$\models A$' if A is a tautology is still appropriate, for if A is a tautology then it is a semantic consequence of the collection consisting of no wffs whatsoever.

We also write a deduction, valid or not, as:

$$\frac{A_1, \ldots, A_n}{B}$$

For example, on the classical interpretation of the connectives:

$$\{p_7 \vee p_{10}, \neg p_{10}\} \models p_7$$

And hence for classical logic:

{Ralph is a dog ∨ Howie is a duck, ¬(Howie is a duck)} ⊨ Ralph is a dog

Arguing that 'either . . . or . . .' should be formalized as inclusive classical disjunction, and 'no way' formalized as classical negation, the following argument is then valid:

Either Ralph is a dog or Howie is a duck
No way is Howie a duck
Therefore:
Ralph is a dog

Note that a valid argument need not have true hypotheses: in this example (under the usual interpretation of the words) 'Either Ralph is a dog or Howie is a duck' is false.

The notion of semantic consequence formalizes the idea that one proposition *follows from* another or collection of others, relative to our interpretation of the connectives. We make the following definition.

Arguments Γ *therefore* A is a (*classically*) *valid deduction* or a (*classically*) *valid argument* just in case $\Gamma \models A$ is a classical semantic consequence; the propositions in Γ are the *hypotheses* or *premises* of the argument, and A is the *conclusion.*

There is not general agreement that this is a good formalization, however, for some say that the premises of an argument must be true for the argument to be valid, and others that the order in which the premises are considered is important. A number of other words and phrases are often used synonomously with 'therefore': *and so, hence, consequently, it follows that,* and even the Latin term *ergo.*

This definition formalizes the following informal characterization of validity.

Valid Deductions A deduction Γ *therefore* B is valid if either there is some proposition in Γ that is false, or else all the propositions in Γ are true and B is true, and one of these alternatives obtains for every (uniform) replacement of the atomic propositions appearing in B and in the propositions in Γ by any other atomic propositions.

A different notion of validity, of ancient lineage, is not formalized by using models in the way we have. Philosophers and logicians until recently adopted some variant of the following:

$A_1, \ldots, A_n \vDash B$ iff it is impossible for $A_1, \ldots, A_n$ all to be true while B is false

This possibility version of validity classifies as valid:

$2 + 2 = 4$

Ralph is a dog $\vDash 2 + 2 = 4$

These are valid because on most accounts the simple truths of arithmetic are necessary truths, impossible to falsify. But according to our notion of validity, these examples would not be valid, for they have the following forms:

p_1

$p_3 \vDash p_1$

The replacement version of validity depends only on form and the interpretations of the connectives. The possibility version of validity depends as well on the meaning of the words appearing in atomic propositions; as such it is outside the scope of formal logic.

The semantic analysis of the propositional connectives we have given here is called *classical propositional logic*: the definition of classical models, of classical validity, and of classical semantic consequence. Sometimes, though, just the classical truth-tables, or just classical semantic consequence, or even just the collection of classical tautologies is referred to as 'classical logic'.

Let's review what we've done. We began by establishing a formal language that would make precise the form of the propositions we would study. We based that formalization on the four connectives : 'and', 'or', 'it's not the case that', and 'if . . . then . . . '. It was the semi-formal language in which we were primarily interested: English with the formal propositional connectives $\neg$, $\rightarrow$, $\wedge$, $\vee$.

Then we made the assumption that the only aspects of a proposition that are of concern to logic are its truth-value and its form as built up from sentence connectives. Making assumptions about the relationship of the semantic properties of a proposition to the properties of its constituents, we analyzed the truth of a complex proposition in terms of its parts by using the truth-tables. Then we explained what it means for a proposition to be true due to its (propositional) form only by reference to the formal language, and then said what it means for one proposition to follow from another.

It's time now to see how to apply some of these ideas to examples in ordinary language.

Exercises for Sections D

1. a. What is a tautology in the formal language?
   b. What is a tautology of the semi-formal language?
   c. What is a tautology in ordinary English?
   d. Relate each of these to the informal characterization of tautologies on p. 28.

2. a. Exhibit 3 formal classical tautologies.
   b. Exhibit 3 formal classical anti-tautologies.
   c. Exhibit a proposition that would be classified as a tautology by the informal characterization that is not a classical tautology.

3. a. What is a classical semantic consequence?
   b. Exhibit 3 formal classical semantic consequences.
   c. What is the relation of classical semantic consequence to 'follows from'?
   d. What is a classically valid argument?
   e. Exhibit 3 classically valid arguments.

E. The Logical Form of a Proposition

1. On logical form

We have chosen to be very stingy with the grammatical tools we allow ourselves in parsing the propositional structure of a proposition. If we confine ourselves to formalizing only 'and', 'or', 'not', 'if . . . then . . .', how are we to analyze 'Ralph barks but George is a duck' or 'Horses eat grass because grass is green'?

In this section I want to discuss some agreements and methods that will allow us to rewrite a wide class of propositions in such a way that the new English sentence is equivalent to the original for all our logical purposes and yet can be easily formalized. Here 'for all our logical purposes' means relative to the assumptions of classical propositional logic. Similar methods and remarks will apply to the assumptions for other logics.

We will have:

English proposition ⇔ rewriting of English ⇔ proposition of semi-formal English

The rewriting in English will exhibit how we've agreed to treat the proposition in logic. We can further abstract the semi-formal equivalent of a proposition into a wff of the formal language, and that is what we call its (*propositional*) *logical form.*

I can't give rules for rewriting every proposition we'll want to use in logic. And often there is more than one way to rewrite a sentence, leading to equally reasonable choices for its logical form. Many discussions in the literature about what is the logical form of a particular sentence or type of sentence seem to me to be analyses of how our assumptions about logic apply to the sentence.

For a platonist there is no question of what the logical form of a proposition is. Propositions are abstract objects that have logical form and the difficulty, instead, is how to determine which proposition is being expressed by a particular sentence.

2. Criteria of formalization

In Chapter I.G, realization (1), I gave an example of an assignment to p_0 and p_1, respectively, 'Ralph is a dog' and 'Four otters are sitting on a log'. Another might be to let p_0 stand for '$2+2=4$' and p_1 stand for 'George is a duck and Ralph is a dog', and p_2 stand for 'Ralph is a dog'. But that assignment would be a bad choice.

A proposition assigned as realization to a propositional variable is *atomic*: indivisible, a single unit whose internal structure cannot be taken into account in a model based on that realization. So though we have the intuitively valid deduction 'George is a duck and Ralph is a dog' *therefore* 'Ralph is a dog', we certainly do not have $p_1 \vDash p_2$. The formalization fails to capture an important logical relationship that we believe can be justified in propositional logic.

But what counts as a good (suitable, correct, appropriate) formalization of an informal proposition? We need criteria.

To begin, we want to observe the assumptions of the logic. For classical logic that means that if we want to formalize a sentence as composed of two propositions, then the composition must be through a truth-functional connective. Thus if we take 'Ralph is a dog' as an atomic proposition in a realization, the truth-functionality of classical logic precludes our formalizing in the same realization:

(13) Anubis wonders whether Ralph is a dog

The sentence (13) contains 'Ralph is a dog' as a part, but the truth-value of (13) does not depend on the truth or falsity of that proposition, so there is no suitable way to model (13) using truth-functional connectives.

We also want to respect truth-values: we don't want to formalize a proposition we consider true as a false proposition, nor a false one as true. That, of course, depends on the particular model we use. But if a proposition is intuitively a tautology, such as 'If Ralph is a dog and George is a duck, then Ralph is a dog', we don't want to formalize it as one that could be false in some models. And intuitively valid deductions should be respected, too. That was why we decided it inappropriate to take both 'George is a duck and Ralph is a dog' and 'Ralph is a dog' as atomic.

Further, we want to be consistent, that is, regular in the way we treat certain words. If we choose to formalize 'but' as $\wedge$ in one case, then we should do so in others, barring good arguments to the contrary.

These are not the only criteria, but they suffice for us to begin analyzing some examples below. In summary, we have the following:

1. The formalization respects the assumptions that govern our choice of primitive connectives and definition of truth in a model. The constraints we work under when we adopt propositional logic must be observed.
Further, the formalization respects the assumptions that govern our choice of a particular logic.

2. If a proposition is informally valid or informally an anti-tautology, then its formalization should be a tautology or anti-tautology, respectively.
If one proposition follows informally from another or a collection of other propositions, then its formalization should be a formal semantic consequence of the formalizations of the other(s) (relative to the logic we adopt).

3. A regular translation of certain words as connectives should be observed. More, formalizations should be regular in the sense that each proceeds in analogy with agreed upon formalizations of other examples, the requirement of *parity of form.*

3. Other propositional connectives

There are some common ways of joining propositions together that are not covered by our four connectives, yet are sufficiently common and regular and, usually, truth-functional for us to establish some conventions about the formalizations of them. Let us adopt the following, barring strong arguments to the contrary in specific cases:

(14)

English connective	*Formalization*
since A, B	$A \to B$
given A, B	$A \to B$
A only if B	$A \to B$
B if A	$A \to B$
if A, B	$A \to B$
B in case A	$A \to B$
B provided that A	$A \to B$
A, so B	$A \to B$

A if and only if B	$(A \rightarrow B) \wedge (B \rightarrow A)$
both A, B	$A \wedge B$
neither A nor B	$\neg A \wedge \neg B$
A or else B	$(A \vee B) \wedge \neg(A \wedge B)$

We usually abbreviate 'A if and only if B' as 'A iff B' and abbreviate its formalization as $A \leftrightarrow B$. We call '$\leftrightarrow$' the *biconditional.* We call 'if A, then B' the *left to right direction* of 'A iff B', and 'if B, then A' the *right to left direction.* We call $B \rightarrow A$ the *converse* of $A \rightarrow B$.

A warning, however: even the four connectives we've agreed are the basis of our logic require care in formalizing. Consider the dictum: 'Spare the rod and spoil the child'. It would be inappropriate to formalize 'and' in this proposition as $\wedge$ interpreted according to the classical table, for what is intended, as George Hughes suggested to me, is: 'If you spare the rod, then you spoil the child'.

F. Examples of Formalization

In this section I present some examples of formalizations of propositions and arguments. For each example I first give an English proposition(s) followed by a semi-formal version, and then (an abbreviation of) a completely formal wff(s) of which that is a realization. I then present reasons for choosing that formalization.

1. **Ralph is a dog or Dusty is a horse and Howie is a cat**
   ***Therefore*:**
   Howie is a cat

   (Ralph is a dog $\vee$ Dusty is a horse) $\wedge$ Howie is a cat
   Therefore:
   Howie is a cat

$$\frac{(p_1 \vee p_2) \wedge p_3}{p_3}$$

Explanation: Without a context there is no preference for reading the hypothesis as $(A \vee B) \wedge C$ or as $A \vee (B \wedge C)$. Regimenting formalizations, say always associating to the left, will simplify, but may not respect the role of propositions in deductions. Here, assuming the deduction is put forward as valid, context does determine a reading for us.

2. **Ralph is a dog and George is a duck and Howie is a cat**

   (Ralph is a dog $\wedge$ George is a duck) $\wedge$ Howie is a cat

   $(p_1 \wedge p_2) \wedge p_3$

Explanation: It doesn't really matter whether we take the formalization I gave or:

Ralph is a dog $\wedge$ (George is a duck $\wedge$ Howie is a cat)

In every model these two formalizations will have the same truth-value: conjunction is *associative.* Our formal language imposes a precision that is not observed or needed in ordinary speech, and, as we want, that further precision does not contradict our intuition that the placement of parentheses (or commas) in the example is really immaterial.

For convenience let us adopt the rule that any conjunction of three or more conjuncts that has no parentheses will be read as *associating the conjuncts to the left,* as in the formalization I gave, and similarly for disjunctions.

3. Ralph is a dog or he's a puppet

Ralph is a dog $\vee$ Ralph is a puppet

$p_1 \vee p_2$

Explanation: First we need to make explicit the reference of 'he'. We rewrite the proposition as:

Ralph is a dog or Ralph is a puppet

We've agreed to formalize 'or' as inclusive disjunction. Here that seems odd, for we suppose the alternatives are mutually exclusive. But that is an observation about the truth-values of the constituent propositions, not about the use of 'or'. When we intend 'or' to be taken exclusively, let us require that the proposition be formulated using 'or else' instead of 'or'.

Richards, 1989, and *Girle, 1989,* debate the formalization of 'or' as exclusive or inclusive disjunction in propositional logic.

4. Ralph is a dog if he's not a puppet

$\neg$(Ralph is a puppet) $\rightarrow$ Ralph is a dog

$\neg p_2 \rightarrow p_1$

Explanation: We rewrite the proposition as 'Ralph is a dog if Ralph is not a puppet'.

Should 'if' be formalized as a conditional? Is our rewrite equivalent to 'If Ralph is not a puppet, then Ralph is a dog'? It seems so, and the formalization then follows.

5. Ralph is a dog although he's a puppet

Ralph is a dog $\wedge$ Ralph is a puppet

$p_1 \wedge p_2$

Explanation: Again we replace 'he' with 'Ralph'. Then we have two propositions: 'Ralph is a dog' and 'Ralph is a puppet'. Can we model 'although' with our propositional connectives? In this example that word seems to function as 'and' plus some signal that we wouldn't normally conjoin these propositions. So in classical logic, which ignores all but truth-values and form, we would be justified in formalizing the proposition using $\wedge$, although that need not be a general convention.

6. Ralph is not a dog because he's a puppet

Not formalizable

Explanation: Rewriting 'he' as 'Ralph', we have two propositions: 'Ralph is not a dog' and 'Ralph is a puppet'. But can we model 'because' as a propositional connective? Compare:

(a) Ralph is a dog because Cedar City is a seaport

(b) Ralph is a dog because Cedar City is not a seaport

I take it that both (a) and (b) are false, and that is independent of whether 'Ralph is a dog' and 'Cedar City is a seaport' are true or false. Hence, 'because' does not act truth-functionally, so in classical logic we cannot formalize these or Example 6. Nor can we take the example as atomic, for from it we can conclude 'Ralph is not a dog', and so an informal deduction that we recognize as valid would not be respected by the formal versions. Other logics attempt to model 'because' with the conditional or other propositional connectives.

A colleague suggested to me that we should construe the example as a deduction: 'Ralph is a puppet' *therefore* 'Ralph is not a dog'. But that fails to capture the intuition that both (a) and (b) should be classified as false.

7. Three faces of a die are even numbered
Three faces of a die are not even numbered
***Therefore*:**
Ralph is a dog

Explanation: It is true that in classical logic anything follows from a contradiction, and the two hypotheses appear to have the form of a contradiction. But this only illustrates that not every use of 'not' creates a negation, for the two hypotheses are both true. The word 'not' applies to 'even numbered' only; it cannot be assimilated to a use of 'It is not the case that . . .'.

8. Ken took off his clothes and went to bed

Explanation: Here 'and' is being used in the sense of 'and then'. The example was true last night, but 'Ken went to bed and took off his clothes' was false. No connective of classical logic can formalize this use of 'and', certainly not $\wedge$.

9. The quotation marks are signals for you to understand what I mean; they are not part of the realization

The quotation marks are signals for the reader to understand what Richard L. Epstein means $\wedge$ $\neg$(the quotation marks are part of the realization)

$p_1 \wedge \neg p_2$

Explanation: There is no word that could be construed as equivalent to 'and' in this example, yet I have used $\wedge$ in the formalization. Semi-colons, commas, and other punctuation are often used in English in place of connectives, and should be formalized accordingly.

This example comes from Chapter I.G and illustrates the difficulties in replacing indexicals and avoiding ambiguity. The 'you' refers to the reader of the text, and more precise than that I cannot be. The 'I' refers to the author, and that we can make precise. There are two plural nouns that could be the antecedent of 'they', but our understanding of English and the context makes 'the quotation marks' and not 'signals' the right choice. The resulting formalization is, I believe, adequate to use in reasoning.

10. Every natural number is even or odd

Every natural number is even or odd

p_1

Explanation: The example is atomic. Propositional logic does not recognize the internal complexity of it. We cannot formalize 'or' by $\vee$ here, for the example is certainly not equivalent to 'Every natural number is even or every natural number is odd.'

11. If Ralph is a dog, then Ralph barks
Ralph barks
***Therefore*:**
Ralph is a dog

Ralph is a dog $\rightarrow$ Ralph barks
Ralph barks
Therefore:
Ralph is a dog

$$\frac{\begin{array}{l} p_1 \rightarrow p_2 \\ p_2 \end{array}}{p_1}$$

Explanation: This is an example of the fallacy of arguing from the consequent. The deduction is not valid: Ralph could be a hyena or a parrot trained to make barking noises, in which case the premisses would be true but the conclusion false.

12. Suppose $\{s_n\}$ is monotonic. Then $\{s_n\}$ converges if and only if it is bounded

$\{s_n\}$ is monotonic $\rightarrow$ ($\{s_n\}$ converges $\leftrightarrow$ $\{s_n\}$ is bounded)

$p_1 \rightarrow (p_2 \leftrightarrow p_3)$

Explanation: In formalizing this theorem from a mathematical textbook on real analysis, *Rudin, 1976* (p. 47), I have changed the grammar significantly. It might seem that we should formalize the example as:

{ $\{s_n\}$ is monotonic }
Therefore: $\{s_n\}$ converges $\leftrightarrow$ $\{s_n\}$ is bounded

That would not be wrong, but it seems to me that when we state a theorem in mathematics we are normally asserting that a proposition is true, not that a deduction is valid. The words 'suppose', 'let', and others that we might take as indicating a deduction are, I believe, best formalized as indicating an antecedent of a conditional when used in mathematical texts.

The context of the discussion in *Rudin, 1964,* indicates what $\{s_n\}$ is. Note that the propositional form of the propositions does not recognize that all three propositions are about $\{s_n\}$.

13. Let A and B be sets of real numbers such that
(a) every real number is either in A or in B;
(b) no real number is in A and in B;
(c) neither A nor B is empty;
(d) if $\alpha \in A$, and $\beta \in B$, then $\alpha < \beta$.
Then there is one (and only one) real number γ such that $\alpha \leq \gamma$ for all $\alpha \in A$, and $\gamma \leq \beta$ for all $\beta \in B$.

{(A is a set of real numbers $\wedge$ B is a set of real numbers) $\wedge$ (every real number is either in A or in B $\wedge$ (no real number is in A and in B $\wedge$ (($\neg$(A is empty) $\wedge$ $\neg$(B is empty)) $\wedge$ ($\alpha \in A \wedge \beta \in B \rightarrow \alpha < \beta$)))) } $\rightarrow$ there is one and only one real number γ such that $\alpha \leq \gamma$ for all $\alpha \in A$ and $\gamma \leq \beta$ for all $\beta \in B$

$((p_1 \wedge p_2) \wedge (p_3 \wedge (p_4 \wedge ((\neg p_5 \wedge \neg p_6) \wedge (p_7 \wedge p_8 \rightarrow p_9))))) \rightarrow p_{10}$

Explanation: I take 'Let' to indicate that the antecedent of a conditional follows; 'such that' to indicate that further conditions are conjoined; the semi-colons to be conjunctions; and 'neither . . . nor . . .' to be formalized according to our earlier convention (p. 34). The grouping of the conjuncts is not entirely arbitrary, though the association to the right in the formalizations of (a)–(d) is. I'll let you convince yourself that 'either . . . or' in (a), and 'and' and 'no' in (b) should not be formalized as propositional connectives here.

Despite the considerable complexity of the propositional form of this example (Dedekind's Theorem from p. 9 of *Rudin, 1964*), note how much is ignored in the formalization. The use of variables, of quantifiers, and of mathematical terms is not recognized in propositional logic. To take account of those we would need to look at the internal structure of propositions, the subject of the volume *Predicate Logics*.

Exercises for Sections E and F

1. What do we mean by 'the logical form of a proposition'? How does the definition given in this text differ from what a platonist would count as logical form?
2. Write down in your own words and justify with examples each of the criteria of formalization.
3. Distinguish between a sentence being vague and a sentence being ambiguous, giving examples of both.
4. a. Present and argue for formalizations of the following English connectives in classical logic:

   not both A and B

   A unless B

   B just in case A

   when A, B

   A, when B

   b. Give at least two examples of English connectives that it would be odd to consider as truth-functional.

   c. Give two examples of apparently complex propositions that must be taken to be atomic if formalized in classical logic.
5. Formalize the following or explain why they cannot be formalized in classical logic according to the format of Section F:

   a. 7 is not even

   b. 7 is not even or odd

   c. If 7 is even, then 7 is not odd

   d. If logic is hard, then art history isn't hard

   e. Either Ralph and Juney will walk to the party on Saturday, or Ralph will drive

   f. If the play is sold out then Don will be here and Laurie will meet us tonight

g. Ralph barks but George is a duck
h. Unless the owner of Milt's restaurant has a dog, Anubis is the best-fed dog in Cedar City
j. Horses eat grass because grass is green
k. It's impossible that $2 + 2 \neq 4$
l. Anubis, you know, eats, sleeps, and barks at night
m. Jack Sprat is fat; his wife Jane is lean
n. If acid and water are mixed and you do not wish to be burned, then you should be careful
o. When acid and water are mixed, you should be careful if you don't wish to get burned
p. You can't make an omelette without breaking eggs

6. Formalize the following arguments and evaluate them for validity in classical logic according to the format of Section F:

a. If strawberries are red, then some colorblind people cannot see strawberries among their leaves
Strawberries are red
Therefore:
Some colorblind people cannot see strawberries among their leaves

b. If cat owners' homes have fleas, then cats are nasty. If cat owners' homes smell bad, then cats are nasty. But cat owners' homes always have fleas or smell bad. So cats are nasty.

c. Dogs are a man's best friend. A friend is loyal. Hence, dogs are loyal.

d. The students are happy if and only if no test is given. If the students are happy, the professor feels good. But if the professor feels good, he won't feel like lecturing, and if he doesn't feel like lecturing a test is given. Therefore, the students are not happy. (*Mates, 1965,* p. 109.)

e. Ralph is Polish, and it's not the case that Ralph is from New York or Virginia
If Ralph is from Syracuse, then Ralph is from New York or Virginia
Therefore:
Ralph is not from Syracuse

f. If you know some logic, you are either very bright or you study very hard
You study very hard
You are very bright
Therefore:
You know some logic

g. If Ralph is a cat, then Ralph meows
Ralph is not a cat
Therefore:
Ralph does not meow

h. Ralph is a dog
Therefore:
Ralph is a dog

j. Either the moon is made of green cheese or $2 + 2 = 4$
The moon is not made of green cheese
Therefore:
$2 + 2 = 4$

k. Either the moon is made of green cheese or $2 + 2 = 5$
The moon is not made of green cheese
Therefore:
$2 + 2 = 5$

l. The government is going to spend less on health and welfare
If the government is going to spend less on health and welfare, then either the government is going to cut the Medicare budget or the government is going to slash spending on housing
If the government is going to cut the medicare budget, the elderly will protest
If the government is going to slash spending on housing, then advocates for the poor will protest
Therefore:
The elderly will protest or the poor will protest

m. All men are mortal
Socrates is a man
Therefore:
Socrates is mortal

n. Socrates exists and Socrates does not exist
Therefore: There is a stick standing in the corner

7. Argue against classifying as valid an argument whose hypotheses are not true.

8. Formalize two theorems (principles) from a mathematics textbook, a biology textbook, a psychology textbook, and an economics textbook.

G. Further Abstractions: The Role of Mathematics in Logic

In classical propositional logic every atomic proposition is abstracted to just its truth-value. Nothing else matters. Thus if we have two models such that for each propositional variable p_i the one model assigns to p_i a proposition which is true if and only if the other model assigns one that is true, then the two models are indistinguishable for the purposes of classical logic. The propositions in one may be about mathematics and in the other about animal husbandry, but that cannot enter into our deliberations if we are using classical logic.

It is appropriate to ignore those differences between models of type **I** that do not matter to classical logic. We can simplify models to:

II

$$L(p_0, p_1, \ldots \neg, \rightarrow, \wedge, \vee) \xrightarrow{\ \upsilon,\ \text{truth-tables}\ } \{T, F\}$$

Here υ assigns truth-values to the variables $p_0, p_1, \ldots$ and is extended to all formal wffs by the classical truth-tables for $\neg$, $\rightarrow$, $\wedge, \vee$ (p. 23). We often name this abstracted form of a model by its valuation, υ.

Our language and semantics have now become formal and abstract. This is a virtue, for with these abstract systems we can analyze many propositions which would have been nearly intractable before because of their complexity. Simplification and abstraction are equally important because they allow us to use mathematics to establish general results about our logic. But these results will have significance for us only if we remember that models of type **II** come from models of type **I**.

I now want to discuss certain abstractions and idealizations that will allow us to apply mathematics more fruitfully to our logic. But first we need to consider the relation between mathematics and logic.

At the beginning of this century *Whitehead and Russell, 1910–1913,* developed a great deal of mathematics from a basis of formal logic. That was part of a program to justify the idea that mathematics is a part of logic. That view, called the *logicist* conception of mathematics, is still an important current in the philosophy of mathematics.

On the other hand there were those such as *Tarski, 1936,* who believed that we are justified in using all of mathematics to develop logic. No work could be called logical if it were not sufficiently mathematicized, made precise by mathematical definitions and axiomatizations. For example, before Tarski defined a formal language he gave a mathematical theory of concatenations of symbols. It would seem from this point of view that logic is part of mathematics.

I believe that there is more of a symbiosis between mathematics and logic. In any advanced parts of mathematics we want to be able to use logic in establishing theorems. For instance, one could argue that you need formal logic in order to give an axiomatization of the theory of symbol concatenation. And we want to be able to claim that the reasoning we do in mathematics is of the same sort as we've formalized in logic, perhaps supplemented by a few special forms of reasoning such as mathematical induction. Yet to develop logic to any degree useful to mathematics and the sciences and to establish any really general results about logic, we need to make logic more mathematical and use mathematics on it. There is no clear division between the disciplines. To my way of thinking both are involved in abstracting a great deal from experience.

What we must do, I believe, is use our common sense, our ordinary reasoning and understanding of language and the world, in establishing formal logic. That same reasoning is what is used in mathematics. An informal mathematics will then be useful in further abstractions and methods of analysis of logic. It is, however, important to see where those further mathematical abstractions and applications are made, so that one who disagrees with them can go with us up to that point and no further.

Propositions, as I've presented them, are written or uttered. There can only be a finite, though perhaps potentially infinite, number of them. We can view any fragment of the formal or semi-formal language as a completed whole, say all wffs using fewer than 47 symbols made up from $p_0, p_1, \ldots, p_{13}$, or the realizations of those in a particular model. We could achieve more generality by establishing results for all wffs using fewer than m symbols made up from $p_0, p_1, \ldots, p_n$ where m and n are any natural numbers 1, 2, 3, ... Already this is an abstraction. Are we to suppose that '¬¬¬¬¬¬¬¬¬¬¬¬¬(Ralph is a dog)' is of significance in reasoning? We would never normally use such a proposition, or if you think we would then come up with your own example of a proposition preceded by so many negations that it seems preposterous to conceive of reasoning with it. Yet by including all such propositions and well-formed-formulas in our logical analyses we may simplify our work considerably. And who is to say with certainty that some mechanical procedure for analyzing the truth-value of a relatively simple compound proposition may never need such a complex proposition? Let us denote:

PV $\equiv_{\text{Def}}$ the collection of propositional variables

It is my preference to treat this collection, as well as the formal and semi-formal languages, as potentially infinite collections, capable of being extended as needed. That assumption, however, is not consonant with some of the logics we study in this text. The mathematics applied to some logics requires that we view these collections as completed infinite wholes. I will try to point out when we will treat these syntactic collections as completed infinities. I will use the symbols of the abstract theory of collections, set theory, throughout this text, without necessarily making infinitistic assumptions. In particular, I will enclose by parentheses of the sort '{ }' either the names of several things or a description of things to indicate that those things are to be taken as a collection; '$\in$' is used for 'is an element of', and '$\notin$' for 'is not an element of', and '$\varnothing$' for the empty collection consisting of no objects whatever; (α, β) indicates an ordered pair, the first element of which is α, the second β.

For the semantics, we now treat a model of type **II** as a mathematical function:

$v: \mathrm{PV} \rightarrow \{\mathrm{T}, \mathrm{F}\}$

This function is extended inductively to all wffs by the truth-tables. Then we make a powerful generalization for classical logic:

The Fully General Classical Abstraction Any function $v: \mathrm{PV} \rightarrow \{\mathrm{T}, \mathrm{F}\}$ which is extended to all wffs by the truth-tables is a model.

Thus not only does it not matter how a model of type **II** arises, we assume any mathematically possible one can arise. In particular, we assume that we can independently assign truth-values to all the variables in PV. All the definitions in Section D are now to be understood in terms of these models and assumptions.

These assumptions are not about logic but about how to do the mathematics of metalogic. They apply to models of type **II**, not type **I**. The simplicity and generality we get by making these assumptions is justified so long as we don't arrive at any contradiction with our previous more fundamental assumptions and intuitions when we apply a metalogical result to actual propositions.

Someone who holds a platonist conception of logic and views propositions as abstract things would most likely take as basic facts about the world what I have called assumptions, idealizations, and generalizations. Words and sentences are understood as abstract objects that sometimes can be represented as inscriptions or utterances; they are not equivalences we impose on the phenomena of our experience: all of them exist whether they are ever uttered or not. The formal and informal language and any model are abstract objects complete in themselves, composed of an infinity of things. Though the distinction between a model of type **I** and **II** can be observed by a platonist, it does not mater whether a model is ever actually expressed or brought to our attention. It simply exists and hence the *Fully General Classical Abstraction* is no abstraction but an observation about the world.

H. Induction

The fundamental mathematical tool used throughout the study of logic is induction. I include a brief description here.

Consider an example. We wish to prove that for every natural number n:

$$1 + 2 + \cdots + n = \tfrac{1}{2}\, n \cdot (n + 1)$$

We test this for some small numbers, say 1, 2, 3, and 4. It looks right. But how are we to proceed?

We've checked the result when $n = 1$. This the *basis of the induction.*

Let's assume now that we've proved it for some number m:

$$1 + 2 + \cdots + m = 1/2\, m \cdot (m + 1)$$

This is called the *induction hypothesis.* Then we try to use this hypothesis to prove that the theorem must be true for the next larger natural number, $m + 1$:

$$1 + 2 + \cdots + m + (m + 1) = (1/2\, m \cdot (m + 1)) + (m + 1)$$

Therefore:

$$1 + 2 + \cdots + m + (m + 1) = 1/2\, (m^2 + m) + 1/2\, (2m + 2)$$

Therefore:

$$1 + 2 + \cdots + m + (m + 1) = 1/2\, (m^2 + 3m + 2)$$

Therefore:

$$1 + 2 + \cdots + m + (m + 1) = 1/2\, (m + 1) \cdot (m + 2)$$

So:

$$1 + 2 + \cdots + m + (m + 1) = 1/2\, (m + 1) \cdot ((m + 1) + 1)$$

That is, we have shown that the theorem is true for $m + 1$ if it is true for m, for the right-hand side of the last equation is exactly what we needed for that. We claim then that the theorem must be true for every natural number.

In summary, the method of induction is:

> We show that the statement is true for the number 1. Then we *assume* it is true for m, an arbitrary but fixed number, and show that it's true for $m + 1$. Then we conclude that it's true for all numbers.

Why can we then claim the statement is true for all numbers? It's true for 1; so it's true for 2; since it's true for 2, it's therefore true for 3; and so on.

Those little words 'and so on' carry a lot of weight. We believe that the natural numbers are completely specified by their method of generation: add 1, starting at 0.

$$0 \quad 1 \quad 2 \quad 3 \quad 4 \quad 5 \quad 6 \quad 7 \ldots$$

To prove a statement A by induction, we first prove it for some starting point in this list of numbers, usually 1, but just as well 0 or 47. We then establish that we have a method of generating proofs which is exactly analogous to the method of generating natural numbers: if A(n) is true, then A($n + 1$) is true. We have the list:

A(0),	if A(0), then A(1), so A(1)	if A(1), then A(2), so A(2)	if A(2), then A(3), ... so A(3)

Then the statement is true for all natural numbers equal to or larger than our initial point, whether the statement is for all numbers larger than 1 or all numbers larger than 47.

In essence we have only one idea: a process for generating objects one after another without end, in one case numerals or numbers, and in the other, proofs. We believe induction is a correct form of proof because the two applications of the single idea are matched up. To deny proof by induction amounts to denying that the natural numbers are completely determined by the process of adding 1 or that we can deduce a proposition C from the propositions $B \to C$ and B.

Generally we will use induction for objects we can number. In the following sections there will be many applications of induction, including examples beginning with basis larger than 1.

An alternative version of the principle of induction that we often use is the following:

> We show that the statement is true for the number 1. Then we *assume* it is true for *all numbers less than* m, an arbitrary but fixed number, and show that it's true for $m + 1$. Then we conclude that it's true for all numbers.

I ask you to show in Exercise 6 that this is equivalent to the first version of induction.

Exercises for Sections G and H

1. a. Explain the difference between a model of type **I** and a model of type **II**.
   b. Give an example of two very different models of type **I** that result in the same model of type **II**.
   c. Why is it appropriate to identify a model with its valuation?
2. Does the *Fully General Classical Abstraction* expand the class of models we must consider in determining validity in classical logic? Respond both in terms of the viewpoint of Section G and from the viewpoint of a platonist.
3. Formulate mathematically and prove by induction:

   The sum of the first n odd numbers is n^2
4. Prove by induction: $1^2 + 2^2 + \cdots + n^2 = \frac{1}{6} n \cdot (n+1) \cdot (2n+1)$.
5. Explain what is wrong with the following proof by induction that in every finite collection of natural numbers all of the numbers are equal.

   > The statement is true for any collection with just one natural number, a, for $a = a$. So suppose it is true for any collection of n natural numbers.
   >
   > Let $a_1, a_2, \ldots, a_n, a_{n+1}$ be any collection of $n+1$ natural numbers. By induction hypothesis, $a_1 = a_2 = \cdots = a_n$. But also we

have $a_2 = \cdots = a_n = a_{n+1}$, because here, too, there are only n numbers. And so $a_1 = a_2 = \cdots = a_n = a_{n+1}$.

6. Prove that the second formulation of induction is equivalent to the first formulation.

J. A Mathematical Presentation of PC

In this section I give a more mathematical treatment of the definitions of the formal logic and the semantics presented earlier, using mathematics to prove some theorems about the syntax and semantics. This section will be used as a reference in all later chapters.

1. The formal language

We have the following primitives:

binary connectives	$\rightarrow, \wedge, \vee$
unary connective	$\neg$
parentheses	(,)
variables	$p_0, p_1, \ldots$

We can use p, q, $q_0, q_1, \ldots$ as metavariables ranging over PV = $\{ p_0, p_1, \ldots \}$. Recall the inductive definition of *wff* from Chapter I.G.

***An Inductive Definition of* wff**

We assign the number:

1	to	(p_i) for each $i = 0, 1, \ldots$
$n+1$	to	$(\neg A)$, $(A \rightarrow B)$, $(A \wedge B)$, $(A \vee B)$ the maximum of the numbers assigned to A and to B is n

A concatenation of symbols is then a *wff* if it is assigned some $n \geq 0$.

The collection of wffs will be denoted *Wffs*.

We can define a formal language using fewer of these connectives by deleting the appropriate clauses in the definition. We may also define a language with any other n-ary connective, γ, by adding:

if the maximum of the numbers assigned to $A_1, \ldots, A_n$ is m then $\gamma(A_1, \ldots, A_n)$ is assigned number $m + 1$

All the definitions and theorems of this section (K.1) can be generalized to any such language, and I will assume those generalizations when needed. Sometimes I refer to a formal language by the primitive connectives it is based on, for example, the language of $\neg$ and $\wedge$.

We need to show that there is only one way to read each wff.

Theorem 1 Unique Readability of Wffs

There is one and only one way to parse each wff.

Proof: To each primitive symbol α of the formal language assign an integer $\lambda(\alpha)$ according to the following chart:

(	-1		$\neg$	0
)	1		$\rightarrow$	0
p_i	0	for each $i \geq 0$	$\wedge$	0
			$\vee$	0

To the concatenation of symbols $\alpha_1 \alpha_2 \cdots \alpha_n$ assign the number:

$$\lambda(\alpha_1 \alpha_2 \cdots \alpha_n) = \lambda(\alpha_1) + \lambda(\alpha_2) + \cdots + \lambda(\alpha_n)$$

First I will show that for any wff A, $\lambda(\text{A}) = 0$. I'll use induction on the number of occurrences of $\neg, \rightarrow, \wedge, \vee$ in A.

If there are no occurrences, then A is atomic, that is, for some $i \geq 0$, A is (p_i). Then $\lambda(\text{'('}) = -1$, $\lambda(p_i) = 0$, and $\lambda(\text{')'}) = 1$. Adding, we have $\lambda(\text{A}) = 0$.

Suppose the lemma is true for every wff that has fewer occurrences of these symbols than A does. Then there are 4 cases, which we cannot yet assume are distinct: A arises as $(\neg\text{B})$, $(\text{B}\rightarrow\text{C})$, $(\text{B}\wedge\text{C})$, or $(\text{B}\vee\text{C})$. By induction $\lambda(\text{B}) = \lambda(\text{C}) = 0$, so adding in each case, we have $\lambda(\text{A}) = 0$.

Now I'll show that, if reading from the left α is an initial segment of a wff other than the entire wff itself, then $\lambda(\alpha) < 0$; and if α is a final segment other than the entire wff itself, then $\lambda(\alpha) > 0$. So no proper initial or final segment of a wff is a wff. To establish this I will again use induction on the number of occurrences of connectives in the wff. I'll let you establish the base case for atomic wffs, where there are no, that is zero, connectives.

Now suppose the lemma is true for any wff that contains $\leq n$ occurrences of the connectives. If A contains $n + 1$ occurrences, then it must have (at least) one of the forms given in the definition of wffs. If A has the form $(\text{B}\wedge\text{C})$, then an initial segment of A must have one of the following forms:

i. (

ii. $(\beta$ where β is an initial segment of B

iii. $(\text{B}\wedge$

iv. $(\text{B}\wedge\gamma$ where γ is an initial segment of C

For (ii), $\lambda(\text{'('}) = -1$ and by induction $\lambda(\beta) < 0$, so $\lambda(\text{'(}\beta\text{'}) < 0$. I'll leave (i), (iii), and (iv) to you.

The other three cases follow similarly, and I'll leave those and the proof for final segments to you.

Now to establish the theorem, we proceed through a number of cases by way of contradiction. Suppose we have a wff that could be read as both $(A \wedge B)$ and $(C \rightarrow D)$. Then $A \wedge B)$ must be the same as $C \rightarrow D)$. Hence either A is an initial part of C or C is an initial part of A. But then $\lambda(A) < 0$ or $\lambda(C) < 0$, which is a contradiction, as we proved above that $\lambda(A) = \lambda(C) = 0$. Hence, A is C. But then we have that $\wedge B)$ is the same as $\rightarrow D)$, which is a contradiction.

Suppose $(\neg A)$ could be read as $(C \rightarrow D)$. Then $(\neg A$ and $(C \rightarrow D$ must be the same. Hence D must be a final segment of A other than A itself, but then $\lambda(D) > 0$, which is a contradiction.

The other cases are similar, and I'll leave them to you. ■

Because each wff has a unique reading, we may define the *principal* or *main connective* of $(A \rightarrow B)$ to be '$\rightarrow$' and the *immediate constituents* to be A and B, and similarly for other wffs. We sometimes refer to a wff which has, for example, '$\rightarrow$' as it's main connective as an '$\rightarrow$-wff'.

By Theorem 1 we may make the following definition of *the length of a wff*:

1. The length of (p_i) is 1.
2. If the length of A is n, then the length of $(\neg A)$ is $n+1$.
3. If the maximum of the lengths of A and B is n, then each of $(A \rightarrow B)$, $(A \wedge B)$, $(A \vee B)$ has length $n+1$.

I leave to you to show that the length of a wff is the same as the number assigned to it in the inductive definition of *wff*.

Now if we want to show that all wffs have some particular property we can use *induction on the length of wffs*: first show that the atomic wffs have the property and then show that if all wffs of length n have it, so do all wffs of length $n+1$. (In *Epstein and Carnielli, 1989,* you can find a discussion of the general notion of induction on inductively defined classes.)

In some proofs by induction it is useful to have an ordering of wffs. We make an ordering in three stages. First, the primitive symbols are alphabetized in the following order:

$$\neg, \rightarrow, \wedge, \vee, (,), p_0, p_1, p_2, \ldots$$

Then for each $n \geq 1$, we give a linear ordering of all wffs of length n:

> If A and B have the same length, and reading left to right at the first place at which they have different symbols A has the prior one in the alphabetization of primitive symbols, then A comes before B

Finally, we order all wffs by dovetailing these orderings:

A_0 is the first wff of length 1
A_1 is the second wff of length 1
A_2 is the first wff of length 2
A_3 is the third wff of length 1
A_4 is the second wff of length 2
A_5 is the first wff of length 3
⋮ ⋮

I leave as Exercise 6 for you to give a precise formulation of this procedure. Finally, we define what we mean by A is a *subformula* of B:

1. A is a subformula of A
2. If A is a subformula of C, then A is a subformula of $(\neg C)$, and for every D, A is a subformula of $(C \rightarrow D)$, $(D \rightarrow C)$, $(C \wedge D)$, $(D \wedge C)$, $(C \vee D)$, and $(D \vee C)$
3. A is a subformula of B only if it is classified as such by repeated application of (1) and (2)

We say that p_i *appears in* A if (p_i) is a subformula of A, and denote by 'PV(A)' the set $\{p_i : p_i \text{ appears in } A\}$.

When I write A(q) I mean that q appears in A ; C(A) means that A is a subformula of C.

Exercises for Section J.1

1. Prove that for any wff in $L(\neg, \rightarrow, \wedge, \vee)$ the length of the wff is the same as the number assigned to it in the inductive definition of *wff*.
2. a. Calculate the length of the following wffs.
   i. $((((p_1) \rightarrow (p_2)) \wedge (\neg(p_2))) \rightarrow (\neg(p_1)))$
   ii. $((((p_4) \wedge (p_2)) \vee (\neg(p_6))) \rightarrow ((p_7) \rightarrow (p_8)))$
   b. Give an unabbreviated example of a wff of length 10 that uses both binary and unary connectives. What is the minimum number of parentheses needed in a wff of length 10? The maximum number?
3. a. Give a full definition of *main connective* and *immediate consitutent(s)* of a wff.
   b. What is the main connective and immediate constituents of the wffs in Exercise 2.a?
4. Write the following wffs without abbreviations.
   a. $(p_1 \vee p_2)$

b. $(p_0 \wedge \neg p_1)$
c. $(p_0 \wedge \neg\neg p_1)$
d. $(p_1 \rightarrow p_2)$
e. $(p_1 \rightarrow (p_2 \rightarrow p_3))$
f. $(\neg p_1)$
g. $(\neg\neg p_1)$
h. $(\neg\neg\neg\neg p_1)$

5. List all subformulas of each of the wffs in Exercises 2 and 4. What propositional variables appear in them?

6. a. What is A_8? A_{11}?
   b. Fill out the description of the ordering of wffs by giving a formula for the *n*th wff being the *m*th wff of length *t*.
   c. Give the number of each wff in Exercise 4.

2. Models and the semantic consequence relation

A *model* for classical logic is a function $v: PV \rightarrow \{T, F\}$ which is extended to all wffs by the classical tables for $\neg$, $\rightarrow$, $\wedge$, $\vee$ (p. 23). The extension is unique since every wff has a unique reading. When it is necessary to distinguish a model of this logic from ones for other logics I will call v a **PC**-model.

We say that a *wff* A *is true or valid in model* v if $v(A) = T$, and notate that $v \models A$, often read 'v validates A'. We say that a wff A *is false in* v if $v(A) = F$, and write $v \not\models A$.

Letting Γ, Σ, Δ stand for collections of wffs, we write $\Gamma \models A$, read as Γ *validates* A or A *is a semantic consequence of* Γ, if given any model v which validates all the wffs in Γ we have that $v(A) = T$. We write $A \models B$ for $\{A\} \models B$, and $\models A$ for $\varnothing \models A$, that is, A is true in every model. If $\models A$, we say that A is a (**PC**-) *tautology* or is (**PC**-) *valid.* We say that v is a *model of* Γ, written $v \models \Gamma$, if $v \models A$ for every A in Γ. We say that a scheme of wffs is tautological if every instance of the scheme is a tautology, and similarly for schema of deductions.

Finally, we say that *two propositions* (or *wffs*) *are semantically equivalent* if each is a semantic consequence of the other: in every model they have the same truth-value. Thus, in classical logic '$\neg$(Ralph is a dog $\wedge$ $\neg$(George is a duck))' is semantically equivalent to 'Ralph is a dog $\rightarrow$ George is a duck', as you can show. Semantic equivalence is *one* formalization of the informal notion of two propositions meaning the same for all our logical purposes.

We denote as **PC** the collection of consequences $\{(\Gamma, A) : \Gamma \models A\}$. The relation $\models$ is called *the semantic consequence relation of* **PC** and is often denoted $\models_{\mathbf{PC}}$. Here are some of the more important properties of $\models_{\mathbf{PC}}$.

Theorem 2 Properties of the Semantic Consequence Relation

a. $A \models A$

b. If $\models A$, then $\Gamma \models A$

c. If $A \in \Gamma$, then $\Gamma \models A$

d. If $\Gamma \models A$ and $\Gamma \subseteq \Delta$, then $\Delta \models A$

e. ***Transitivity***
If $\Gamma \models A$ and $A \models B$, then $\Gamma \models B$

f. ***Transitivity, the Cut Rule***
If $\Gamma \models A$ and $\Delta, A \models B$, then $\Gamma \cup \Delta \models B$

g. If $\Gamma \cup \{A_1, \ldots, A_n\} \models B$ and $\Gamma \models A_i$ for $i = 1, \ldots, n$, then $\Gamma \models B$

(Parts (a)–(g) will apply to the semantic consequence relation of every logic presented in this book, whereas those that follow may not.)

h. $\models A$ iff for every nonempty Γ, $\Gamma \models A$

j. ***The Semantic Deduction Theorem for* PC**
$\Gamma, A \models B$ iff $\Gamma \models A \rightarrow B$

k. ***The Semantic Deduction Theorem for* PC *for Finite Consequences***
$\Gamma \cup \{A_1, \ldots, A_n\} \models B$ iff $\Gamma \models A_1 \rightarrow (A_2 \rightarrow (\cdots \rightarrow (A_n \rightarrow B) \cdots))$

l. ***Alternate version of* (k)**
$\Gamma \cup \{A_1, \ldots, A_n\} \models B$ iff $\Gamma \models (A_1 \wedge \cdots \wedge A_n) \rightarrow B$

m. ***Substitution***
If $\models A(p)$, then $\models A(B)$, where $A(B)$ is the result of substituting B uniformly for p in A (i.e., B replaces every occurrence of p in A)

Proof: I'll leave the proofs of all of these as exercises except for (h) and (m).

Part (h): The left to right direction follows from (b). For the other direction, if every nonempty Γ validates A, then in particular $\neg A \models A$. So by (j), $\models \neg A \rightarrow A$ and hence $\neg A \rightarrow A$ is true in every model. Therefore A is true in every model.

Part (m): The only place that p can occur in A(B) is within B itself. Hence given any model v, once we have evaluated $v(B)$ the evaluation of $v(A(B))$ proceeds as if we were evaluating A(p) in a model w, where w and v agree on all propositional variables except p and $w(p) = v(B)$. Since A(p) is true in every model, $w(A(p)) = T$ and so $v(A(B)) = T$. Hence, $\models A(p)$. ∎

Part (m) says that propositional variables are really variables that can stand for any proposition, a consequence of the *Division of Form and Content.*

Exercises for Section J.2

1. What does it mean to say that v validates a wff A?

2. What does '$v \models \Gamma$' mean?

3. What does '$\Gamma \models A$' mean?

4. Let $v: PV \to \{T, F\}$ be a function. Show that there is one and only one way to extend v to all wffs via the classical truth-tables.

5. a. Prove parts (a)–(g) and (j)–(l) of Theorem 2.
   b. Show that part (m) of Theorem 2 is false if the replacement is not uniform.

6. Prove that the following pairs are semantically equivalent.
   a. $(A \wedge A)$ $\quad$ A
   b. $(A \vee A)$ $\quad$ A
   c. $(A \wedge B) \wedge C$ $\quad$ $A \wedge (B \wedge C)$
   d. $(A \vee B) \vee C$ $\quad$ $A \vee (B \vee C)$
   e. $A \vee (B \wedge C)$ $\quad$ $(A \vee B) \wedge (A \vee C)$
   f. $A \wedge (B \vee C)$ $\quad$ $(A \wedge B) \vee (A \wedge C)$

3. The truth-functional completeness of the connectives

Suppose we add to the language a new connective $\gamma(A_1, \ldots, A_n)$ which we wish to interpret as a truth-function. Then we must specify its interpretation by a truth-table, which for every sequence of n T's and F's assigns either T or F. Doing this we say that we have added a *truth-functional connective* to the language.

I will show in this section that every truth-functional connective can be defined in classical logic in terms of $\neg, \to, \wedge, \vee$, that is: $\{\neg, \to, \wedge, \vee\}$ is *truth-functionally complete*. More precisely, given a truth-functional connective γ, we can find a schema $S(A_1, \ldots, A_n)$ that uses only $\neg, \to, \wedge, \vee$ such that for all wffs $A_1, \ldots, A_n$ we have that $S(A_1, \ldots, A_n)$ is semantically equivalent to $\gamma(A_1, \ldots, A_n)$. Thus, except for the convenience of abbreviation, further truth-functional connectives add nothing of significance to our language and semantics.

Theorem 3 In **PC** every truth-functional connective can be defined in terms of $\neg, \to, \wedge, \vee$.

Proof: Let γ be a truth-functional connective. If γ takes only the value F, then $\gamma(A_1, \ldots, A_n)$ is semantically equivalent to $(A_1 \wedge \neg A_1) \vee \cdots \vee (A_n \wedge \neg A_n)$. Otherwise let the following be those sequences of T's and F's that are assigned T by the table for γ ⊬:

$$\alpha_1 = (\alpha_{11}, \ldots, \alpha_{1n})$$
$$\vdots$$
$$\alpha_k = (\alpha_{k1}, \ldots, \alpha_{kn})$$

Define for $i = 1, \ldots, k$ and $j = 1, \ldots, n$:

$$B_{ij} = \begin{cases} A_{ij} & \text{if } \alpha_{ij} = \mathsf{T} \\ \neg A_{ij} & \text{if } \alpha_{ij} = \mathsf{F} \end{cases}$$

And $D_i = B_{i1} \wedge \cdots \wedge B_{in}$, associating to the left (this corresponds to the ith row of the table). Then $D_1 \vee D_2 \vee \cdots \vee D_n$, associating to the left, is semantically equivalent to $\gamma(A_1, \ldots, A_n)$. ■

As an example of this procedure, suppose we wish to define in **PC** a ternary connective γ with the following table:

A_1	A_2	A_3	$\gamma(A_1, A_2, A_3)$
T	T	T	T
T	T	F	F
T	F	T	F
T	F	F	T
F	T	T	T
F	T	F	T
F	F	T	F
F	F	F	F

Then:

$$\alpha_1 = (\mathsf{T},\mathsf{T},\mathsf{T}) \qquad \alpha_2 = (\mathsf{T},\mathsf{F},\mathsf{F}) \qquad \alpha_3 = (\mathsf{F},\mathsf{T},\mathsf{T}) \qquad \alpha_4 = (\mathsf{F},\mathsf{T},\mathsf{F})$$

and

$$D_1 = (A_1 \wedge A_2) \wedge A_3 \qquad D_2 = (A_1 \wedge \neg A_2) \wedge \neg A_3$$

$$D_3 = (\neg A_1 \wedge A_2) \wedge A_3 \qquad D_4 = (\neg A_1 \wedge A_2) \wedge \neg A_3$$

So $((D_1 \vee D_2) \vee D_3) \vee D_4$ is semantically equivalent to $\gamma(A_1, A_2, A_3)$.

The proof of Theorem 3 actually shows that $\{\neg, \wedge, \vee\}$ is truth-functionally complete. By noting the following semantic equivalences we can show that *each of* $\{\neg, \rightarrow\}$, $\{\neg, \wedge\}$, $\{\neg, \vee\}$ *is truth-functionally complete.*

(15) *Equivalences in* **PC**

(a) $A \rightarrow B \qquad \neg(A \wedge \neg B)$

(b) $A \rightarrow B \qquad \neg A \vee B$

(c) $A \wedge B \qquad \neg(\neg A \vee \neg B)$

(d) $A \wedge B \qquad \neg(A \rightarrow \neg B)$

(e) $A \vee B \qquad \neg(\neg A \wedge \neg B)$

(f) $A \vee B \qquad \neg A \rightarrow B$

However, $\{\rightarrow, \wedge, \vee\}$ *is not truth-functionally complete*: any wff made from these connectives using only p_0 will take value T if p_0 has value T, so $\neg$ cannot be defined.

There are two truth-functional connectives each of which is, by itself, truth-functionally complete. The first is the *Sheffer stroke* ('nand'):

$A | B$ is T iff not both A and B are T

Then $\neg A$ is semantically equivalent to $A | A$, and $A \rightarrow B$ to $A | \neg B$.

The other connective formalizes 'neither . . . nor . . .':

$A \downarrow B$ is true iff both A and B are false

Then $\neg A$ is semantically equivalent to $A \downarrow A$, and $A \wedge B$ to $\neg A \downarrow \neg B$.

4. The choice of language for PC

Since every truth-functional connective is definable in **PC** from $\neg$ and $\rightarrow$, the connectives $\wedge$ and $\vee$ are in some sense superfluous. We could take $L(\neg, \rightarrow, p_0, p_1, \ldots)$ as the language of classical logic and then define connectives $\wedge$ and $\vee$ as in table (15). We feel we have the same logic because we have a translation from the language $L(\neg, \rightarrow, \wedge, \vee, p_0, p_1, \ldots)$ to $L(\neg, \rightarrow, p_0, p_1, \ldots)$:

$$(p_i)^{\dagger} = p_i$$
$$(\neg A)^{\dagger} = \neg(A^{\dagger})$$
$$(A \rightarrow B)^{\dagger} = A^{\dagger} \rightarrow B^{\dagger}$$
$$(A \wedge B)^{\dagger} = \neg(A^{\dagger} \rightarrow \neg B^{\dagger})$$
$$(A \vee B)^{\dagger} = \neg(A^{\dagger}) \rightarrow B^{\dagger}$$

This translation is faithful to our models: given any $v: PV \rightarrow \{T, F\}$, if we first extend it to all wffs of $L(\neg, \rightarrow, \wedge, \vee, p_0, p_1, \ldots)$ and call that v_1, and then extend it to $L(\neg, \rightarrow, p_0, p_1, \ldots)$ and call that v_2, we have that $v_1 \vDash A$ iff $v_2 \vDash A^{\dagger}$.

Similarly, we feel we have the same logic if we formalize it using $\{\neg, \wedge\}$ or $\{\neg, \vee\}$ or any other truth-functionally complete set of connectives. We recognize this by saying, loosely, that **PC** *may be formalized in any of several languages*. And we may speak of *a formalization of* **PC** *in the language of* $L(\neg, \rightarrow, p_0, p_1, \ldots)$. These phrases are meant only to refer to the translations and equivalences we've established, and are not meant to suggest that there is some formless body of truths comprising **PC**. Similar comments will apply to any other logic we study in which some of the primitive connectives may be defined in terms of the others.

5. Normal forms

Given any wff we can find another wff that has a particularly simple form and that is semantically equivalent to it in **PC** using the proof of Theorem 3.

We say that a wff is in *disjunctive normal form* if it is a disjunction of conjunctions of variables or negations of variables; it is in *conjunctive normal form* if it is a conjunction of disjunctions of variables or negations of variables. For example:

$((p_1 \wedge p_2) \wedge (\neg p_1)) \vee ((\neg p_6 \wedge \neg p_8) \wedge p_9) \vee p_2)$ is in disjunctive normal form

$((p_1 \vee p_2) \vee (\neg p_1)) \wedge ((\neg p_6 \vee \neg p_8) \vee p_9) \wedge p_2)$ is in conjunctive normal form

In Exercise 6 above I asked you to show:

$(A \wedge A)$ is semantically equivalent to A

$(A \vee A)$ is semantically equivalent to A

So we can view either (p) or $(\neg p)$ as a disjunction or a conjunction. Thus a wff A is in disjunctive normal form if A is $B_1 \vee B_2 \vee \cdots \vee B_n$ where each B_i is, for some m, of the form $C_i \wedge \cdots \wedge C_m$, where each C_i is either a propositional variable or the negation of one. Further, the conjuncts or disjuncts may be associated in any way, since:

$(A \wedge B) \wedge C$ is semantically equivalent to $A \wedge (B \wedge C)$

$(A \vee B) \vee C$ is semantically equivalent to $A \vee (B \vee C)$

Theorem 4 ***The Normal Form Theorem for* PC** Given any wff A there is an effective procedure for finding a wff B which is in disjunctive normal form, contains exactly the same propositional variables as A, and is semantically equivalent to A in **PC**; similarly, there is an equivalent wff in conjunctive normal form which contains the same propositional variables.

Proof: Any wff A determines a truth-function: if $PV(A) = \{q_1, \ldots, q_n\}$, then we can make up a table with 2^n rows corresponding to the ways we can assign T or F to each of the variables. Then we can calculate the truth value of A under each assignment. The proof of Theorem 3 then establishes a disjunctive normal form for A. By repeated use of the following equivalences we can obtain a conjunctive normal form for A:

$A \vee (B \wedge C)$ is semantically equivalent to $[(A \vee B) \wedge (A \vee C)]$

$A \wedge (B \vee C)$ is semantically equivalent to $[(A \wedge B) \vee (A \wedge C)]$ ■

6. The principle of duality for PC

Consider the truth-conditions for $\wedge$ and $\vee$:

$A \wedge B$ is T iff both A, B are T; otherwise F

$A \vee B$ is F iff both A, B are F; otherwise T

These are dual to each other in the sense that one arises by reversing the roles of T and F in the other. It's no accident that the symbol for conjunction is an upside down $\vee$.

Let A be any wff that contains no occurrence of '$\rightarrow$'. The *dual of* A, notated here as A*, is the wff we get by replacing every occurrence of $\wedge$ by $\vee$ in A and every occurrence of $\vee$ by $\wedge$. For instance, the dual of $\neg(A \vee B)$ is $\neg(A \wedge B)$; the dual of $\neg A \wedge \neg B$ is $\neg A \vee \neg B$. The evaluation of A* proceeds exactly as for A except that the roles of T and F are reversed. So if A has the same truth-value as B in all models of **PC**, then A* will have the same truth-value as B* in all models of **PC**. Using the definition of '$\leftrightarrow$' (p. 34), we have the *principle of duality for* **PC**:

$$\vDash_{\mathbf{PC}} A \leftrightarrow B \text{ iff } \vDash_{\mathbf{PC}} A^* \leftrightarrow B^*$$

Exercises for Sections J.3–J.6

1. Define the following ternary connective in **PC** in the language $L(\neg, \rightarrow, \wedge, \vee)$. Use no abbreviations in your schema.

A_1	A_2	A_3	$\gamma(A_1, A_2, A_3)$
T	T	T	F
T	T	F	F
T	F	T	T
T	F	F	F
F	T	T	T
F	T	F	T
F	F	T	T
F	F	F	T

2. Show that each of $\{\neg, \rightarrow\}$, $\{\neg, \wedge\}$, $\{\neg, \vee\}$ is truth-functionally complete.

3. a. Show that no binary connective except $|$ and $\downarrow$ is truth-functionally complete.
   b. Exhibit a single ternary connective that is truth-functionally complete.

4. Put the following wffs in disjunctive normal form; then put each in conjunctive normal form.
   a. $p_1 \rightarrow (\neg p_2 \rightarrow p_3)$
   b. $(p_1 \rightarrow p_2) \rightarrow ((p_1 \rightarrow \neg p_2) \rightarrow \neg p_1)$
   c. $p_1 \rightarrow \neg\neg p_1$
   d. $\neg(p_1 \rightarrow (p_2 \rightarrow p_2))$

5. Write out the dual of each of the following wffs.
   a. $\neg(p_1 \wedge \neg\neg(p_2 \vee p_3))$

b. $\neg p_1 \wedge (\neg\neg p_2 \vee p_3))$

c. $p_1 \rightarrow \neg(p_2 \vee p_3)$

7. The decidability of tautologies

It is fairly easy to decide whether a given wff is a tautology, but it's worth pointing out exactly what that depends on. In doing so we'll see that our infinitistic methods have not fouled up what seems obvious.

a. In any model, the truth-value of any wff depends only on the truth-values of the propositional variables appearing in it.

b. Given any assignment of truth-values to a finite number of propositional variables, the assignment can be extended to all variables (for example, take F to be the value assigned to all other propositions).

c. Therefore, a wff is not a tautology iff there is an assignment of truth-values to the variables appearing in it for which it takes the value F. For if there is such an assignment, it can be extended to be a model in which the wff is false. And if there is no such assignment, then the wff must be a tautology.

d. For any assignment of truth-values to the variables appearing in a wff we can effectively calculate the truth-value of the wff by repeatedly using the classical tables for $\neg$, $\rightarrow$, $\wedge$, $\vee$ (p. 23).

Thus we can decide whether a wff is a **PC**-tautology by listing all possible truth-value assignments to the variables appearing in it—if there are n variables then there are 2^n assignments—and for each assignment mechanically checking the resulting truth-value for the wff. The wff is a tautology iff it always receives the value T. See *Epstein and Carnielli, 1989,* Chapter 19, for a formal presentation of this procedure, showing that it is computable.

Note that this decision procedure for the validity of wffs also allows us to decide whether any two given wffs are semantically equivalent: A is semantically equivalent to B iff $A \leftrightarrow B$ is a tautology.

This decision procedure for validity, though effective, is practically unusable for wffs containing, say, 47 or more propositional variables, for we would have to check some 2^{47} assignments. There are many shortcuts we can take, quite a few of which are spelled out in *Quine, 1950.* But to date no one has come up with a method of checking validity that is substantially faster and could be run on a computer in less than (roughly) exponential time relative to the number of variables appearing in the wff; it has been conjectured that there is no such faster method.

One particular way of using (a)–(d) above is worth noting: we attempt to falsify the wff. If we can, it's not a tautology, otherwise it is. This is particularly useful for $\rightarrow$-wffs because only one line of the table yields F: $A \rightarrow B$ can be falsified iff there is an assignment that makes both A true and B false. Generally we consider schema rather than particular wffs. For example:

	$\neg(A \wedge B) \rightarrow (B \rightarrow A)$		
is false iff	T	F	
which is iff	$A \wedge B$	B	A
are	F	T	F

And this is the falsifying assignment. Hence the scheme is not tautological.

Similarly,	$((A \wedge B) \rightarrow C) \rightarrow (A \rightarrow (B \rightarrow C))$			
is false iff	T	F		
iff		A	$B \rightarrow C$	
are		T	F	
iff			B	C
are			T	F

But if A is T, B is T, and C is F, then $(A \wedge B) \rightarrow C$ is F, so there is no way to falsify the scheme. Hence it is a tautology.

Exercises for Sections J.7

1. Determine whether the following are classical tautologies. Here p, q, and r are used in place of p_1, p_2, and p_3.

   a. $(p \wedge q) \rightarrow (p \vee q)$

   b. $(p \vee q) \rightarrow (p \wedge q)$

   c. $(p \rightarrow q) \rightarrow (\neg p \vee q)$

   d. $(p \wedge q) \rightarrow p$

   e. $[\neg(p \wedge q) \wedge p] \rightarrow \neg q$

   f. $\neg q \rightarrow [\neg(p \wedge q) \wedge p]$

   g. $[(p \rightarrow q) \wedge (\neg p \rightarrow r)] \rightarrow (q \vee r)$

   h. $\neg(p \wedge \neg\neg p)$

   j. $\neg p \vee \neg\neg p$

   k. $\neg\neg p \vee \neg\neg\neg p$

   l. $\neg\neg\neg p \vee \neg\neg\neg\neg\neg p$

   m. $((p \vee q) \wedge \neg p) \rightarrow q$

   n. $(q \vee \neg q) \rightarrow q$

   o. $q \rightarrow (q \vee \neg q)$

   p. $q \rightarrow \neg q$

   r. $\neg\neg q \rightarrow q$

   s. $\neg(\neg p \vee \neg\neg p) \vee p$

t. $p \rightarrow (p \vee q)$

u. $\neg p \rightarrow (p \rightarrow q)$

v. $p \rightarrow (q \rightarrow p)$

w. $((p \leftrightarrow \neg q) \leftrightarrow \neg p) \leftrightarrow q$

x. $[q \leftrightarrow (r \rightarrow \neg p)] \vee [(\neg q \rightarrow p) \leftrightarrow r]$

y. $[(p \rightarrow q) \wedge (\neg p \rightarrow r)] \rightarrow (q \vee r)$

2. Evaluate whether the following are classically valid deductions.

a. $\dfrac{A \wedge B}{A \vee B}$

b. $\dfrac{A \vee B,\ \neg A}{B}$

c. $\dfrac{A \wedge \neg A}{B}$

d. $\dfrac{(A \wedge \neg\neg A) \vee B}{\neg B}$

e. $\dfrac{A \rightarrow \neg B,\ B \wedge \neg C}{A \rightarrow C}$

f. $\dfrac{A \rightarrow \neg\neg B,\ \neg C \vee A,\ C}{B}$

8. Some PC-tautologies

Here are some **PC**-tautologies, given in the form of schema. These schema can be seen as the syntactic codification of many of our assumptions, especially since we may identify the validity of an $\rightarrow$ -scheme with an assertion about semantic consequence in **PC**.

1.	$A \vee \neg A$	*excluded middle, tertium non datur* (a third way is not given)
2.	$\neg(A \wedge \neg A)$	*law of noncontradiction*
3.	$\neg\neg A \rightarrow A$ $A \rightarrow \neg\neg A$	*laws of double negation*
4.	$(A \wedge B) \rightarrow A$ $(A \wedge B) \rightarrow B$	*laws of simplification for conjunction*
5.	$(A \wedge A) \leftrightarrow A$	*principle of tautology for conjunction*
6.	$(A \wedge B) \rightarrow (B \wedge A)$	*commutativity of conjunction*

7.	$((A \wedge B) \wedge C) \leftrightarrow (A \wedge (B \wedge C))$	*associativity of conjunction*
8.	$A \rightarrow (A \vee B)$	*laws of addition for disjunction*
	$B \rightarrow (A \vee B)$	
9.	$(A \vee A) \leftrightarrow A$	*principle of tautology for disjunction*
10.	$(A \vee B) \rightarrow (B \vee A)$	*commutativity of disjunction*
11.	$((A \vee B) \vee C) \leftrightarrow (A \vee (B \vee C))$	*associativity of disjunction*
12.	$A \vee (B \wedge C) \leftrightarrow ((A \vee B) \wedge (A \vee C))$	*distribution laws for*
	$A \wedge (B \vee C) \leftrightarrow ((A \wedge B) \vee (A \wedge C))$	*conjunction and disjunction*
13.	$\neg(A \wedge B) \leftrightarrow (\neg A \vee \neg B)$	*De Morgan's laws*
	$\neg(A \vee B) \leftrightarrow (\neg A \wedge \neg B)$	
14.	$((A \vee B) \wedge \neg A) \rightarrow B$	(a propositional form of) *disjunctive syllogism*
15.	$A \rightarrow A$	*identity*
16.	$(\neg A \rightarrow A) \rightarrow A$	*Clavius' law, consequentia mirabilis*
17.	$A \rightarrow (B \rightarrow A)$	*paradoxes of material implication*
	$\neg A \rightarrow (A \rightarrow B)$	
18.	$A \rightarrow (B \rightarrow B)$	*paradoxes of strict implication*
	$(A \wedge \neg A) \rightarrow B$	
19.	$(A \rightarrow B) \rightarrow ((\neg A \rightarrow B) \rightarrow B)$	
20.	$((A \rightarrow B) \wedge (B \rightarrow C)) \rightarrow (A \rightarrow C)$	(propositional forms of)
	$(A \rightarrow B) \rightarrow ((B \rightarrow C) \rightarrow (A \rightarrow C))$	*transitivity of* $\rightarrow$
21.	$(A \wedge (A \rightarrow B)) \rightarrow B$	(propositional forms of)
	$A \rightarrow ((A \rightarrow B) \rightarrow B)$	*modus ponens*
22.	$(\neg B \wedge (A \rightarrow B)) \rightarrow \neg A$	(propositional forms of)
	$\neg B \rightarrow ((A \rightarrow B) \rightarrow \neg A)$	*modus tollens*
23.	$(A \rightarrow B) \rightarrow ((A \rightarrow \neg B) \rightarrow \neg A)$	*principle of contradiction; reductio ad absurdum*
24.	$((A \wedge B) \rightarrow C) \rightarrow (A \rightarrow (B \rightarrow C))$	*exportation*
25.	$(A \rightarrow (B \rightarrow C)) \rightarrow ((A \wedge B) \rightarrow C)$	*importation*
26.	$(A \rightarrow (B \rightarrow C)) \rightarrow (B \rightarrow (A \rightarrow C))$	*interchange of premisses*
27.	$(A \rightarrow B) \leftrightarrow (\neg B \rightarrow \neg A)$	*contraposition*

$\neg B \rightarrow \neg A$ is called the *contrapositive* of $A \rightarrow B$

28. $(A \to B) \leftrightarrow \neg(A \wedge \neg B)$
29. $(A \to (B \wedge C)) \to ((A \to B) \wedge (A \to C))$
30. $(A \to (B \vee C)) \to ((A \to B) \vee (A \to C))$
31. $(A \to (B \to C)) \to ((A \to B) \to (A \to C))$
32. $((A \to B) \to A) \to A$ *Pierce's law*

Exercises for Section J.8

1. Prove that in **PC**: $\models A \to B$ iff $A \models B$.
2. Use Exercise 1 to reformulate as valid deductions all the tautologies you can from the list above.
3. Prove that in **PC**: $A \leftrightarrow B$ is a tautology iff A is semantically equivalent to B.
4. Use the principle of duality to deduce the second De Morgan law from the first.
5. Give semi-formal examples of *exportation, importation, De Morgan's laws,* and the propositional form of *disjunctive syllogism.*

K. Formalizing the Notion of Proof

The material in this part is motivated by our work on **PC**, but it is not tied to classical logic and will apply to every logic in this volume. *Nothing in this section presupposes any of the mathematical abstractions* discussed in Section G; all the notions here make perfectly good sense in a finitistic, constructive context, though unless I state otherwise later I will follow the abstractions of Section G.

1. Reasons for formalizing

Historically, logic proceeded from observations that there were propositions true solely because of their form, to an investigation of those forms, then to symbolization and generality, and finally to laying out a few forms from which all other logically acceptable ones could be mechanically derived. That approach dates back at least to the Stoics, and to Aristotle too, though he was concerned more with the form of logically acceptable arguments than valid wffs. The modern versions came to fruition in the work of *Frege, 1879,* and *Whitehead and Russell, 1910–13.*

Logic in this tradition was seen as one language for all reasoning. It was only in the nineteenth century, with the use of models for non-euclidean geometry, that mathematicians considered the idea that a set of formal propositions could have more than one interpretation. It is revealing to read the correspondence between Hilbert, who had combined the notion of models for geometry with rigorous formal logic, and Frege (*Frege, 1980,* pp. 31–52) to see how novel the idea of many models for one language appeared at that time.

Still, the tables for the classical connectives had been implicit in the writings of logicians since antiquity:

> Philo said that the conditional is true whenever it is not the case that its antecedent is true and its consequent is false; so that according to him, the conditional is true in three cases and false in one case. . . . they say that a conjunction holds when all the conjuncts are true, but is false when it has at least one false conjunct . . .
>
> Sextus Empiricus, *Against the Mathematicians,* VIII, 112, 125
> translation from *Mates, 1953,* pp. 97–98

Post, 1921 showed that the propositional wffs derivable as theorems in Whitehead and Russell's system of logic coincide with the classical tautologies (see Section J.1 below). Since we know now that we can use the semantics to decide whether a wff is a **PC**-tautology, why should we bother to show that we can derive the **PC**-tautologies from a few acceptable forms?

For one thing, the decision procedure takes an unrealistic amount of time, and so for all practical purposes we have no decision procedure for wffs of even moderate length. Worse, when we consider the internal form of propositions we find that classical predicate logic, which deals with quantification and is based on **PC**, has no decision method (see, for example, *Epstein and Carnielli, 1989*). In that case, and also for other propositional logics for which we have no decision procedure, it is essential to have a way to derive tautologies syntactically.

Moreover, formalizing the notion of derivability, of when one proposition can be proved from others, is itself important not only for logic but for mathematics and science. The semantics of **PC** gave us a notion of semantic consequence. We would like a syntactic counterpart of semantic consequence based on formalizing the notion of a proof.

Finally, by approaching logic in terms of the forms and not the meanings of wffs we will have another way to isolate our assumptions. We start with an informal notion of validity and take as fundamental the validity of a few simple wffs and the fact that some few basic rules lead always from valid wffs to valid wffs. Then we can say: if you accept these you will accept our logic. Unless, that is, you accept other wffs that are not derivable in our system. To convince you that there are no other acceptable wffs we must return to our formalization of the meaning of wffs and prove that the valid wffs, the tautologies, can all be derived within our system. That proof, which links the syntax and semantics, will be a central point of our investigations of a logic.

My emphasis in this book is on semantics for logic; I will present only one approach to the theory of proofs, usually called Hilbert-style proof theory, which I will use throughout. For a general survey of formalizations of the notion of proof see Chapter IX of *Kneale and Kneale, 1962.*

2. Proof, syntactic consequence, and theories

We have the formal language $L(\neg, \rightarrow, \wedge, \vee, p_0, p_1, \ldots)$. Whenever I say 'proposition' below I mean one in the semi-formal object language of a model of L.

A *rule* is a direct consequence relation: given a collection of propositions of specified form, called the hypotheses, another proposition is taken to be a consequence. Since it's the forms that are significant, rules are given in the formal language and are usually presented as schema, where any collection of wffs of the form $\{A_1, \ldots, A_n\}$ can be taken as hypotheses, and of the form B as consequence:

$$\frac{A_1, \ldots, A_n}{B}$$

For instance, we have:

modus ponens $\qquad \dfrac{A, A \rightarrow B}{B}$

adjunction $\qquad \dfrac{A, B}{A \wedge B}$

Sometimes the hypotheses are required to have an additional property, dependent in some way on their form, as we'll see with the rule of substitution below (Section L.4).

When we choose a rule, we believe it's self-evident that if the hypotheses are instantiated with true propositions, the conclusion will be true, too.

The notion of proof is formalized as follows. We begin with a collection of propositions that we take to be self-evidently true, called the *axioms*. We then set out a collection of rules that together with the axioms comprise an *axiom system*.

Proof There is a *proof* or *derivation of* A *from the axioms* if there is a sequence $A_1, \ldots, A_n$ such that A_n is A and each A_i is either an axiom or is a direct consequence of some of the preceding A_j's by one of the rules. In that case we say that A is a *theorem* of that system, and write '$\vdash A$'.

We usually give a name to the system and add that as a subscript to '$\vdash$'.

At this point we have a choice whether to let a nonconstructive element enter into our formalization. We may decide that 'there is a proof of A' means that we can actually produce one, as is sometimes suggested by constructivists, or we may take it nonconstructively. Usually the choice depends on the underlying semantic assumptions motivating the development of the logic. I'll discuss the difference with respect to classical logic in Section L.2,3 and Section M.1.

In axiomatizing logic itself, we'll want to take as axioms propositions that we believe are self-evidently true due to their form only. The best way to describe those is by using formal wffs or schema. The axiom system is then in the formal language.

Once we have an axiom system for our logic we may wish to add further hypotheses specific to some discipline, say geometry or physics. A *proof of a proposition* A *from a collection of propositions* Σ is a sequence just as above except that we are also allowed to use the propositions of Σ as hypotheses in the rules. That is, each A_i in a proof either is an axiom, is in Σ, or is derivable by a rule from some of the preceding A_j's. If there is a proof of A from Σ, then we say that A *is a syntactic consequence of* Σ and write $\Sigma \vdash A$. We sometimes read this as 'A is deducible from Σ'. We will write $A \vdash B$ for $\{A\} \vdash B$, and $\Sigma, A \vdash B$ for $\Sigma \cup \{A\} \vdash B$. Note that $\varnothing \vdash A$ is the same as $\vdash A$. A shorthand notation for 'there is not a proof of A from Σ' is $\Sigma \nvdash A$.

With these definitions of 'consequence' and 'theorem', once we prove a theorem we can use it to prove further theorems. I'll let you convince yourself that if Γ is a collection of theorems and $\Gamma \vdash A$, then $\vdash A$.

The relation '$\vdash$' is our syntactic formalization of the notion of a proposition following from one or more propositions. In this we are again primarily interested in the syntactic forms, so we will generally take Σ to be a collection of formal wffs and then pass back and forth between wffs and propositions as we did for the notions of truth and validity in Section C.2. So I'll let Σ, Γ, Δ and subscripted versions of these *stand variously for collections of wffs, semi-formal propositions, or propositions* depending on the context.

Note that I have not defined the notions of proof and consequence here. What I've done, relative to any particular axiom system, is give a definition of '$\vdash$', which we read colloquially as 'is a theorem' or 'has as a consequence' or 'proves'. I must assume that you already have an idea of what it means to prove something and it's that which we are formalizing, as I need to use the informal notion in proving theorems about the formal or semi-formal languages. I prove those (informal) theorems in the language of this book, the *metalanguage,* which is English supplemented with various technical notions.

We may take any particular syntactic consequence, or more usually scheme of consequences $\{A_1, \ldots, A_n\} \vdash B$, and use it as a further *derived rule* in proofs. If we can prove A from Σ using a derived rule as well as the original rules, then we also have $\Sigma \vdash A$; hence, derived rules may shorten proofs, but don't allow any new consequences (I'll demonstrate this in Theorem 5 below). For example, if we have $\{\neg(A \wedge \neg B), A\} \vdash B$ then we may use the derived rule:

$$\frac{A, \neg(A \wedge \neg B)}{B}$$

We say that Σ is *closed under* a rule if whenever the hypotheses of the rule are in Σ, then so is the consequence. The *closure* of Σ under a rule is the smallest collection of wffs Δ such that $\Delta \supseteq \Sigma$ and Δ is closed under the rule.

We say Σ is a *theory* if it is closed under syntactic deduction, that is, if $\Sigma \vdash A$, then A is in Σ. This, roughly, is how we think of scientific theories: not just their hypotheses, but their consequences too. In general, we call the collection of all syntactic consequences of Σ the *theory of* Σ, denoted $\text{Th}(\Sigma)$. That is,

$$\text{Th}(\Sigma) = \{ A : \Sigma \vdash A \}$$

And Σ is a *theory* if $\text{Th}(\Sigma) = \Sigma$. We say that Δ is the *closure of* Γ *under a rule* to mean that $\Delta = \text{Th}(\Gamma)$, where that rule is the only one used in derivations.

Theorem 5 Properties of Syntactic Consequence Relations

a. $A \vdash A$

b. If $\vdash A$, then $\Gamma \vdash A$

c. If $A \in \Gamma$, then $\Gamma \vdash A$

d. If $\Gamma \vdash A$ and $\Gamma \subseteq \Delta$, then $\Delta \vdash A$

e. ***Transitivity***
If $\Gamma \vdash A$ and $A \vdash B$, then $\Gamma \vdash B$

f. ***Transitivity, the Cut Rule***
If $\Gamma \vdash A$ and $\Delta, A \vdash B$, then $\Gamma \cup \Delta \vdash B$

g. If $\Gamma \cup \{A_1, \ldots, A_n\} \vdash B$ and $\Gamma \vdash A_i$ for $i = 1, \ldots, n$, then $\Gamma \vDash B$

h. ***Compactness***
$\Gamma \vdash A$ iff there is some finite collection $\Delta \subseteq \Gamma$ such that $\Delta \vdash A$

j. If there is a sequence $A_1, \ldots, A_n$ where each A_i is either an axiom, in Σ, or a consequence of some of the previous A_n's using a rule of the system or a derived rule, and $A_n = A$, then $\Sigma \vdash A$

Proof: With the exception of part (j), I will leave these as an exercise.

Part (j): Given such a sequence, each time a derived rule is employed, say to obtain A_k from $A_{i_1}, \ldots, A_{i_m}$, we can insert into the proof the derivation of A_k from those propositions (wffs) using only the axioms, the rules of the system, and the propositions (wffs) in Σ; the use of a derived rule to obtain A_k is just a shorthand way of saying that A_k is a consequence of $\{A_{i_1}, \ldots, A_{i_m}\}$. In this way we can obtain a (longer) proof of A using no derived rules. ■

Parts (e) and (f) are generalizations of the remark I made above that once we prove a proposition we may use it as if it were an axiom. Part (h) is of significance only if we make the assumption that we are dealing with completed infinite totalities of propositions.

We could, if we wish, rephrase several parts of the previous theorem and other statements about '$\vdash$' in terms of theories, and this is what many texts do. For example, (c) can be stated as: $\Gamma \subseteq \text{Th}(\Gamma)$. But that sort of description is usually linked to infinitistic assumptions about the nature of collections and may not be suited to some of our discussions.

Some theories are better than others, as any scientist will tell you. As logicians we are especially interested in theories that are *consistent*, that is, ones from which no contradiction can be deduced. A collection Σ that is not consistent we call *inconsistent*. The notion of contradiction that we employ will depend on the formal or informal semantics we are using, how we understand truth. For instance, in **PC** a consistent theory is one that for no A contains both A and $\neg$A. For a 3-valued logic we'll study in Chapter VIII.C.1 the notion of a consistent theory is one which for no A contains all three of A, $\neg$A, A$\leftrightarrow\neg$A.

A *complete theory* is one which is as full a description as possible of "the way the world is" relative to the atomic propositions we've assumed and the semantics, formal and informal, that we are using. It might be inconsistent; if not, it will contain as many complex propositions as possible relative to those atomic ones while still being consistent. So a complete and consistent theory corresponds to a possible description of the world, relative to our semantic intuitions and choice of atomic propositions. Therefore, *if we have a formal semantics we will want the notion of a complete and consistent theory to correspond to the notion of the collection of propositions true in a model.*

Post, 1921, devised general criteria for completeness and consistency that do not depend on the particular language in which a system is formulated. Often, though not always (see Section M.8), they correspond to the semantic notions of completeness and consistency. Relative to a particular syntactic consequence relation, they are:

Γ is *Post-consistent* iff there is some A such that $\Gamma \nvdash$ A

Γ is *Post-complete* iff for every B not in Γ and for every C, $\Gamma \cup \{B\} \vdash$ C

Theorem 6 ***a.*** If Γ is Post-consistent and $\Delta \subseteq \Gamma$, then Δ is Post-consistent.

b. Γ is Post-consistent iff every finite subset of Γ is Post-consistent.

c. If Γ is Post-complete and consistent, then Γ is a theory.

Proof: If Δ is not Post-consistent, then any collection $\Gamma \supseteq \Delta$ is not Post-consistent, since the same proof from Δ can be made from Γ. And indeed, since every proof is finite, we only need consider finite collections of Γ.

For (c), suppose Γ is Post-complete and consistent, and $\Gamma \vdash$ A. If A is not in Γ, then for every C, $\Gamma \cup \{A\} \vdash$ C. But then by the *Cut Rule* (Theorem 5.f) for every C, $\Gamma \vdash$ C. But this is a contradiction, since Γ is Post-consistent. So A is in Γ. ■

3. What is a logic?

Note that the first seven properties of syntactic consequence relations correspond to the first seven properties for semantic consequence relations listed in Theorem 2.

Given a logic, we would like the semantic and syntactic consequence relations to coincide, so that we have only one formal metalogical notion of 'follows from'. This is not always the case, so we distinguish various possibilities:

Methods of Presentation of a Logic

1. Define a notion of model and truth in a model for a formal language. The logic is then the collection of wffs that are true in all models.
2. Define a notion of model and truth in a model for a formal language. Semantic consequence is then defined in the manner it was for **PC**, and the logic is considered to be the semantic consequence relation.
3. Present an axiom system in a formal language. Then the collection of all theorems is the logic.
4. Present an axiom system in a formal language. The logic is then the syntactic consequence relation.

I will consider any one of these to be *a presentation of a logic*; examples of each of them can be found in what follows and in the literature of modern logic. I'll say a logic is *presented semantically* if it is given by (1) or (2) and *presented syntactically* if it is given by (3) or (4). Sometimes we say that (3) or (4) is the *syntax of the logic*.

Given a semantics and a syntax for the same language, if every theorem is a tautology then we say that *the axiomatization is sound* or that we have *provided sound semantics for the axiomatization*, depending on which was presented first. In that case, for all A, if $\vdash A$ then $\models A$. This is the minimal condition we impose on any pair of semantics and syntax we want to claim characterize the same pre-formal logical notions.

Given a sound axiomatization of a semantics, if we also have that every tautology is a theorem, that is for all A, if $\models A$ then $\vdash A$, we say that *the axiomatization is complete for the semantics* or that *we have provided complete semantics for the axiomatization*. Sometimes we say simply that *the* (*axiomatized*) *logic has a semantics*. A proof of that is called a *completeness theorem*.

The strongest correlation we can have is when the semantic and syntactic consequence relations are the same. That is, for all Γ and A, $\Gamma \models A$ iff $\Gamma \vdash A$. In that case we say that we have *strongly complete semantics for the axiomatization*, or a *strongly complete axiomatization of the semantics*. A proof of the equivalence is called a *strong completeness theorem* for the logic.

If we can prove a strong completeness theorem for a logic, then we can conclude by Theorem 5.h that $\Gamma \models A$ iff there is some finite collection of wffs $\Delta \subseteq \Gamma$ such that $\Delta \models A$. For most logics this compactness property cannot be proved without infinitistic assumptions (compare Section L.3 and Section M.1) that may contradict the semantic ideas upon which the logic is based (see, for example, Chapter VII.B.2 on intuitionistic logic). The practitioners of the logic may, therefore, choose to prove only a *finite* strong completeness theorem: for any finite Γ and any A, $\Gamma \vDash A$ iff $\Gamma \vdash A$.

Exercises for Section K

1. What does '$\Gamma \vdash A$' mean? Prove that if Γ is a collection of theorems and $\Gamma \vdash A$, then $\vdash A$. Does it matter whether Γ is finite or not?
2. Why is it right to identify $\vdash A$ and $\varnothing \vdash A$?
3. a. What is a theory?
   b. Show that if Γ is a theory, then every axiom is in Γ.
4. a. Is $\{p_1, p_2 \rightarrow p_3, p_4 \rightarrow p_5\}$ closed under *modus ponens*?
   b. Is $\{p_1, p_2, p_1 \wedge p_2, p_2 \wedge p_1\}$ closed under the rule of adjunction?
   c. Is $\{p_1, p_1 \rightarrow p_2, p_2, p_2 \rightarrow p_3\}$ closed under *modus ponens*?
5. a. What is the closure of the set {p, q} under the rule $A \vdash A \vee B$?
   b. Give an inductive definition of the closure of a set Γ under the rule of *modus ponens*. (Compare the inductive definition of 'wff', p. 47.)
6. Give an inductive definition of *theorem.*
7. Prove Theorem 5.
8. Reformulate Theorem 5 in terms of theories.
9. What does it mean to say an axiomatization is sound? Complete? Strongly complete?
10. In what ways does the formal definition of '$\vdash A$' differ from your own intuitive notion of proof?
11. Distinguish the syntactic notion of 'follows from' from the semantic notion.
12. Distinguish the four ways of presenting a logic.

L. An Axiomatization of PC

1. The axiom system

How are we to come up with axioms for **PC**? Generally we have some idea, some strategy for how to prove an axiom system complete or strongly complete. So we begin a proof with no axioms in hand and, say, one rule, usually *modus ponens.* Then each time we need an axiom (scheme) or an additional rule to make the completeness proof work, we add it to the list. The list might get a bit long that way, but we can be sure it's complete if we've added enough axioms to make the proof go through. We can always go back and try to simplify the list by showing that one of the axioms is superfluous by deducing it from the others. And this is how I got the axiomatization for **PC** in the language $L(\neg, \rightarrow)$ that I present in this section. I'll highlight uses of the axioms by printing them in boldface so you can see where they are needed. I'll discuss other axiomatizations in Section M.

The syntactic consequence relation (see pp. 63–64) for **PC** will be defined with respect to the following axiomatization.

PC

in $L(\neg, \rightarrow)$

axiom schema

Every formal wff which is an instance of one of the following schema is an axiom:

1. $\neg A \rightarrow (A \rightarrow B)$
2. $B \rightarrow (A \rightarrow B)$
3. $(A \rightarrow B) \rightarrow ((\neg A \rightarrow B) \rightarrow B)$
4. $(A \rightarrow (B \rightarrow C)) \rightarrow ((A \rightarrow B) \rightarrow (A \rightarrow C))$

rule $\dfrac{A, A \rightarrow B}{B}$ *modus ponens*

Here are two examples of derivations, both of which we will need in the completeness proof below.

Lemma 7 ***a.*** $\vdash A \rightarrow A$

b. $\{A, \neg A\} \vdash B$

Proof: On the right-hand side I give a justification for each step of the derivation.

a.
1. $\vdash A \rightarrow ((A \rightarrow A) \rightarrow A)$ — an instance of **axiom 2**
2. $\vdash A \rightarrow (A \rightarrow A)$ — an instance of **axiom 2**
3. $\vdash (A \rightarrow ((A \rightarrow A) \rightarrow A)) \rightarrow ((A \rightarrow (A \rightarrow A)) \rightarrow (A \rightarrow A))$ — an instance of **axiom 4**
4. $\vdash (A \rightarrow (A \rightarrow A)) \rightarrow (A \rightarrow A)$ — *modus ponens* using (1) and (3)
5. $\vdash A \rightarrow A$ — *modus ponens* using (2) and (4)

b.
1. $\vdash \neg A \rightarrow (A \rightarrow B)$ — **axiom 1**
2. $\neg A$ — premise
3. $(A \rightarrow B)$ — *modus ponens* using (1) and (2)
4. A — premise
5. B — *modus ponens* using (3) and (4) ■

Next we show the syntactic version of the deduction theorem.

Theorem 8 ***a. The Syntactic Deduction Theorem***

$\Gamma, A \vdash B$ iff $\Gamma \vdash A \rightarrow B$

b. The Syntactic Deduction Theorem for Finite Consequences

$\Gamma \cup \{A_1, \ldots, A_n\} \vdash B$ iff $\Gamma \vdash A_1 \rightarrow (A_2 \rightarrow (\cdots \rightarrow (A_n \rightarrow B) \cdots)$

Proof: a. From right to left is immediate, since *modus ponens* is our rule.

To show that if $\Gamma, A \vdash B$, then $\Gamma \vdash A \rightarrow B$, suppose that $B_1, \ldots, B_n$ is a proof of B from $\Gamma \cup \{A\}$. I'll show by induction that for each i, $\Gamma \vdash A \rightarrow B_i$.

Either $B_1 \in \Gamma$ or B_1 is an axiom, or B_1 is A. In the first two cases the result follows by using **axiom 2**. If B is A, we could just add $A \rightarrow A$ to our list of axioms. But I showed in Lemma 7.a, using **axioms 2** and **4**, that $\vdash A \rightarrow A$.

Now suppose that for all $k < i$, $\vdash A \rightarrow B_k$. If B_i is an axiom, or $B_i \in \Gamma$, or B_i is A, then we have $\vdash A \rightarrow B_i$ as before. The only other case is when B_i is a consequence by *modus ponens* of B_m and $B_j = B_m \rightarrow B$ for some $m, j < n$. Then by induction $\Gamma \vdash A \rightarrow (B_m \rightarrow B_i)$ and $\Gamma \vdash A \rightarrow B_m$, so by **axiom 4**, $\Gamma \vdash A \rightarrow B$. Part (b) I leave as an exercise. ■

Note that this proof depends only on axiom schema 2 and 4, and these involve just the one connective '$\rightarrow$'.

Exercises for Section L.1

1. Exhibit derivations that establish the following.

   a. $\vdash A \rightarrow (B \rightarrow B)$

   b. $\vdash A \rightarrow \neg\neg A$

   c. $\vdash \neg\neg A \rightarrow A$

   d. $\vdash (\neg A \rightarrow A) \rightarrow A$

   e. $\vdash (A \rightarrow B) \rightarrow (\neg B \rightarrow \neg A)$

   f. $\vdash A \rightarrow ((A \rightarrow B) \rightarrow B)$

   g. $\vdash (A \rightarrow (B \rightarrow C)) \rightarrow (B \rightarrow (A \rightarrow C))$

2. Prove the *Syntactic Deduction Theorem for Finite Consequences.*

3. a. Establish $A \vdash \neg A \rightarrow B$

   b. Establish $\{A \rightarrow B, B \rightarrow C\} \vdash A \rightarrow C$

   c. Use the algorithm implicit in the proof of the *Syntactic Deduction Theorem* to convert the derivations in (a) and (b) to proofs of $\vdash A \rightarrow (\neg A \rightarrow B)$ and $\vdash (A \rightarrow B) \rightarrow ((B \rightarrow C) \rightarrow (A \rightarrow C))$.

4. What changes would need to be made to the material in this section if we replace axiom schema 1 by $A \rightarrow (\neg A \rightarrow B)$?

2. A completeness proof

In axiomatizing a logic we usually need to find formulas whose presence or absence in a theory is equivalent to Post-consistency and Post-completeness (p. 67). For classical logic we have:

Γ is *classically consistent* $\equiv_{\text{Def}}$ for every A, not both $\Gamma \vdash A$ and $\Gamma \vdash \neg A$

Γ is *classically complete* $\equiv_{\text{Def}}$ for every A, at least one of A, $\neg A$ is in Γ

To connect these definitions to those of Post-consistency and completeness we need to look at collections that are theories.

Theorem 9 Γ is classically consistent iff Γ is Post-consistent
If Γ is a theory, then
Γ is classically complete iff Γ is Post-complete

Proof: Suppose that Γ is classically inconsistent. Then by Lemma 7.b, $\Gamma \vdash B$ for every B, so Γ is Post-inconsistent. If for every B, $\Gamma \vdash B$, then in particular for some B, $\Gamma \vdash B$ and $\Gamma \vdash \neg B$.

If Γ is classically complete, suppose that B is not in Γ. Then $\neg B$ is in Γ, so by Lemma 7.b, $\Gamma \cup \{B\} \vdash C$ for every C. If Γ is a Post-complete theory, suppose B is not in Γ. We must show that $\neg B$ is in Γ. First, $\Gamma \cup \{B\} \vdash \neg B$. So by the *Syntactic Deduction Theorem,* $\Gamma \vdash B \rightarrow \neg B$. By Lemma 7.a, $\Gamma \vdash \neg B \rightarrow \neg B$. So by **axiom 3,** $\Gamma \vdash \neg B$. As Γ is a theory, $\neg B$ is in Γ. ■

Now we will show that these notions of consistency and completeness correctly correlate to the semantics. By Theorem 9 we may use the two definitions of consistency and the two definitions of completeness interchangeably. Below I will simply write 'consistent' or 'complete', usually invoking classical consistency and completeness.

Lemma 10 Γ is complete and consistent
iff there is a **PC** model v such that $\Gamma = \{A : v(A) = \mathsf{T}\}$

Proof: I leave to you to show that the set of wffs true in a **PC** model is complete and consistent.

Suppose now that Γ is complete and consistent. Consider v: Wffs $\rightarrow \{\mathsf{T}, \mathsf{F}\}$ defined by:

$v(A) = \mathsf{T}$ if $A \in \Gamma$
$v(A) = \mathsf{F}$ if $A \notin \Gamma$

This definition is acceptable because for every A exactly one of A, $\neg A$ is in Γ. We will show that v is a model by showing that it evaluates the connectives correctly.

if $v(\neg A) = \mathsf{T}$,

then $\neg A \in \Gamma$,
so $A \notin \Gamma$ by consistency,
so $v(A) = \mathsf{F}$

if $v(A) = \mathsf{F}$,
then $A \notin \Gamma$,
so $\neg A \in \Gamma$ by completeness,
so $v(\neg A) = \mathsf{T}$

Suppose $v(A \rightarrow B) = \mathsf{T}$. Then $A \rightarrow B \in \Gamma$. By Lemma 6.c, Γ is a theory. If $v(A) = \mathsf{T}$, we have $A \in \Gamma$, so by *modus ponens,* $B \in \Gamma$, and hence $v(B) = \mathsf{T}$. Conversely, suppose $v(A) = \mathsf{F}$ or $v(B) = \mathsf{T}$. If the former then $A \notin \Gamma$, so $\neg A \in \Gamma$, and by **axiom 1**, $A \rightarrow B \in \Gamma$; so $v(A \rightarrow B) = \mathsf{T}$. If the latter then $B \in \Gamma$, and as Γ is a theory, by **axiom 2**, $A \rightarrow B \in \Gamma$; so $v(A \rightarrow B) = \mathsf{T}$. ■

The next lemma is the crucial nonconstructive step in this method of proving completeness.

Lemma 11 If $\nvdash D$, then there is some complete and consistent collection of wffs Γ such that $D \notin \Gamma$.

Proof: Let $A_0, A_1, \ldots$ be a numbering of the wffs of the formal language (see pp. 49–50). Define:

$$\Gamma_0 = \{\neg D\}$$

$$\Gamma_{n+1} = \begin{cases} \Gamma_n \cup \{A_n\} & \text{if this is consistent} \\ \Gamma_n & \text{otherwise} \end{cases}$$

$$\Gamma = \bigcup_n \Gamma_n$$

First, we have that Γ_0 is Post-consistent, since if $\{\neg D\} \vdash C$ for every C, then in particular, $\{\neg D\} \vdash D$, and as in the proof of Theorem 9, we would have $\vdash D$. So by construction, each Γ_n is consistent. Hence Γ is consistent, for if not some finite $\Delta \subseteq \Gamma$ is inconsistent by Theorem 6.b, and Δ being finite, $\Delta \subseteq \Gamma_n$ for some n. In that case Γ_n would be inconsistent, a contradiction.

Γ is (Post-) complete, because if B is not in Γ, then by construction, $\Gamma \cup \{B\}$ is inconsistent, so for every C, $\Gamma \cup \{B\} \vdash C$. So Γ is a theory, too, and since $\neg D$ is in Γ, D is not in Γ. ■

Theorem 12 ***a. Completeness of the axiomatization of* PC**

$\vdash A$ iff $\models A$

b. Finite Strong Completeness

For finite Γ, $\Gamma \vdash A$ iff $\Gamma \models A$

Proof: a. First, the axiomatization is sound: every axiom is a tautology, as you can check, and if A and $A \to B$ are tautologies, then so is B. Hence if $\vdash A$, then A is a tautology.

Now suppose that $\nvdash A$. We will show that A is not a tautology. Since $\nvdash A$, by Lemma 11 there is a complete and consistent Γ such that $A \notin \Gamma$. By Lemma 10 there is some **PC**-model υ such that Γ is the set of wffs true in υ. So $\upsilon(A) = \mathsf{F}$ and A is not a tautology. Thus if A is a tautology, it is a theorem.

b. Suppose that $\Gamma = \{A_1, \ldots, A_n\}$. If $\{A_1, \ldots, A_n\} \vdash A$ then by the *Syntactic Deduction Theorem for Finite Consequences* (Theorem 8.b), $\vdash A_1 \to (A_2 \to (\cdots \to (A_n \to A) \cdots)$. Hence by part (a), $\vDash A_1 \to (A_2 \to (\cdots \to (A_n \to A) \cdots)$, and so $\{A_1, \ldots, A_n\} \vDash A$ (Theorem 2.k). ■

We have shown that for **PC** the syntactic and semantic formalizations of 'follows from' result in the same relation on wffs and hence on propositions. Moreover, these metalogical notions can be identified with the theoremhood or validity of the corresponding conditional.

Corollary 13 The following are equivalent:

a. $A \vDash B$

b. $A \vdash B$

c. $\vDash A \to B$

d. $\vdash A \to B$

Proof: The equivalence of (a) and (b) is Theorem 12.a. I remarked earlier on the equivalence of (c) and (a) (Theorem 2.j). The equivalence of (d) and (b) is the *Syntactic Deduction Theorem.* ■

The nonconstructive nature of this proof of completeness has a serious disadvantage: given a tautology such as $(p_1 \to (p_2 \to p_3)) \to (p_2 \to (p_1 \to p_3))$, Theorem 12 tells us that a derivation for it exists in our system, but it doesn't tell us how to produce it. In Section M.2 below I will give a constructive proof of completeness that will show how to produce the derivations, but the results that we next prove depend heavily on the nonconstructive proof given above.

3. The *Strong Completeness Theorem* for PC

It is not possible to give a constructive proof of the strong completeness of **PC** (see *Henkin, 1954*). Therefore, a version of Lemma 12 is essential.

Lemma 14

a. If Σ is consistent and $\Sigma \nvdash D$, then there is a complete and consistent set Γ such that D is not in Γ and $\Sigma \subseteq \Gamma$.

b. Every consistent collection of wffs has a model.

Proof: Part (a) follows just as for Lemma 11, except that we take Γ_0 to be $\Sigma \cup \{\neg D\}$. Part (b) then follows by Lemma 10. ■

***Theorem 15 Strong completeness of the axiomatization of* PC**
For all Γ, $\Gamma \vdash A$ iff $\Gamma \vDash A$

Proof: If $\Gamma \vdash A$, then given any model of Γ all the wffs in Γ are true, as are the axioms. Since the rule is valid, A must be true in the model, too. Hence $\Gamma \vDash A$. Suppose $\Gamma \nvdash A$. Then proceed as in the proof of Theorem 12 using Lemma 14. ■

Here is a striking consequence of these infinitistic methods.

Theorem 16 Compactness of the semantic consequence relation

a. $\Gamma \vDash A$ iff there is some finite collection $\Delta \subseteq \Gamma$ such that $\Delta \vDash A$.

b. Γ has a model iff every finite subset of Γ has a model.

Proof: Part (a) follows from Theorem 15 and Theorem 5.h.

For (b), if Γ has a model, then so does every finite subset of it. In the other direction, if every finite subset of Γ has a model, then every finite subset of Γ is consistent by the soundness part of Theorem 15. Hence by Lemma 6.b, Γ is consistent; so by Lemma 14 it has a model. ■

Exercises for Sections L.2 and L.3

1. Prove that if v is a model and Γ is the collection of wffs true in v, then Γ is complete and consistent.

2. If Γ is a theory, show without using the completeness theorem that Γ is classically complete and consistent iff exactly one of $\neg A \to A$, $A \to \neg A$ is in Γ.

3. Prove without using the completeness theorem:
   a. If Γ is consistent, then $\Gamma \cup \{A\}$ or $\Gamma \cup \{\neg A\}$ is consistent
   b. $\Gamma \nvdash A$ iff $\Gamma \cup \{\neg A\}$ is consistent
   c. $\Gamma \nvdash \neg A$ iff $\Gamma \cup \{A\}$ is consistent

4. Use the completeness theorem to establish:
   a. $\vdash \neg(A \to B) \to (A \to \neg B)$
   b. $\vdash ((A \to B) \to A) \to A$

5. Prove Lemma 11 using Exercise 3 and the construction:

$$\Gamma_0 = \{\neg D\}$$

$$\Gamma_{n+1} = \begin{cases} \Gamma_n \cup \{A_n\} & \text{if this is consistent} \\ \Gamma_n \cup \{\neg A_n\} & \text{otherwise} \end{cases}$$

$$\Gamma = \bigcup_n \Gamma_n$$

6. Show that axiom scheme 3 can be replaced by $(\neg A \rightarrow A) \rightarrow A$.

7. Given a wff D such that $\nvdash D$, does the complete and consistent collection of wffs Γ that contains $\neg D$ given in the proof of Lemma 11 depend on the particular ordering of wffs we use?

8. Prove $\{A_1, \ldots, A_n\} \vdash_{PC} B$ iff $\vdash_{PC} \neg((A_1 \wedge \ldots \wedge A_n) \wedge \neg B)$.

9. Show that in the construction in the proof of Lemma 11 we can replace A_n by p_n (*Silver, 1978*).

10. Prove that if $\vdash_{PC} A$, then there is a proof $A_1, \ldots, A_n$ of A in **PC** such that all the propositional variables that appear in $A_1, \ldots, A_n$ appear in A. (Hint: Restrict the language.)

4. Derived rules: substitution

Consider:

$$\frac{A \vee B,\ \neg A}{B}$$

By the *Finite Strong Completeness Theorem,* to show this is a derived rule (p. 65) we need only show that $\{A \vee B, \neg A\} \vDash A$.

Two derived rules about how we may substitute into wffs in **PC** are worth noting. First, define $\wedge$ in terms of $\neg$ and $\rightarrow$ as in the table of equivalences (15), p. 54. Then set:

$$A \leftrightarrow B \ \equiv_{Def}\ (A \rightarrow B) \wedge (B \rightarrow A)$$

In every model $\upsilon(A \leftrightarrow B) = \mathsf{T}$ iff $\upsilon(A) = \upsilon(B)$.

Corollary 17 The following are derived rules in **PC**:

a. Substitution If A(B) is A(p) with every occurrence of p replaced by B, then

$$\frac{\vdash A(p)}{\vdash A(B)}$$

b. Substitution of Logical Equivalents If C(B) is the result of substituting B for some but not necessarily all occurrences of the subformula A in C, then

$$\frac{A \leftrightarrow B}{C(A) \leftrightarrow C(B)}$$

Proof: Part (a) follows from its semantic version, Theorem 2.m (p. 52), and the finite strong completeness of the axiomatization.

For part (b), given evaluations of $\upsilon(A)$ and $\upsilon(B)$, the evaluations of $\upsilon(C(A))$

and $v(C(B))$ can differ only if $v(A) \neq v(B)$. So if $v \vDash A \leftrightarrow B$, then $v \vDash C(A) \leftrightarrow C(B)$. Hence, $A \leftrightarrow B \vDash C(A) \leftrightarrow C(B)$, and the corollary follows. ■

Thus propositional variables are really variables. If $\vdash A(p)$, then no matter what proposition 'p' is to stand for, simple or complex, we have a theorem. For this reason, axiomatizations are sometimes given using the rule of substitution rather than schema, as I discuss further in Section M.3.

Exercises for Section L.4

1. Reformulate as a derived rule each tautology in Section J.8 that is of the form $A \rightarrow B$ (compare Exercise 2 for Section J.8).
2. Exhibit formulas A and B such that $\vdash A(p)$, but if C arises by replacing some but not all occurrences of p in A by B, then $\nvdash C$.

M. Other Axiomatizations and Proofs of Completeness of PC

In this section I'll present other ways to axiomatize **PC** and survey alternate proofs of the completeness theorem.

1. History and Post's proof

The first axiomatization of classical logic in terms that are recognizably the same as ours was by *Russell and Whitehead, 1910–13,* in their *Principia Mathematica.* Their work was indebted to an earlier axiomatization by *Frege, 1879.* For Frege, Whitehead, and Russell, and many others at the time, the question of completeness simply did not arise. Their goal was to show that all logic, and indeed all mathematics, could be developed *within* their system; hence they did not look outside their formal systems for justification in terms of meanings of their formulas. *Dreben and van Heijenoort, 1986,* pp. 44–47, and *Goldfarb, 1979,* discuss their views.

The first use of semantic notions to justify Whitehead and Russell's system, giving a completeness proof, was by Bernays, though that was not published until *Bernays, 1926* (see *Dreben and van Heijenoort, 1986*). The first published completeness proof for Whitehead and Russell's system, and long the most influential, was by *Post, 1921.* He showed that every wff is semantically equivalent to one in disjunctive normal form. Supplementing the system of *Principia Mathematica* with further axioms, he then showed how given any tautology A it was possible to produce a derivation of the equivalent disjunctive normal form B of A and a derivation of $B \rightarrow A$. Joining those derivations he then had a proof of A. Only later was it shown that his further axioms could be proved in the system of *Principia Mathematica.*

Post's proof has a very clear advantage over the proof I gave above (Section L.2): given a tautology it shows how to produce a derivation for it in the formal system. However, Post's proof has a disadvantage: it cannot be generalized easily to other logics. Moreover, the strong completeness of **PC** cannot be deduced from it, since the proof of strong completeness requires substantial infinitistic nonconstructive assumptions (see *Henkin, 1954*) as in the proof I gave, which is due to *Łos, 1951,* based on the work of Lindenbaum (see *Tarski, 1930,* Theorem 12). There are several other constructive proofs of completeness, surveyed by *Surma, 1973 B,* all of which share the same advantage and disadvantages. Here is one.

2. A constructive proof of the completeness of PC

The constructive proof of the completeness of **PC** that I give here is due to *Kalmár, 1935.* I will use Lemma 7 and *The Syntactic Deduction Theorem* from Section L.1, since the proofs of those show how to produce the derivations.

Lemma 18 ***a.*** $\vdash A \to \neg\neg A$
b. $\{A, \neg B\} \vdash \neg(A \to B)$

Proof: a. As before, I'll put the justification to the right of each step of the derivation.

1. $(\neg A \to (A \to \neg\neg A)) \to [(\neg\neg A \to (A \to \neg\neg A)) \to (A \to \neg\neg A)]$ — an instance of axiom 3
2. $\neg A \to (A \to \neg\neg A)$ — an instance of axiom 1
3. $(\neg\neg A \to (A \to \neg\neg A)) \to (A \to \neg\neg A)$ — *modus ponens* on (1) and (2)
4. $\neg\neg A \to (A \to \neg\neg A)$ — an instance of axiom 2
5. $A \to \neg\neg A$ — *modus ponens* on (3) and (4)

b. It's immediate that $\{A, \neg B, A \to B\} \vdash \neg B$ and $\{A, \neg B, A \to B\} \vdash B$. Hence by Lemma 7.b and Theorem 5.f, $\{A, \neg B, A \to B\} \vdash \neg(A \to B)$. So by the *Syntactic Deduction Theorem* (Theorem 8), $\{A, \neg B\} \vdash (A \to B) \to \neg(A \to B)$. By Lemma 7.a and Theorem 5.b, $\{A, \neg B\} \vdash \neg(A \to B) \to \neg(A \to B)$. An instance of axiom 3 is $[(A \to B) \to \neg(A \to B)] \to [(\neg(A \to B) \to \neg(A \to B)) \to \neg(A \to B)]$. Hence by using *modus ponens* twice, we get $\{A, \neg B\} \vdash \neg(A \to B)$. ■

In the proof of part (b) I have shown that our previous work allows us to conclude that there is such a derivation. The proofs of the earlier lemmas on which that depends show how to produce one.

Lemma 19 Let C be any wff and $q_1, \ldots, q_n$ the propositional variables appearing in C. Let v be any valuation. Define, for all $i \leq n$:

$$Q_i = \begin{cases} q_i & \text{if } v(q_i) = \mathsf{T} \\ \neg q_i & \text{if } v(q_i) = \mathsf{F} \end{cases}$$

If $v(C) = \mathsf{T}$, then $\{Q_1, \ldots, Q_n\} \vdash C$.

If $v(C) = \mathsf{F}$, then $\{Q_1, \ldots, Q_n\} \vdash \neg C$.

Proof: The proof is by induction on the length of C. If C is (p_i), then we need to show that $\vdash p_i \to p_i$ and $\vdash \neg p_i \to \neg p_i$, which we did in Lemma 7.a. Now suppose it is true for all wffs shorter than C.

Now let $\Gamma = \{Q_1, \ldots, Q_n\}$. If C is $\neg A$, suppose $\nu(C) = \mathsf{T}$. Then $\nu(A) = \mathsf{F}$. So by induction $\Gamma \vdash \neg A$. If $\nu(C) = \mathsf{F}$, then $\nu(A) = \mathsf{T}$, so by induction $\Gamma \vdash A$. By the previous lemma and Lemma 5.b, $\Gamma \vdash A \to \neg\neg A$, so using *modus ponens*, $\Gamma \vdash \neg\neg A$ as we wish.

If C is $A \to B$, first suppose $\nu(C) = \mathsf{T}$. Then $\nu(A) = \mathsf{F}$, or $\nu(B) = \mathsf{T}$. If $\nu(A) = \mathsf{F}$, then $\Gamma \vdash \neg A$, so by axiom 1, $\Gamma \vdash A \to B$. If $\nu(B) = \mathsf{T}$, then axiom 2 allows us to conclude that $\Gamma \vdash A \to B$. Finally, if $\nu(C) = \mathsf{F}$, then $\nu(A) = \mathsf{T}$ and $\nu(B) = \mathsf{F}$. So $\Gamma \vdash A$ and $\Gamma \vdash \neg B$. Hence by the previous lemma, $\Gamma \vdash \neg(A \to B)$. ■

Theorem 20 ***a. Completeness of the axiomatization of*** **PC**

$\vdash A$ iff $\models A$

b. Finite Strong Completeness

For finite Γ, $\Gamma \vdash A$ iff $\Gamma \models A$

Proof: a. As in the previous proof (Theorem 12), if $\vdash A$, then $\models A$. For the converse, suppose A is valid and $q_1, \ldots, q_n$ are the propositional variables appearing in A. For every valuation ν, $\nu(A) = \mathsf{T}$.

By Lemma 19, for any collection $\{Q_1, \ldots, Q_n\}$ where for each *i*, Q_i is either q_i or $\neg q_i$, $\{Q_1, \ldots, Q_n\} \vdash A$. Hence by the *Syntactic Deduction Theorem* (Theorem 9), $\{Q_1, \ldots, Q_{n-1}\} \vdash Q_n \to A$. In particular, for any $Q_1, \ldots, Q_{n-1}$, $\{Q_1, \ldots, Q_{n-1}\} \vdash q_n \to A$ and $\{Q_1, \ldots, Q_{n-1}\} \vdash \neg q_n \to A$. Hence, via axiom 3, for any $Q_1, \ldots, Q_{n-1}$, we have $\{Q_1, \ldots, Q_{n-1}\} \vdash A$. Repeating this argument $n-1$ times, we have $\vdash A$.

b. This follows by the *Syntactic Deduction Theorem,* as in the proof of Theroem 12.b. ■

In the proof of Theorem 20 I didn't exhibit the derivation of the given tautology, but you can trace through the earlier proofs on which it depends to produce one.

Exercises for Sections M.1 and M.2

1. Use the algorithm implicit in Kalmar's proof to establish derivations of each of the following.

   a. $\vdash A \to (B \to B)$

   b. $\vdash A \to \neg\neg A$

   c. $\vdash (\neg A \to A) \to A$

   d. $\vdash (A \to B) \to (\neg B \to \neg A)$

2. Use the method of Post described in Section M.1 to give a constructive completeness proof for **PC**.

3. Schema vs. the rule of substitution

One well-known axiomatization of **PC** is due to Łukasiewicz (see *Łukasiewicz and Tarski, 1930*). It uses the rule of substitution rather than schema.

Łukasiewicz's axiomatization of **PC**
in $L(\neg, \rightarrow)$

axioms

$$(p_1 \rightarrow p_2) \rightarrow ((p_2 \rightarrow p_3) \rightarrow (p_1 \rightarrow p_3))$$

$$(\neg p_1 \rightarrow p_1) \rightarrow p_1$$

$$p_1 \rightarrow (\neg p_1 \rightarrow p_2)$$

rules *modus ponens* substitution

$$\frac{A, A \rightarrow B}{B} \qquad \frac{\vdash A(p)}{\vdash A(B)}$$

We could give a proof from scratch that this system is strongly complete for **PC**, perhaps using Łukasiewicz's methods (see *Surma, 1973 B*). Or we could rely on our previous proof by showing that there is a derivation in this system of each axiom of each scheme of the system in Section L.1. I'll let you choose in Exercise 3.

Any axiom system that uses the rule of substitution can be presented using schema instead of that rule. However, only if substitution is a derived rule can an axiom system that uses schema be converted to one using the rule of substitution. There is a drawback to using the rule of substitution as a proof method: we must formulate it to apply only to the derivation of theorems and not to the syntactic consequence relation. That is, we take the rule to be:

$$\frac{\vdash A(p)}{\vdash A(B)}$$

and not

$$\frac{A(p)}{A(B)}$$

If we were to use the latter, every wff would be a syntactic consequence of $\{p_1\}$, which we certainly do not want. Moreover, the rule of substitution only makes sense for the formal language and not for the semi-formal language of propositions. For these reasons I've presented every axiom system in this book using schema rather than the rule of substitution, even if the originators of the logic did otherwise.

Originally, all axiom systems for propositional logic were given using the rule of substititution. See *Church, 1956,* Section 29, for a history of early systems of axioms for classical propositional logic.

4. Independent axiom systems

Given an axiom system, we'd like to know if any of the axioms are superfluous—that is, can some be proved from others? We want the simplest system we can get, for then the syntactic agreements necessary to establish the logic are as persuasive and as few as possible.

Given an axiom system we say that an *axiom is independent* of the others if we can't prove it from the system that results by deleting that axiom. An *axiom scheme is independent* if there is at least one instance of that scheme that cannot be proved if the scheme is deleted. A *rule* is independent if there is at least one wff that can be proved using the rule that cannot be proved if the rule is deleted. *An axiom system is independent* if each axiom (scheme) and rule is independent. I will often say that an axiom is independent when I mean that the scheme is.

In Chapter VIII.G I show that the axiomatization of **PC** of Section L.1 is independent. *Łukasiewicz, 1929,* proves that his axiomatization is independent also. See *Church, 1956*, p. 163, for a history of the earliest independence proofs.

5. Proofs using rules only

By using the *Deduction Theorem* and the *Strong Completeness Theorem* we can obtain a derived rule from each of our axioms: take the antecedent as hypothesis and consequent as conclusion. Thus we have, for example, the derived rule:

$$\frac{\neg A}{A \to B}$$

Some of the tautologies of Section J.8 are more familiar in the form of rules:

modus tollens $$\frac{\neg B,\ A \to B}{\neg A}$$

reductio ad absurdum $$\frac{A \to B,\ A \to \neg B}{\neg A}$$

With the Hilbert-style notion of proof we use, it is customary to minimize the number of rules taken as primitive at the expense of additional axioms. There are other formalizations of the notion of proof, however, that use only rules and no axioms, and in some cases these give clearer derivations. *Kneale and Kneale, 1962,* Chapter IX, Section 3, give a general history and discussion of this method, and *Kleene, 1952,* presents a formalization of both classical and intuitionistic logic (Chapter VII below) using rules.

Exercises for Sections M.3–M.5

1. a. Distinguish between the use of schema and the use of the rule of substitution in presenting axiom systems.
   b. Reformulate Łukasiewicz's axiomatization using schema instead of the rule of substitution.
   c. When can schema replace the rule of substitution in an axiomatization?
   d. When can the rule of substitution replace schema in an axiomatization?

2. Reformulate the axiomatization of **PC** in $L(\neg, \rightarrow)$ of Section L.1 to maximize the number of rules and minimize the number of axioms; prove that your system is complete.

3. a. Prove that Łukasiewicz's axiomatization of **PC** is complete.
   b. Prove or disprove: For Łukasiewicz's axiomatization of **PC**, the collection of axioms is complete iff Th(the collection of axioms) is Post-complete.

6. An axiomatization of PC in $L(\neg, \rightarrow, \wedge, \vee)$

We originally defined **PC** in $L(\neg, \rightarrow, \wedge, \vee)$, and it behooves us to give an axiomatization of it in that language. We can use our earlier proof from Section L to accomplish this. The only place the other connectives need enter is in the proof of Lemma 10: we need to add axioms to ensure that the connectives $\wedge$ and $\vee$ are evaluated correctly.

PC
in $L(\neg, \rightarrow, \wedge, \vee)$

axiom schema

1. $\neg A \rightarrow (A \rightarrow B)$
2. $B \rightarrow (A \rightarrow B)$
3. $(A \rightarrow B) \rightarrow ((\neg A \rightarrow B) \rightarrow B)$
4. $(A \rightarrow (B \rightarrow C)) \rightarrow ((A \rightarrow B) \rightarrow (A \rightarrow C))$
5. $A \rightarrow (B \rightarrow (A \wedge B))$
6. $(A \wedge B) \rightarrow A$
7. $(A \wedge B) \rightarrow B$
8. $A \rightarrow (A \vee B)$
9. $B \rightarrow (A \vee B)$
10. $(A \rightarrow C) \rightarrow ((B \rightarrow C) \rightarrow ((A \vee B) \rightarrow C))$

rule $\dfrac{A,\ A \rightarrow B}{B}$

Theorem 21 This axiomatization is strongly complete for **PC**.

Proof: The proof is as for the axiomatization in $L(\neg, \rightarrow)$ in Section L, except that we supplement the proof of Lemma 10:

$$\begin{aligned} v(A \wedge B) = \mathsf{T} \ &\text{iff} \ v(A) = \mathsf{T} \text{ and } v(B) = \mathsf{T} \\ &\text{iff} \ A \in \Gamma \text{ and } B \in \Gamma \\ &\text{iff} \ (A \wedge B) \in \Gamma \text{ using } \textbf{axioms 5, 6} \text{ and } \textbf{7} \end{aligned}$$

For disjunction, we have that if $A \in \Gamma$ or $B \in \Gamma$, then $(A \vee B) \in \Gamma$ by **axioms 8** and **9**. So if $v(A) = \mathsf{T}$ or $v(B) = \mathsf{T}$, then $v(A \vee B) = \mathsf{T}$. If $v(A \vee B) = \mathsf{T}$, then $(A \vee B) \in \Gamma$. If both $A \notin \Gamma$ and $B \notin \Gamma$, then by Post-consistency, for any C, $\Gamma \cup \{A\} \vdash C$ and $\Gamma \cup \{B\} \vdash C$. So by the *Syntactic Deduction Theorem,* $\Gamma \vdash A \rightarrow C$ and $\Gamma \vdash B \rightarrow C$. Hence by **axiom 10**, $\Gamma \vdash (A \vee B) \rightarrow C$. By the *Cut Rule* (Theorem 5.f), $\Gamma \vdash C$ for every C, a contradiction on the consistency of Γ. Hence $A \in \Gamma$ or $B \in \Gamma$, so that $v(A) = \mathsf{T}$ or $v(B) = \mathsf{T}$. ■

As a corollary to Theorem 21 I'll let you prove:

Corollary 22 ***The Conjunctive Form of the Syntactic Deduction Theorem***

$\Gamma \cup \{A_1, \ldots, A_n\} \vdash B$ iff $\Gamma \vdash (A_1 \wedge \cdots \wedge A_n) \rightarrow B$

For future reference we also need the following, which is immediate from the proof above.

Theorem 23 The axiom system consisting of axiom schema **1–7** and the rule of *modus ponens* is strongly complete for **PC** in the language $L(\neg, \rightarrow, \wedge)$.

7. An axiomatization of PC in L(¬, ∧)

Later in this volume we shall need an axiomatization of **PC** in the language of $\neg$ and $\wedge$. First, we set:

$$A \supset B \equiv_{\text{Def}} \neg(A \wedge \neg B)$$

We call this connective *material implication* whenever $\neg$ and $\wedge$ are interpreted classically. This will be the only way I use the symbol '$\supset$' in this book.

PC

in $L(\neg,\wedge)$

$$A \supset B \equiv_{Def} \neg(A \wedge \neg B)$$

axiom schema

1. $B \supset (A \supset B)$
2. $(A \supset (B \supset C)) \supset ((A \supset B) \supset (A \supset C))$
3. $(A \wedge B) \supset A$
4. $(A \wedge B) \supset B$
5. $A \supset (B \supset (A \wedge B))$
6. $\neg A \supset (A \supset B)$
7. $(A \supset B) \supset ((\neg A \supset B) \supset B)$

rule $\dfrac{A,\ A \supset B}{B}$ (*material detachment*)

I leave for you to prove that this system is strongly complete for **PC**.

Exercises for Sections M.6 and M.7

1. a. Show that axiom scheme 10 of the axiomatization of **PC** in the language $L(\neg,\rightarrow,\wedge,\vee)$ can be replaced by either of the following:

   $((A \vee B) \wedge \neg A) \rightarrow B$

   $(A \vee B) \rightarrow (\neg A \rightarrow B)$

   b. Why are these inferior choices?

2. Without using the completeness theorem, prove the *Conjunctive Form of the Syntactic Deduction Theorem* for $L(\neg,\rightarrow,\wedge,\vee)$.

3. Axiomatize **PC** in $L(\neg,\rightarrow,\vee)$.

4. What is wrong with the following argument?

   To obtain an axiomatization of **PC** in $L(\neg,\wedge)$ just take the axiomatization in $L(\neg,\rightarrow)$ from Section L and replace '$\rightarrow$' with '$\supset$' everywhere. After all, $A \supset B$ and $A \rightarrow B$ are evaluated the same in every model.

5. Prove the strong completeness of the axiomatization of **PC** in $L(\neg,\wedge)$. (Hint: Use the proof in Section L, reading '$\supset$' for '$\rightarrow$' whenever possible.)

6. Show that axiom schema 6 and 7 of the axiomatization of **PC** in $L(\neg,\wedge)$ can be replaced by: $(\neg A \supset A) \supset A$ $(A \wedge \neg A) \supset B$.

7. *Hilbert, 1922,* gives the following axiom system.

   PC *in* $L(\neg, \rightarrow)$

   1. $A \rightarrow (B \rightarrow A)$
   2. $(A \rightarrow (A \rightarrow B)) \rightarrow (A \rightarrow B)$
   3. $(A \rightarrow (B \rightarrow C)) \rightarrow (B \rightarrow (A \rightarrow C))$
   4. $(B \rightarrow C) \rightarrow ((A \rightarrow B) \rightarrow (A \rightarrow C))$
   5. $A \rightarrow (\neg A \rightarrow B)$
   6. $(A \rightarrow B) \rightarrow ((\neg A \rightarrow B) \rightarrow B)$

   rule $\dfrac{A, A \rightarrow B}{B}$

   Prove that this system is strongly complete for **PC**.

8. *Frege, 1879,* presents (in our notation) the following axiom system.

   PC *in* $L(\neg, \rightarrow)$

   1. $A \rightarrow (B \rightarrow A)$
   2. $(A \rightarrow (B \rightarrow C)) \rightarrow ((A \rightarrow B) \rightarrow (A \rightarrow C))$
   3. $(A \rightarrow (B \rightarrow C)) \rightarrow (B \rightarrow (A \rightarrow C))$
   4. $(A \rightarrow B) \rightarrow (\neg B \rightarrow \neg A)$
   5. $\neg\neg A \rightarrow A$
   6. $A \rightarrow \neg\neg A$

   rule $\dfrac{A, A \rightarrow B}{B}$

   a. Prove that this system is not independent.
   b. Prove that this system is strongly complete for **PC**.

9. *Rosser, 1953,* gives the following axiomatization of **PC**.

   axiom schema in $L(\neg, \wedge)$

   $A \supset (A \wedge A)$

   $(A \wedge B) \supset A$

   $(A \supset B) \supset (\neg(B \wedge C) \supset \neg(C \wedge A))$

   rule $\dfrac{A,\ A \supset B}{B}$

   Prove that this system is strongly complete.

10. (Open) Give an axiomatization of **PC** in $L(\neg, \vee)$ that uses no defined connectives and uses only rules that appear natural for these connectives, such as $A \vee B, \neg A \vdash B$.

11. (Open) Prove or disprove that the axiomatization of **PC** in $L(\neg, \wedge)$ is independent. (See Chapter VIII.G for possible methods.)

12. Give an axiomatization and constructive proof of completeness for **PC** in $L(\neg, \wedge)$.

8. Classical logic without negation: the positive fragment of PC

Classical logic as we have formulated it includes negation as an integral part. We have seen that each of $\{\neg, \wedge\}$, $\{\neg, \vee\}$, $\{\neg, \rightarrow\}$ is functionally complete, but that $\neg$ cannot be defined from $\{\rightarrow, \wedge, \vee\}$.

The fragment of **PC** formulated in $L(\rightarrow, \wedge, \vee)$ is called the *positive part of classical logic.* I shall give an axiomatization of this fragment, first considering only $L(\rightarrow)$, to illustrate the use of Post-completeness and Post-consistency in the completeness theorems and also in order to allow for comparisons with the positive part of other logics we will study.

The fragment of **PC** *in* $L(\rightarrow)$

axiom schema

1. $B \rightarrow (A \rightarrow B)$
2. $(A \rightarrow (B \rightarrow C)) \rightarrow ((A \rightarrow B) \rightarrow (A \rightarrow C))$
3. $(B \rightarrow A) \rightarrow [(C \rightarrow A) \rightarrow (((B \rightarrow C) \rightarrow A) \rightarrow A)]$

rule $\dfrac{A,\ A \rightarrow B}{B}$

First note that due to **axioms 1** and **2**, we have $\vdash A \rightarrow A$, as in the proof of Lemma 7, and the *Syntactic Deduction Theorem* (Theorem 8).

Lemma 24 If Γ is a Post-complete theory, then there is a model υ for $L(\rightarrow)$ such that $\Gamma = \{A\colon \upsilon(A) = \mathsf{T}\}$.

Proof: Define υ:Wffs $\rightarrow \{\mathsf{T}, \mathsf{F}\}$ by:

$\upsilon(A) = \mathsf{T}$ iff $A \in \Gamma$

To show that υ evaluates $\rightarrow$ correctly, suppose $\upsilon(B) = \mathsf{T}$. Then $B \in \Gamma$. So by **axiom 1**, $A \rightarrow B \in \Gamma$, as Γ is a theory. And hence $\upsilon(A \rightarrow B) = \mathsf{T}$. If $\upsilon(A) = \mathsf{F}$, then $A \notin \Gamma$, so by Post-completeness $\Gamma \cup \{A\} \vdash B$. So by the *Syntactic Deduction Theorem,* $\Gamma \vdash A \rightarrow B$, and hence since Γ is a theory, $A \rightarrow B \in \Gamma$. So $\upsilon(A \rightarrow B) = \mathsf{T}$.

If $\upsilon(A \rightarrow B) = \mathsf{T}$, suppose $\upsilon(A) = \mathsf{T}$ and $\upsilon(B) = \mathsf{F}$. Then A, $A \rightarrow B$ are in Γ, but B is not in Γ, a contradiction on Γ being Post-complete. Hence $\upsilon(A \rightarrow B) = \mathsf{T}$ iff $\upsilon(A) = \mathsf{F}$ or $\upsilon(B) = \mathsf{T}$. ∎

I ask you to prove in Exercise 5 below that the converse to Lemma 24 is false. That is, there is a collection of wffs that has a model and yet is not Post-complete and Post-consistent.

Lemma 25 If $\Sigma \nvdash A$, then there is a Post-complete and Post-consistent theory Γ such that $\Sigma \subset \Gamma$ and $A \notin \Gamma$.

Proof: Define

$$\Gamma_0 = \Sigma$$

$$\Gamma_{n+1} = \begin{cases} \Gamma_n \cup \{A_n\} & \text{if } \Gamma_n \cup \{A_n\} \nvdash A \\ \Gamma_n & \text{otherwise} \end{cases}$$

$$\Gamma = \bigcup_n \Gamma_n$$

By construction $\Gamma \nvdash A$ (Lemma 5.h). So Γ is Post-consistent. And Γ is a theory, for if $\Gamma \vdash B$ and $B \notin \Gamma$, then we must have that for some n, $\Gamma_n \cup \{B\} \vdash A$. But then $\Gamma \vdash A$, which is a contradiction.

To show that Γ is Post-complete, suppose $B \notin \Gamma$. Let C be any wff. If $C \in \Gamma$, then $\Gamma \cup \{B\} \vdash C$. Suppose $C \notin \Gamma$. If $B \rightarrow C \in \Gamma$, we are done. So suppose $B \rightarrow C \notin \Gamma$. Then $\Gamma \cup \{B\} \vdash A$, $\Gamma \cup \{C\} \vdash A$, and $\Gamma \cup \{B \rightarrow C\} \vdash A$. Hence by the *Syntactic Deduction Theorem*, $\Gamma \vdash B \rightarrow A$, $\Gamma \vdash C \rightarrow A$, and $\Gamma \vdash (B \rightarrow C) \rightarrow A$. Hence by **axiom 3**, $\Gamma \vdash A$, a contradiction. Hence $B \rightarrow C \in \Gamma$ and $\Gamma \cup \{B\} \vdash C$. ■

Theorem 26 ***Strong Completeness of the Axiomatization of* PC *in* L($\rightarrow$)**

For all Γ, $\Gamma \vdash A$ iff $\Gamma \vDash A$

Proof: I leave to you to show that the axiomatization is sound.

Suppose $\Gamma \nvdash A$. Then by Lemma 25 there is some Post-complete theory that contains Γ and excludes A. So by Lemma 24 there is a model of Γ in which A is false. So $\Gamma \nvDash A$. ■

Now we turn to the full positive part of **PC**. In order to use easily the postitive axioms of the axiomatization of **PC** in L($\neg, \rightarrow, \wedge, \vee$), p. 82, note that in Lemma 25 we obtain a Post-consistent theory Σ, so we may assume in Lemma 24 that Γ is Post-consistent, too.

The fragment of **PC** *in* L($\rightarrow, \wedge, \vee$)

axiom schema

1. $B \rightarrow (A \rightarrow B)$
2. $(A \rightarrow (B \rightarrow C)) \rightarrow ((A \rightarrow B) \rightarrow (A \rightarrow C))$
3. $(B \rightarrow A) \rightarrow [(C \rightarrow A) \rightarrow (((B \rightarrow C) \rightarrow A) \rightarrow A)]$
4. $A \rightarrow (B \rightarrow (A \wedge B))$
5. $(A \wedge B) \rightarrow A$
6. $(A \wedge B) \rightarrow B$
7. $A \rightarrow (A \vee B)$
8. $B \rightarrow (A \vee B)$
9. $(A \rightarrow C) \rightarrow ((B \rightarrow C) \rightarrow ((A \vee B) \rightarrow C))$

rule $\dfrac{A,\ A \rightarrow B}{B}$

***Theorem 27 Strong Completeness of the Axiomatization of* PC *in* L($\rightarrow,\wedge,\vee$)**
For all Γ, $\Gamma \vdash A$ iff $\Gamma \models A$

Proof: The proof is as for the fragment of **PC** in L($\rightarrow$), only now we must prove that $\wedge$ and $\vee$ are evaluated correctly in the model of Lemma 24. But for that we may use the proof of Theorem 22. ■

Corollary 28 Axiom schema **1–6** and the rule of *modus ponens* give a strongly complete axiomatization of the fragment of **PC** in L($\rightarrow,\wedge$).
Axiom schema **1–3**, **7–9** and the rule of *modus ponens* give a strongly complete axiomatization of the fragment of **PC** in L($\rightarrow,\vee$).

Exercises for Section M.8

1. Show that the axiomatization of **PC** in L($\rightarrow,\wedge,\vee$) is sound.
2. Prove that the axiomatization of the positive part of **PC** is independent. (Hint: Use Theorem 26 and Corollary 28.)
3. Prove Corollary 28.
4. Prove that the first four axiom schema of Hilbert's axiomatization of **PC** in L($\neg,\rightarrow$) (p. 85) plus Pierce's Law, $((A \rightarrow B) \rightarrow A) \rightarrow A$, and the rule of *modus ponens* give a strongly complete axiomatization of the fragment of **PC** in L($\rightarrow$).
5. Show that for L($\rightarrow,\wedge,\vee$) there is a valuation v that validates all wffs. Conclude that the converse of Lemma 24 is false.
6. Give a constructive proof of completeness for the positive part of **PC**.
7. (Open) Prove or disprove: A strongly complete axiomatization of **PC** in L($\rightarrow$) is given by replacing schema 3 of the axiomatization of **PC** in L($\rightarrow$) by Pierce's Law, $((A \rightarrow B) \rightarrow A) \rightarrow A$.
8. (Open?) Axiomatize the fragment of **PC** in L($\wedge,\vee$).

N. The Reasonableness of PC

1. Why classical logic is classical

When Whitehead and Russell wrote *Principia Mathematica* (*1910–13*) they codified what they saw as the laws of logic in order to use them as a basis for the foundations of mathematics. When *Post, 1921,* showed that the propositional logic they set out axiomatically was characterized by the truth-tables of **PC** there was overwhelming reason for logicians and mathematicians to accept it: they could feel confident using it because they could check their work with truth-tables, and this (semantically) simple logic seemed to be all that was needed to reason about propositions as wholes (without quantification), since so much mathematics had been developed from it. It was simple, easy to use, and applicable in science and mathematics. It came to be called 'classical,' though until then the truth-functional reading of the connectives had been only one of several competing logical traditions dating back to the ancient Greeks, as you can read in *Bochenski, 1970,* or *Mates, 1953.*

Here are some of the major features of classical propositional logic that recommend it to us as a good, useful model of reasoning:

a. It's simple. Indeed, given the general assumptions of propositional logic of Chapter I, I believe it has the simplest semantics we can devise.

b. It's easy to use both formally and in applications because we can check our work with the truth-tables. And this simplicity is essential, since we want to use our logic in reasoning on any subject. If a logic were harder to use and understand than, say, mathematics, we'd hardly want to use it to do mathematics.

c. It has broad scope: it can be used on any atomic propositions, any sentences that we agree to view as having truth-values.

d. We can give it a simple axiomatization and the resulting syntactic consequence relation is equivalent to the semantic consequence relation.

What more could there be to propositional logic? We know from Examples 6 and 8 of Section F that there are propositional connectives we can't deal with in **PC**, but so much the worse for them we might argue. Logic is only concerned with truth and consequence. But does **PC** do a good job of formalizing those? There is one more criterion that **PC** must be measured against:

e. Is **PC** a reasonably accurate model or our pre-formal notions? Does it require us to give up a great deal of our intuitions about what is true, or what follows from what?

2. The paradoxes of PC

If the moon is made of green cheese, then $2+2=4$

PC tells us this is true. That clashes with almost everyone's (naive) intuitions, assuming we're not using the sentence as hyperbole.

A classical logician might argue that this is a confusion. **PC** doesn't tell us that this proposition is true, but only that its formal version using '$\rightarrow$' is. And '$\rightarrow$' captures only those aspects of 'if . . . then . . .' that depend solely on the truth-values of propositions. So in this logic the formalization of any 'if . . . then . . .' proposition with false antecedent or true consequent is true. There's nothing paradoxical about that.

To lessen the shock of having to accept as true the formal version of propositions such as the one above, many people read '$\rightarrow$' in **PC** as 'materially implies' and call a wff formed using it a 'material implication'. At least they do so when they're being careful: most of the time they forget and read it as 'if . . . then . . .', because that is what they set out to model. There's just a bad match here that doesn't get better by saying we weren't really trying to formalize our understanding of 'if . . . then . . .', 'implies', or 'follows from'. These observations were made by *Lewis, 1912* even before the last volume of *Principia Mathematica* was published. Due to his critique the formulas $A \rightarrow (B \rightarrow A)$ and $\neg A \rightarrow (A \rightarrow B)$ have come to be known as the *paradoxes of classical logic*.

PC is only partially successful at capturing our informal notions, our intuitions about what follows from what, about which 'if . . . then . . .' propositions are true. That's because in almost all reasoning some other property of propositions and some relationship between propositions other than that between their truth-values matters. It seems to me that **PC** is applicable and reasonable only in those areas of propositional reasoning where it can be argued that the truth-values and forms of propositions are all that matter, such as in logic itself or perhaps mathematics and science. But even in mathematics there are serious arguments given by the intuitionists against using classical logic (see Chapter VII).

Still, there are many who disagree and claim that **PC** is the right logic. Let's consider two arguments to that effect.

First, consider the proposition, 'If triangles had four sides, then bananas would be purple'. It seems unreasonable to us to accept this as true. *Bennett, 1969,* in a sustained defense of **PC**, says that's because any proposition that begins with 'If triangles had four sides . . .' will, in his words, 'strike one as implausible, weird, unsatisfactory' because it's antecedent is false.

I don't think that's what makes the conditional weird; rather, it's the total lack of connection between the antecedent and consequent. Nor is the subjunctive mood indicating a counterfactual the point. I'd venture to say that almost everyone would accept as true and not weird or implausible, 'If triangles had four sides, then rectangles would have five sides'.

Bennett's reply to this is that we shouldn't or can't incorporate any notion of connection of meanings into a logical analysis of 'if . . . then . . .'. Hughes and Cresswell make the same point:

> . . . to insist on it [some connection of "content" or "meaning"] seems to introduce into an otherwise clear and workable account of deducibility a gratuitously vague element which will make it impossible to determine whether a formal system is a correct logic of entailment or not.
>
> *Hughes and Cresswell, 1968,* pp. 336–337

Here they are defending classical modal logic (see Chapter VI below), which they see as an extension of classical logic. In classical modal logic any proposition with impossible antecedent or necessarily true consequent is adjudged true. Since '2+2 = 4' is considered to be a necessary truth, the initial proposition of this section would be true under their analysis, too. Comparable to the paradoxes of **PC** we have the *paradoxes of strict implication*: $A \to (B \to B)$ and $(A \wedge \neg A) \to B$.

The argument seems to me that classical logic (or its extension, classical modal logic) is the right logic because only the truth or falsity (and the possibility or impossibility) of an atomic proposition should matter to logic. These are clear and precise notions. Classical logic is right, and where it clashes with our intuitions about reasoning, our intuitions are defective. Logic is primarily prescriptive.

But outside a platonic heaven, propositions are full of imprecision and ambiguity. If we are to use logic—and I want to use logic in arguments ethical, mathematical, and political—we must face the fact that in almost all contexts some content of a proposition other than its truth-value matters to reasoning. And I will argue in Chapter VI.G.3 that even in classical modal logic an imprecise notion of content is crucial.

We need some general methods for modeling such contents. Before I present those in Chapter IV, I want to go to the other extreme from classical logic and show how to incorporate into a formal model of reasoning a very imprecise notion of content: subject matter.

III Relatedness Logic: The Subject Matter of a Proposition – S and R –

In this chapter we will see how to incorporate a particular aspect of propositions, subject matter, into a formal analysis of reasoning. This example provides a simple introduction to the general methods used throughout the text. Nothing in later chapters, however, depends on accepting the analysis of subject matters presented here.

A. An Aspect of Propositions: Subject Matter

If the moon is made of green cheese, then $2+2=4$

This isn't true. It's not that it's odd. What possible connection does 'the moon is made of green cheese' have with '$2+2=4$' ? They have nothing in common, so how can the latter follow from the former? In any normal discourse, in any normal reasoning and use of logic, something other than the truth-values of propositions matters in establishing the truth-value of an 'if . . . then . . .' proposition. But are other aspects of propositions too ambiguous to take account of in our reasoning? Or would it be hopelessly complicated to take account of them?

Even if I can't precisely specify and define, say, the subject matter of a proposition, I can nonetheless tell you in many examples if two propositions have some common subject matter, in what fashion subject matter affects the truth of an 'if . . . then . . .' proposition, and how the subject matter of a proposition is related to its parts. I might not be able to tell you what the subject matter of '$2+2=4$' is, or of 'the moon is made of green cheese'. But I can tell you that in any reasonable discussion we'll say that they are unrelated. And '$2+2=4$ and the moon is made of green cheese' does have some subject matter in common with both. If I can formulate rules that you'll agree are faithful to some commonly held notion of subject matter, then we can investigate together whether subject matter is a coherent notion and whether it's suitable to be taken account of in logical argumentation.

The approach I'm going to take is that the subject matter of a proposition isn't so much a property of it, as a relationship it has to other propositions. We may agree that 'Ralph is a dog' is, say, true, and that the truth-value is a property of that proposition. But the subject matter of 'Ralph is a dog' is an aspect of it related to other propositions. For example:

'Ralph is a dog' is related to {
'George is a duck'
'Dogs are faithful'
'Cats are nasty'
⋮

'Ralph is a dog' is not related to {
'2+2=4'
'The action of an internal combustion engine can be described in physics'
⋮

Suppose you stop me here and say, 'Look, no one's going to use both '2+2=4' and 'Ralph is a dog' in the same argument, in some logical deduction, so why worry? When we actually make up some particular model of classical logic, some assignment for the propositional variables, we won't include both of these propositions, so classical logic will do fine for us.'

> Only those conditionals are worth affirming which follow from some manner of relevance between antecedent and consequent—some law, perhaps, connecting the matters which these two components describe. But such connection underlies the useful application of the conditional without needing to participate in its meaning.
>
> *Quine, 1950,* p. 24

Then classical logic isn't universal? It's only applicable if all the propositions in the model have the same subject matter? That's exactly my point.

If I'm going to convince you of something, I must assume some shared knowledge. For example, if I want to convince you of the Pythagorean Theorem, I'll need to assume some mathematical propositions are true and accepted so by you, such as 'the area of a square with sides of length s is s^2'. Similarly, if I want to convince you of the truth of some proposition in a ordinary language setting, say a law court, then I'll have to assume a common stock of true propositions; but I will also have to assume that we share a common notion of subject matter or related propositions. Otherwise you're likely to dismiss my reasoning on the ground that I've introduced unrelated propositions.

Usually the common stock of propositions we accept as true is given implicitly, with only an occasional explicit truth-value stated, sometimes as a working hypothesis. Similarly, it's only with an occasional proposition that we may find disagreement about whether it's related to certain others. In that case we need a way to survey the consequences of whether it is or is not related to those others.

Suppose we want to take some unproved mathematical conjecture, such as Riemann's Hypothesis, as an atomic proposition in a model. We can't agree whether it's true or false, yet we want to see what its logical consequences are. We can do this because we know how to survey the assignments in which it's assumed true, and those in which it's assumed false. Similarly, I don't know whether 'Riemann's Hypothesis holds' is related in subject matter to 'Light bulbs have copper filaments', but I can work out their consequences if I can survey assignments in which they are assumed to be related and those in which they are not.

The notion of the subject matter of an atomic proposition is apparently both more dependent on context and more holistic than truth. In a particular model we may decide that 'Ralph is a dog' and 'George is a duck' are related because one of the topics under discussion is living creatures. But were our topics restricted to mammals and birds we would, perhaps, take these propositions to be unrelated. The holism of this approach is in part a consequence of the fact that the content of a sentence can only be given by considering all other atomic sentences in the model to decide which it's related to and which not, for in that survey we may find that certain topics are under discussion. However, in practice we can deal with the notion molecularly by either postulating the topics to begin with or by considering only a small number of propositions in any one disputation. Hence we can give a determinate content to each proposition and build up the contents in a molecular pattern thereafter.

To further explain this notion of common subject matter I need to be able to talk about the internal structure of propositions. I understand two propositions as being related if they share, either explicitly or implicitly, some predicate or both refer to some common object. Thus 'Ralph is a dog' and 'Ralph barks' share a common reference; 'Don is a bachelor' and 'Don is a man' share, implicitly, a common predicate. Agreements should be established in terms of predicates and names, and that is how I have developed the notion with Stanisław Krajewski, as I will present in *A General Framework for Semantics for Predicate Logics* (see *Krajewski, 1986*).

Since we're not using the tools of predicate logic here, I have to hope that you find the notion of subject matter plausible and have gained some insight into it from this discussion. Based on that assumption I will motivate structural properties of this aspect of propositions.

Relatedness Assumption A proposition can be viewed as being related in subject matter to some propositions and unrelated to all others.

The more aspects of a proposition we model, the better we'll be able to satisfy the criterion that a logic concur with our intuitions about what's true and what follows from what, until we finally reach the complexity of natural language itself. But the more we model, the less simple our logic will be and the harder it will be to apply. So let's look at only this one new aspect of propositions.

The Relatedness Abstraction The only properties of a proposition that matter to (this) logic are its form, its truth-value, and what propositions it is related to in subject matter.

But which forms?

B. The Formal Language

We have been discussing the same English language connectives that we considered for classical logic. Should we choose some new formal symbols to use in modeling them? That would have the advantage of reminding us that in one case relatedness of propositions matters, and in another it doesn't. But it hides a basic point: both here and in classical logic we are formalizing the same English connectives. A comparison of how different aspects of propositions matter in using these connectives in logic will be obscured if we choose different symbols. Therefore, the formal language I'll use is $L(\neg, \rightarrow, \wedge, p_0, p_1, \ldots)$.

I don't include $\vee$ because here, even more than for classical logic, there are many good choices for how to formalize 'or', and arguments about the suitability of one over another are secondary to the main ideas of this chapter. We'll see in Section J that all of the choices can be defined from the other primitives.

Why, you may ask, don't we take 'relatedness' as a primitive in the language? The answer is that the binary relation of subject matter relatedness, call it R, is not a connective. Consider:

R(George is a duck, Ralph is a dog)

This is read as:

'George is a duck' is related to 'Ralph is a dog'

The original propositions are not used in this example, but rather their names. It's an unnecessary confusion of objects and their names to formalize 'is related to' as a sentence connective. It makes no sense to iterate it, for how could we read R(R(George is a duck, Ralph is a dog), Dusty is a horse)?

C. Properties of the Primitive: Relatedness Relations

Did you realize that by making the *Relatedness Abstraction* I'm continuing to assume that every proposition can be viewed as being either true or false? If 'If Ralph is a dog, then $2+2=4$' is not true, then it's false. Not odd, or unacceptable, or some third truth-value. Just false. There are two truth-values: one for propositions we accept and use in deducing "what is the case", and one for those we reject. And part of what is the case is how we understand 'if . . . then . . .'. Between affirming and denying there is no third choice: *tertium non datur*. Throughout this book any oddness, uncertainty, or other aspect of a proposition will be taken account of in terms of its content, in this case subject matter.

But then what is the relation between the subject matter of a proposition and its truth-value? When I suggested that 'Ralph is a dog' is unrelated to '$2+2=4$' I'm fairly sure you didn't ask whether 'Ralph is a dog' is true. The use or assertion of a proposition carries with it the subject matter of the proposition. But it cannot also carry the truth of it. We may assert a sentence to be true, but that's not to say that it is true. That is, I am assuming:

> The subject matter of a proposition is independent of its truth-value

As I see it, a virtual consequence of this assumption is that the logical connectives are neutral with respect to subject matter: they are syncategorematic, without any content. Roughly speaking, 'Ralph is a dog and $2+2=4$' is related to the same propositions as 'If Ralph is a dog, then $2+2=4$' because the predicate implicit in and things referred to by the former are the same as for the latter. There are, however, other ways to view the logical connectives, two of which I discuss in Chapter V.A.2. Therefore, I will list this as an additional assumption:

> The logical connectives are *syncategorematic*: they are neutral with respect to subject matter

These assumptions rule out the view that a tautology has no subject matter, as propounded by *Prior, 1960,* p. 89. Whatever the notion of content behind that view, it surely can't be based on how or to what propositions refer or the predicates involved in the propositions.

The assumption that the logical connectives are syncategorematic tells us that A and $\neg$A are related to the same propositions, and that $A \to B$ and $A \wedge B$ have the same subject matter. Using '$\mathsf{R}(A, B)$' to stand for 'A is related to B' we can symbolize these observations as

$\mathsf{R}(A,B)$ iff $\mathsf{R}(\neg A,B)$

$\mathsf{R}(A,B \wedge C)$ iff $\mathsf{R}(A,B \to C)$

I should also list $\mathsf{R}(B,A)$ iff $\mathsf{R}(B,\neg A)$, and similarly for the second. But that's not necessary, for I believe you'll agree, subject matter relatedness is symmetric. It doesn't matter in which order we consider 'Ralph is a dog' and 'George is a duck' in determining whether they have some common subject matter. That is, the relation should be *symmetric*:

$\mathsf{R}(A,B)$ iff $\mathsf{R}(B,A)$

It seems in keeping with our understanding of this notion to further agree that every proposition is related to itself, that is, the relationship R is *reflexive*:

$\mathsf{R}(A, A)$

So the only question left is: what is $A \wedge B$, and hence $A \rightarrow B$, related to?

To begin with, is 'Ralph is a dog' related to 'Ralph is a dog $\wedge$ 2+2=4' ? Surely yes, as the former is a part of the latter. Roughly speaking, the former proposition refers to some of the same things and involves some of the same predications as the latter. Don't confuse truth and subject matter here: we aren't saying that because the latter proposition is a conjunction then if it's true so is the former.

If A is related to $A \wedge B$, shouldn't also A be related to $C \wedge D$ if it's related to C or to D? And how else can A have something in common with $C \wedge D$ unless it's related to C or to D? Thus:

$\mathsf{R}(A, B \wedge C)$ iff $\mathsf{R}(A, B)$ or $\mathsf{R}(A, C)$

$\mathsf{R}(A, B \rightarrow C)$ iff $\mathsf{R}(A, B)$ or $\mathsf{R}(A, C)$

You may have answered my last question differently, claiming that something transcendent happens when we conjoin propositions, so that $C \wedge D$ can be related to propositions other than those that C or D are related to. I don't agree, but you can view this last condition as a simplification if you like, along the lines of a Fregean assumption (Chapter II.A, p. 20).

To summarize, we have:

R1. $\mathsf{R}(A, A)$

R2. $\mathsf{R}(A, B)$ iff $\mathsf{R}(\neg A, B)$

R3. $\mathsf{R}(A, B)$ iff $\mathsf{R}(B, A)$

R4. $\mathsf{R}(A, B \rightarrow C)$ iff $\mathsf{R}(A, B)$ or $\mathsf{R}(A, C)$

R5. $\mathsf{R}(A, B \wedge C)$ iff $\mathsf{R}(A, B \rightarrow C)$

Let's give the name *subject matter relatedness relation* to any binary relation on propositions of the semi-formal language that we take as establishing subject matter relatedness and that satisfies R1–R5.

R1–R5 allow us to define inductively a unique subject matter relatedness relation from a reflexive and symmetric relation on atomic propositions. The following lemma establishes that there is a simple functional relation between the subject matter relations of a complex proposition and the subject matter relations of its parts. Thus, as with truth-values, we may take subject matter relatedness as primitive on just the atomic propositions.

Lemma 1 $\mathsf{R}(A,B)$ iff there is some atomic proposition p that appears in A and some atomic proposition q that appears in B such that $\mathsf{R}(p, q)$.

Proof: The proof of this lemma is an example of double induction on the length of propositions.

The lemma is immediate if both A and B have length 1. Suppose it's true for all A of length $\leq n$ and for all B of length 1, i.e., B is q. I will show that it holds for all A of length $\leq n+1$ and all B of length 1.

Case 1: A is $C \wedge D$. Then $R(C \wedge D, q)$ iff $R(C, q)$ or $R(D, q)$ by R5 and R3. So by induction $R(C \wedge D, q)$ iff some p appears in C and $R(p, q)$ or some p appears in D and $R(p, q)$, which is iff some p appears in $C \wedge D$ and $R(p, q)$.

Case 2: A is $C \rightarrow D$. Then proceed as in Case 1 using R2.

Case 3: A is $\neg C$. Then proceed as in Case 1 using R1 .

Now suppose the lemma is true for all A when B has length $\leq n$. I will show it holds for all A, when B has length $\leq n+1$.

Case 1: B is $C \wedge D$. Then $R(A, C \wedge D)$ iff $R(A, C)$ or $R(A, D)$, which by induction is iff for some p in A, and some q in C or q in D, $R(p, q)$, which is iff for some p in A, some q in $C \wedge D$, $R(p, q)$.

The other cases follow similarly. ■

D. Subject Matters As Set-Assignments

We've taken as primitive the notion of two propositions being related. Can we derive a definite content for each proposition from that relationship, one which we could call 'the subject matter of the proposition'? David Lewis has shown how. Though I'll use mathematical notation, no infinitistic assumptions are necessary.

Given a subject matter relatedness relation, R, set the subject matter of a proposition, A, to be:

$$s(A) = \{\{A, B\} : R(A, B)\}$$

We call s the *subject matter set-assignment associated with* R.

Lemma 2 Given a relatedness relation R and s the subject matter set-assignment associated with R, then: $R(A, B)$ iff $s(A) \cap s(B) \neq \varnothing$.

I leave the proof as an exercise.

Alternatively, we could take the notion of a definite content for a proposition, a subject matter, as primitive and then define two propositions as being related if they have some subject matter in common. To do that we postulate a set of topics that we assign to the propositions under consideration. That is, we have:

a set of topics S

an assignment s that for every atomic proposition p gives
$s(p) \subseteq S$ and $s(p) \neq \varnothing$

We need $s(p) \neq \varnothing$ so that each proposition has some subject matter in common with itself.

We extend s to complex propositions by taking the *subject matter of a proposition to be the sum of the subject matter of its parts*:

$$s(A) = \bigcup \{s(p) : \text{p appears in A}\}$$

Any such s and S we call a *subject matter assignment.*

Define the *relatedness relation associated with* s to be:

$$R(A, B) \text{ iff } s(A) \cap s(B) \neq \varnothing$$

Example: For a thoroughly developed example we would need to consider the internal structure of propositions, where subject matter is explained in terms of the subject matter of predicates and names. Even so, a simple example may help.

Let the set of topics of our discussion be S = { mammals, birds, dogs, cats, ducks, eagles, wolves, humans, eats, sleeps, runs, flies, swims, barks, quacks, iron, rust, car, drives, squashes, plastic, rubber, tires, engines, Ralph, George, Don }. Then we might take s(Ralph is a dog) = {mammals, dogs, eats, sleeps, runs, barks} and s(George is a duck) = {birds, ducks, flies, swims, quacks}, so that R(George is a duck, Ralph is a dog) fails. I'll leave to you to determine s(Don drives a car $\wedge$ cars have tires) and whether R(Don drives a car $\wedge$ cars have tires, George is a duck).

In this example I understand the elements of S to be bits of language, inscriptions, in the tradition of Buridan (see the Introduction to *T. K. Scott, 1966,* especially pp. 46–47). Someone working in the tradition of *Frege, 1892,* might choose to interpret S to be a collection of nonsensible objects such as meanings or senses.

Lemma 3 The relatedness relation associated with a subject matter assignment s satisfies R1–R5.

Proof: First, because for every p, $s(p) \neq \varnothing$, we have $s(A) \neq \varnothing$ for every A, so $R(A, A)$ holds. And R is symmetric. We also have $s(A) = s(\neg A)$, so $R(A, B)$ iff $R(\neg A, B)$. And $s(A \wedge B) = s(A \rightarrow B) = s(A) \cup s(B)$ so $R(A, B \rightarrow C)$ iff $R(A, B \wedge C)$. Finally,

$$\begin{aligned} R(A, B \rightarrow C) \text{ holds} \quad & \text{iff} \quad s(A) \cap [s(B) \cup s(C)] \neq \varnothing \\ & \text{iff} \quad [s(A) \cap s(B)] \cup [s(A) \cap s(C)] \neq \varnothing \\ & \text{iff} \quad s(A) \cap s(B) \neq \varnothing \text{ or } s(A) \cap s(C) \neq \varnothing \\ & \text{iff} \quad R(A, B) \text{ or } R(A, C) \end{aligned}$$

■

Can we use these two approaches interchangeably? The next lemma shows that for every R there is a unique s, and for every s a unique R such that we can pass back and forth between them without losing any significant information.

Lemma 4 ***a.*** Given a relatedness relation R, let s be the subject matter assignment associated with R. Let $\mathsf{R_s}$ be the relatedness relation associated with s. Then $\mathsf{R} = \mathsf{R_s}$.

b. Given a subject matter assignment s, let R be the relatedness relation associated with s. Let $\mathsf{s_R}$ be the subject matter assignment associated with R. Then $\mathsf{s}(A) \cap \mathsf{s}(B) \neq \varnothing$ iff $\mathsf{s_R}(A) \cap \mathsf{s_R}(B) \neq \varnothing$.

Proof: a. $\mathsf{R_s}(A, B)$ iff $\mathsf{s}(A) \cap \mathsf{s}(B) \neq \varnothing$

iff there are C, D such that $\{A, C\} \in \mathsf{s}(A)$ and $\{B, D\} \in \mathsf{s}(B)$ and $\{A, C\} = \{B, D\}$ and $\mathsf{R}(A, C)$ and $\mathsf{R}(B, D)$

iff either $A = B$, so $\mathsf{R}(A, B)$, or $A \neq B$, so $C = B$ and $\mathsf{R}(A, B)$.

b. This follows from part (a) by passing to the relatedness relation associated with $\mathsf{s_R}$. ■

We don't claim in Lemma 4 that $\mathsf{s} = \mathsf{s_R}$. We only have that $\mathsf{s_R}$ preserves two-way overlap from s. The following two subject matter assignments result in the same $\mathsf{s_R}$ in Lemma 4:

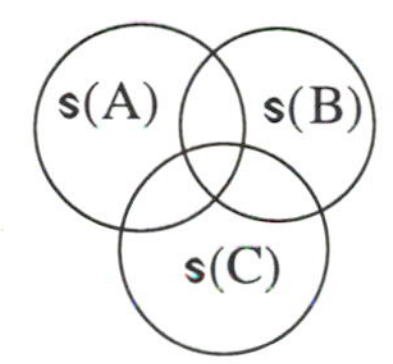

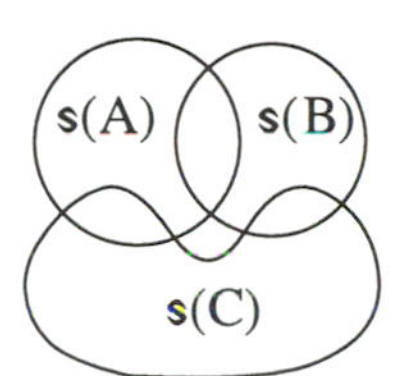

In the left one there is three-way overlap of $\mathsf{s}(A)$, $\mathsf{s}(B)$, and $\mathsf{s}(C)$, whereas in the right hand one there isn't. Because we are investigating only unary and binary connectives at this stage, the difference between these two is inessential.

Exercises for Sections A–D

1. Argue that classical logic is universal and the notion of subject matter is irrelevant to reasoning in logic. In a particular discussion how would you respond to an objection that a proposition you have introduced is irrelevant to what has gone before?

2. What is the *Relatedness Assumption*? What is the *Relatedness Abstraction*? Compare both to the *Classical Abstraction.*

3. Why do we continue to classify 'If the moon is made of green cheese, then $2 + 2 = 4$' as true or false, rather than nonsensical?

4. What is a subject matter relatedness relation? Argue for each condition. Are we correct in requiring it to be symmetric? Can you think of reasoning in which it might not be symmetric?

5. a. What is the subject matter set-assignment associated with a subject matter relatedness relation?
   b. Why do we require the set of topics to be non-empty?
   c. Explain the different assumptions needed for taking set-assignments as primitive versus taking relatedness relations as primitive.

6. Using the set of topics from the example in Section D, assign subject matter to:

Ralph is a dog	Cars squash dogs
Dogs bark	Engines run cars
Ducks quack	Humans run cars
Ducks are mammals	Don squashed a duck
Don drives a car	Cars fly
Cars have tires	Dogs fly
Cars kill ducks	Ducks fly
Cars squash ducks	Ducks squash cars

   Using your assignment, calculate the subject matter of all propositions of the form $\neg p$, $p \wedge q$, and $p \rightarrow q$, where p, q are from the list above. Then calculate which of those propositions are related to which others.

   How do you respond to someone who says that 'Cars have tires' is related to 'Ducks are mammals' because both are related to 'Cars kill ducks'?

7. Prove Lemma 2.

8. Prove Lemma 4.b.

E. Truth-and-Relatedness Tables

How are we to determine the truth-value of a complex proposition? For the same reasons as for classical logic (Chapter II.A) I am going to invoke the *Fregean Assumption*:

> The truth-value of a complex proposition is determined by its form and the properties of its constituents

And for the same reasons as before (Chapter II.B) we should invoke the *Division of Form and Content*:

> If two propositions have the same semantic properties, then they are indistinguishable in any semantic analysis, regardless of their form.

We have already decided which semantic properties of the constituents are to be taken into consideration with the *Relatedness Abstraction.* So we may conclude that each connective is interpreted as a function of the truth-values of and relatedness between its parts. That is, the connectives will be interpreted as *truth-and-relatedness functions*.

What is the function appropriate for negation? The arguments I gave for the independence of subject matter from truth-value should convince you that the table should be the same as for classical negation:

A	$\neg A$
T	F
F	T

It also seems to me that we should be allowed to conjoin any two propositions without regard to their subject matters: using $\wedge$ is like making a list, for example, 'Ralph is a dog $\wedge$ $2+2=4$ $\wedge$ light bulbs have copper filaments'. So I will take the classical table for conjunction:

A	B	$A \wedge B$
T	T	T
T	F	F
F	T	F
F	F	F

I believe that these tables are the best choices, but you can view them as an application of a simplicity constraint: the simplest way in which we can take account of subject matter relatedness is to let it affect only the one connective that originally worried us, the conditional.

When is $A \to B$ true? If A and B are related, then the usual classical truth-table should apply, as suggested in Section A: classical logic models are ones in which all propositions are taken to be related. If A and B are unrelated, then $A \to B$ is false: when we wanted to reject 'If Ralph is a dog, then $2+2=4$' the question of whether 'Ralph is a dog' is true didn't arise. The lack of subject matter overlap in itself disqualifies $A \to B$ from being true, regardless of the truth-values of A and B. Thus the *truth-table for the* (*subject matter*) *related conditional* is:

A	B	$\mathsf{R}(A, B)$	$A \to B$
any	value	fails	F
T	T	holds	T
T	F		F
F	T		T
F	F		T

Should we take account of relatedness in formalizing 'or'? If we want a related disjunction we can use $\neg A \to B$, which is true iff A is related to B and at least one of A, B is true. If we believe 'or' should be formalized as classical disjunction, we can use $\neg(\neg A \wedge \neg B)$. Because our other primitives suffice, I will avoid introducing a controversy here by making a definitive choice for how to interpret $\vee$.

F. The Formal Semantics for S

1. Models based on relatedness relations

We've made enough agreements to present a logic.

A *subject matter relatedness model* is:

I

$$L(\neg, \rightarrow, \wedge, p_0, p_1, \ldots)$$

$\downarrow$ realization

$\{p_0, p_1, \ldots,$ complex propositions formed from these using $\neg, \rightarrow, \wedge\}$

$\downarrow$ ν, R, and truth-tables

$$\{T, F\}$$

Here $p_0, p_1, \ldots$ are the realizations of $p_0, p_1, \ldots$, which we take to be atomic; ν is a way of assigning truth-values to these; and R is a relatedness relation on pairs of atomic propositions. We extend R to all complex propositions inductively by R1–R5, or equivalently by: R(A,B) iff for some p_i in A, some p_j in B, $R(p_i, p_j)$. Then ν is extended inductively to all propositions by using the classical tables for $\neg$ and $\wedge$ and the table for the related conditional for $\rightarrow$ from the previous section. A proposition A of the semi-formal language is *true* if $\nu(A) = T$, *false* if $\nu(A) = F$.

Since every atomic proposition is abstracted in our model to only its truth-value and subject matter (in terms of its place in a relatedness relation) we can simplify a model of type **I** in the same way we did for classical logic:

II

$$L(\neg, \rightarrow, \wedge, p_0, p_1, \ldots)$$

$\downarrow$ ν, R, and truth-tables

$$\{T, F\}$$

Here ν is a way to assign truth-values to the propositional variables; R is a symmetric and reflexive relation on the variables that is extended to all wffs by R1–R5, or equivalently by R(A, B) iff for some p_i in A, some p_j in B, $R(p_i, p_j)$. And ν is extended to all wffs by the tables for classical $\neg$ and $\wedge$ and the related conditional for $\rightarrow$. Thus any difference between two models of type **I** is obliterated if they result in the same model of type **II**.

I will usually refer to a relatedness model as $\langle \nu, R \rangle$, where R is either specified on all wffs or only on the propositional variables. Then a *wff* A *is true in* $\langle \nu, R \rangle$ if $\nu(A) = T$, false if $\nu(A) = F$. We often write $\langle \nu, R \rangle \vDash A$, read '$\langle \nu, R \rangle$ validates A', for $\nu(A) = T$, and $\langle \nu, R \rangle \nvDash A$ for $\nu(A) = F$.

Up to this point we have made no infinitistic assumptions, either about the formal language or the semantics. To have full generality and to freely use mathematics in the study of this logic, and for all the reasons and with all the provisos for mathematizing classical logic (Chapter II.G), we will view the collection of propositional variables, PV, and the formal language as completed wholes. And we will further abstract our models. First, given two collections U and V, write $U \times V$ for the collection of ordered pairs whose first element is from U and second element is from V.

The Fully General Relatedness Abstraction Any function $\nu: PV \rightarrow \{T, F\}$ and any symmetric, reflexive relation $R \subseteq PV \times PV$ together form a model $<\nu, R>$ when R is extended to all ordered pairs of wffs by R1–R5 and ν is extended to all wffs by the truth-tables.

Thus we not only obliterate any difference between models of type **I**, we no longer care where a model of type **II** comes from.

The notions of validity, semantic consequence, and semantic equivalence can now be defined with respect to these models for both propositions and wffs in the same manner as they were for the classical case (Chapter II.J.2). We say that a wff is a *subject matter relatedness tautology,* or *relatedness tautology* for short, if it is valid: for every $<\nu, R>$, $<\nu, R> \models_{\mathbf{S}} A$. And $\Gamma \models_{\mathbf{S}} A$ if whenever all the wffs in Γ are true in a subject matter relatedness model, so, too, is A. This consequence relation is what we call (*subject matter*) *relatedness logic,* denoted as **S**. When we want to remind ourselves that these notions are defined with respect to relatedness models (as opposed to, say, classical ones) we'll preface them with '**S**-' as in '**S**-validity', or '**S**-tautology'. Throughout this chapter, though, I will use '$\models$' for this semantic consequence relation.

As for classical logic, we say that two wffs A and B are *semantically equivalent* if in every model they take the same truth-value: $A \models B$ and $B \models A$. However, this notion of semantic equivalence does not fully reflect the roles of propositions in reasoning in relatedness logic, for it classifies as semantically equivalent:

'¬(Ralph is a dog ∧ ¬ Ralph is a dog)' '$\neg(2+2=4 \wedge \neg(2+2=4))$'

And these may be related to entirely different wffs in a model. For relatedness logic we define two wffs A and B to be *strongly semantically equivalent* if:

For every model $<\nu, R>$,
$\nu(A) = T$ iff $\nu(B) = T$, and for every wff C, $R(A, C)$ iff $R(B, C)$

Then the two wffs above are not strongly semantically equivalent, though the following are:

'¬(Ralph is a dog ∧ ¬ Ralph is a dog)' 'Ralph is a dog → Ralph is a dog'

2. Models based on subject matter assignments

Using the observations of Section D we can describe **S** in terms of subject matters, too. I will skip directly to a fully general abstraction.

Given any countable set $\mathsf{S} \neq \varnothing$ together with any function $\mathsf{s}: \mathrm{PV} \to$ subsets of S such that $\mathsf{s}(p) \neq \varnothing$ for all variables p, and any function $\upsilon: \mathrm{PV} \to \{\mathsf{T}, \mathsf{F}\}$, we will call $\langle \upsilon, \mathsf{s} \rangle$ a *set-assignment model for* **S** if s is extended to all wffs via $\mathsf{s}(A) = \bigcup\{\mathsf{s}(p)\text{: p appears in A}\}$ and υ is extended to all wffs by the classical tables for $\neg$ and $\wedge$ and the relatedness table for $\to$, reading '$\mathsf{s}(A) \cap \mathsf{s}(B) \neq \varnothing$' for '$\mathsf{R}(A, B)$'. We define $\langle \upsilon, \mathsf{s} \rangle \vDash A$ iff $\upsilon(A) = \mathsf{T}$. Then from Lemma 2, given any model $\langle \upsilon, \mathsf{R} \rangle$, if s is the subject matter set-assignment associated with R, we have $\langle \upsilon, \mathsf{s} \rangle \vDash A$ iff $\langle \upsilon, \mathsf{R} \rangle \vDash A$. And similarly, by Lemma 3, given $\langle \upsilon, \mathsf{s} \rangle$, if R is the relatedness relation associated with s, then $\langle \upsilon, \mathsf{R} \rangle \vDash A$ iff $\langle \upsilon, \mathsf{s} \rangle \vDash A$. Hence if we define the notions of validity, semantic consequence, etc. with respect to set assignment models for **S**, they will coincide with those we define for relatedness models. Therefore, I will use both kinds of models as the occasion arises, calling them collectively *models for* **S**.

3. Nonsymmetric relatedness logic, R

A variation on subject matter relatedness logic was designed to analyze arguments about actions. We can say that one proposition about an action, such as 'Ralph barked', is related to another, say, 'Richard killed a duck', if the action it describes, were it to have occurred, is close in space to, and slightly preceding in time, the action described by the latter proposition. In this case we do not want the relation on propositions to be symmetric. But as before, a complex proposition should be related to another if any one part of it is related to any part of the other.

At least that was the idea when the logic was proposed (see Section L below). But this motivation now seems rather weak: we want, it would seem, the relation on propositions to be anti-symmetric: if $\mathsf{R}(A, B)$ and $\mathsf{R}(B, A)$, then $B = A$. And do we really want a proposition describing an action and its negation to be related to exactly the same propositions? But as a start anyway, let's consider just lifting the requirement that the relatedness relation be symmetric.

Call a binary relation R on Wffs$\times$Wffs a *nonsymmetric relatedness relation* if it satisfies R1, R2, R4, and R5 (p. 91) and:

R6. $\mathsf{R}(B, A)$ iff $\mathsf{R}(B, \neg A)$

R7. $\mathsf{R}(B \wedge C, A)$ iff $\mathsf{R}(B \to C, A)$

R8. $\mathsf{R}(B \wedge C, A)$ iff $\mathsf{R}(B, A)$ or $\mathsf{R}(C, A)$

Define the logic **R** to be the consequence relation of the models $\langle \upsilon, \mathsf{R} \rangle$ where: $\mathsf{R} \subseteq \mathrm{PV} \times \mathrm{PV}$ is a reflexive relation, $\upsilon: \mathrm{PV} \to \{\mathsf{T}, \mathsf{F}\}$, and R is extended to all wffs by R1, R2, R4, and R5–R8, and υ is extended to all wffs by the classical tables for $\neg$

and $\wedge$, and the relatedness table for $\rightarrow$. I call **R** *nonsymmetric relatedness logic.* Note that every subject matter relatedness relation also is a nonsymmetric relatedness relation, since we do not require the relation to be anti-symmetric, just not necessarily symmetric. Hence, if $\Gamma \vDash_{\mathbf{R}} A$ then $\Gamma \vDash_{\mathbf{S}} A$, as I will let you show.

In the rest of this chapter I discuss only **S**, unless I explicitly state otherwise.

Exercises for Sections E and F

1. What is the *Division of Form and Content* for relatedness logic? Are we justified in invoking it? Is the *Fregean Assumption* as plausible in relatedness logic as in classical logic? Show that these two assumptions are independent. (Compare Exercise 8, p. 26.)

2. a. What is a truth and relatedness functional connective?
   b. Which of $\neg$, $\rightarrow$, $\wedge$ are truth and relatedness functional connectives?
   c. Why do we take falsity as the default truth-value in the table for $\rightarrow$? That is, why do we say and $\rightarrow$-wff is false if the antecedent and consequent are not related?

3. Give two examples of propositions or deductions in which it seems reasonable to formalize 'or' as related disjunction.

4. Prove that if $v: PV \rightarrow \{T, F\}$, and $R \subseteq PV \times PV$ is a symmetric and reflexive relation, then there is a unique way to extend R to a relatedness relation on all pairs of wffs and to extend v to all wffs according to the relatedness table for $\rightarrow$ and the classical tables for $\neg$ and $\wedge$.

5. What are set-assignment semantics for **S**? Prove that they give rise to the same logic as the original semantics for **S**.

6. Show that $\wedge$ can be defined from $\neg$ and $\rightarrow$ in **S**. Is there a single truth-and-relatedness connective from which all three of $\neg$, $\rightarrow$, and $\wedge$ can be defined?

7. a. Give a precise definition of the logic **R**.
   b. Show that every **R**-consequence is an **S**-consequence.
   c. Give three examples of **S**-tautologies that are not **R**-tautologies.
   d. (Open) Devise a notion of content that can be assigned to every wff that can replace the notion of a non-symmetric relatedness relation in the definition of the logic **R**.

G. Examples of Formalization

In this section I present some examples of formalizations of propositions and arguments relative to the assumptions that motivated subject matter relatedness logic. The format will be the same as in Chapter II.F, and the criteria that govern the acceptability of formalizations will be those discussed in Chapter II.E.2.

1. If the moon is made of green cheese, then 2 + 2 = 4

The moon is made of green cheese $\to 2+2=4$

$p_1 \to p_2$

Explanation: This fails to be a relatedness tautology, but of course it fails to be a classical tautology. This is despite the claim that it is impossible for '2 + 2 = 4' to be false: logical validity, whether relatedness or classical, does not take into account the meaning of words, which is what that claim depends on.

2. 2 + 2 = 4
***Therefore*:**
If the moon is made of green cheese, then 2 + 2 = 4

$2+2=4$
Therefore:
The moon is made of green cheese $\to 2+2=4$

$$\frac{p_1}{p_2 \to p_1}$$

Explanation: The example is a classically valid deduction: we have explicitly added '2 + 2 = 4' as a premiss. But the example is not a relatedness tautology: just because '2 + 2 = 4' is true, we need not have that 'the moon is made of green cheese' is related to '2 + 2 = 4'.

The "paradoxes" of classical logic likewise fail to be relatedness tautologies, as exemplified here:

$2+2=4 \to$ (Ralph is a dog $\to 2+2=4$)

$\neg$(Ralph is a dog) $\to$ (Ralph is a dog $\to 2+2=4$)

In both examples the hypothesis can be true in a model while the consequent is false, since 'Ralph is a dog' may be unrelated to '2 + 2 = 4'.

3. If Ralph is a dog and if 1 =1 then 1 = 1, then 2 + 2 = 4 or 2 + 2 ≠ 4

[Ralph is a dog $\wedge$ $(1=1 \to 1=1)] \to (2+2=4 \vee \neg (2+2=4))$

$(p_1 \wedge (p_2 \to p_2)) \to (p_3 \vee \neg p_3)$

Explanation: The example is a classical tautology; it is not a relatedness tautology, since $\mathsf{R}(p_1 \wedge (p_2 \to p_2),\ p_3 \vee \neg p_3)$ may fail. But following the motivation I gave for relatedness relations, if we examine the internal structure of propositions we should have $\mathsf{R}(1=1, 2+2=4)$, since both propositions involve the predicate '='. And in that case we would have that the example would be a relatedness tautology. Or at least we can say that in every reasonable model, '1 = 1' and '2 + 2 = 4' will be related. And then the example will be formalized as a true proposition.

Some say this example illustrates why relatedness is not a significant addition to the notions of logic that establish classical logic. They say that this example is just as counterintuitive as Example 1, yet is classified as true. After all, if 'Ralph is a dog' is unrelated to '$2+2=4$', then surely 'Ralph is a dog $\wedge$ $(1=1 \vee \neg\,(1=1))$' should be unrelated to '$2+2=4$'.

That 'surely' seems to be based on the view that a tautology has no subject matter and shouldn't contribute to establishing relatedness of propositions. Whether you agree that a tautology has no subject matter, it is clear that such a view is not compatible with the analysis of subject matter that I gave and must be fully motivated along quite different lines.

4. If John loves Mary, then Mary has 2 apples
If Mary has 2 apples, then 2 + 2 = 4
***Therefore*:**
If John loves Mary, then 2 + 2 = 4

John loves Mary $\rightarrow$ Mary has 2 apples
Mary has 2 apples, $\rightarrow$ $2+2=4$
Therefore:
John loves Mary $\rightarrow$ $2+2=4$

$$\frac{p_1 \rightarrow p_2 \qquad p_2 \rightarrow p_3}{p_1 \rightarrow p_3}$$

Explanation: The example is not valid. We may have that each of the premisses is true in a model while the conclusion fails because $\mathsf{R}(p_1, p_3)$ fails.

According to the motivation given for relatedness logic, this seems sensible: intuitively 'John loves Mary' and '$2+2=4$' have nothing in common. And the premisses do not establish any common subject matter for them. The transitivity of the conditional fails. And that is right, for relatedness is not transitive.

But how can we reason if chains of conditionals do not yield the conclusions we want? Isn't this a defect of our analysis?

Perhaps we could simply require every relatedness relation to be transitive. But then the only relatedness relation would be the universal relation on pairs of wffs: for every A, B we have $\mathsf{R}(A, A \rightarrow B)$ and $\mathsf{R}(A \rightarrow B, B)$, so transitivity would yield $\mathsf{R}(A, B)$. Note that we may have that R is transitive on atomic propositions, yet fails to be transitive on complex wffs:

$\models_{\mathbf{S}} (p_1 \wedge \neg p_1) \rightarrow (p_1 \wedge (p_2 \wedge \neg p_2))$ and $\models_{\mathbf{S}} (p_1 \wedge (p_2 \wedge \neg p_2)) \rightarrow (p_2 \wedge p_3)$
but $\not\models_{\mathbf{S}} (p_1 \wedge \neg p_1) \rightarrow (p_2 \wedge p_3)$

If subject matter relatedness is transitive, we are back to classical logic, since every two propositions will be related. And indeed the argument from transitivity of relatedness to every proposition being related is the one that is often given by classical logicians to establish that there is no nontrivial notion of subject matter relatedness. As in the example, 'John loves Mary' is related to 'Mary has 2 apples', which in turn is related to '2+2=4'. By the transitivity of relatedness, 'John loves Mary' would be related to '2+2=4'. Not surprisingly, if you think that these two propositions are related, then you'll probably believe that any two propositions are related. It may be comforting to mathematicians to think that love and mathematics have something in common, but for most of us this sounds like sophistry.

It has been argued that one cannot do logic if '$\rightarrow$' is not transitive, and I think this is what lies at the bottom of the insistence that relatedness be transitive. Yet here we have a logic based on some (reasonable) intuitions about truth and language in which '$\rightarrow$' is not transitive. *T. J. Smiley, 1959,* has also developed a logic in which '$\rightarrow$' is not transitive. He says:

> The need for an unrestrictedly transitive entailment-relation for serious logical work is no reason at all against accepting a relation which is not unrestrictedly transitive as being a satisfactory reconstruction of an intuitive idea of entailment. But the need itself is undeniable: the whole point of logic as an instrument, and the way it brings us new knowledge, lies in the contrast between the transitivity of 'entails' and the non-transitivity of 'obviously entails,' and all this is lost if transitivity cannot be relied on. Of course if there is an effective way of predicting when transitivity will hold then most of the objection vanishes.
>
> *Smiley, 1959,* p. 242

In relatedness logic we do have that the consequence relation is transitive, for it takes into account only truth-values: if $A \vDash B$ and $B \vDash C$, then $A \vDash C$.

E. P. Specker, 1960, also discusses the problem of a nontransitive '$\rightarrow$', in relation to a logic of quantum mechanics. He suggests that a further condition needed to infer $A \rightarrow C$ from $A \rightarrow B$ and $B \rightarrow C$ is that A and C be simultaneously decidable.

5. If Don squashed a duck and Don drives a car, then a duck is dead
***Therefore*:**
If Don squashed a duck, then if Don drives a car, then a duck is dead

(Don squashed a duck $\wedge$ Don drives a car) $\rightarrow$ a duck is dead
Therefore:
Don squashed a duck $\rightarrow$ (Don drives a car $\rightarrow$ a duck is dead)

$$\frac{(p_1 \wedge p_2) \rightarrow p_3}{p_1 \rightarrow (p_2 \rightarrow p_3)}$$

Explanation: The example is classically valid, an instance of the deductive form of *exportation.* But the example is not valid in relatedness logic: we may have a model in which all the atomic propositions above are true, 'Don squashed a duck' is related to ‘a duck is dead’, yet ‘Don drives a car’ is unrelated to ‘a duck is dead’. In relatedness logic deductions are sensitive to the placement of hypotheses.

Exercises for Section G

1. Formalize the following and discuss according to the format of Section G.
   a. If the moon is made of green cheese, then if green cheese stinks, we have that no one would eat the moon
   b. Ralph is a dog or not both $2 + 2 = 4$ and $2 + 2 \neq 4$
   c. If Ralph is a dog then dogs bark, so if dogs bark, Ralph is a dog
   d. If Ralph is a dog then dogs bark, so if dogs don’t bark, then Ralph is not a dog
   e. Ralph is a dog
      Ralph is not a dog
      Therefore:
         $2 + 2 = 4$
   f. Ducks quack or dogs quack
      Dogs don’t quack
      Therefore:
         Ducks quack
   g. If dogs bark, then Ralph barks and ducks quack; so if dogs bark then Ralph barks, and if dogs bark then ducks quack
   h. Ralph is not a dog because he’s a puppet

H. Relatedness Logic Compared to Classical Logic

1. The decidability of relatedness tautologies

We can decide for any wff whether it is a relatedness tautology or not. The method is a variation on the one for classical logic (Chapter II.J.7).

Given a wff A in $L(\neg, \rightarrow, \wedge, p_0, p_1, \ldots)$, the following is a computable procedure (though for long wffs unfeasible): list each of the finite number of ways (a) to assign truth-values to the variables appearing in A and (b) to symmetrically and reflexively relate those variables; then evaluate the wff for each assignment by the tables for $\neg$, $\wedge$, $\rightarrow$. If there is no assignment for which the wff comes out false, then A is a relatedness tautology; if there is one, then A is not a relatedness tautology, for any such valuation and binary symmetric reflexive relation can be extended to all variables, for example, assigning F to all other variables, and, except for relating each variable to itself, relating no others. In that model A is false.

I ask you in an exercise to expand on this description and devise a falsification procedure for verifying subject matter relatedness tautologies comparable to the one for classical logic described in Chapter II.J.7.

2. Every relatedness tautology is a classical tautology

I said earlier that in the face of the objections that motivated subject matter relatedness logic, some logicians justify **PC** by saying that every pair of propositions are related (see, for example, *Bennett, 1969*). Or at least, as Quine (p. 94 above), they will argue that in any useful model of classical logic every two propositions are related. The semantics for subject matters reflect this.

Let U denote the *universal relation* on Wffs$\times$Wffs, that is, for every A and B we have $\mathsf{U}(A,B)$. This is a subject matter relatedness relation. In any model $<\mathsf{v},\mathsf{U}>$ the relation contributes nothing to the evaluation of the truth-value of a wff: $<\mathsf{v},\mathsf{U}>\models_{\mathbf{S}} A$ iff $\mathsf{v}\models_{\mathbf{PC}} A$. Hence the collection of wffs true in all models $<\mathsf{v},\mathsf{U}>$ is exactly **PC**. So for a wff to qualify as a relatedness tautology it must also be a classical tautology. And if $\Gamma\models_{\mathbf{S}} A$, then $\Gamma\models_{\mathbf{PC}} A$. In this sense we can view relatedness logic as contained in classical logic. And this is as we want, for we have put further conditions for the formalization of an 'if . . . then . . .' proposition for it to be true in the semantics for **S**. In the next section we will see classical tautologies that fail to be relatedness tautologies.

However, if we think of **PC** as formalized in the language of $\neg$ and $\wedge$, then **S** can be viewed as an extension of **PC** in the language of $\neg$, $\wedge$, and $\rightarrow$. If we had chosen a different symbol for the relatedness formalization of the conditional, then our language would have suggested this point of view. I believe it is the wrong way to think of **S**, for both **S** and **PC** are formalizing the same English propositions, interpreting 'if . . . then . . .' differently, **S** more stringently than **PC**, so it is proper to think of **S** as contained in **PC**. In Chapter X I will discuss translations of **S** into **PC**.

3. Classical tautologies that aren't relatedness tautologies

Many familiar classical tautologies fail to be relatedness tautologies due to a failure of relatedness between antecedent and consequent. In Section G (Example 2) we saw that the paradoxes of classical logic are not valid in relatedness logic:

$$A \rightarrow (B \rightarrow A)$$

$$\neg A \rightarrow (A \rightarrow B)$$

Similarly, the paradoxes of strict implication fail to be relatedness tautologies:

$$(A \wedge \neg A) \rightarrow B$$

$$A \rightarrow (B \rightarrow B)$$

Each is true in a model $<\mathsf{v},\mathsf{R}>$ iff $\mathsf{R}(A, B)$ holds in that model.

In Example 4 of Section G we saw that transitivity of the conditional fails for deductions, but equally it fails in its propositional forms:

$$[(A \rightarrow B) \wedge (B \rightarrow C)] \rightarrow (A \rightarrow C)$$

$(A \to B) \to [(B \to C) \to (A \to C)]$

Simply consider a model in which A is related to B, B is related to C, and A is not related to C, and all of A, B, and C are true.

We also saw, in Example 5 of Section G, that relatedness logic is sensitive to the placement of premisses in conditionals. The following fail to be relatedness tautologies:

exportation $((A \wedge B) \to C) \to (A \to (B \to C))$

This is false if A, B, C are all true, $\mathsf{R}(A, C)$ holds, and $\mathsf{R}(B, C)$ fails.

importation $(A \to (B \to C)) \to ((A \wedge B) \to C)$

This is false if A is false, $\mathsf{R}(A, B)$ holds, and both $\mathsf{R}(A, C)$ and $\mathsf{R}(B, C)$ fail.

But some **PC**-tautologies are relatedness tautologies:

$(A \wedge (A \to B)) \to B$

$\neg B \to ((A \to B) \to \neg A)$

$(A \to B) \to ((A \to \neg B) \to \neg A)$

And if we take $\vee$ to be interpreted as either the classical or related version of 'or' (see p. 103), then a propositional form of *disjunctive syllogism* is a tautology:

$((A \vee B) \wedge \neg A) \to B$

In Chapter V.E I'll give some syntactic criteria for when **PC**-tautologies in which $\to$ is the main connective are **S**-tautologies.

Exercises for Section H

1. a. State in detail a procedure for deciding whether a formal wff (scheme) is a relatedness tautology (scheme), making explicit the assumptions on which the procedure is based (compare the procedure for classical logic, Chapter II.J.7).
   b. Devise a falsification procedure for establishing whether a formal $\to$-wff is a relatedness tautology (cf. the falsification procedure for classical logic, p. 59).
2. Decide for each scheme of classical tautologies listed in Chapter II.J.8 (pp. 60–62) whether it is a relatedness tautology scheme.

J. Functional Completeness of the Connectives and the *Normal Form Theorem* for S

Though we didn't take relatedness as a connective, the paradoxes of strict implication in the previous section show how we can define a formula that plays the same role. I use the following, which involves the conditional only:

$$\mathrm{R}(\mathrm{A},\mathrm{B}) \equiv_{\mathrm{Def}} \mathrm{A}\rightarrow(\mathrm{B}\rightarrow\mathrm{B})$$

In any model, $<\upsilon,\mathsf{R}>\models \mathrm{R}(\mathrm{A},\mathrm{B})$ iff $\mathsf{R}(\mathrm{A},\mathrm{B})$.

This abbreviation is useful for axiomatizing **S**. But I do not view 'R' as a connective; it is only an abbreviation useful in metalogical investigations.

We can reduce the number of primitives of **S** by defining $\wedge$ in terms of $\neg$ and $\rightarrow$. In **S** we have:

$\mathrm{A}\wedge\mathrm{B}$ is strongly semantically equivalent to
$$\neg(\mathrm{A}\rightarrow(\mathrm{B}\rightarrow \neg((\mathrm{A}\rightarrow\mathrm{B})\rightarrow(\mathrm{A}\rightarrow\mathrm{B}))))$$

Indeed, it is possible to define every truth-and-relatedness connective from $\neg$ and $\rightarrow$, as I'll now demonstrate.

Informally, a truth-and-relatedness functional connective is some connective $\gamma(\mathrm{A}_1, \ldots, \mathrm{A}_n)$ that is evaluated semantically by a truth table assigning T or F to the entire formula for each of the various ways T or F can be assigned to the A_i's and the various A_i's related, subject to the condition that they are related reflexively and symmetrically. More formally, γ is an n-ary *truth-and-relatedness functional connective* if it is interpreted semantically as a function from $\{\mathsf{T},\mathsf{F}\}^n \times \{\mathsf{T},\mathsf{F}\}^m$ to $\{\mathsf{T},\mathsf{F}\}$ where $m = n(n-1)/2$, and the ith entry of the sequence for $i \leq n$ indicates whether $\upsilon(\mathrm{A}_i) = \mathsf{T}$ or $\upsilon(\mathrm{A}_i) = \mathsf{F}$, while the succeeding entries stipulate whether the following hold (notated T), or fail (notated F):

$$\mathsf{R}(\mathrm{A}_1, \mathrm{A}_2),\ \mathsf{R}(\mathrm{A}_1, \mathrm{A}_3), \ldots, \mathsf{R}(\mathrm{A}_1, \mathrm{A}_n); \mathsf{R}(\mathrm{A}_2, \mathrm{A}_3), \ldots, \mathsf{R}(\mathrm{A}_2, \mathrm{A}_n);$$
$$\ldots; \mathsf{R}(\mathrm{A}_{n-1}, \mathrm{A}_n)$$

This covers all cases, since R must be reflexive and symmetric.

Theorem 5 $\{\neg,\rightarrow\}$ is truth-and-relatedness functionally complete for **S** in a strong sense. That is, in **S** every truth-and-relatedness functional connective is strongly semantically equivalent to a scheme in which only $\neg$ and $\rightarrow$ are used.

Proof: We've already shown how to define $\wedge$ and R in terms of $\neg$ and $\rightarrow$. For the purposes of this proof I'll take $\mathrm{A}\vee\mathrm{B}$ as an abbreviation of $\neg(\neg\mathrm{A}\wedge\neg\mathrm{B})$, the truth-functional inclusive 'or'.

If $\gamma(\mathrm{A}_1, \ldots, \mathrm{A}_n)$ always takes the value F, then it is semantically equivalent to $(\mathrm{A}_1\wedge\neg\mathrm{A}_1)\vee(\mathrm{A}_2\wedge\neg\mathrm{A}_2)\vee\cdots\vee(\mathrm{A}_n\wedge\neg\mathrm{A}_n)$, associating to the left. Otherwise let those sequences that are assigned T by the function for γ be:

$$\delta_1 = (\alpha_{11}, \ldots, \alpha_{1n};\ \beta_{11}, \ldots, \beta_{1m})$$
$$\vdots$$
$$\delta_k = (\alpha_{k1}, \ldots, \alpha_{kn};\ \beta_{k1}, \ldots, \beta_{km})$$

Define for $i = 1, \ldots, k$:

for $j = 1, \ldots, n$:

$$B_{ij} = \begin{cases} A_{ij} & \text{if } \alpha_{ij} = \mathsf{T} \\ \neg A_{ij} & \text{if } \alpha_{ij} = \mathsf{F} \end{cases}$$

for $j = 1, \ldots, m$, where the jth entry refers to whether $\mathsf{R}(A_s, A_t)$ holds:

$$C_{ij} = \begin{cases} R(A_s, A_t) & \text{if } \beta_{ij} = \mathsf{T} \\ \neg R(A_s, A_t) & \text{if } \beta_{ij} = \mathsf{F} \end{cases}$$

Lastly, associating the conjuncts to the left, define:

$$D_i = (B_{i1} \wedge B_{i2} \wedge \cdots \wedge B_{in}) \wedge (C_{i1} \wedge C_{i2} \wedge \cdots \wedge C_{im})$$

Then $\gamma(A_1, \ldots, A_n)$ is strongly semantically equivalent to $D_1 \vee \cdots \vee D_k$, associating to the left. ■

Example Consider a binary connective $*$ interpreted by the following table:

A_1	A_2	$\mathsf{R}(A_1, A_2)$	$A_1 * A_2$
T	T	T	F
T	T	F	F
T	F	T	T
T	F	F	F
F	T	T	T
F	T	F	T
F	F	T	F
F	F	F	F

Then $\delta_1 = (\mathsf{T}, \mathsf{F}; \mathsf{T})$, $\delta_2 = (\mathsf{F}, \mathsf{T}; \mathsf{T})$, and $\delta_3 = (\mathsf{F}, \mathsf{T}; \mathsf{F})$. And $A_1 * A_2$ is strongly semantically equivalent in **S** to:

$$((A_1 \wedge \neg A_2) \wedge R(A_1, A_2)) \vee ((\neg A_1 \wedge A_2) \wedge R(A_1, A_2)) \vee ((\neg A_1 \wedge A_2) \wedge \neg R(A_1, A_2))$$

The proof of Theorem 5 also yields a normal form for every wff with respect to the semantics for **S**.

***Corollary 6 A Normal Form Theorem for* S**

Given any wff A there is a wff B that contains exactly the same propositional variables and is a truth-functional disjunction of truth-functional conjunctions of either propositional variables or their negations or wffs of the form $R(p, q)$ or $\neg R(p, q)$, such that:

$\langle v, \mathsf{R} \rangle \vDash A$ iff $\langle v, \mathsf{R} \rangle \vDash B$

and for every C, $\mathsf{R}(A, C)$ iff $\mathsf{R}(B, C)$

It is possible to reduce the primitives further by defining them all from just one, a relatedness version of the Sheffer stroke:

$A | B$ is true in $<v, R>$ iff at least one of A, B is F and $R(A, B)$ holds

Then $\neg A$ is equivalent to $A | A$, and $A \rightarrow B$ is equivalent to $A | \neg B$.

Relative to the subject matter assignment models there is a broader notion of a truth-and-subject matter assignment connective that is sensitive to *n*-ary overlap between subject matters for $n > 2$. For that notion $\{\neg, \rightarrow\}$ is not functionally complete (consider the diagram on p. 101).

There is one particular connective, *material implication,* we will need below. We define it as for classical logic (p. 83):

$$A \supset B \equiv_{Def} \neg(A \wedge \neg B)$$

This is true in a model iff either A is false or B is true.

Exercises for Section J

1. Explain why the scheme in Theorem 5 is *strongly* semantically equivalent to the given connective.
2. Devise a table for a ternary truth-and-relatedness connective and give an explicit scheme that defines it in **S** in the language $L(\neg, \rightarrow, \wedge)$.
3. a. Show that $\{\neg, \wedge\}$ is not functionally complete for **S**.
   b. Show that $\{\neg, \vee\}$ is strongly functionally complete for **S** if $\vee$ is interpreted as related disjunction.
4. Prove the *Normal Form Theorem* for **S**.
5. Give a normal form in **S** for the following wffs:
   a. $p_1 \rightarrow (\neg p_2 \rightarrow p_3)$
   b. $(p_1 \rightarrow p_2) \rightarrow ((p_1 \rightarrow \neg p_2) \rightarrow \neg p_1)$
   c. $p_1 \rightarrow \neg\neg p_1$
   d. $\neg(p_1 \rightarrow (p_2 \rightarrow p_2))$
6. Can you define a notion of the dual of a wff in **S** as for **PC** (pp. 56–57)?
7. (Open)
   a. Define a notion of strong functional completeness for a collection of connectives in **R**.
   b. Prove or disprove that $\{\neg, \rightarrow, \wedge\}$ is strongly functionally complete for **R**.
   c. Prove or disprove that $\wedge$ is definable from $\neg$ and $\rightarrow$ in **R**.
   d. Prove or disprove that there is a single connective that is strongly functionally complete in **R**.

K. An Axiom System for S

1. S in L(¬, →)

In this section I will present a strongly complete axiomatization of **S**. I will begin with L(¬,→), since ∧ can be defined in that language and the proof will be simpler.

S, Subject Matter Relatedness Logic
in L(¬,→)

$$R(A,B) \equiv_{Def} A\to(B\to B)$$
$$A\wedge B \equiv_{Def} \neg(A\to(B\to \neg((A\to B)\to(A\to B))))$$
$$A \supset B \equiv_{Def} \neg(A\wedge\neg B)$$

axiom schema

1. $R(A, B) \to R(\neg A, B)$
2. $R(\neg A, B) \to R(A, B)$
3. $R(A, B) \to R(B, A)$
4. $R(A, B) \to R(A, B\to C)$
5. $R(A, C) \to R(A, B\to C)$
6. $R(A, B\to C) \to (\neg R(A, B) \to R(A, C))$
7. $(A\to B) \to R(A, B)$
8. $(A\to B) \to (A\supset B)$
9. $A\to A$
10. $(A\supset B) \to (R(A, B) \to(A\to B))$
11. $\neg A \to (A\supset B)$
12. $B \to (A\supset B)$
13. $[A \supset (C\to D)] \to [(A\supset C) \to (A\supset D)]$
14. $(\neg A\supset\neg B) \to ((\neg A\supset B) \to A)$

rule $\dfrac{A,\ A\to B}{B}$

The syntactic consequence relation is defined in the usual manner (see Chapter II.K.2, pp. 64–65). For the duration of this section, rather than writing '$\vdash_S$' for this consequence relation, I will use '$\vdash$'.

The proof that this axiom system is strongly complete follows the lines of the proof for classical logic in Chapter II.L.2. I will highlight the use of each axiom. I'll leave to you to show that the system is sound.

We define, as for classical logic:

Γ is *consistent* iff for every A not both $\Gamma\vdash A$ and $\Gamma\vdash\neg A$

Γ is *complete* iff for every A at least one of A, ¬A is in Γ

We cannot prove a syntactic deduction theorem as we did for classical logic, for we may have both $A \vDash_S B$ and $\nvDash_S A \rightarrow B$. In particular, if p and q are distinct variables, $p \vDash_S q \rightarrow q$ but $\nvDash_S p \rightarrow (q \rightarrow q)$. For our purposes, however, the following will suffice:

Lemma 7 If $\Gamma, A \vdash B$ then $\Gamma \vdash A \supset B$.

Proof: We proceed as for classical logic (p. 71) by inducting on the length of a proof. If the proof is just one wff long, then B is an axiom or B is A. If an axiom, we have $\Gamma \vdash A \supset B$ by **axiom 12**. If B is A, we have $\Gamma \vdash A \supset A$ by **axioms 9** and **8**.

Suppose the lemma true for all proofs of length less than *n*, and $B_1, B_2, \ldots,$ $B_n = B$ is a proof of B from $\Gamma \cup \{A\}$. If B_n is an axiom or A, we are done as before. Otherwise, for some *j*, B_j is $B_k \rightarrow B$, where $j, k < n$. Then by the induction hypothesis, $\Gamma \vdash A \supset (B_k \rightarrow B)$ and $\Gamma \vdash A \supset B_k$. Hence by **axiom 13**, we have $\Gamma \vdash A \supset B$. ■

Though we could rely now on showing that consistency and completeness are the same as Post-consistency and Post-completeness as for classical logic, there is an easier route.

Lemma 8 If $\Gamma \nvdash A$, then $\Gamma \cup \{\neg A\}$ is consistent.

Proof: Suppose not. Then for some B, $\Gamma \cup \{\neg A\} \vdash B$ and $\Gamma \cup \{\neg A\} \vdash \neg B$. So via Lemma 7 and **axiom 14**, $\Gamma \vdash A$. ■

Lemma 9 Γ is complete and consistent iff there is an **S** model $\langle \upsilon, \mathsf{R} \rangle$ such that $\Gamma = \{A: \upsilon(A) = \mathsf{T}\}$

Proof: The right to left direction is an exercise.

Now suppose Γ is complete and consistent. Note that Γ is a theory. Define:

$\upsilon(A) = \mathsf{T}$ iff $A \in \Gamma$
$\mathsf{R}(A, B)$ holds iff $R(A, B) \in \Gamma$

We must show that $\langle \upsilon, \mathsf{R} \rangle$ is an **S**-model.

First I will show that R is a relatedness relation. From **axiom 9** we have that for every A, $A \rightarrow A \in \Gamma$, so by **axiom 7**, $R(A, A) \in \Gamma$. So $\mathsf{R}(A, A)$.

If $\mathsf{R}(A, B)$, then $R(A, B) \in \Gamma$, so by **axiom 3**, $R(B, A) \in \Gamma$. Hence $\mathsf{R}(B, A)$.

Suppose now that $\mathsf{R}(A, B)$. Then by **axiom 1**, $\mathsf{R}(\neg A, B)$, and by **axiom 2**, if $\mathsf{R}(\neg A, B)$, then $\mathsf{R}(A, B)$.

We have that if $\mathsf{R}(A, B)$ or $\mathsf{R}(A, C)$, then $\mathsf{R}(A, B \rightarrow C)$ by **axioms 4** and **5**. If $\mathsf{R}(A, B \rightarrow C)$, then $R(A, B \rightarrow C) \in \Gamma$. Now suppose that $\mathsf{R}(A, B)$ does not hold. Then $R(A, B) \notin \Gamma$, so by completeness, $\neg R(A, B) \in \Gamma$. So by **axiom 6**, $R(A,C) \in \Gamma$, and hence $\mathsf{R}(A, C)$. This shows that for the language $L(\neg, \rightarrow)$, R is a relatedness relation.

To show that v evaluates wffs correctly, as for classical logic we have:

If $\mathsf{v}(\neg A) = \mathsf{T}$, then $\neg A \in \Gamma$,
so $A \notin \Gamma$ by consistency,
so $\mathsf{v}(A) = \mathsf{F}$

If $\mathsf{v}(A) = \mathsf{F}$, then $A \notin \Gamma$,
so $\neg A \in \Gamma$ by completeness,
so $\mathsf{v}(\neg A) = \mathsf{T}$

Suppose $\mathsf{v}(A \rightarrow B) = \mathsf{T}$. Then $A \rightarrow B \in \Gamma$. So by **axiom 7**, $R(A, B) \in \Gamma$, so $\mathsf{R}(A, B)$. If $\mathsf{v}(A) = \mathsf{T}$, then $A \in \Gamma$ and so $B \in \Gamma$, as Γ is a theory. So $\mathsf{v}(B) = \mathsf{T}$.

If $\mathsf{v}(A) = \mathsf{F}$ or $\mathsf{v}(B) = \mathsf{T}$, and $\mathsf{R}(A, B)$, then $\neg A \in \Gamma$ or $B \in \Gamma$, and $R(A, B) \in \Gamma$. So by **axioms 10**, **11**, and **12**, $A \rightarrow B \in \Gamma$, and hence $\mathsf{v}(A \rightarrow B) = \mathsf{T}$. Hence v evaluates all $\rightarrow$-wffs correctly, too. ■

Lemma 10 If $\Sigma \nvdash D$ then there is some complete and consistent collection of wffs Γ such that $\Sigma \subseteq \Gamma$ and $D \notin \Gamma$.

Proof: The construction is the same as for classical logic, Lemmas II.11 and II.14.

$$\Gamma_0 = \Sigma \cup \{\neg D\}$$

$$\Gamma_{n+1} = \begin{cases} \Gamma_n \cup \{A_n\} & \text{if this is consistent} \\ \Gamma_n & \text{otherwise} \end{cases}$$

$$\Gamma = \bigcup_n \Gamma_n$$

To show that Γ is consistent, we invoke Lemma 8. To show that Γ is complete, suppose $\neg A \notin \Gamma$. Then $\Gamma \cup \{\neg A\}$ must be inconsistent. So as in the proof of Lemma 8, $\Gamma \vdash A$. So $\Gamma \cup \{A\}$ is consistent, and hence A must be in Γ. ■

***Theorem 11 Strong completeness of the axiomatization of* S**
For all Γ, $\Gamma \vdash A$ iff $\Gamma \vDash A$

Proof: The proof is exactly as for classical logic, Theorem II.15, p. 75. ■

And again, as for classical logic, we have the following.

Theorem 12 Compactness of the semantic consequence relation
a. $\Gamma \vDash A$ iff there is some finite collection $\Delta \subseteq \Gamma$ such that $\Delta \vDash A$
b. Γ has a model iff every finite subset of Γ has a model

Exercises for Section K.1

1. Show that the axiomatization of **S** in $L(\neg, \rightarrow)$ is sound.

2. Fill in all the details in Lemma 10.
3. Prove: Γ is consistent iff Γ is Post-consistent
   Γ is complete iff Γ is Post-complete
4. Exhibit derivations of:
   a. $\vdash R(A, A)$
   b. $\vdash \nvdash (A \to B) \to (\neg B \to \neg A)$
5. Prove that if there is an **S** model $<\mathsf{v}, \mathsf{R}>$ such that $\Gamma = \{A: \mathsf{v}(A) = \mathsf{T}\}$, then Γ is complete and consistent.
6. Explain why the axiomatization of **S** in $L(\neg, \to)$ guarantees that $\wedge$ as a defined connective is evaluated by the classical table.
7. (Open) Give an independent axiomatization of **S** in $L(\neg, \to)$ or show that the axiomatization above is independent.
8. (Open) Show that if there is a proof of A, then there is a proof of A that uses only wffs that share a variable with A.

2. S in L(¬, →, ∧)

To devise an axiomatization of **S** in $L(\neg, \to, \wedge)$ we need to add to our previous system axioms that ensure that $\wedge$ is evaluated correctly by the v in Lemma 9.

S , Subject Matter Relatedness Logic
in $L(\neg, \to, \wedge)$

$R(A,B) \equiv_{\text{Def}} A \to (B \to B)$ $\qquad$ $A \supset B \equiv_{\text{Def}} \neg(A \wedge \neg B)$

axiom schema

Schema 1–14 as for **S** in $L(\neg, \to)$, using these definitions, plus:

15. $R(A, B \wedge C) \to R(A, B \to C)$
16. $R(A, B \to C) \to R(A, B \wedge C)$
17. $A \wedge B \to A$
18. $A \wedge B \to B$
19. $A \to (B \to (A \wedge B))$

rule $\dfrac{A,\ A \to B}{B}$

I leave to you to prove that this axiomatization is strongly complete.

3. R in L(¬, →)

An axiomatization of **R** in $L(\neg, \to)$ is a simple modification of the one for **S**.

R, Nonsymmetric Relatedness Logic
in $L(\neg, \rightarrow)$

axiom schema

As for **S**, except delete axiom scheme 3 and add:

$R(A, B) \rightarrow R(A, \neg B)$

$R(A, \neg B) \rightarrow R(A, B)$

$R(B, A) \rightarrow R(B \rightarrow C, A)$

$R(C, A) \rightarrow R(B \rightarrow C, A)$

$R(B \rightarrow C, A) \rightarrow (\neg R(B, A) \rightarrow R(C, A))$

rule $\dfrac{A,\ A \rightarrow B}{B}$

Alternatively, we could begin with an axiomatization of **R**. Then:

$\mathbf{S} = \text{Th}(\ \mathbf{R} \cup \{\text{all instances of } R(A, B) \rightarrow R(B, A)\})$

4. Substitution

In **S** the role of variables continues to allow for substitutions.

Theorem 13 Substitution

If A(B) is A(p) with every occurrence of p replaced by B, then $\dfrac{\vdash A(p)}{\vdash A(B)}$

Proof: The proof is similar to the one for classical logic. The only place that p can occur in A(B) is within B itself. Hence given any model $<\nu, R>$, consider a model $<w, R^*>$ in which for every q different from p, $w(q) = \nu(q)$, while $w(p) = \nu(B)$, and $R^*(q_1, q_2)$ iff $R(q_1, q_2)$ if both q_1 and q_2 are different from p, while $R^*(p, q)$ iff $R(B, q)$. Then the evaluation of A(B) in $<\nu, R>$ is the same as the evaluation of A(p) in $<w, R^*>$. So $<\nu, R> \vDash A(B)$. The result then follows by finite strong completeness. ■

Recall that in classical logic we also had the derived rule of substitution of logical equivalents (Corollary II.17, p. 76):

$$\frac{A \leftrightarrow B}{C(A) \leftrightarrow C(B)}$$

This fails in **S**: take a model in which $p_1 \leftrightarrow p_2$ is true, $R(p_1, p_3)$ holds, but $R(p_2, p_3)$ fails. Then $(p_1 \rightarrow p_3) \leftrightarrow (p_2 \rightarrow p_3)$ is false, so the rule leads from a true wff to a false one.

Even the rule of substitution of provable logical equivalents fails in **S**:

$$\frac{\vdash A \leftrightarrow B}{\vdash C(A) \leftrightarrow C(B)}$$

For example, take A to be $\neg(p_1 \wedge (p_2 \wedge \neg p_2))$ and B to be $\neg(p_2 \wedge \neg p_2)$ and C to be $p_1 \rightarrow p_3$ where the substitution is for p_3. To ensure that a substitution of B for A in C leads from a true wff to a true wff in a model we need that A and B always have the same truth-value *and* that they are related to precisely the same subformulas of C, which is more than logical equivalence can guarantee. This insensitivity to semantic values other than truth-values accounts for the failure of the deduction theorem.

5. The *Deduction Theorem*

For **S**, unlike classical logic, the metalogical formalization of 'follows from' as semantic consequence does not coincide with the validity of the corresponding conditional. For every A, B we have that if $\vDash A \rightarrow B$ then $A \vDash B$. But the converse fails if p and q are distinct variables: $p \vDash q \rightarrow q$, but $\nvDash p \rightarrow (q \rightarrow q)$. The semantic consequence relation is sensitive only to the relative truth-values of the wffs, while we have formalized an additional property of 'if . . . then . . .' with '$\rightarrow$'. Because we designed the relation of syntactic consequence for **S** to be the same as the relation of semantic consequence, the same break occurs between syntactic consequence and the theoremhood of the corresponding conditional. To summarize:

Theorem 14 ***a.*** If $\vDash A \rightarrow B$, then $A \vDash B$
b. If $\vdash A \rightarrow B$, then $A \vdash B$
c. The converses of (a) and (b) fail.

The connective whose validity corresponds to semantic consequence is the material conditional. We showed in Lemma 7 that if $\Gamma, A \vdash B$, then $\Gamma \vdash A \supset B$. Since $\Gamma, A \vDash B$ iff $\Gamma \vDash A \supset B$, by the strong completeness of the axiomatization we have:

Theorem 15 ***The Material Implication Form of the Deduction Theorem***
$\Gamma, A \vdash B$ iff $\Gamma \vdash A \supset B$

For the connective '$\rightarrow$' we can prove a deduction theorem in a restricted form:

Theorem 16 ***a.*** $\Gamma, A \vDash B$ and $\Gamma \vDash R(A, B)$ iff $\Gamma \vDash A \rightarrow B$
$\Gamma, A \vdash B$ and $\Gamma \vdash R(A, B)$ iff $\Gamma \vdash A \rightarrow B$
b. If A and B share a propositional variable, then:
$\Gamma, A \vDash B$ iff $\Gamma \vDash A \rightarrow B$
$\Gamma, A \vdash B$ iff $\Gamma \vdash A \rightarrow B$

Theorem 14 might suggest that there is a serious break between our logic and metalogic. But Theorem 16 may be all we need for our metalogic. We could also argue that in our metalogic, being the study of logic, we can rightly assume that all the propositions under consideration share the common subject matter of logic.

Or one could argue, though I wouldn't, that the deducibility of B from A is properly modeled by $\vdash A \rightarrow B$. Then the semantic and syntactic consequence relations are only curiosities, a viewpoint taken by modal logicians (Chapter VI.C.1.b).

If $A \vdash B$, then there is a chain of related implications that leads from A to B, for there is a proof each step of which is of the form C, $C \rightarrow D$, D where C and D are related in any model in which A is true. Classical logicians argue that such a chain establishes the relatedness of A to B; hence we should have $\vdash A \rightarrow B$ (see for example *Bennett, 1969*). However, this assumes that relatedness must be transitive, which fails in our analysis (see Example 4 of Section G). In relatedness logic the deduction is not denied, but exhibiting the chain matters. For any wffs that involve only $\neg$ and $\wedge$ we do have $A \vdash_{\mathbf{PC}} B$ iff $A \vdash_{\mathbf{S}} B$, for both kinds of models evaluate $\neg$ and $\wedge$ classically.

Is there some way we could modify the formal notions of consequence to make them coincide with the validity or provability of our formalization of the corresponding 'if . . . then . . .' proposition? Here are some possibilities:

a. We could change the definition of truth in a model to:

$$\langle v, s \rangle \models A \text{ iff } v(A) = \mathsf{T} \text{ and } s(A) = \mathsf{S}$$

Then $\models A \rightarrow B$ iff $A \models B$. This is similar to the approach of both modal logic (Chapter VI) and intuitionistic logic (Chapter VII). But I can't see how to reconcile this idea with the notion of subject matter.

b. We could change the definition of 'proof', requiring further relatedness conditions to enter in. We have that $R(A,B)$ is deducible if we have deduced $A \rightarrow B$. If we want to detach B we might also require that $R(C,B)$ be deduced for every C preceding B in the proof. That will guarantee a deduction theorem. But I don't see how to provide a complete axiomatization using this notion of proof. Moreover, strengthenings of the definition of 'proof' along this line have the flaw that we can't automatically use a theorem we've proved in the proof of another theorem.

c. Newton da Costa suggested that we keep the definition of $\vdash A$ and $\models A$ as before, and define:

$$\mathrel{/\!\!\!\vdash} A \text{ iff } \vdash A$$
$$\Gamma \mathrel{/\!\!\!\vdash} A \text{ iff there are } A_1, \ldots, A_n \text{ in } \Gamma \text{ such that } \vdash (A_1 \wedge \cdots \wedge A_n) \rightarrow A$$

Then we have $A \mathrel{/\!\!\!\vdash} B$ iff $\mathrel{/\!\!\!\vdash} A \rightarrow B$. The appropriate notion of semantic consequence to correlate to $\mathrel{/\!\!\!\vdash}$ is:

$$A \mathrel{/\!\!\!\models} B \text{ iff in every model A is related to B, and if A is true, then B is true}$$

Then $\Gamma \mathrel{/\!\!\!\models} A$ should mean that in every model there is some C in Γ such that C is related to A and if all the propositions in Γ are true then A is true. In that case we have $\Gamma \mathrel{/\!\!\!\models} A$ iff $\Gamma \mathrel{/\!\!\!\vdash} A$.

But the full form of the *Deduction Theorem* fails: because *importation* fails (see Section H.3 above) we may have $\Gamma \vdash A \rightarrow B$, but $\Gamma, A \nvdash B$. For example, $\{p \wedge \neg p\} \vdash p \rightarrow q$, but if p and q are distinct $\{p \wedge \neg p, p\} \nvdash q$. Similarly, because *exportation* fails (Section H.3 above) we may have $\Gamma, A \vdash B$ but $\Gamma \nvdash A \rightarrow B$.

Exercises for Sections K.2–K.5

1. Exhibit derivations of the following in the axiomatization of **S** in $L(\neg, \rightarrow, \wedge)$.
   a. $A \wedge \neg A \vdash B$
   b. $\vdash R(A, B) \rightarrow R(A, B \wedge C)$
2. Verify that the axiomatization of **S** in $L(\neg, \rightarrow, \wedge)$ is strongly complete.
3. Assuming that $\vee$ is evaluated classically in **S**, prove that the fragment of **S** in $L(\neg, \wedge, \vee)$ is the same as **PC** in $L(\neg, \wedge, \vee)$.
4. a. Prove that the following is a strongly complete axiomatization of **S**. (Hint: Compare the axiomatization of **PC** in $L(\neg, \wedge)$, p. 80.)

S *in* $L(\neg, \rightarrow, \wedge)$

$R(A, B) \equiv_{Def} A \rightarrow (B \rightarrow B) \qquad A \supset B \equiv_{Def} \neg(A \wedge \neg B)$

axiom schema

1. $B \supset (A \supset B)$
2. $(A \supset (B \supset C)) \supset ((A \supset B) \supset (A \supset C))$
3. $(A \wedge B) \supset A$
4. $(A \wedge B) \supset B$
5. $A \supset (B \supset (A \wedge B))$
6. $(\neg A \supset A) \supset A$
7. $(A \rightarrow B) \rightarrow R(A, B)$
8. $(A \wedge \neg B) \supset \neg(A \rightarrow B)$
9. $\neg A \supset (R(A,B) \supset (A \rightarrow B))$
10. $B \supset (R(A, B) \supset (A \rightarrow B))$
11. $R(A, A)$
12. $R(A, B) \supset R(\neg A, B)$
13. $R(\neg A, B) \supset R(A, B)$
14. $R(A, B) \supset R(B, A)$
15. $R(A, B) \supset R(A, B \rightarrow C)$
16. $R(A, C) \supset R(A, B \rightarrow C)$
17. $R(A, B \rightarrow C) \supset (\neg R(A, B) \supset R(A, C))$
18. $R(A, B \rightarrow C) \supset R(A, B \wedge C)$
19. $R(A, B \wedge C) \supset R(A, B \rightarrow C)$

rule $\dfrac{A,\ A \supset B}{B}$

b. Compare this axiomatization to the one given in Section K.1, and explain how they codify different views of the relation of **S** to **PC**.

5. On p. 121 we saw that we could take a model in which $p_1 \leftrightarrow p_2$ is true and $(p_1 \to p_3) \leftrightarrow (p_2 \to p_3)$ is false. Why does this show that the rule of substitution of logical equivalents does not hold in **S**?

6. Verify that the axiomatization of **R** in L(¬, →, ∧) is strongly complete.

7. Show that substitution is a valid rule in **R**.

8. Prove Theorem 15.

9. Prove $\{A_1, \ldots, A_n\} \vdash_{\mathbf{S}} B$ iff $\vdash_{\mathbf{S}} \neg((A_1 \wedge \ldots \wedge A_n) \wedge \neg B)$.

10. (Open) Show that in **S** there is no schema S such that:

$$\frac{S(A, B)}{C(A) \leftrightarrow C(B)}$$

11. (Open) Devise a more natural axiomatization of **S** in L(¬, →) or L(¬, →, ∧) that does not rely on axioms involving ⊃. (Suggestion: Devise axioms that allow a proof of Theorem 15.b, which is all that is needed in the proof of Lemma 9).

12. (Open) Give a constructive proof of a completeness theorem for **S** (you may prefer to give a different axiomatization).

L. Historical Remarks

I developed the logics **S** and **R** in Wellington, New Zealand with D. Walton and R. Goldblatt in 1977. Those logics were first presented in *Epstein, 1979*. *Walton, 1982* and *1985*, has given applications of them, while *Iseminger, 1986*, has discussed whether it is appropriate to view **S** as a formalization of entailment.

At the same time, *B. J. Copeland, 1978 and 1984*, working from somewhat different motivations independently arrived at a notion of relevance that obeys almost the same laws as R1–R5.

Nonsymmetric relatedness logic was devised by D. Walton and me to use in action theory, see *Walton, 1979*. *Walton, 1985*, has also applied it to analyses of *ad hominem* fallacies. I use the term *relatedness logic* to refer to both systems and the semantic analyses of this chapter. Only when both systems **R** and **S** are under consideration at the same time do I use the full names 'subject matter relatedness logic', 'subject matter relation', etc. In *Epstein, 1979*, I called **S** 'symmetric relatedness logic'.

The axiom system for **S** is a correction to the one I gave in *Epstein, 1979* and also differs from the one in the first edition of this book. *Carnielli, 1987 A*, has produced an elegant syntactic formulation of **S** in terms of tableaux.

IV A General Framework for Semantics for Propositional Logics

---

In this chapter I present a general form of semantics that unifies the propositional logics in this volume. Each logic, other than classical logic, is based on some aspect of propositions in addition to form and truth-value; different aspects give rise to different structural conditions on the semantics, yielding a spectrum of semantics.

Classical logic and relatedness logic serve as examples and motivation for this overview, though I will repeat definitions I made in the discussion of those logics. The succeeding chapters present further illustrations of the general framework, and you may, if you prefer, read those first before returning to this chapter.

I will begin by discussing propositions and then present the general framework without making mathematical generalizations. I follow with some open questions that can serve as a guide to suceeding chapters. I then consider a different view of logics that does not see them as a spectrum but rather as divided into those that extend and those that compete with classical logic. Thereafter, I present a fully formal, mathematical treatment of the framework along with an axiomatization of a fragment of the valid deductions for the general framework.

A. Aspects of Sentences

1. Propositions

In many kinds of reasoning some aspect of a proposition beyond truth-value and form is taken into consideration. For instance, in this volume we study:

subject matter
the possible ways in which a proposition could be true
the constructive mathematical content of a proposition

the consequences of a proposition in some particular logic
the likelihood of a proposition being true
the connotation or sense of a proposition
the ways in which we could come to know whether a proposition is true

If we wish to take into account some aspect of propositions in our formal models of logic, how are we to incorporate it as a new primitive?

We should not change our notion of a proposition. Perhaps this other aspect seems significant only for some restricted class of propositions: after all, who would be concerned with the constructive mathematical content of 'Ralph is a dog'? We might wish to specify that restricted class when we first set out a logic, but just as often we discover that class by using the logic. We can do so only if we retain the same meaning for a *proposition*: a written or uttered declarative sentence that we agree to view as being either true or false, but not both (Chapter I) .

You may balk here, thinking of some of the aspects I've mentioned. Why should a complex assertion be true or false? Why not simply nonsensical, or meaningless, or unacceptable? Why? Because there are only two classes of propositions: those that are true, that correspond to the case, part of which is how we understand the connectives, and those that are not true. Third truth-values, undefined truth-values, dual truth-values, levels of plausibility, all these can be taken into account as the content of a proposition. There are only two mutually exclusive truth-values, and every logician ascribes to something like this in that, in the end, he parcels out propositions into those that are acceptable to proceed on as the basis of reasoning in determining what is the case, and those that are not. Between affirming and denying there seems to be no third choice.

Given a background of agreements, which may be adopted for metaphysical, psychological, physical, or pragmatic reasons (and it might not be possible to know which of these reasons prevails), then I agree with C. I. Lewis and C. H. Langford, if by 'fact' we understand a true proposition:

> We make an inference upon observation of a certain relation between facts. Whether the facts have that relation or not we do not determine. But whether we shall *be observant of* just this particular relation of facts, and whether we shall *make that relation the basis of our inference* are things which we do determine.
>
> *Lewis and Langford, 1932,* p. 258

If you wish, you can view this dichotomy of propositions as a simplification, a simplicity constraint. But if it is, then it seems to be one that is embedded in the very way we (currently?) perceive the world.

In Chapter I.B and Chapter XI I discuss more fully the notion of a proposition, how we can view a proposition as a type, and why truth-values should be viewed hypothetically. In Chapter I.B and throughout I have explained how the technical work that I present could be interpreted in platonist terms, viewing a proposition as an abstract object, which sometimes can be correlated to sentences.

2. The logical connectives

We should also retain the same formal language when incorporating another aspect of propositions into our logic. Almost all reasoning involves *and, not,* and to a lesser extent *or*. And if our logic is to have any notion of *if . . . then . . .*, or *implies*, any notion whose converse is *follows from*, then we should use the symbol '$\rightarrow$' for it. There are two reasons for this.

First, it will allow us to compare how these various aspects affect our logics. And second, the comparisons are valid because these aren't different notions of *follows from,* but the same notion taking into account different aspects of propositions.

> Now all the Dialecticians agree in asserting that a conditional holds whenever its consequent follows from its antecedent; but as to when and how it follows, they disagree with one another and set forth conflicting criteria for this "following."
>
> Sextus Empiricus, *Against the Mathematicians,* VIII, 112
> translated in *Mates, 1953,* p.97

We aren't talking different languages; we are paying attention to different things. In that sense a translation of one logic into another either is, or ought to be accompanied by a reduction of one aspect of propositions, for example constructive mathematical content, to another, say the possible ways in which a proposition could be true.

So each formal logic owes us an explanation of these four connectives, though by defining some in terms of the others fewer may be taken as primitive. Other connectives may become important relative to a particular aspect of propositions we are studying, and the general semantics I present in the next section can easily be extended to accommodate them.

So I am going to assume that we will be working throughout with the same formal language $L(\neg, \rightarrow, \wedge, \vee, p_0, p_1, \ldots)$. In Chapter I.G I've given a definition of this language and discussed the relation of it to reasoning in ordinary language.

3. Two approaches to semantics

There are two ways we can view how an aspect of propositions can affect the truth-values of compound propositions. If what is of primary interest to us is how the aspect affects the truth-value of '$\rightarrow$' propositions, then we can view it as a primitive relationship between propositions, as when we say that one proposition is relevant to another. That's how relatedness logic was first developed in Chapter III.

Alternatively, we can view the primitive as a property of propositions. Each proposition has some content. That's how subject matters were treated in Chapter III.D. This latter approach seems a bit more intuitive to me, so I'll present it first, then explain the binary relation approach and the connection between the two.

Exercises for Section A

1. What is a proposition? (See Chapter I.B.)
2. If we accept 'Ralph is a dog' and '$2 + 2 = 4$' as propositions, why should we accept 'If Ralph is a dog, then $2 + 2 = 4$' as a proposition? Can you explain how to distinguish between meaningful and meaningless/nonsensical compound wffs?
3. Why do we use the same symbol '$\rightarrow$' to symbolize the formalization of 'if . . . then . . .' in all the logics we study?

B. Set-Assignment Semantics

1. Models

To give a *model*, or interpretation, of the formal language we first realize $p_0, p_1, \ldots$ as particular sentences in English that we take to be atomic propositions. That is, we agree to view each as either true (T) or false (F) but not both, and the internal structure of each will not be taken into account in this model. We will symbolize this assumption for a proposition p by writing $v(p) = \mathsf{T}$ or $v(p) = \mathsf{F}$. The assignment v is called a *valuation*.

From these we can form compound *semi-formal propositions* using the formal connectives $\neg, \wedge, \vee, \rightarrow$ on the pattern of the formal language. Thus, 'Ralph is a dog' might be assigned to p_0 and '$2 + 2 = 4$' could be assigned to p_1. Then the interpretation of $p_0 \wedge p_1$ is 'Ralph is a dog $\wedge$ $2 + 2 = 4$' (cf. Chapter I.G). The purpose of the model is to assign truth-values to all of these compound propositions.

To do that we first choose for this model a nonempty collection S to represent the least bits of *content* in terms of the aspect being formalized. Then every proposition, complex or atomic, is assigned a collection from within S, the method of assignment being denoted s. For example, the modal logics of Chapter VI take S to be a collection of possible worlds and the content of a proposition to be those worlds in which it is true. Often we specify $\mathsf{s}(A)$ for every A, and then the collection S is given implicitly as the union of those. Depending on what aspect is being formalized, there may be only one S for all models (see the logic **DPC** of Chapter V.F), or S may be any of a range of specified sets. In Section D below I discuss the relation between the content of atomic propositions and the content of compound ones formed from them and to what extent that should be a regular functional relationship.

Note that we do not assume that s is independent of v. In some logics it is, for example, relatedness and dependence logics (Chapters III and V), in some logics not, for example, Heyting's intuitionistic logic (Chapter VII).

In the model we are giving here, the only properties of a proposition that matter are its form, its truth-value, and its content relative to the one aspect being considered. So the truth-value of a complex proposition must be a function of these properties of its parts. For if not, what else would determine the truth-value of the whole? If something transcendent occurs when a connective joins two propositions, then how are we to reason? Reasoning requires some regularity. We explicitly make the following assumption.

The Fregean Assumption The truth-value of a compound proposition is determined by its form and the truth-values and contents of its parts.

Here 'parts' is to be understood as 'proper parts' and not the whole.

Clearly the truth-value of a compound proposition must depend on its form, but all the semantic properties of the constituents should be accounted for by their truth-values and contents. Certainly that's true when the constituents are atomic, but to ensure that that is also the case when the constituents are compound, we have to invoke the following.

The Division of Form and Content If two propositions have the same semantic properties, then they are indistinguishable in any semantic analysis, regardless of their form. So if A is part of C, then the truth-value of C depends only on the semantic properties and not the form of A, except insofar as the form of A determines the semantic properties of A.

This rules out, for example, the evaluation of a negation, ¬A, by first taking into account the truth-value and content of A, and then counting how many negation signs A begins with, using one analysis for less than three negations and another for more than three. For example, we might argue that any excessive use of 'no' or 'not' in English serves only to emphasize the initial use, as in 'No, no, no, cat's aren't nice'. So the truth-value of ¬¬¬¬p should simply be T if p is F, and F if p is T. In that case we would be assigning semantic content to the appearance of the negation signs, and therefore their presence or absence should be noted as part of the content of A.

So we can conclude from the *Fregean Assumption* and the *Division of Form and Content* that the connectives will operate semantically as truth-and-content functions in all circumstances, regardless of whether their constituents are compound or atomic; the only semantic properties of propositions we are concerned with are truth-value and content. It only remains to decide which truth-and-content functions correspond to the connectives. Consider first the conditional.

Where there is the correct "connection of meaning" between antecedent and consequent then only the truth-values matter; in that case the classical table is appropriate. Where there fails to be the connection then the conditional is false, for we may not infer truths from it.

What is this connection? It will vary depending on the aspect being modeled. For example, in relatedness logic (Chapter III) we require that the antecedent, A, and consequent, B, have some subject matter in common, and hence the relation is $\mathsf{s}(A) \cap \mathsf{s}(B) \neq \varnothing$. For example, we could have a model in which 'Ralph is a dog $\rightarrow$ $2+2=4$' is false because the antecedent and consequent are viewed as having no common subject matter. For most modal logics (Chapter VI) the connection is that the ways in which the antecedent could be true are all ways that the consequent could be true, and so in structural terms it is $\mathsf{s}(A) \subseteq \mathsf{s}(B)$. For a logic of equality of contents designed as a logic of sense and reference (Chapter V.D), it is $\mathsf{s}(A) = \mathsf{s}(B)$. In each case the connection is taken to be some fixed relationship, B, on contents of propositions. We will write $\mathsf{B}(A,B)$, though it will be clear from context that we mean $\mathsf{B}(\mathsf{s}(A), \mathsf{s}(B))$. Thus the truth-table for the conditional is:

(1)

A	B	$\mathsf{B}(A,B)$	$A \rightarrow B$
any	value	fails	F
T	T	holds	T
T	F		F
F	T		T
F	F		T

Often content is not considered significant in evaluating negations, so that the classical table for $\neg$ applies. However, in some logics the content of a proposition must have some particular property, symbolized here by N, or else its negation cannot be taken as true. For example, in Heyting's intuitionistic logic (Chapter VII) the property is $\mathsf{s}(A) = \varnothing$. Thus the general table for negation is:

(2)

A	$\mathsf{N}(A)$	$\neg A$
any value	fails	F
T	holds	F
F		T

Here, too, we write $\mathsf{N}(A)$ when $\mathsf{N}(\mathsf{s}(A))$ is meant.

I have not encountered a logic in which conjunction cannot be interpreted classically, but I see no reason why the contents of the conjuncts should not be taken into consideration. For example, we might want to distinguish between 'Ken took off his clothes and went to bed' and 'Ken went to bed and took off his clothes'. Thus the general table for conjunction is:

(3)

A	B	C(A,B)	A∧B
any	value	fails	F
T	T	holds	T
T	F		F
F	T		F
F	F		F

Here C is a fixed relation between contents of propositions, where we write C(A, B) for C(s(A), s(B)). Unless specified otherwise, however, C will be assumed to be the universal relation holding between every two propositional contents.

Also unless specified otherwise ∨ will be a defined connective. When it isn't, its truth-table will be governed by a relation A between contents of propositions: when the relation holds the table for classical inclusive ∨ is used; when the relation fails the alternation is false:

(4)

A	B	A(A,B)	A∨B
any	value	fails	F
T	T	holds	T
T	F		T
F	T		T
F	F		F

I've chosen the letters 'N', 'C', 'A' as mnemonics for 'the relations governing the table for negation, conjunction, alternation,' and 'B' to stress the binary aspect of the relation governing the table for the conditional. However, I often use other symbols for these relations in specific contexts to reflect the aspect being considered. These are collectively referred to as the *relations governing the truth-tables.*

Using these tables the truth-value of each proposition of the semi-formal language of the model may be uniquely evaluated. We write v(A) = T or v(A) = F according to whether A is true or false in the model.

Schematically then, a model for our formal language is:

I

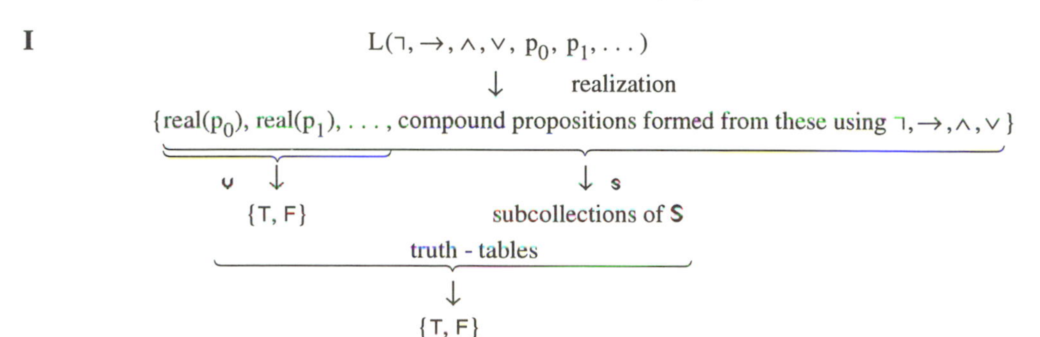

By taking the relation governing a truth-table to be universal, holding between (the contents of) any two propositions, we obtain a truth-functional connective, one that does not depend on content. However, we have not made provision for *wholly intensional* versions of $\neg$, $\rightarrow$, $\wedge$, $\vee$, that is, models in which the truth-value of compounds formed from (one of) these depends solely on the contents and not the truth-values of the parts.

The role of wholly intensional connectives in logic is not clear. Perhaps for subjunctive conditionals, such as 'If dogs could meow, then more cats would be killed', in which a false antecedent does not affect the truth-value of the whole, an intensional interpretation of $\rightarrow$ should be used. In Chapter VI I discuss subjunctive conditionals briefly, and also find that a minimal modal system, **K**, requires an intensional interpretation of $\rightarrow$ in the form:

$$A \rightarrow B \text{ is true iff } \mathsf{s}(A) \subseteq \mathsf{s}(B)$$

But I also point out there that there is no indication what reasoning that system might be modeling.

In Chapter VII a minimal logic of intuitionism is presented that also can be given semantics using a wholly intensional connective, in this case $\neg$:

$$\neg A \text{ is true iff } \mathsf{s}(A) \subseteq \mathsf{s}(\neg A)$$

But there the set-assignment semantics are parisitic on other semantics that are not in keeping with the motivation of the logic. In Chapter IX a paraconsistent logic is presented in which some notion of negation can be interpreted intensionally, though only in the presence of another negation that is modeled classically.

I am unsure, then, of the role of wholly intensional versions of $\neg$, $\rightarrow$, $\wedge$, and $\vee$ as formalizations of our usual understandings of 'not', 'if . . . then . . .', 'and' and 'or'. Though I consider wholly intensional versions of $\neg$ and $\rightarrow$ in the text, in the formal framework below I consider those as extensions of the general framework. However, I do make provision in the general framework for other wholly intensional connectives, such as a necessity operator.

Exercises for Section B.1

1. What is a semi-formal language?
2. What is the *Fregean Assumption*? Why do we make it?
3. What is the assumption of the *Division of Form and Content*? Give an example in which it could be violated. Compare its application in classical logic and its application in relatedness logic.
4. Why do the *Fregean Assumption* and the *Division of Form and Content* yield that the evaluation of the connectives are truth-and-content functions?
5. If the correct "connection of meaning" holds between antecedent and consequent,

why should we then use the classical table to evaluate the conditional?

6. Diagram schematically a general model and explain the parts of your diagram.
7. What is a wholly intensional connective?

2. Abstract models

In any model the only properties of an atomic proposition that are considered are its truth-value and the content assigned to it. So we can simplify our models:

II

$$L(\neg, \rightarrow, \wedge, \vee, p_0, p_1, \ldots)$$

$\downarrow$ v, s, and truth-tables governed by B, N, C, and A

$$\{T, F\}$$

A wff is true or false in this model according to whether $v(A) = T$ or $v(A) = F$. We have explicitly recognized that all other aspects of the propositions are to be ignored.

What is the simplest, most general way to conduct investigations of these models? Up to this point models have been built from propositions that are English sentences, and content sets are motivated and argued for with respect to the aspect of propositions under consideration, as are the relations B, N, C, and A. Even in diagram **II** the assumption is that the model arose from one as in diagram **I**.

We are not obliged to carry out our logical investigations with any greater abstraction than this. But to achieve full generality, to simplify our proofs and to isolate the structural nature of these models, it is useful and now customary to make a number of assumptions relative to models of type **II**. These assumptions allow us to apply mathematics in our studies.

The first assumption we make is to view the set of propositional variables as a completed whole, PV. We also view the collection of well-formed-formulas as a completed totality, Wffs. We treat these collections mathematically as sets. Then we make the following abstraction, writing 'Sub S' for 'subsets of S'.

The Fully General Abstraction

1. Any function $v: PV \rightarrow \{T, F\}$ is suitable to use as a truth-value assignment to the variables.
2. Any set S and any function $s: Wffs \rightarrow \text{Sub } S$ which (together with v) satisfy the structural rules for modeling the aspect under consideration are suitable to use as a content assignment.
3. Given v and a pair s and S as in (1) and (2), then any relations B, N, C, and A on the sets assigned to the variables that satisfy the structural rules for modeling the relations governing the truth-tables are suitable to use in a model.

To use parts (2) and (3) we need to have proceeded beyond examples or the ability to produce models of type **I** to a structural analysis of contents of propositions.

The *Fully General Abstraction* obliterates the differences between aspects of propositions that satisfy the same formal rules. For example, the same structural analysis, $\mathsf{s}(A) = \bigcup \{ \mathsf{s}(p) : p \text{ appears in } A \}$, is used for the referential content of a proposition (Chapter V.A) and the content of a proposition taken to be its consequences in classical logic (Chapter V.F).

We could assume the *Fully General Abstraction* and use mathematical tools on our models without assuming anything about infinite completed totalities such as PV, Wffs, or $\mathbf{S}$. That is fairly unusual, and in that case I would say that we make a *Finitistic Fully General Abstraction*. More usual is to restrict the sets $\mathbf{S}$ to be *countable*, which is the assumption I make in this book unless noted otherwise.

If we make these abstractions, the results we obtain using them when applied to propositions in fully motivated models of type **I** shouldn't contradict our original assumptions and intuitions, that is, our semantics. In Chapter II.G I have discussed more fully the role of mathematical abstraction in logic as well as its relation to a platonist view of propositions.

3. Semantics and logics

The *Fully General Abstraction* reduces the semantics to a translation from one language, $L(\neg, \rightarrow, \wedge, \vee, p_0, p_1, \ldots)$, to another, the (informal) language of mathematics. That by itself is no semantics at all, it is just a translation. The fully general abstract semantics are rules governing the form of meanings, comparable to an axiomatization of the logic in the more formal language $L(\neg, \rightarrow, \wedge, \vee, p_0, p_1, \ldots)$. Another syntactic characterization, from a different point of view, does not give meaning.

Semantics are only given when arguments are made, agreements based on common understandings are reached, our intuitions sharpened, so that we can give an analysis in terms of models of type **I**. Questions of meaning, truth, reference, and our relation to "the world" must be dealt with. Otherwise we have no semantics, only an empty formalism.

To summarize, what I have presented here is the framework for semantics for a propositional logic. It is fleshed out into a logic when:

1. We choose an aspect of propositions to formalize or choose to formalize none except truth-value.

2. We explain that aspect in simple terms, giving examples, developing an intuition, justifying the assumption that *a proposition can be viewed as having this aspect*. In doing so we may come to restrict ourselves to a certain class of propositions to which this assumption

applies, either ordinary language propositions or ones from some technical scientific language. We may also choose to add further connectives to the formal language.

3. We argue that in the context of the deductions we are formalizing it is reasonable to ignore all other aspects of propositions, justifying the abstraction that *the only properties of a proposition that matter to (this) logic are its form, its truth-value, and this one additional aspect.*

4. We stipulate a class of content-assignments that are appropriate to model this aspect, usually through rules they must obey; that is, a structural analysis of content is given. In particular, we decide whether content assignments are independent of the assignment of truth-values to the atomic propositions.

5. We explain the connections of meaning that govern the truth-tables and stipulate a class of relations B, C, A, and property N that are appropriate to model them. These classes, too, are usually presented on the basis of a structural analysis.

6. We choose whether to make the *Fully General Abstraction* of models.

7. We return to our examples and show that the formalization of them and the resulting semantic analysis are reasonable in terms of the original motivation of (2). When our pre-formal intuitions clash with the formal analysis we may explain why our semantics are appropriate to act prescriptively.

The *logic* then is this analysis: the syntax and the formal semantics developed through (1)–(7).

For the rest of this chapter I am going to discuss structural questions of propositional logics under the assumption that in each case we do make the *Fully General Abstraction.* The models I will refer to are of type **II**. Indeed, *throughout this volume I will assume the* Fully General Abstraction *for every logic unless I specify otherwise.*

4. Semantic and syntactic consequence relations

There are two major concerns of formal logic that result in other uses of the term *a logic*.

As logicians we cannot specify any particular model as being an accurate interpretation of "the world". We can only say which compound propositions are true given a specific truth- and content-assignment. However, certain propositions will be true regardless of the assignment. They are true (in this logic) due to their (propositional) form only. Their schematic form as wffs will be true in every model.

These are the "universal truths" (of this logic). We call the class of wffs true in every model the *tautologies* of this logic and say that they are *valid*. The tautologies are the forms of what we must accept due to our assumptions about how we (should) speak and reason. I have discussed more fully the notion of validity in Chapter II.E, and that discussion applies generally if references to classical logic are deleted. When the classification of propositions true due to their form only becomes the main focus of an investigation, then the class of tautologies is sometimes referred to as the logic, and on occasion I will use the name of a logic to refer to that class.

But it is not just what is true due to form only that is of interest, but what, due to form, follows from what. That is, given certain propositions as hypotheses, what can we conclude solely due to our assumptions about how we (should) speak and reason? There are two distinct semantic ways to analyze this.

First, we may argue that our formalization of 'if . . . then . . .' was meant to capture exactly the idea that 'B follows from A': the validity of $A \rightarrow B$ means that B follows from A.

Alternatively, we may say that what we are concerned with is whether B follows from the assumption of the truth of A, or from the truth of a collection of propositions Γ. Given a logic **L**, we write $A \vDash_{\mathbf{L}} B$ to mean that in every model of **L** in which A is true, so is B. We write $\Gamma \vDash_{\mathbf{L}} B$ to mean that in every model in which every A in Γ is true, so is B. This is the *semantic consequence relation for* **L**. In this notation $\vDash_{\mathbf{L}} A$ means that A follows from no propositions, and hence is true in all models; that is, A is a tautology. I have discussed these notions as well as semantic equivalence and the relation of them to deduction in ordinary language in Chapter II.D.2 in a way that generalizes if references to classical logic are deleted. In Chapter II.J.2 I have given a more formal presentation, and Theorem II.2.(a)–(g) (p.52) applies to the semantic consequence relation of any logic in this book.

Which is the more appropriate way to formalize the notion of B following from A? The answer depends on the logic being considered. For modal logics (Chapter VI) it is usually the validity of an '$\rightarrow$' wff. For classical logic (Chapter II) it is arguably the semantic consequence relation. For some logics no choice need be made, for the two formalizations are equivalent: $A \vDash_{\mathbf{L}} B$ iff $\vDash_{\mathbf{L}} A \rightarrow B$. This is a version of the *Semantic Deduction Theorem,* which I discuss in Section E below.

The phrase 'A *implies* B' or 'A *entails* B' is often used to indicate that B follows from A. Whether that phrase is an appropriate reading of $A \vDash_{\mathbf{L}} B$ or of $\vDash_{\mathbf{L}} A \rightarrow B$ will depend on the logic in question. There has been much discussion of whether implication is a metalogical notion or a connective and whether formalizing it as a connective involves a use-mention confusion, a topic I treat in Chapter II.E.2, Chapter III.K.5, Chapter VI.B.2.

Given a class of models and associated notion of validity, there may be other ways to define a semantic consequence relation peculiar to the logic in question. In Chapter III.K.5, for instance, other proposals for a semantic consequence relation for relatedness logic are discussed; *Smiley, 1976,* discusses three different consequence relations for a particular many-valued logic. Usually the intention is to ensure a semantic deduction theorem. Hence the term 'a logic' is sometimes reserved for the chosen semantic consequence relation. Throughout this book the semantic consequence relation for a logic will be derived from the notion of a wff being true in a model, as above. *Wójcicki, 1988,* discusses consequence relations more generally and abstractly.

There is yet another conception of logic, concerned primarily with form and not meaning. As I discussed in Chapter II.L.1 and Chapter II.N.1, earlier in this century the notion of a logic was understood in terms of formal systems of axioms and rules meant to formalize the notion of proof. In Chapter II.L I present the general form of one such formalization, which I use throughout this book. On this conception the notion of B following from A is understood to mean that we can prove B on the hypothesis of A. A logic is then either the collection of theorems—that is wffs that can be proved from the axioms—or the syntactic consequence relation. It will be clear from context whenever I use the term in one of these senses.

In Chapter II.K.3 I have summarized these four ways to present a logic.

Exercises for Sections B.2–B.4

1. a. How does a model of type **I** differ from a model of type **II**?
   b. Why is it legitimate to reduce a model of type **I** to a model of type **II**?
2. a. Explain how the *Fully General Abstraction* applies to models of type **II**.
   b. Why might we have more models of type **II** if we use the *Fully General Abstraction*?
   c. Why do we not apply the *Fully General Abstraction* to models of type **I**?
   d. Why do we make the *Fully General Abstraction*? Can you think of a context of formalizing reasoning in which it would be inappropriate?
   e. What is the *Finitistic Fully General Abstraction*?
3. In what way are formal abstract semantics only a translation from one language to another?
4. What steps are needed to give semantics for a logic, in the sense of an analysis of meaning?
5. a. What is a valid wff?
   b. What is a valid deduction?
   c. What is a semantic consequence relation?
   d. When do we not need to choose between the validity of $\rightarrow$-wffs and valid deductions to model the notion of *follows from*?
6. a. What is a logic?
   b. Distinguish and compare the four ways to present a logic (cf. Chapter II.K.3).

C. Relation-Based Semantics

In a set-assignment model of type **II** the content of a proposition is of significance only in establishing what other propositions it is related to by the relations on contents B, C, A, and whether the content has property N. Hence a model could be given by stipulating these relations and property of propositions, ignoring completely the contents.

Given any binary relations B, C, A on propositions (wffs), and N a property of propositions (wffs), and assignment v of truth-values to the atomic propositions (propositional variables) we have a model:

II

$$L(\neg, \rightarrow, \wedge, \vee, p_0, p_1, \ldots)$$

$\downarrow$ v and truth-tables governed by B, N, C, and A

$$\{T, F\}$$

The truth-value assignment is extended to all propositions (wffs) in the model by tables (1)–(4), pp. 132–133.

If models are stipulated by giving relations B, C, A and property N only, then we say we have *relation-based semantics*. That is how relatedness logic was first presented in Chapter III. The *Fregean Assumption* and the *Division of Form and Content* yield that the truth-value of a compound proposition is determined by its principal form, the truth-values of its parts, and the relations holding between its parts. That is, the connectives are *truth-and-relation functions* (cf. Chapter III.J). I will let you modify the summary of how to give semantics ((1)–(7) of Section B.3 above) and the *Fully General Abstraction* so as to apply here.

Every set-assignment semantics gives rise to a relation-based semantics. And as Donald Pigozzi points out, any relation-based semantics gives rise to a set-assignment semantics: given a relation-based model, we can create a set-assignment model that validates exactly the same wffs by assigning to each wff A the set {A}, setting the relation on contents to be the same as the relation on wffs. At least formally, then, the two types of presentations are equivalent. But this formal set-assignment semantics is hardly a semantics in its own right, lacking any clarity or simplicty of its presentation. These are what I call *trivial* set-assignments.

When should we say that a set-assignment semantics and a relation-based semantics *give rise to the same logic*, or simply *are the same logic*? Certainly their semantic consequence relations must be the same, but it is not clear that this is enough. In Chaper II.K.4 and Chapter X.B.6 I look at criteria for calling two different presentations 'the same logic'.

Exercises for Section C

1. a. What are relation-based semantics?
   b. Give an example in which a relation-based semantics approach to formalizing reasoning seems more natural than a set-assignment approach.
2. Modify the summary of how to give semantics in Section B.3 to apply to relation-based semantics.
3. How do the *Fregean Assumption* and the *Division of Form and Content* for relation-based semantics differ from those for set-assignment semantics?
4. State the *Fully General Abstraction* for relation-based semantics.
5. Given a relation-based semantics, show that is there a set-assignment semantics that has the same consequence relation.

D. Semantics Having a Simple Presentation

In this section I put forward some tentative definitions of what it means for a semantics to have a simple presentation, based on the examples in this book.

Let's first look at set-assignment semantics in which only the relation governing the table for '$\rightarrow$', B, is not universal, and s and B are independent of v. An example is subject matter relatedness logic of Chapter III.

What should we mean by 'a simple presentation for B'? B should be a relation on sets assigned to formulas satisfying some equation using the boolean operations union, intersection, and complementation, as well as possibly some (small) finite number of designated subsets of S, that is, some parameters. For example:

$\mathsf{B}(A,B)$ iff $\mathsf{s}(A) \cap \mathsf{s}(B) \neq \varnothing$	relatedness logic, Chapter III
$\mathsf{B}(A,B)$ iff $\mathsf{s}(A) \supseteq \mathsf{s}(B)$	dependence logic, Chapter V
$\mathsf{B}(A,B)$ iff $\mathsf{s}(A) = \mathsf{s}(B)$	**Eq**, Chapter V
$\mathsf{B}(A,B)$ iff $(\mathsf{s}(A) \cap \mathsf{T}) \subseteq (\mathsf{s}(B) \cap \mathsf{T})$	for a parameter $\mathsf{T} \subseteq \mathsf{S}$ modal logic **T**, Chapter VI

Note that containments can be defined using equations and the Boolean operations.

The rules that pick out the class of set-assignments should also be a finite number, or possibly a simple recursive set of equations between sets assigned to formulas and their subformulas, using the boolean operations and a finite number of parameters, or truth-functional conditions involving such equations. For instance, for relatedness logic in Chapter III we used: $\mathsf{s}(A) = \bigcup \{\mathsf{s}(p) : p \text{ appears in } A\}$; many modal logics in Chapter VI use: if $\mathsf{s}(A) = \mathsf{S}$ then $\mathsf{s}(A \rightarrow (A \rightarrow A)) = \mathsf{S}$. Truth-functional connectives are the appropriate ones here because working in the metalogic it's reasonable to make the mathematician's assumption that all propositions about wffs of the formal language and semantics are "connected in meaning."

For comparable relation-based semantics in which only B is not universal and B is independent of v, a simple presentation of the semantics ought to be a finite number or a simple recursive set of formulas that determine the class of relations. Each formula should be built from the symbol B for the relation, variables for wffs and their subformulas, truth-functional propositional connectives, and possibly a finite (or recursive) set of parameter wffs. As an example, one of the formulas establishing the class of relations B for relatedness logic (Chapter III) is B(A, B→C) iff B(A, B) or B(A, C).

If the semantics allow v to enter into the determination of s or of B, then 'v(A) = T' and 'v(A) = F' should also be allowed into the conditions, as it is for Heyting's intuitionistic logic (Chapter VII) which uses: if v(A) = T, then s(A) = S. In that case it may be difficult to present simple relation-based semantics, since this condition may not be expressible in terms of the relations governing the truth-tables.

I'll let you generalize my remarks to semantics in which C, N, or A is not universal.

Should we require a version of the *Fregean Assumption* for contents of wffs? For example,

(C) The content of a complex proposition is determined by its form and the contents of its proper parts, as well as possibly the truth-values of the parts.

And the comparable assumption for relation-based semantics:

(R) Whether two propositions are related is determined by their forms and the relations between their proper parts, as well as possibly the truth-values of the parts.

In Chapter III, for non-symmetric relatedness logic, we saw an example of a simple relation-based semantics that satisfies (R), yet I know of no simple set-assignment semantics for that logic, much less ones satisfying (C).

Further, in coming to grips with some difficult notion, some aspect of propositions that is vague or hard to formalize, we may wish to begin our investigations using a logic in which there is no, or at least no simple functional relation between the set assigned to a complex proposition and the sets assigned to its parts. Here I'm thinking of some preliminary investigations I made of the referential content of negated propositions, where I could at best say that $s(\neg A) \subseteq \overline{s(A)}$.

And sometimes, though the semantics do not satisfy (C) or (R), the notion of content may nonetheless be reducible to some other primitive that does satisfy a functional relation between the parts and the whole. For example, in the modal logics of Chapter VI s(A) is understood as the possible worlds in which A is true, and a simple translation to Kripke models is given. I would like to think that where the Fregean assumptions (C) and (R) both fail there is (or should be) some underlying, more primitive notion to which contents can be reduced and that satisfies a functional relation between the part and the whole.

Note that I haven't allowed for quantification in the formulas or equations of semantics having a simple presentation, other than implicit universal quantification over all wffs. This has to do with the possibility of representing the relation syntactically and axiomatizing the logic, as I discuss in questions Q4 and Q5 below.

E. Some Questions

This general framework for semantics raises many questions, and many more questions arise when we try to apply the framework to the logics in the chapters that follow. Here I present a number of questions that can serve as a guide to succeeding chapters. Some of these questions are admittedly vague, but the best way to make them more precise will come from trying to answer them.

Q1 Simply presented semantics

If the set-assignment semantics have a simple presentation, do the relation-based semantics have a simple presentation, too?

Given a relation-based semantics having a simple presentation, does it arise from some set-assignment semantics having a simple presentation?

Q2 The Deduction Theorem

In some logics we can establish that the two ways of semantically formalizing that B follows from A are equivalent, *The Semantic Deduction Theorem*:

$$\mathrm{A} \vDash \mathrm{B} \text{ iff } \vDash \mathrm{A} \rightarrow \mathrm{B}$$

Relatedness logic (Chapter III) does not satisfy this, but does satisfy $\mathrm{A} \vDash_{\mathrm{S}} \mathrm{B}$ iff $\vDash_{\mathrm{S}} \neg(\mathrm{A} \wedge \neg \mathrm{B})$. In general we would like to know when there is a connective, primitive or defined, whose validity corresponds to semantic consequence. Can structural conditions be given for when relation-based or set-assignment semantics determine a logic that satisfies the following?

A Semantic Deduction Theorem

There is a scheme $\theta(\mathrm{A}, \mathrm{B})$ such that $\mathrm{A} \vDash \mathrm{B}$ iff $\vDash \theta(\mathrm{A}, \mathrm{B})$.

If we have a semantic deduction theorem, we can represent any finite semantic consequence in terms of validity. By induction we can show that for every $n \geq 1$ there is a scheme θ_n such that for all $\mathrm{A}_1, \ldots, \mathrm{A}_n, \mathrm{B}$:

$$\{\mathrm{A}_1, \ldots, \mathrm{A}_n\} \vDash \mathrm{B} \text{ iff } \vDash \theta_n(\mathrm{A}_1, \ldots, \mathrm{A}_n, \mathrm{B})$$

Namely, let θ_1 be θ, and set

$$\theta_n(\mathrm{A}_1, \ldots, \mathrm{A}_{n+1}, \mathrm{B}) \equiv_{\mathrm{Def}} \theta(\mathrm{A}_1, \theta_n(\mathrm{A}_2, \ldots, \mathrm{A}_{n+1}, \mathrm{B}))$$

This reduction of finite consequences to validity is a *Semantic Deduction Theorem for Finite Consequences.*

We say that *semantics are compact* if for all Γ and A, $\Gamma \vDash A$ iff there is some finite $\Delta \subseteq \Gamma$ such that $\Delta \vDash A$. For compact semantics a semantic deduction theorem reduces the entire semantics consequence relation to the set of tautologies.

The standard way to show that semantics are compact is to prove a strong completeness theorem for a particular axiomatization, since our notion of syntactic consequence is compact (Chapter II.K.2). The proof of strong completeness, as opposed to completeness or finite strong completeness (Chapter II.L.3), usually requires nonconstructive infinitistic assumptions (cf. Chapter II.L.3 and II.M.1).

In demonstrating that a particular axiomatization is complete it is often useful to show a syntactic counterpart to these theorems:

A Syntactic Deduction Theorem

There is a scheme $\theta(A, B)$ such that $A \vdash B$ iff $\vdash \theta(A, B)$.

I reserve the term '*The* Syntactic Deduction Theorem' for the case where $\theta(A, B)$ is $A \rightarrow B$.

Blok and Pigozzi, 1982 and *1988,* have studied these questions from an algebraic viewpoint (see Q6 below), and *Porte, 1982,* gives an extensive bibliography in his survey of deduction theorems.

Q3 Functional completeness of the connectives

We began by formalizing 'not', 'and', 'or', 'if . . . then . . .' as $\neg$, $\wedge$, $\vee$, $\rightarrow$. In classical logic we showed that every truth-functional connective could be defined in terms of these (Chapter II.J.3). For relatedness logic these four interpreted connectives sufficed to define every truth-and-relatedness connective.

Given a semantics determined by a class of quadruples of relations **B**, **C**, **A**, and **N** governing the truth-tables, we say that θ *is a truth-and-relation functional connective* for these semantics if there is a table that determines the truth-value of $\theta(A_1, \ldots, A_m)$ in a model solely on the basis of the truth-values of the A_i's and whether or not the various A_i's are related by **B**, **C**, **A**, or have property **N**. When the relations arise from set-assignments, we say that θ is a *truth-and-content functional connective.*

Given a logic (a language plus semantics) we say that a collection of connectives of the language, $\{\theta_1, \ldots, \theta_n\}$, is *functionally complete* if every truth-and-relation (truth-and-content) functional connective can be defined in terms of these. That is, given a truth-and-relation functional connective θ (if necessary extending the language to include it) there is a scheme S using only $\theta_1, \ldots, \theta_n$ such that $\theta(A_1, \ldots, A_m)$ and $S(A_1, \ldots, A_m)$ have the same truth-value in every model. That is, $\theta(A_1, \ldots, A_m)$ is *semantically equivalent* to $S(A_1, \ldots, A_m)$. The primitive connectives for Łukasiewicz's three-valued logic (Chapter VIII.C.1) are not functionally complete. What structural conditions on the semantics of a logic ensure that the connectives chosen as primitive are functionally complete?

There are other notions of functional completeness specific to the original semantics of other logics studied in this volume. For example, in the paraconsistent logic $\mathbf{J_3}$ a connective that always takes value $\frac{1}{2}$ cannot be defined in terms of the original semantics (Chapter IX.B.3); nonetheless, we can find a scheme that is semantically equivalent to it. A correspondingly stronger sense of functional completeness for the general framework requires that $\theta(A_1, \ldots, A_m)$ and $S(A_1, \ldots, A_m)$ have the same truth-value and content (are related to the same wffs) in every model. What structural conditions on the semantics of a logic ensure that the connectives chosen as primitive are functionally complete in this stronger sense?

In Chapter X.B.6 I discuss a stronger notion of definability of connectives in terms of semantically faithful translations. *Smiley, 1962,* has a general discussion of what is meant by the definability of a connective in a logic.

Q4 Representing the relations governing the truth-tables within the formal language
For relatedness logic (Chapter III) we demonstrated the functional completeness of the connectives and gave an axiomatization by first exhibiting a formula $R(A,B)$ in the language such that $R(A,B)$ is true in a model iff $\mathsf{R}(A,B)$ holds, where R is the relation governing the truth-table for '$\rightarrow$'.

Given a logic, that is, a language with relation or set-assignment semantics, when is there a formula $\theta(A,B)$ that is true in a model iff $\mathsf{B}(A,B)$ holds, and similarly for the other relations?

Q5 Characterizing the relations governing the truth-tables in terms of valid schema
Even if we can represent via defined connectives the relations governing the truth-tables, a further problem often remains before we can axiomatize the logic: can the stipulations that determine the class of relations be given in propositional form? Sometimes it's not hard: in relatedness logic (Chapter III) conditions such as $\mathsf{R}(A,\neg B)$ iff $\mathsf{R}(A,B)$ can be represented via the defined connective R as $R(A,\neg B) \leftrightarrow R(A,B)$. Sometimes it's surprisingly complicated, even when the set-assignment semantics are simple, as in the logic of equality of contents of Chapter V.D. Can some general criteria be given for when there is a positive answer? Better yet, is there some general way to translate from rules governing the relations into propositional formulas?

I suspect that if the equations or formulas defining the relations of the semantics necessarily involve existential quantifiers over propositions, it will not be possible to define the class of relations using propositional schema.

Q6 Translating other semantics into the general framework
Some of the logics considered in this volume are given set-assignment semantics by a translation from Kripke-style possible-world semantics or many-valued matrices.

Are there general translations of these kinds of semantics into set-assignment or relation-based semantics? And conversely, when can set-assignment or relation-based semantics be translated into one of these? (Compare Exercise 7, p. 373.)

Relation-based semantics seem to be more general than algebraic semantics. For instance, a Lindenbaum–Tarski algebra cannot be formed for **S**, subject matter relatedness logic (Chapter III), for neither semantic equivalence, $A \models_S B$ iff $B \models_S A$, nor provable equivalence, $\vdash_S A \leftrightarrow B$, is a congruence relation on the algebra of formulas. Can conditions be given for a logic presented by relation-based semantics to have algebraic semantics?

In *Epstein, 1987*, I give algebraic semantics for dependence logic (Chapter V.A), but I also show that that logic is not algebraizable. That is, not every consequence relation on the base algebra corresponds to a theory in the logic. So a separate question is: What conditions guarantee that the logic is algebraizable? *Blok and Pigozzi, 1982* and *1989,* discuss this question.

Q7 Decidability

A logic is *decidable* if there is a computable procedure to determine for any wff whether it is valid, or, if the logic is presented syntactically, whether it is a theorem.

If the logic has a *Deduction Theorem*, then the decidability of validity implies that the finite consequence relation, $\{A_1, \ldots, A_n\} \models B$, is decidable, too. There is no corresponding notion of decidability for the full consequence relation, $\Gamma \models A$, for in general it is not decidable whether a given wff belongs to an infinite collection Γ.

We say that a relation-based semantics is *compactly decidable* if given any wff we can determine if it is valid by surveying the finite number of ways to assign truth-values to the variables in it and ways to relate its parts by the relations governing the truth-tables: the wff is valid if and only if it comes out true for each assignment, which can be evaluated using the truth-tables. An analogous definition can be given for set-assignment semantics, where the finite number of ways to assign sets to its parts in terms of Venn diagrams are surveyed. The logics of Chapters III and V have compactly decidable set-assignment semantics.

For some logics it may be that an assignment that makes a wff false cannot be extended to an assignment for all wffs, and hence there is no model corresponding to it. In that case it seems that the assumptions about infinite totalities have entered more strongly into the semantics. Can structural conditions on relation-based or set-assignment semantics be found that guarantee the resulting logic is decidable? That the resulting logic is compactly decidable?

Q8 Extensionally equivalent propositions and the rule of substitution

Two propositions are said to be extensionally equivalent relative to a particular logic if they have the same semantic properties in a model, regardless of their form. They mean the same.

In classical logic the only semantic property under consideration is truth-value, so if $A \leftrightarrow B$ is true (in a model) then A and B are extensionally equivalent. From this we have the valid rule of substitution:

$$\frac{A \leftrightarrow B}{C(A) \leftrightarrow C(B)}$$

Here B is substituted for some but not necessarily all occurrences of the subformula A of C.

For relatedness logic there is no scheme $S(A, B)$ whose truth guarantees that A is extensionally to B. In particular, $A \leftrightarrow B$ does not. Nor is there any scheme S for which the following is valid:

$$\frac{S(A, B)}{C(A) \leftrightarrow C(B)}$$

Given a logic is there is a scheme $S(A, B)$ such that in every model $S(A, B)$ is true iff A and B are extensionally equivalent? If there is such a scheme, then the *Division of Form and Content* should guarantee the following is valid, too:

$$\frac{S(A, B)}{S(C(A), C(B))}$$

I suspect that finding such a schema is related to whether there is a congruence relation on the algebra of formulas and whether algebraic semantics can be given for the logic.

F. On the Unity and Division of Logics

1. Quine on deviant logical connectives

The views of *Quine, 1970*, Chapter 6, on logical connectives are in marked contrast to those of the previous sections.

Suppose we were to say that subject matters affect conjunction. And suppose further that we denied that 'Ralph is a dog $\wedge$ $2 + 2 = 4$' is true, while assenting to each conjunct. Quine would, I believe, argue that the use of the symbol '$\wedge$' is inappropriate.

> If a native is prepared to assent to some compound but not to a constituent, this is a reason not to construe the construction as conjunction.
>
> *Quine, 1970*, p. 82

(I assume Quine would say similarly that if "the native" were to assent to both constituents but not the compound, we should not construe the construction as conjunction.)

> Here, evidently, is the deviant logician's predicament: when he tries to deny the doctrine he only changes the subject.
>
> *Quine, 1970*, p. 81

I certainly agree that we would no longer be talking of classical conjunction. But he stresses that we are, first of all, discussing natural language 'and'. No native speaks using $\wedge$, $\neg$, $\rightarrow$ or words like 'conjoin', unless he is a logician. He speaks an ordinary language, using words such as 'and' or 'et'. Moreover, I don't believe that anyone invariably uses 'and' truth-functionally. Yet it would only be on the basis of everyone doing so that Quine could claim that logical truths are obvious:

> Naturally, the native's unreadiness to assent to a certain sentence gives us reason not to construe the sentence as saying something which would be obvious to the native at the time. ... I must stress that I am using the word 'obvious' in an ordinary behavioral sense, with no epistemological overtones. When I call '$1 + 1 = 2$' obvious to a community I mean only that everyone, nearly enough, will unhesitatingly assent to it, for whatever reason; ... Logic is peculiar; every logical truth is obvious, actually or potentially.
>
> *Quine, 1970*, p. 82

Simple logical "truths" involving the formal '$\wedge$' which we explicitly agree to interpret truth-functionally are certainly obvious, but not in the sense of Quine, for no one speaks a language with that "word" in it. Logical truths, whatever they may be, are not obvious in his sense if it means only that, nearly enough, people will unhesitatingly assent to them. For 'nearly enough' conceals an immense problem of explication; for example, we know that 'or' is not used nearly always in the truth-functional inclusive sense. Moreover, subjects of a tyrant may unhesitatingly assent to 'The tyrant is good'; logic is not so peculiar. The reason for assent matters.

There is evidence in our language and reasoning to support many different models of the same connectives. And, you may recall, we chose the connectives to investigate and built our formal language first, then decided to use the classical interpretation in Chapter II. We are speaking the same language; we are paying attention to different aspects or uses of it. There is a change of subject in that sense, but we may still have real disagreements about which model represents the way people do reason (if you're listening to the natives) or should reason. Quine's sense of obviousness is accommodated and better understood as agreement between, archetypally, two participants in a logical discussion. That explanation, and how agreements are founded upon a background understanding of the world, I explain further in Chapter XI.

2. Classical vs. nonclassical logics

Haack, 1974, has developed what I take to be Quine's viewpoint, dividing propositional logics according to whether they are rivals or supplements to classical logic, based on syntactic comparisons. In this book I hope to have shown that the same viewpoint underlies both kinds of logics. Classical logic stands out because it has the simplest semantics, resulting from the greatest abstraction from ordinary propositions.

Within the framework here, a division of the sort Haack suggests might be justified on the basis of some semantic division of aspects of propositions into those that extend classical logic and those that are deviant. Here are some possibilities:

1. If the aspect of propositions of significance to the logic can be defined with reference to, or is reducible to the semantics of classical logic, one might classify the logic as nondeviant. For example, the modal logics of Chapter VI would not be deviant, for the content of a proposition is taken to be the ways in which it could be true, and that is explained in terms of classical models. However, the originator of modern modal logics, *C. I. Lewis, 1912* and *1932,* formulated them as rivals to classical logic.

Another logic that would be nondeviant by this criteria is the one discussed in Chapter V.F in which the content of a proposition is taken to be its consequences in classical logic.

2. One might argue that a logic is compatible with classical logic if it has relation-based semantics in which any valuation v can be paired with the universal relations for B, C, A, and the universal property for N to yield a model. In that case the collection of tautologies of the logic will be contained in those of classical logic.

3. If $\neg$ and $\wedge$ are interpreted classically according to some relation-based semantics for the logic, then the logic could be viewed as extending classical logic. Or one could take the logic as a classical extension if $\neg$ and at least one of $\wedge$, $\vee$, $\rightarrow$ is interpreted classically.

By both criteria (3) and (4) subject matter relatedness logic (Chapter III) would be nondeviant, yet the aspect of propositions on which it is based doesn't seem reducible to the semantics of classical logic.

4. One might also try to divide logics by using the notion of a translation discussed in Chapter X. Unfortunately, the obvious divisions yield counterintuitive results: if we say a logic is nondeviant if there is a semantically faithful translation of classical logic into it, then Łukasiewicz's three-valued logic (Chapter VIII.C.1) is not deviant. If we say a logic is compatible with classical logic if it can be translated grammatically into classical logic, then I know of no example of a nondeviant logic other than (fragments of) classical logic itself.

H. A Mathematical Presentation of the General Framework

with the assistance of **Walter Carnielli**

1. Languages

We consider languages $L(\gamma_0, \gamma_1, \ldots, \gamma_n; p_0, p_1, \ldots)$ where p_i is a propositional variable and γ_i is an i_j-ary *connective* for a natural number $i_j \geq 0$. If I write '$\gamma_j(A_1, \ldots, A_r)$', then I assume that $i_j = r$. Note that we allow γ_j to be a 0-ary connective, that is, a *constant*. The well-formed-formulas (*wffs*) of the language are defined inductively:

Assign the number:

1 to (p_i) for each $i = 0, 1, \ldots$

$n+1$ to $\gamma_j(A_1, \ldots, A_{i_j})$, where the maximum of the numbers assigned to $A_1, \ldots, A_{i_j}$ is n

A concatenation of symbols is then a *wff* if it is assigned some $n \geq 0$.

For a proof of the unique readability of wffs and further definitions concerning formal languages see Chapter II.K.1. The set of propositional variables is denoted 'PV', and the set of wffs of the language 'Wffs'.

I restrict attention to languages with a finite number of connectives. This excludes from consideration certain modal logics of belief, knowledge, action, and so on, in which there is a distinct modal operator for each of an infinite number of individuals, for example, $K_x A$ read as 'x knows that A'. The definitions below could be extended to such languages.

2. Formal set-assignment semantics

I use the word 'formal' here to distinguish these technical mathematical structures from semantics for logics as described in Section B.3 above.

Let L be a fixed language throughout the following.

Given an r-ary connective γ_j, a *truth-function* for it is a function:

$$f_j : \{T, F\}^{i_j} \rightarrow \{T, F\}$$

We reserve the symbols $\neg$, $\rightarrow$, $\wedge$, and $\vee$ for connectives that are assigned the classical truth-functions (tables (3)–(6) of Chapter II.B, p. 23). We call these, respectively, *negation*, *the conditional*, *conjunction*, and *disjunction*. For example, if γ_j is $\wedge$, then $i_j = 2$ and

$$f_j : \{T, F\} \times \{T, F\} \rightarrow \{T, F\}$$

is given by:

$$f_j(T,T) = T \quad f_j(T,F) = F \quad f_j(F,T) = F \quad f_j(F,F) = F$$

We reserve the symbol $\perp$ for 0-ary connectives that are assigned the constant truth-function F.

A *formal set-assignment model* for L with respect to truth-functions $f_0, \ldots, f_n$ for the connectives $\gamma_0, \ldots, \gamma_n$ is

$$M = \langle v, s, S, R_j : 0 \leq j \leq n \rangle$$

where:

v is a *valuation*, $v: PV \to \{T, F\}$

S is a set

$s: \text{Wffs} \to \text{Sub}(S)$ is a *set-assignment*, where, unless otherwise noted, $S \neq \varnothing$

$R_j \subseteq \text{Sub}(S)^{i_j}$ is the *relation governing the truth-table for* γ_j

The valuation v is extended inductively to all wffs by:

$$v(\gamma_j(A_1, \ldots, A_{i_j})) = T \quad \text{iff} \quad \begin{cases} R_j(s(A_1), \ldots, s(A_{i_j})) \\ \quad \text{and} \\ f_j(v(A_1), \ldots, v(A_{i_j})) = T \end{cases}$$

This is the *truth-table* for γ_j.

If in every model the relation R_j is the *universal* relation, that is, $R_j(s(A_1), \ldots, s(A_{i_j}))$ holds for every sequence of wffs $A_1, \ldots, A_{i_j}$, then we say that γ_j is *truth-functional*. If $\neg$, $\to$, $\wedge$, or $\vee$ is truth-functional, we say it is a *classical connective*.

If $v(A) = T$, we say A is *true in the model* and write $M \vDash A$. We write $M \vDash \Gamma$ if for every $B \in \Gamma$, $v(B) = T$.

A *formal set-assignment semantic structure* for a language L is a choice of truth-functions $f_0, \ldots, f_n$ for $\gamma_0, \ldots, \gamma_n$ and a collection of models for L with respect to them. A formal semantic structure will be denoted $\mathcal{M}$.

Given a formal semantic structure $\mathcal{M}$ for L, the class of *tautologies with respect to* $\mathcal{M}$ is $\{A : M \vDash A \text{ for every } M \in \mathcal{M}\}$. The *consequence relation with respect to* $\mathcal{M}$ is the relation $\vDash_{\mathcal{M}}$ on $\text{Sub(Wffs)} \times \text{Wffs}$ given by: $\Gamma \vDash_{\mathcal{M}} A$ iff for every $M \in \mathcal{M}$, if $M \vDash \Gamma$ then $M \vDash A$. Here 'if . . . then . . .' is understood classically.

Denote by $\mathcal{SA}_L$ the class of all formal set-assignment semantic structures for the language L, and by $\mathcal{SA}$ the class of all formal set-assignment structures for all languages.

3. Formal relation-based semantics

A *formal relation-based model* for $L \in \mathbf{L}$ with respect to truth-functions $f_0, \ldots, f_n$ for $\gamma_0, \ldots, \gamma_n$ is:

$$M = \langle v, R_j : 0 \leq j \leq n \rangle$$

where:

v is a *valuation*, $v: \mathrm{PV} \rightarrow \{\mathsf{T},\mathsf{F}\}$

$\mathsf{R}_j \subseteq \mathrm{Sub}(\mathrm{Wffs})^{i_j}$ is the *relation governing the truth-table for* γ_j

and v is extended inductively to all wffs by:

$$v(\gamma_j(\mathrm{A}_1, \ldots, \mathrm{A}_{i_j})) = \mathsf{T} \text{ iff } \begin{cases} \mathsf{R}_j(\mathrm{A}_1, \ldots, \mathrm{A}_{i_j}) \\ \quad \text{and} \\ \mathrm{f}_j(v(\mathrm{A}_1), \ldots, v(\mathrm{A}_{i_j})) = \mathsf{T} \end{cases}$$

A *formal relation-based semantic structure* for L, the class of tautologies, and the consequence relation for such a semantic structure are defined as in the previous section but with respect to relation-based models.

I denote by RB_{L} the class of all formal relation based semantic structures for the language L, and by RB the class of all formal relation-based structures for all languages.

Denote by $\mathrm{C}(\mathit{SA})$ and $\mathrm{C}(\mathit{RB})$ the class of consequence relations of all set-assignment semantics structures, and of all relation-based semantic structures respectively. As noted on p. 140, $\mathrm{C}(\mathit{SA}) = \mathrm{C}(\mathit{RB})$.

Exercises for Sections G.1–G.3

1. a. What is an r-ary connective for $r \geq 1$?
   b. What is a 0-ary connective?
   c. What 0-ary connective interprets $\perp$?
2. Define the formal languages without induction (see Chapter I.G.1).
3. Give the truth-function for all interpretations of $\rightarrow$ (not as a table).
4. What is a classical connective?
5. What does $\mathsf{M} \vDash \Gamma$ mean?
6. a. What is a semantic structure?
   b. What is the class of tautologies for a semantic structure?
   c. What is the consequence relation with respect to a semantic structure?
7. Why does $\mathrm{C}(\mathit{SA}) = \mathrm{C}(\mathit{RB})$?
8. Axiomatize the class of tautologies of all semantic structures. (Hint: It's easy.)

4. Specifying semantic structures

To specify a semantic structure we have two steps:

1. We first specify the truth-functions associated with the connectives.

For example, for the conditional we adopt the classical truth-function (table (6)

p. 23). That is how we know that in each of the models we are talking about the same connective, the conditional.

Specifying the truth-functions first accords with our hypothesis that every proposition is to be viewed as having a truth-value and that an additional aspect of propositions can be factored into the semantics in terms of content or relations between propositions. Then:

> **2.** We specify the class of models comprising the semantic structure by specifying the class of relations governing the truth-tables (and the class of set-assignments for structures of *SA*).

In this section I want to examine how to do (2).

a. Set-assignments and relations for SA

A *connective* γ_j *is uniformly interpreted in* $\mathcal{M}$, where $i_j = r$, if there is a single set-theoretic condition $C(x_1, \ldots, x_r)$ such that for each formal model $M \in \mathcal{M}$,

$$R_j(s(A_1), \ldots, s(A_{i_j})) \text{ iff } C(s(A_1), \ldots, s(A_{i_j}))$$

For example, for most of the modal logics of Chapter VI the relation governing the table for '$\rightarrow$' is $x_1 \subseteq x_2$; for relatedness logic (Chapter III) the relation governing the table for '$\rightarrow$' is $x_1 \cap x_2 \neq \varnothing$.

Here 'a set-theoretic condition' is meant to be any formula of first-order set theory with exactly i_j free variables. The complexity, variety, and pathological possibilities for such set-theoretic conditions make it important to restrict our attention to some class of "simple" uniform interpretations, as suggested in Section D.

The *class of set-assignments of* $\mathcal{M}$ *is uniformly presented* if there is a finite number or recursive set of set-theoretic conditions on contents of wffs that determine the class. The class of assignments is *equational* if each condition can be expressed as a universally quantified equation. For example, the conditions on contents of wffs specifying the class of union set-assignments used by the logics of Chapter V are: $s(\neg A) = s(A)$, $s(A \wedge B) = s(A) \cup s(B)$, $s(A \rightarrow B) = s(A) \cup s(B)$.

The language in which such conditions are to be expressed should contain function symbols, variables ranging over Wffs, variables ranging over PV, a symbol for the relation 'is a subformula of', as well as the language of set theory with equality. But to be more specific than this requires further study.

The class of set-assignments of $\mathcal{M}$ is *presented uniformly depending on truth-values* if the conditions on set-assignments are allowed to include expressions of the form '$v(A) = T$' and '$v(A) = F$'. For example, one of the conditions for the modal logic **S5** (Chapter VI.G.4) is: if $v(A) = T$, then $s(\Diamond A) \neq \varnothing$.

It seems to me that only uniformly presented relations and set-assignments are of importance to logic: an infinite collection of *ad hoc* cases does not amount to a logic, for reasoning requires uniformity.

b. Relations for RB

A *connective* γ_j *is uniformly interpreted* in a relation-based semantic structure if the class of relations governing its truth-table is uniformly presented. That is, there is a finite number or recursive collection of conditions on relations that determine the class. For example, the conditions defining the class of relations governing the table for '$\rightarrow$' for dependence logic (Chapter V.A.4) are: $\mathsf{D}(A,A)$; $\mathsf{D}(\neg A,A)$; $\mathsf{D}(A,\neg A)$; if $\mathsf{D}(A,B)$ and $\mathsf{D}(B,C)$, then $\mathsf{D}(A,C)$; $\mathsf{D}(A,B\wedge C)$ iff $\mathsf{D}(A,B)$ and $\mathsf{D}(A,C)$; $\mathsf{D}(A,B\rightarrow C)$ iff $\mathsf{D}(A,B)$ and $\mathsf{D}(A,C)$.

The language in which such conditions are to be expressed should have variables ranging over Wffs and variables ranging over PV, a relation to be interpreted as 'is a subformula of', and relation symbols, though further study may suggest a broader language.

The class of relations governing the truth-table is *uniformly presented depending on truth-values* if the conditions on set-assignments are allowed to include expressions of the form '$v(A) = \mathsf{T}$' and '$v(A) = \mathsf{F}$'.

5. Wholly intensional connectives

The semantic structures above allow us to consider truth-functional interpretations of connectives, but do not provide for an evaluation of a connective solely in terms of content, for example, $v(\Box A) = \mathsf{T}$ iff $s(A) = \mathsf{S}$. Such connectives are wholly intensional.

Wholly intensional connective can be modeled in the semantics of the previous sections by taking the associated truth-function for it to have constant output T, though this may be counterintuitive. Thus, except for relaxing the conditions on interpretations of $\neg$, $\rightarrow$, $\wedge$, and $\vee$, semantic structures with wholly intensional connectives give rise to no new consequence relations. Their role should be seen as providing more intuitively acceptable presentations.

A *formal model with wholly intensional connectives* is defined as above except that a connective γ_j need not have a truth-function assigned to it. Then γ_j is called a *wholly intensional connective* in the model and is evaluated by the table:

$$v(\gamma_j(A_1, \ldots, A_{i_j})) = \mathsf{T} \quad \text{iff} \quad \mathsf{R}_j(A_1, \ldots, A_{i_j})$$

In set-assignment models R_j is understood to be a relation on the sets assigned to $A_1, \ldots, A_{ij}$.

A model with wholly intensional connectives may use the symbols $\neg$, $\rightarrow$, $\wedge$, $\vee$ for wholly intensional connectives, but only if there is no other connective that is assigned the classical truth-function for that symbol.

A *formal semantic structure* for $L(\gamma_0, \gamma_1, \ldots, \gamma_n; p_0, p_1, \ldots)$ *with intensional connectives* is defined as a choice of truth-functions $f_{i_1}, \ldots, f_{i_m}$ for $\gamma_{i_1}, \ldots, \gamma_{i_m}$ and a collection of models with respect to those. That is, if γ_j is interpreted wholly intensionally in one model, it is interpreted wholly intensionally in all.

6. Truth-default semantic structures

The models outlined above are *falsity-default*: $\gamma_j(A_1, \ldots, A_{i_j})$ is evaluated to be *false* if $\mathsf{R}_j(A_1, \ldots, A_{i_j})$ does not hold; otherwise it is evaluated by the truth-function for γ_j. I have argued for this two-valued falsity-weighted analysis of semantics throughout this volume, partly by presenting a wide variety of logics in that form. I have also developed that theme in *Epstein, 1992*.

One could, however, devise a mirror image of these semantic structures by taking truth to be the default value. That is, in a model $\gamma_j(A_1, \ldots, A_{i_j})$ is evaluated to be *true* if $\mathsf{R}_j(A_1, \ldots, A_{i_j})$ does not hold; otherwise it is evaluated by the truth-function for γ_j. Thus $\gamma_j(A_1, \ldots, A_{i_j})$ is true unless it passes two tests to be false, instead of false unless it passes two tests to be true.

$$v(\gamma_j(A_1, \ldots, A_{i_j})) = \mathsf{F} \quad \text{iff} \quad \begin{cases} \mathsf{R}_j(A_1, \ldots, A_{i_j}) \\ \quad \text{and} \\ f_j(v(A_1), \ldots, v(A_{i_j})) = \mathsf{F} \end{cases}$$

This is a *truth-default truth-table for* γ_j. See Chapter IX.G for the truth-default tables for $\neg, \rightarrow, \wedge, \vee$.

Using these tables we may define *formal truth-default semantic models*, structures, consequence relations, and so on, as in the previous sections.

Given any such table there is another table that uses the negation of that relation. So the same class of models, etc., can be defined by using the following, which I also call *truth-default tables*:

$$v(\gamma_j(A_1, \ldots, A_{i_j})) = \mathsf{T} \quad \text{iff} \quad \begin{cases} \mathsf{R}_j(A_1, \ldots, A_{i_j}) \\ \quad \text{or} \\ f_j(v(A_1), \ldots, v(A_{i_j})) = \mathsf{T} \end{cases}$$

Here 'or' is to be understood as classical inclusive 'or'.

From this viewpoint truth-default semantics take the relation governing the table for '$\rightarrow$' as an additional way for the conditional to be true even if the antecedent is true and consequent false.

It is my opinion that such an interpretation of $\neg, \rightarrow, \wedge$, or $\vee$ is not of a piece with our usual understandings of 'not', 'if . . . then . . .', 'and', 'or'. I discuss an example in Chapter IX.G, the paraconsistent logic $\mathbf{J_3}$, where I suggest that '$\neg$' interpreted by a truth-default table is better understood as an intensional connective.

However, within a falsity-default semantic structure it may be possible to define truth-default connectives, and vice versa; certainly this is so if the logic has a functionally complete set of connectives. One example is $\Diamond A$ in various modal logics (table (14) of Chapter VI.G.1 using $\mathsf{s}(A) = \varnothing$ in place of $\mathsf{s}(A) \neq \varnothing$). Further research may suggest a greater reliance on truth-default semantics or a mixing of truth-default with falsity-default connectives in the definition of a logic.

Exercises for Sections G.4–G.6

1. When is a connective interpreted uniformly?
2. What does it mean to say that a class of set-assignments is presented uniformly?
3. Why are uniformly presented set-assignments and relations and uniformly interpreted connectives important for logic? Can you give an example of reasoning that seems to require nonuniformly interpreted connectives or nonuniformly presented relations to model it?
4. What is a wholly intensional connective?
5. a. Why are the models of the general framework called 'falsity-default'?
   b. What is a truth-default connective?
   c. What is a truth-default semantic structure?
   d. Can you give an example of reasoning in which truth rather than falsity seems to be the appropriate truth-value? (Consider a principle of charity in interpreting what someone tells you?)
   e. Show that the classical interpretations of $\neg$, $\rightarrow$, $\wedge$, $\vee$ can be viewed as either falsity-default or truth-default.

7. Tautologies of the general framework

A wff of the language $L(\neg, \rightarrow, \wedge, \vee)$ is a *falsity-default tautology* if it is true in every falsity-default semantic structure, a *falsity-default anti-tautology* if it is always false. Similarly we can define *truth-default tautologies* and *truth-default anti-tautologies*.

There are no falsity-default tautologies: for a wff to be true it must pass two tests, one of which refers to its content (or place in a relation). And the relation on contents may be made to fail.

There are, however, falsity-default anti-tautologies. For example:

$A \wedge \neg A$

if A, B are falsity-default anti-tautologies, then so is $A \vee B$

if A is a falsity-default anti-tautology, then for any wff C so are $A \wedge C$, $C \wedge A$

I suspect that these conditions completely characterize the class of falsity-default anti-tautologies. In particular, note that for no A can $\neg A$ be an anti-tautology, since that would require A to be a tautology; and $A \rightarrow B$ cannot be an anti-tautology, since that would require A to be a tautology.

There are no truth-default anti-tautologies. There are, however, truth-default tautologies:

$A \vee \neg A$ is a truth-default tautology.

if A, B are truth-default tautologies, then so is $A \wedge B$

if A is a truth-default tautology, then for any wff C, so are $A \vee C$, $C \vee A$, $C \rightarrow A$

I suspect that these conditions completely characterize the class of truth-default tautologies. In particular, note that for no A can $\neg A$ cannot be a truth-default tautology, since that would require A to be an anti-tautology.

8. Valid deductions of the general framework

A pair Γ, A from the language $L(\neg, \rightarrow, \wedge, \vee)$ is a *falsity-default valid deduction* if any falsity-default model that validates Γ also validates A.

a. Examples

Here are some valid deductions:

$\dfrac{A}{A}$ $\qquad \dfrac{A \wedge B}{B}$ $\qquad \dfrac{A \vee B,\ \neg A}{B}$ $\qquad \dfrac{\neg B,\ A \rightarrow B,\ C \rightarrow A,\ C \vee D}{D}$ $\qquad \dfrac{A \wedge \neg A}{B}$

Here are some deductions that are not valid:

$\dfrac{A \wedge B}{B \wedge A}$ This can fail if C(A, B) holds, but C(B, A) fails

$\dfrac{A}{\neg\neg A}$ A may be true, yet $N(\neg A)$ fails

$\dfrac{\neg\neg A}{A}$ A may be false, and N(A) fails, and $N(\neg A)$ holds

$\dfrac{\neg\neg A \vee A}{A}$ A may be false, N(A) fails, and $N(\neg A)$ holds

$\dfrac{A \rightarrow B,\ \neg A \rightarrow B}{B}$ Both $A \rightarrow B$ and $\neg A \rightarrow B$ can be true if A is false and N(A) fails, while B is false

Can we characterize syntactically the class of all valid deductions?

b. The subformula property

I shall prove the following theorem for the language $L(\neg, \rightarrow, \wedge, \vee)$, from which it will follow for sublanguages of that language, too.

Theorem 1 If Γ has a model and $\Gamma \vDash A$, then A is a subformula of some wff in Γ.

Proof: Suppose Γ has a model and $\Gamma \vDash A$.

Let M be a model of Γ where $\mathsf{M} = \langle \mathsf{v}_{\mathsf{M}}, \mathsf{N}_{\mathsf{M}}, \mathsf{A}_{\mathsf{M}}, \mathsf{B}_{\mathsf{M}}, \mathsf{C}_{\mathsf{M}} \rangle$. We can construct further models of Γ, $\langle \mathsf{v}, \mathsf{N}, \mathsf{A}, \mathsf{B}, \mathsf{C} \rangle$ by:

- for each q that appears in some formula in Γ, set $\mathsf{v}(q) = \mathsf{v}_{\mathsf{M}}(q)$
- for every B such that $\neg B$ is a subformula of some formula of Γ, set $\mathsf{N}(B)$ iff $\mathsf{N}_{\mathsf{M}}(B)$
- for every pair B, C such that $B \to C$ is a subformula of some formula in Γ, set $\mathsf{B}(B, C)$ iff $\mathsf{B}_{\mathsf{M}}(B, C)$
- for every pair B, C such that $B \wedge C$ is a subformula of some formula in Γ, set $\mathsf{C}(B, C)$ iff $\mathsf{C}_{\mathsf{M}}(B, C)$
- for every pair B, C such that $B \vee C$ is a subformula of some formula in Γ, set $\mathsf{A}(B, C)$ iff $\mathsf{A}_{\mathsf{M}}(B, C)$

Then any extension of $\langle \mathsf{v}, \mathsf{N}, \mathsf{A}, \mathsf{B}, \mathsf{C} \rangle$ to all wffs will be a model of Γ.

Suppose now that A is a wff that is not a subformula of any formula in Γ. To show that there is a model of Γ in which A is false, consider the form of A. If A is a propositional variable then there is an extension of $\langle \mathsf{v}, \mathsf{N}, \mathsf{A}, \mathsf{B}, \mathsf{C} \rangle$ in which A is false. Suppose A is of the form $B \to C$. Since $B \to C$ is not a subformula of a wff in Γ, any extension of the construction above to all wffs in which $\mathsf{B}(B, C)$ fails is also a model of Γ. In such a model, $B \to C$ is false. Similar observations apply to wffs of the form $B \wedge C$, $B \vee C$, $\neg B$.

Hence, if $\Gamma \vDash A$, then A is a subformula of some wff in Γ. ∎

***Note*:** The proof of this theorem is easy because we are considering all falsity-default models; for particular logics not all extensions as above need be models of the logic.

c. *Axiomatizing deductions in* L($\neg, \to, \wedge$)

To axiomatize the valid deductions in L($\neg, \to, \wedge$) we use a Hilbert-style proof system with no axioms.

Falsity-Default Valid Deductions
in L($\neg, \to, \wedge$)

rules

1. $\dfrac{A}{A}$

2. $\dfrac{A, \neg A}{B}$

3. $\dfrac{A \wedge B}{A}$

4. $\dfrac{A \wedge B}{B}$

5. $\dfrac{A \to B, A}{B}$

A proof of A from Γ, notated $\Gamma \vdash A$, is a sequence of instances of these schema:

$$\frac{\Gamma_1}{A_1} \quad \frac{\Gamma_2}{A_2} \quad . \; . \; . \quad \frac{\Gamma_n}{A_n}$$

where A_n is A, $\Gamma_1 \subseteq \Gamma$, and for each i, $\Gamma_{i+1} \subseteq \Gamma \cup \{A_1, \ldots, A_i\}$.

I leave to you to show that the system is sound:

Theorem 2 If $\Gamma \vdash A$, then $\Gamma \vDash A$.

Define Γ to be *consistent* if for no A do we have $\Gamma \vdash A$ and $\Gamma \vdash \neg A$.

Theorem 3 If Γ is consistent and not empty, then there is a model M of Γ such that for every A, $\Gamma \vdash A$ iff $\mathsf{M} \vDash A$.

Proof: Suppose Γ is consistent and not empty. Define a model M by requiring for all propositional variables p and all wffs B, C:

$$\mathsf{v}(p) = \mathsf{T} \quad \text{iff} \quad \Gamma \vdash p$$

$$\mathsf{B}(B, C) \text{ holds iff } \Gamma \vdash B \rightarrow C$$

$$\mathsf{C}(B, C) \text{ holds iff } \Gamma \vdash B \wedge C$$

$$\mathsf{N}(B) \text{ holds iff } \Gamma \vdash \neg B$$

First we show that if $\mathsf{M} \vDash A$, then $\Gamma \vdash A$ by showing that if $\Gamma \nvdash A$, then $\mathsf{M} \nvDash A$. So suppose that $\Gamma \nvdash A$. Consider the form of A. If A is a variable, we are done by construction. If A is of the form $B \rightarrow C$, then since $\Gamma \nvdash B \rightarrow C$, we have that $\mathsf{B}(B, C)$ fails in M, and hence $\mathsf{M} \nvDash B \rightarrow C$. The case where A is $B \wedge C$ or $\neg B$ follow similarly.

Next I will show by induction on the length of A that if $\Gamma \vdash A$ then $\mathsf{M} \vDash A$. If A is a propositional variable, we are done by construction. So suppose the claim is true for all wffs shorter than A.

If A is $B \wedge C$, then if $\Gamma \vdash B \wedge C$, then $\mathsf{C}(B, C)$ holds by construction. By **rules 3** and **4**, $\Gamma \vdash B$ and $\Gamma \vdash C$, so by induction $\mathsf{M} \vDash B$ and $\mathsf{M} \vDash C$. Hence $\mathsf{M} \vDash B \wedge C$.

If A is $B \rightarrow C$, then if $\Gamma \vdash B \rightarrow C$, then $\mathsf{B}(B, C)$ holds. If $\mathsf{M} \nvDash B$, then $\mathsf{M} \vDash B \rightarrow C$. If $\mathsf{M} \vDash B$, then by above, $\Gamma \vdash B$, hence by **rule 5**, $\Gamma \vdash C$. By induction, then, $\mathsf{M} \vDash C$, so $\mathsf{M} \vDash B \rightarrow C$.

If A is $\neg B$, then if $\Gamma \vdash \neg B$ we have that $\mathsf{N}(B)$ holds. If $\mathsf{M} \vDash B$, then by above $\Gamma \vdash B$, a contradiction on the consistency of Γ. Hence $\mathsf{M} \nvDash B$, and so $\mathsf{M} \vDash \neg B$.

Finally, to show that M is a model of Γ, note that by **rule 1**, if $A \in \Gamma$ then $\Gamma \vdash A$, and hence $\mathsf{M} \vDash A$. ■

Theorem 4 In $L(\neg, \rightarrow, \wedge)$, $\Gamma \vdash A$ iff $\Gamma \vDash A$.

Proof: As noted above, there are no falsity-default tautologies. Hence if Γ is empty, the theorem is true, since there are no theorems of this proof system.

If Γ is inconsistent, then for some A, $\Gamma \vdash A$ and $\Gamma \vdash \neg A$. Hence by Theorem 2, Γ has no model. Hence for all B, $\Gamma \vDash B$. And by **rule 2**, for all B, $\Gamma \vdash B$.

If Γ is consistent and not empty, then by Theorem 2, if $\Gamma \vdash A$ then $\Gamma \vDash A$. If $\Gamma \nvdash A$, then by Theorem 3 there is a model of Γ in which A is false, so $\Gamma \nvDash A$. ■

d. Deductions in languages containing disjunction

Theorem 3 fails in the presence of disjunction: in every model of $\{ p \vee q \}$ one of p, q must be true, yet $\Gamma \nvDash p$ and $\Gamma \nvDash q$.

Straightforward adaptations of the usual construction of a complete consistent extension of a consistent theory do not appear to work. For consider $\Gamma = \{ \neg p \to p \}$. We have both $\Gamma \nvDash p$ and $\Gamma \nvDash \neg p$, the former because **N**(p) may fail. Yet $\Gamma \cup \{\neg p\}$ has no model. So if $\vDash$ and $\vdash$ are coextensive, Γ consistent and $\Gamma \nvdash A$ does not imply $\Gamma \cup \{\neg A\}$ is consistent.

I can only list a number of rules that are sound and may be sufficient for axiomatizing the valid deductions in $L(\neg, \to, \wedge, \vee)$.

Rules 1–5 above, plus:

6. $\dfrac{A \vee A}{A}$

7. $\dfrac{A \vee B,\ \neg A}{B} \qquad \dfrac{A \vee B,\ \neg B}{A}$

$\dfrac{A \vee \neg B,\ B}{A} \qquad \dfrac{\neg A \vee B,\ A}{B}$

8. $\dfrac{(A \wedge B) \vee (A \wedge C)}{A} \qquad \dfrac{(B \wedge A) \vee (A \wedge C)}{A}$

$\dfrac{(A \wedge B) \vee (C \wedge A)}{A} \qquad \dfrac{(B \wedge A) \vee (C \wedge A)}{A}$

9. $\dfrac{\neg A \to A,\ A \vee \neg A}{A} \qquad \dfrac{\neg A \to A,\ \neg A \vee A}{A}$

$\dfrac{A \to \neg A,\ A \vee \neg A}{\neg A} \qquad \dfrac{A \to \neg A,\ \neg A \vee A}{\neg A}$

10. $\dfrac{\neg B,\ A_1 \to B,\ A_2 \to A_1,\ \ldots,\ A_{n+1} \to A_n,\ A_{n+1} \vee C}{C}$

$\dfrac{\neg B,\ A_1 \to B,\ A_2 \to A_1,\ \ldots,\ A_{n+1} \to A_n,\ C \vee A_{n+1}}{C}$

$$\frac{B,\ A_1 \to \neg B,\ A_2 \to A_1,\ \ldots,\ A_{n+1} \to A_n,\ A_{n+1} \vee C}{C}$$

$$\frac{B,\ A_1 \to \neg B,\ A_2 \to A_1,\ \ldots,\ A_{n+1} \to A_n,\ C \vee A_{n+1}}{C}$$

Exercises for Sections G.7–G.8

1. a. Why are there no falsity-default tautologies?
   b. Give three falsity default anti-tautologies not mentioned in the text.
2. What is a falsity-default valid deduction?
3. What is the subformula property for deductions?
4. What is a proof in the axiomatic system of valid deductions for $L(\neg, \to, \wedge)$?
5. Show that the axiomatization of valid falsity-default deductions is sound.
6. Axiomatize the valid deductions in $L(\to, \wedge)$.
7. (Open) Axiomatize the falsity-default anti-tautologies.
8. (Open) Axiomatize the truth-default tautologies.
9. (Open) Axiomatize the valid falsity-default deductions in $L(\neg, \to, \wedge, \vee)$.

V Dependence Logics – D, Dual D, Eq, DPC –

Dependence logic is a case study of how to develop a logic based on a particular aspect of propositions using the general framework of the last chapter, though it is not necessary to have read that chapter. The aspect of propositions studied is the referential content of a proposition, a notion closely allied to the idea of analyticity in logic. It is similar to the notion of the subject matter of a proposition (Chapter III) and differs perhaps only in applications: subject matters divide predicates and propositions roughly according to the traditional categories of philosophy, with the concern being whether the subject matter of one proposition overlaps that of another; analyticity arranges propositions hierarchically according to what is implicit in them, and containment of analytic content is the significant relationship between propositions.

After formulating dependence logic, I will consider in Sections B–F variations on its semantics. For dependence logic and these other semantics I will give Hilbert-style axiomatizations. *Walter Carnielli, 1987,* has produced natural syntactic formulations of dependence and dual dependence logic in terms of analytic tableaux.

A. Dependence Logic

1. The consequent is contained in the antecedent

One of the most persistent ideas concerning the conditional is that $A \rightarrow B$ can hold only if the consequent is contained in the antecedent.

> And those who judge by "suggestion" [that is, 'the power of signifying more than is expressed', i.e., what is implicit] declare that a conditional is true if its consequent is in effect included in its antecedent. According to these, 'If it is day, then it is day,' and every repeated conditional will probably be false, for it is impossible for a thing to be included in itself.
>
> *Sextus Empiricus, Outlines of Pyrrhonism*, II, 111
> translated in *Mates, 1953,* p. 109

There have been many explanations of what it is that's contained and the manner of containment. Sometimes it's said that the 'information conveyed by B is included in A' (*Hanson, 1980,* p. 659); Leibniz spoke of the terms of the consequent being implicitly contained in those of the antecedent. However, there seems to be general agreement now that containment need not be strict.

Could classical logic somehow be invoked to model this containment notion of the conditional? Consider two examples.

Smullyan, 1980, says that there must be a person, call him a, such that if a drinks then everyone drinks. That is, 'if a drinks, then everyone drinks' is true. For if everyone drinks, then anyone will do for a. And if there's someone who doesn't drink, then choose one, call him a, and the proposition is true as the antecedent is false.

Also, one of the following must be true:

If $2+2=4$, then there are natural numbers $n \geq 2$ and x, y, z such that $x^n + y^n = z^n$

If $2+2=4$, then there are no natural numbers $n \geq 2$ and x, y, z such that $x^n + y^n = z^n$

These examples seem odd because in classical logic we identify an atomic proposition with its truth-value. That the consequent of each example "conveys more information" than the antecedent is irrelevant to the classical analysis of truth. We need another analysis.

Consider how Prior explains the nature of containment:

> One proposition entails another . . . means rather that the fact expressed by the consequent is identical with that expressed by the antecedent, or at least is "not a new fact in addition to" the first. . . . The proposition that John owns a mare entails the proposition that John owns a horse, while the *fact* that John owns a horse does not entail but contains the fact that John owns a horse.
>
> *Prior, 1948,* p. 62

Here facts are understood by Prior not as true propositions nor as abstract objects, but as arrangements or configurations of the world. And ways in which the world is contain other ways in which the world is. But if that is how we understand 'facts', then we have to account for negative facts and possible facts—a daunting journey into the abstract.

It seems to me that propositions with which we describe the world suffice; beyond those there is only the world itself. There is no need of a further kind of object called 'facts'. For 'Ralph is a dog' to be true the world must be some way, but the way the world must be is: Ralph is a dog. "The way the world is" is what our propositions are meant to describe.

Therefore, I would rather say that the *content* of the proposition 'Trina owns a horse' is contained in the content of 'Trina owns a mare.' I explain that further by saying that anything referred to or any predicate implicit in the one proposition is referred to or implicit in the other. This is the notion of the *referential content of a proposition* or its *conceptual framework.*

Compare Prior's explanation:

> The fact that John does not own a mare "contains" the sex of the animal that John does not own (not in the sense in which a fact contains other facts, but in the sense in which a fact contains the objects it is about).
>
> *Prior, 1948,* p. 66

But what objects are contained in possible facts or negative facts: is the sex of an animal a thing?

I prefer to say that certain predicates, for example, 'is a mare', have implicit in them certain others, in this case 'is a horse', 'is female', 'is an animal' . . . . Whether we wish to include on our list 'has lungs' or 'has over 5,000,000 red blood cells' will depend on the context established by our discussion. There is no fixed eternal body of predicates that constitutes the content of a proposition unless predicates and propositions are taken to be eternal, timeless abstract objects, for ordinary language is fluid, and therein lie our predicates.

This aspect of propositions is apparently more holistic than truth, depending on the range of propositions under discussion; but in practice it is molecular, since only a finite number of predicates are usually under discussion (compare the discussion of subject matters in Chapter III.A). But a fuller explanation of the referential content of propositions requires an analysis of predicates, predication, and the entire apparatus of predicate logic, which I shall defer to *A General Framework for Semantics for Predicate Logics.* I hope to have done enough here, however, to convince you that the following assumption is reasonable.

Referential Content Assumption

A proposition can be viewed as having referential content.

Now suppose you were to criticize this notion by saying that the only way we could discern if the content of one proposition contains that of another is by recognizing whether the corresponding entailment holds. Even granting that, we may be able to set out how referential content functions in reasoning by making certain observations or agreements concerning how some containments are derivative from others. Thus we might decide that if the content of B is contained in that of A then it is contained in that of $A \wedge C$ for any C. In the process of reckoning the logical structure of referential content we may sharpen that notion, for as it stands now there are a number of different interpretations we could make.

2. The structure of referential content

We want to examine the view that B is *dependent* on A if, and only if, B does not go beyond the conceptual framework established by A.

This is a weaker view than that A must analytically imply B. Consider: 'Ralph is a bachelor' analytically implies 'Ralph is a man' if 'bachelor' and 'man' have their usual meanings. But 'Ralph is a bachelor' does not analytically imply 'Ralph is not a man', though the former does establish the conceptual framework for the latter. This holds, though, only if we take the view that the logical connectives are neutral in establishing what is referred to by a proposition. We assume, then,

> The logical connectives are syncategorematic:
> they have no referential content

Thus, informally, 'Ralph is a dog' and 'Ralph is not a dog' have the same content; and 'Ralph is a dog and George is a duck' has the same content as 'If Ralph is a dog, then George is a duck'.

Under this view the truth of a proposition is not part of its content. The predicates implicit in a proposition and the truth-value of the proposition are distinct, for the referential content remains unchanged whether the proposition is true or false. The assertion or use of a proposition carries with it the contents of the proposition, but it cannot also carry the truth of it. We may say that when we assert a sentence then we assert that it is true, but that is not to say it is true. Yet what we take as its contextual frame of reference does stay constant. Our assumption is thus:

> The referential content of a proposition is independent of its truth-value

Based on these assumptions the simplest and, I believe, the most reasonable agreement we can make about the content of a compound is that what the whole refers to and predicates is exactly what its parts refer to and predicate. This accords with the *Fregean Assumption* (Chapter IV.D, p. 142) that the content of the whole is a function of the content of its parts.

The Structure of Referential Content The referential content of a compound proposition is the union of the contents of its parts.

Thus the content of 'Ralph is a bachelor' does contain that of 'Ralph is not a man'. But in any reasonable model 'If Ralph is a bachelor, then Ralph is not a man' fails, because the truth-value of the parts will matter in establishing the truth-value of the compound sentence. Hence we begin to approximate a notion of analytic implication when we evaluate 'if . . . then . . .' propositions using both the referential content and truth-values of their parts. And these will be the only aspects of propositions we'll consider.

Dependence Logic Abstraction The only properties of a proposition that matter to (this) logic are its form, its truth-value, and its referential content.

But consider two other views about the referential content of a compound sentence. First, one could argue that the connective 'not', for example, does not change the content of 'Ralph is a dog' but rather augments it. That is, 'Ralph is not a dog' speaks of the content of 'Ralph is a dog' and of negation. Similarly, the content of an 'if . . . then . . .' sentence might be said to include hypothetical assertion. If, on this view, the logical connectives have distinct contents, then we would never have any conditional evaluated as true if the consequent had a connective not appearing in the antecedent—unless, that is, we adopt a hierarchy of contents of logical connectives. Alternatively, if the content of each of the connectives is the same, then it's clear that nothing new would be added to our logic except that 'if A, then B' would fail if A is atomic and B isn't. For an argument that relies on the logical connectives having content see *Prior, 1964,* pp. 161–162, where Prior says, 'I cannot see how the sense of a sentence can ever be identical with a logical complication of itself.' *K. Fine, 1979,* discusses that idea, too.

The second view is that negating a sentence distorts the reference of it. Asserting 'Ralph is not a dog' includes a wider reference than that of 'Ralph is a dog.' This view is difficult to accommodate into the motivation I've given, as it seems to confute the truth of a sentence with its content.

There is also a view that tautologies have no content: see *Prior, 1960.* Whatever the motivation of such a view, it cannot be that the predications expressed by the parts of the proposition contribute uniformly to the content of the whole. *Searle, 1970,* p. 124, discusses this point.

3. Set-assignment semantics

We can now present a logic that incorporates the assumptions we've made; we only need fill out the general form of semantics for a propositional logic.

Our assumptions dictate the class of set-assignments appropriate to model the notion of the referential content of a proposition.

Union Set-Assignments

(U) $\mathsf{s}(A) = \bigcup \{\mathsf{s}(p) : p \text{ is atomic and } p \text{ appears in } A\}$

It might be in keeping with our motivation to require that $\mathsf{s}(A) \neq \varnothing$ for all A. But that would not affect the resulting logic, so I do not suppose that.

For a sketch of such an assignment, compare the example for subject matters in Chapter III.D, p. 100, reading 'predicates' for 'topics'.

What remains to specify are the connections that govern the truth-tables. The relation governing the conditional is containment, as we first stated.

Dependent Implication

A	B	$s(A) \supseteq s(B)$	$A \to B$
any	values	fails	F
T	T		T
T	F	holds	F
F	T		T
F	F		T

The referential content of a proposition is independent of its truth. So, it seems, we should take negation to be classical. And barring any good reason why content should affect conjunction, I'll assume that conjunction, too, is classical.

Negation and conjunction are classical

If you wish, you can view these last two choices as a simplicity constraint: we allow the contextual framework of a proposition to affect only the conditional, as that is the main concern of the motivating discussion. However, I think that these are the natural interpretations of the connectives in terms of the notions of referential content and analyticity.

There are at least three plausible ways to formalize 'or'. We could take $\vee$ to be classical, and hence $A \vee B$ would be definable as:

$$\neg(\neg A \wedge \neg B)$$

Or we could interpret $A \vee B$ to be true iff at least one of A, B is true and the content of A contains that of B, and that of B contains the content of A, what I call *strongly dependent disjunction.* This is definable as:

$$(\neg A \to B) \wedge (\neg B \to A)$$

Or we could interpret $A \vee B$ as being true iff either A or B is true and either the content of B is contained in A or vice versa, what I call *weakly dependent disjunction.* This is definable as:

$$\neg(\neg(\neg A \to B) \wedge \neg(\neg B \to A))$$

Arguments for each of these could be made, but they would be inessential to the main analysis of this chapter. So I leave to you the choice of one of these, or some other interpretation of $\vee$.

Therefore, the formal language we will use is $L(\neg, \to, \wedge, p_0, p_1, \ldots)$. For this language we can specify the union set-assignments as those that satisfy:

U1. $s(\neg A) = s(A)$

U2. $s(A \wedge B) = s(A) \cup s(B)$

U3. $s(A \to B) = s(A) \cup s(B)$

With the general framework for semantics for a propositional logic filled in with these choices, we can specify the models for our logic, recalling that we have agreed that truth-values and contents are independent.

A *dependence model* is:

I

$$L(\neg, \rightarrow, \wedge, p_0, p_1, \ldots)$$

$\downarrow$ realization

$$\{p_0, p_1, \ldots, \text{complex propositions formed from these using } \neg, \rightarrow, \wedge\}$$

$\downarrow$ v, s, and truth-tables

$$\{\mathsf{T}, \mathsf{F}\}$$

Here v is a valuation for the atomic propositions; s is a union set-assignment on the atomic propositions that is extended to all propositions of the semi-formal language by (**U**) or inductively by **U1–U3**. And v is extended to all propositions of the semi-formal language by using the classical tables for $\neg$ and $\wedge$ and the table for dependent implication for $\rightarrow$.

By virtue of the *Dependence Logic Abstraction*, nothing matters about an atomic proposition except its truth-value and content. So we can simplify a model:

II

$$L(\neg, \rightarrow, \wedge, p_0, p_1, \ldots)$$

$\downarrow$ v, s, and truth-tables

$$\{\mathsf{T}, \mathsf{F}\}$$

And we make a *Fully General Abstraction* that any function $\mathsf{v}: PV \rightarrow \{\mathsf{T}, \mathsf{F}\}$ together with any *countable* set $\mathbf{S}$ and map $\mathsf{s}: PV \rightarrow \text{Sub}\,\mathbf{S}$ can comprise a model $<\mathsf{v},\mathsf{s}>$. Equivalently we can take $\mathsf{s}: \text{Wffs} \rightarrow \text{Sub}\,\mathbf{S}$ satisfying **U1–U3**. Unless previously specified otherwise, when I write '$<\mathsf{v},\mathsf{s}>$', the set $\mathbf{S}$ is to be understood to be $\bigcup\{\mathsf{s}(A) : A \text{ is a wff}\}$.

As usual, $<\mathsf{v},\mathsf{s}> \vDash A$ means that $\mathsf{v}(A) = \mathsf{T}$, and $<\mathsf{v},\mathsf{s}> \vDash \Gamma$ means that $\mathsf{v}(A) = \mathsf{T}$ for every A in Γ. The semantic consequence relation is the usual one: $\Gamma \vDash A$ iff for any model $<\mathsf{v},\mathsf{s}>$, if $<\mathsf{v},\mathsf{s}> \vDash \Gamma$ then $<\mathsf{v},\mathsf{s}> \vDash A$. A proposition is a *dependence tautology* if it is true in every dependence model (under the *Fully General Abstraction*). I'll refer to this consequence relation as **D**, or **dependence logic**. Sometimes I use that name for just the tautologies, as the context will make clear. When other logics are under consideration I will write $\vDash_{\mathbf{D}}$ for $\vDash$.

As for classical logic, we say that two wffs A and B are *semantically equivalent* if in every model they take the same truth-value: $A \vDash B$ and $B \vDash A$. As for relatedness logic (p. 105), however, we want a stronger notion of semantic equivalence, one that more fully reflects assumptions of this logic about the role of a proposition in reasoning. We say two wffs A and B are *strongly semantically equivalent* if for every model $<\mathsf{v},\mathsf{s},\mathsf{S}>$, $\mathsf{s}(A) = \mathsf{s}(B)$ and $\mathsf{v}(A) = \mathsf{T}$ iff $\mathsf{v}(B) = \mathsf{T}$.

4. Relation-based semantics

The set-assignment semantics of dependence logic give rise to a relation-based semantics. Call a binary relation D on wffs a *dependence relation* if there is a union set-assignment s such that $\mathsf{D}(A,B)$ iff $\mathsf{s}(A) \supseteq \mathsf{s}(B)$. Dependence logic can be viewed as the class of all tautologies for all models $<\mathsf{v},\mathsf{D}>$ where D is a dependence relation, $\neg$ and $\wedge$ are interpreted classically, and the table for $\rightarrow$ is governed by D. We can characterize the class of dependence relations as follows.

Lemma 1 D is a dependence relation iff all of the following hold:

1. D is reflexive
2. D is transitive
3. $\mathsf{D}(A,B\wedge C)$ iff $\mathsf{D}(A,B)$ and $\mathsf{D}(A,C)$
4. $\mathsf{D}(A,B\rightarrow C)$ iff $\mathsf{D}(A,B)$ and $\mathsf{D}(A,C)$
5. $\mathsf{D}(\neg A,A)$
6. $\mathsf{D}(A,\neg A)$

Proof: If D is a dependence relation then (1)–(6) hold. If (1)–(6) hold, set $\mathsf{t}(A) = \{B\colon \mathsf{D}(B,A)\}$. Then by (1) and (2), $\mathsf{D}(A,B)$ iff $\mathsf{t}(A) \subseteq \mathsf{t}(B)$. By (3) and (4), $\mathsf{t}(A\rightarrow B) = \mathsf{t}(A\wedge B) = \mathsf{t}(A) \cap \mathsf{t}(B)$, and by (5) and (6), $\mathsf{t}(\neg A) = \mathsf{t}(A)$. Set $\mathsf{s}(A) = \text{Wffs} - \mathsf{t}(A)$, that is, $\mathsf{s}(A) = \{B\colon B \text{ is a wff and } B \notin \mathsf{t}(A)\}$. Then s is a union set-assignment and $\mathsf{D}(A,B)$ iff $\mathsf{s}(A) \supseteq \mathsf{s}(B)$. ■

Despite the simplicity of the conditions of Lemma 1, dependence relations do not satisfy a version of the *Fregean Assumption* (see (R) of Chapter IV.D, p. 142). We may have a binary relation on the propositional variables and extend it to a dependence relation on the collection of all wffs in two distinct ways. For instance, we may have that both $\mathsf{D}(p,q)$ and $\mathsf{D}(p,r)$ fail, yet either $\mathsf{D}(p\wedge r, q)$ holds, or $\mathsf{D}(p\wedge r, q)$ fails, as these diagrams illustrate:

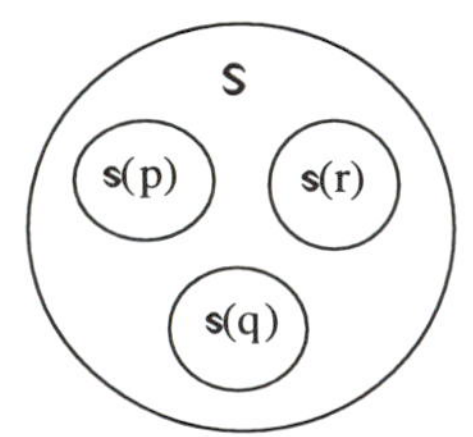

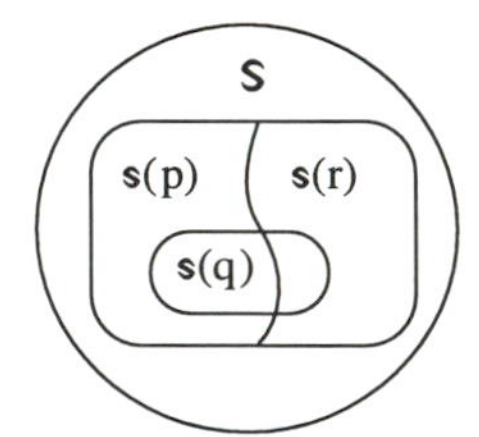

A dependence relation cannot be specified locally; it is a global notion. However, we have the following, which you can prove using the previous lemma.

Lemma 2 D is a dependence relation iff all of the following hold:

1. D is reflexive

2. D is transitive

3. D(A,B) iff D(A,p) for every p in B

Exercises for Section A.1–A.4

1. Give an example of a realization and an assignment of referential content to the atomic propositions in it comparable to the example for relatedness logic, p. 100.
2. Prove that s is a union set-assignment iff s satisfies **U1–U3**.
3. What is the table for dependent implication? Why is negation chosen to be classical in dependence logic? Should we choose conjunction to be classical?
4. Argue for one of the choices of interpretation of $\vee$ given in Section A.3, providing examples; or give examples of all three interpretations that are compatible with the motivation of dependence logic.
5. What is the *Fully General Abstraction* for dependence logic?
6. Explain why the *Fregean Assumption* holds for contents but not for relations in dependence logic.
7. Verify that given a valuation, v, on atomic propositions and a union set-assignment, s, on wffs, there is one and only one way to extend v to all wffs in accord with the interpretations of the connectives for dependence logic.
8. Prove Lemma 2.
9. (Open) Compare and contrast notion of the referential content of a proposition with themedieval notion of the signification of a proposition.
10. (Open) Give a proof of Lemma 2 by building a set-assignment s corresponding to a given D without recourse to Lemma 1 or its proof. Each s(p) should be built up corresponding to whether, for each A, D(A,p) holds and whether D(p,A) holds. (A rough sketch is given in *Epstein and Maddux, 1980.*)

    Such a construction is important for giving a set-assignment that has a good reading in terms of the given D, which the s produced in the proof of Lemma 1 does not (that proof was supplied by an anonymous referee). More importantly, such a construction can reconcile the holism of the notion of referential content with some regular inductive procedure for producing contents, showing that our infinitistic assumptions have not led to counterintuitive results.

5. The decidability of dependence logic tautologies

We can decide for any wff whether it is a dependence logic tautology, much as we did for relatedness logic. Given a wff we first list each of the finite number of ways (a) to assign truth-values to the variables appearing in A and (b) to create distinct

Venn diagrams assigning sets to those variables. For each assignment of truth-values and Venn diagram, calculate first the content of each subformula in the wff and whether for each two subformulas the content of the one contains the content of the other. Then determine the truth-value of the whole wff using the classical tables for $\neg$ and $\wedge$ and the dependent implication table for $\rightarrow$. If for each assignment the wff takes value T, then the wff is a tautology, for if not a falsifying assignment could be used to create a model by extending it to all other variables, assigning T and the empty set to those. That is, dependence logic is compactly decidable (p. 146).

For example, consider a propositional form of *modus ponens*:

$$p \rightarrow [(p \rightarrow q) \rightarrow q]$$

Assign p and q the value T, and take the following Venn diagram to assign content to the variables:

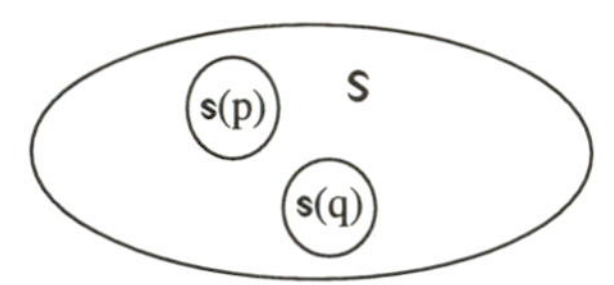

We calculate that $s((p \rightarrow q) \rightarrow q) = (s(p) \cup s(q)) \not\subseteq s(p)$, so the wff fails to be true. This assignment could be extended to a model for the entire language; hence the wff is not a tautology.

6. Examples of formalization

Let us look at some examples of formalizations of propositions and arguments, comparing the analyses to those for classical logic and for relatedness logic.

1. Ari doesn't drink
***Therefore*:**
If Ari drinks, then everyone drinks

$\neg$(Ari drinks)
Therefore:
Ari drinks $\rightarrow$ everyone drinks

$$\frac{\neg p_1}{p_1 \rightarrow p_2}$$

Explanation: The deduction is not valid: $\neg p_1$ may be true, yet if $s(p_1) \not\supseteq s(p_2)$, then $p_1 \rightarrow p_2$ will be false.

The "paradoxes" of material implication fail to be valid in dependence logic, just as for relatedness logic. The following fail to be tautologies:

$\neg$(Ari drinks) $\rightarrow$ (Ari drinks $\rightarrow$ everyone drinks)

Everyone drinks $\rightarrow$ (Ari drinks $\rightarrow$ everyone drinks)

The antecedents may be true while the consequents false, since the referential content of ‘Ari drinks’ need not contain that of ‘everyone drinks’.

Note that in any reasonable model of relatedness logic these examples would be true, since they share some subject matter (they both involve the predicate ‘drinks’). Indeed, in a predicate logic analysis they would be tautological.

The paradoxes of strict implication fail to be dependence logic tautologies, too. For example:

Ralph is a dog $\rightarrow$ (cats are nasty $\rightarrow$ cats are nasty)

(Ralph is a dog $\wedge$ $\neg$(Ralph is a dog)) $\rightarrow$ cats are nasty

The content of ‘Ralph is a dog’ need not contain that of ‘cats are nasty’: no “information” about dogs and Ralph can lead us to the referential content, the information of ‘cats are nasty’ (even though the latter may be a necessary truth).

2. Not both Ralph is a dog and cats aren’t nasty
***Therefore*:**
If Ralph is a dog, then cats are nasty

$\neg$(Ralph is a dog $\wedge$ $\neg$(cats are nasty))
Therefore:
Ralph is a dog $\rightarrow$ cats are nasty

$$\frac{\neg(p_1 \wedge \neg p_2)}{(p_1 \rightarrow p_2)}$$

Explanation: This fails to be valid in dependence logic, again since the content of ‘Ralph is a dog’ need not contain that of ‘cats are nasty’. But the following is valid:

Ralph is a dog $\rightarrow$ cats are nasty
Therefore:
$\neg$(Ralph is a dog $\wedge$ $\neg$(cats are nasty))

For a dependent implication to hold it is necessary but not sufficient that the corresponding material implication be true.

3. If Ralph is a bachelor, then Ralph is a man

Ralph is a bachelor $\rightarrow$ Ralph is a man

$p_1 \rightarrow p_2$

Explanation: The example is not a tautology. The content of ‘Ralph is a bachelor’ need not contain that of ‘Ralph is a man’. However, noting the meaning of the words in the proposition, we may wish to restrict ourselves to models in which the content of ‘Ralph is a bachelor’ does contain that of ‘Ralph is a man’. But the wff need not be true even in those models. Rather, we would have to restrict ourselves to models in which, perhaps, the following is true:

Ralph is a bachelor → (Ralph is a man ∧ ¬(Ralph is married))

Or, were we to look at the internal structure of propositions, we might require that the following be true in a model:

For every x (x is a bachelor → x is a man ∧ ¬(x is married))

With such formal hypotheses classical logic can often model the assumptions about meanings of words that lead us to classify some propositions as analytic, as I have discussed in Chapter V.K of *Predicate Logic.* But classical logic, even utilizing the internal structure of propositions, is not always adequate to model a notion of containment of referential content, as the first two examples illustrate.

4. If dogs bark and Juney is a dog, then Juney barks
***Therefore*:**
If dogs bark, then if Juney is a dog, then Juney barks

Dogs bark ∧ Juney is a dog → Juney barks
Therefore:
Dogs bark → (Juney is a dog → Juney barks)

$$\frac{(p_1 \wedge p_2) \rightarrow p_3}{p_1 \rightarrow (p_2 \rightarrow p_3)}$$

Explanation: The example is not valid, even were we to take into consideration the internal structure of propositions. *Exportation* of premisses can delete content, making a conditional false: we may have each of 'Dogs bark', 'Juney barks', and 'Juney barks' true, and, quite reasonably, s('Dogs bark) ∪ s('Juney is a dog') ⊇ s('Juney barks), while s('Dogs bark) ⊉ s('Juney is a dog') ∪ s('Juney barks), since 'Dogs bark' does not speak (explicitly) of Juney.

However, *importation* of premisses is allowed, which is not the case in relatedness logic. For example, the following is a dependence tautology:

[Juney barks → (Juney is a dog → Juney howls)]
→ [Juney barks ∧ Juney is a dog → Juney howls]

5. If dogs bark, then Juney barks
If Juney barks, then a dog has scared a thief
***Therefore*:**
If dogs bark, then a dog has scared a thief

Dogs bark → Juney barks
Juney barks → a dog can scare a thief
Therefore:
Dogs bark → a dog can scare a thief

$$\frac{\begin{array}{l} p_1 \rightarrow p_2 \\ p_2 \rightarrow p_3 \end{array}}{p_1 \rightarrow p_3}$$

Explanation: The deduction is valid. In dependence logic, unlike relatedness logic, the conditional is transitive, for containment of contents is transitive.

6. If Ralph is a dog, then Ralph barks
***Therefore*:**
If Ralph doesn't bark, then Ralph is not a dog

Ralph is a dog $\rightarrow$ Ralph barks
Therefore:
$\neg$(Ralph barks) $\rightarrow$ $\neg$(Ralph is a dog)

$$\frac{p_1 \rightarrow p_2}{\neg p_2 \rightarrow \neg p_1}$$

Explanation: This deduction is not valid, though it is valid in relatedness logic. We may have a model in which s(Ralph is a dog) $\supseteq$ s(Ralph barks), but s(Ralph barks) $\not\supseteq$ s(Ralph is a dog), since we may wish to say that included in being a dog is barking, but barking need not contain the information that the object is a dog. *Contraposition,* $(A \rightarrow B) \rightarrow (\neg B \rightarrow \neg A)$, fails to be a scheme of dependence tautologies.

7. Ralph is not a dog because he's a puppet

Not formalizable

Explanation: Dependent implication will not do to formalize 'because'. In any reasonable model 'Ralph is a puppet $\rightarrow$ $\neg$(Ralph is a dog)' will be false: the content of the antecedent does not contain that of the consequent, even though the example is arguably true. We would need a different notion of content for the conditional to model 'because', one in which, apparently, the content of a $\neg A$ is (contained in?, contains?) the complement of the content of A.

7. Dependence logic tautologies compared to classical tautologies

Every dependence logic tautology must be a classical tautology, for every classical model is a dependence logic model: assign the same content to all atomic propositions. For relatedness logic, models in which the universal relation governs the truth-table for '$\rightarrow$' could be claimed by a classical logician to represent his view that all propositions are related. But here the notion of referential content does not so easily evaporate: under what intuition can one claim that all propositions should have the same content?

Many familiar classical tautologies fail to be dependence logic tautologies. As noted in Example 1, the paradoxes of material implication fail in dependence logic.

And, also as noted in Example 1, the paradoxes of strict implication fail to be dependence tautologies:

$(A \wedge \neg A) \rightarrow B$ $\qquad$ $A \rightarrow (B \rightarrow B)$

Each of these is true in a model $<\mathsf{v},\mathsf{s}>$ if, and only if $\mathsf{s}(A) \supseteq \mathsf{s}(B)$.

As noted in Example 3, *exportation* fails:

$((A \wedge B) \rightarrow C) \rightarrow (A \rightarrow (B \rightarrow C))$

We may have $\mathsf{s}(A) \cup \mathsf{s}(B) \supseteq \mathsf{s}(C)$ and $\mathsf{s}(A) \not\supseteq \mathsf{s}(B) \cup \mathsf{s}(C)$. But *importation* is tautological:

$(A \rightarrow (B \rightarrow C)) \rightarrow ((A \wedge B) \rightarrow C))$

Dependence logic is sensitive to the placement of premisses. For example, some propositional forms of *modus ponens* and *modus tollens* fail to be dependence tautologies:

$A \rightarrow ((A \rightarrow B) \rightarrow B)$ $\qquad$ $\neg B \rightarrow ((A \rightarrow B) \rightarrow \neg A)$

But if we pack the premisses together into the antecedent, we have tautologies:

$(A \wedge (A \rightarrow B)) \rightarrow B$ $\qquad$ $(\neg B \wedge (A \rightarrow B)) \rightarrow \neg A$

As noted in Example 5, dependence logic is unlike relatedness logic in that dependent implication is transitive. We have both:

$\{A \rightarrow B,\ B \rightarrow C\} \vDash A \rightarrow C$ $\qquad$ $((A \rightarrow B) \wedge (B \rightarrow C)) \rightarrow (A \rightarrow C)$

Consider also:

$(A \rightarrow (B \wedge C)) \rightarrow ((A \rightarrow B) \wedge (A \rightarrow C))$

This is a dependence tautology, but fails to be a relatedness tautology. Relatedness logic isn't suitable to model the idea that for an implication to hold the antecedent must contain the consequent. But the first example of this chapter already showed that, for in any plausible model the subject matter of '*a* drinks' overlaps that of 'everyone drinks' (see also Example 1 above).

If we take '$\vee$' to be either classical, weakly dependent, or strongly dependent disjunction (see p. 168 of Section 3 above), then we cannot add disjuncts to a consequent; the following fail to be schema of dependence tautologies:

$A \rightarrow (A \vee B)$ $\qquad$ $(A \rightarrow B) \rightarrow (A \rightarrow (B \vee C))$

But both of the following would be dependence tautologies:

$((A \vee B) \wedge \neg A) \rightarrow B$ $\qquad$ $(A \rightarrow (B \vee C)) \rightarrow ((A \rightarrow B) \vee (A \rightarrow C))$

In Section E I'll discuss the form of dependence tautologies in which '$\rightarrow$' is the principal connective.

8. Functional completeness of the connectives

Define:

$$D(A,B) \equiv_{\text{Def}} A \rightarrow (B \rightarrow B)$$

Then $D(A,B)$ is true in a model $\langle v, s \rangle$ iff $s(A) \supseteq s(B)$.

This is the same formula we called 'R(A,B)' in relatedness logic. It's right to give it a new name here because we're not formalizing the same connective of English: 'D' is not a connective at all, but an abbreviation useful for the metalogic. We read 'D(Ralph is a dog, Ralph barks)' as 'the proposition 'Ralph barks' is dependent on the proposition 'Ralph is a dog''. It makes no sense to iterate 'D'.

We also have:

$A \wedge B$ is strongly semantically equivalent to

$$\neg((((A \rightarrow B) \rightarrow (A \rightarrow B)) \rightarrow A) \rightarrow \neg B)$$

With this observation it is possible to define every truth-and-content functional connective from just $\neg$ and $\rightarrow$: the proof is exactly the same as for subject matter relatedness logic (pp. 114–115) reading 'D' for 'R'. A normal form theorem also follows as for relatedness logic (Corollary III.6, p. 115).

Exercises for Sections A.5–A.8

1. Devise a falsification procedure for establishing whether a formal $\rightarrow$-wff is a dependence logic tautology (cf. the falsification procedure for classical logic, pp. 58–59, and Exercise 1, p. 113).
2. For each scheme of classical tautologies listed in Chapter II.J.8 (pp. 60–62) decide if it is a scheme of dependence logic tautologies.
3. Formalize and discuss the following:
   a. If $2 + 2 = 4$, then $4 + 4 = 8$
   b. If $2 + 2 = 4$, then $2^{32} > 4^{12}$
   c. If $2 + 2 = 4$, then there are nondifferentiable functions on the real line
   d. If Ralph is a bachelor and every bachelor is unmarried, then Ralph is not married
   e. If dogs bark, then Anubis barks; so if Anubis doesn't bark, then dogs don't bark
   f. If Ralph is a dog, then Ralph is a dog
   g. If Ralph is a dog, then Ralph is a dog or Howie is a cat
   h. If dogs bark, then Ralph barks and ducks quack; so if dogs bark then Ralph barks, and if dogs bark then ducks quack
   j. If Ralph is a dog and if $1 = 1$ then $1 = 1$, then $2 + 2 = 4$

k. Ralph is a dog
   Ralph is not a dog
   Therefore: $2+2=4$

4. Verify that $A \wedge B$ is strongly semantically equivalent to:
   $\neg((((A \rightarrow B) \rightarrow (A \rightarrow B)) \rightarrow A) \rightarrow \neg B)$

5. State and prove a normal form theorem for dependence logic.

6. Show that *The Semantic Deduction Theorem* fails for **D**: $\Gamma, A \vDash B$ iff $\Gamma \vDash A \rightarrow B$.

9. An axiom system for D

In this section I will give a strongly complete axiomatization for **D**. The proof is similar to the one for subject matter relatedness logic, **S**, Chapter III.K.1.

D, Dependence Logic
in $L(\neg, \rightarrow, \wedge)$

$D(A,B) \equiv_{Def} A \rightarrow (B \rightarrow B)$ $\qquad$ $A \supset B \equiv_{Def} \neg(A \wedge \neg B)$

axiom schema

1. $D(A, B) \wedge D(B, C) \rightarrow D(A, C)$
2. $D(A, B \wedge C) \rightarrow D(A, B) \wedge D(A, C)$
3. $D(A, B) \wedge D(A, C) \rightarrow D(A, B \wedge C)$
4. $D(A, B \wedge C) \rightarrow D(A, B \rightarrow C)$
   $D(A, B \rightarrow C) \rightarrow D(A, B \wedge C)$
5. $D(A, \neg A)$
   $D(\neg A, A)$
6. $(A \wedge B) \rightarrow A$
   $(A \wedge B) \rightarrow B$
7. $(A \rightarrow B) \rightarrow D(A, B)$
8. $(A \wedge \neg B) \rightarrow \neg(A \rightarrow B)$
9. $(D(A, B) \wedge \neg A) \rightarrow (A \rightarrow B)$
10. $(D(A, B) \wedge B) \rightarrow (A \rightarrow B)$
11. $(B \wedge D(A, A)) \rightarrow (A \supset B)$
12. $A \supset A$
13. $[(A \supset (C \rightarrow D)) \wedge (A \supset C)] \rightarrow (A \supset D)$
14. $((\neg A \supset \neg B) \wedge (\neg A \supset B)) \rightarrow A$

rules $\quad$ *modus ponens* $\quad$ adjunction

$$\frac{A,\ A \rightarrow B}{B} \qquad \frac{A,\ B}{A \wedge B}$$

The definition of proof is the usual one from Chapter II.K. Throughout this section I will use $\vdash$ for the syntactic consequence relation of this system.

This system differs from the ones in previous chapters in using two rules. We need adjunction because it seems that without it we cannot derive $A \wedge B$ from $\{A, B\}$; the usual wff that allows that derivation, $A \rightarrow (B \rightarrow (A \wedge B))$, is not a tautology in **D**. So I have not taken $\wedge$ as a defined symbol; the definability of $\wedge$ in terms of $\neg$ and $\rightarrow$ in this axiom system will follow from the completeness theorem.

We begin by choosing appropriate notions of consistency and completeness. First of all, Γ is a *dependence theory* if Γ is closed under deduction. That is, if $\Gamma \vdash A$, then $A \in \Gamma$. In particular note that every theorem of this system is in every dependence theory.

Γ is a *complete dependence theory* if Γ is a dependence theory and for every A, one of $A, \neg A$ is in Γ.

Γ is *consistent* if there is no A such that $\Gamma \vdash A$ and $\Gamma \vdash \neg A$.

A *model* of Γ is a dependence model in which every wff in Γ is true.

These are the same definitions as for classical logic and relatedness logic.

The proof of strong completeness now follows the same lines as for **S**. I will use the relation-based semantics for **D**, invoking Lemma 1 above.

Lemma 3 Γ is complete and consistent iff there is a **D** model $\langle \mathsf{v}, \mathsf{D} \rangle$ such that $\Gamma = \{A: \mathsf{v}(A) = \mathsf{T}\}$

Proof: The right to left direction is an exercise.

Now suppose Γ is complete and consistent. Note that Γ is a theory. Define:

$\mathsf{v}(A) = \mathsf{T}$ iff $A \in \Gamma$

$\mathsf{D}(A, B)$ holds iff $D(A, B) \in \Gamma$

We must show that $\langle \mathsf{v}, \mathsf{D} \rangle$ is a **D**-model.

First I will show that D is a dependence relation. From **axiom 5** we have that for every A, $D(A, \neg A)$ and $D(\neg A, A) \in \Gamma$. So we have $\mathsf{D}(A, \neg A)$ and $\mathsf{D}(\neg A, A)$. And also by adjunction, $D(A, \neg A) \wedge D(\neg A, A) \in \Gamma$. So by **axiom 1**, $D(A, A) \in \Gamma$. And hence, $\mathsf{D}(A, A)$. The rest of the proof that D is a dependence relation follows by invoking **axioms 1–4** and Lemma 1.

To show that v evaluates wffs correctly, as for classical logic we have that v evaluates $\neg$ correctly (see p. 70).

Suppose now that $\mathsf{v}(A \wedge B) = \mathsf{T}$. Then $A \wedge B \in \Gamma$. Hence by **axiom 6**, $A \in \Gamma$ and $B \in \Gamma$, so $\mathsf{v}(A) = \mathsf{T}$ and $\mathsf{v}(B) = \mathsf{T}$. On the other hand, if $\mathsf{v}(A) = \mathsf{T}$ and $\mathsf{v}(B) = \mathsf{T}$, then $A \in \Gamma$ and $B \in \Gamma$, so by adjunction, $A \wedge B \in \Gamma$. And hence $\mathsf{v}(A \wedge B) = \mathsf{T}$.

Suppose now $\mathsf{v}(A \rightarrow B) = \mathsf{T}$. Then $A \rightarrow B \in \Gamma$. So by **axiom 7**, $D(A, B) \in \Gamma$ and hence $\mathsf{D}(A, B)$. If $\mathsf{v}(A) = \mathsf{T}$ and $\mathsf{v}(B) = \mathsf{F}$, then $A \in \Gamma$ and $\neg B \in \Gamma$, so by adjunction, $A \wedge \neg B \in \Gamma$. And hence by **axiom 8**, $\neg(A \rightarrow B) \in \Gamma$, a contradiction on the consistency of Γ. So either $\mathsf{v}(A) = \mathsf{F}$ or $\mathsf{v}(B) = \mathsf{T}$.

Lastly, suppose $v(A) = F$ or $v(B) = T$ and $D(A, B)$. Then $D(A, B) \in \Gamma$ and either $\neg A \in \Gamma$ or $B \in \Gamma$. Hence by **axioms 9** and **10**, $A \to B \in \Gamma$. ■

The following will suffice in place of a deduction theorem here.

Lemma 4 If $\Gamma, A \vdash B$ then $\Gamma \vdash A \supset B$.

Proof: The proof is similar to the one for relatedness logic, p. 118, using **axioms 11**, **12**, and **13**, and I leave it as an exercise (use $\vdash D(A, A)$ as established in the proof of Lemma 3). ■

Lemma 5 ***a.*** If Γ is consistent, then $\Gamma \cup \{A\}$ or $\Gamma \cup \{\neg A\}$ is consistent.
b. $\Gamma \nvdash A$ iff $\Gamma \cup \{\neg A\}$ is consistent

Proof: As for relatedness logic, p. 118, using **axiom 14**. ■

Lemma 6 If $\Gamma \nvdash D$, then there is some complete and consistent collection of wffs Σ such that $\Sigma \cup \{\neg D\} \in \Gamma$.

Proof: The proof is as for relatedness logic, Lemma III.10, p. 119. ■

Theorem 7 ***Strong completeness of the axiomatization of* D**
For all Γ, $\Gamma \vdash A$ iff $\Gamma \vDash A$

Proof: The proof is exactly as for classical logic, Theorem II.15, p. 75. ■

And again, as for classical logic, we have the following.

Theorem 8 ***Compactness of the semantic consequence relation***
a. $\Gamma \vDash A$ iff there is some finite collection $\Delta \subseteq \Gamma$ such that $\Delta \vDash A$
b. Γ has a model iff every finite subset of Γ has a model

The Semantic Deduction Theorem fails: $(A \wedge \neg A) \vDash B$ for every A and B, yet $\nvDash (A \wedge \neg A) \to B$ if B contains a propositional variable that does not appear in A. But we have the *Material Implication Form of the Deduction Theorem*:

$\Gamma, A \vDash B$ iff $\Gamma \vDash \neg(A \wedge \neg B)$

And, much as for relatedness logic, we do have restricted forms of a semantic/syntactic deduction theorem.

Theorem 9 $\Gamma, A \vDash B$ and $\Gamma \vDash R(A, B)$ iff $\Gamma \vDash A \to B$
$\Gamma, A \vdash B$ and $\Gamma \vdash R(A, B)$ iff $\Gamma \vdash A \to B$

If every propositional variable that appears in B also appears in A, then:

$\Gamma, A \vDash B$ iff $\Gamma \vDash A \to B$
$\Gamma, A \vdash B$ iff $\Gamma \vdash A \to B$

Unlike for relatedness logic, the rule of substitution of logical equivalents is a derived rule in dependence logic:

$$\frac{A \leftrightarrow B}{C(A) \leftrightarrow C(B)}$$

If $<\mathsf{v},\mathsf{s}> \models A \leftrightarrow B$, then $\mathsf{v}(A) = \mathsf{v}(B)$ and $\mathsf{s}(A) = \mathsf{s}(B)$, so that A and B play exactly the same role in the model.

Except for the question of whether **D** has possible-world semantics, we have answered all the questions of Chapter IV as they apply to dependence logic.

Exercises for Section A.9

1. Derive the following in the axiom system for **D**.
   a. $A \wedge B \rightarrow B \wedge A$
   b. $A \rightarrow A$
   c. $A \rightarrow B \rightarrow \neg(A \wedge \neg B)$
   d. $(A \rightarrow B \wedge D(B, A)) \rightarrow (\neg B \rightarrow \neg A)$
   e. $A \wedge \neg A \vdash B$

2. Prove without recourse to completeness that if $\Gamma, A \vdash B$, then $\Gamma \vdash A \supset B$.

3. Prove without recourse to completeness:
   a. If Γ is consistent, then $\Gamma \cup \{A\}$ or $\Gamma \cup \{\neg A\}$ is consistent.
   b. $\Gamma \nvdash A$ iff $\Gamma \cup \{\neg A\}$ is consistent.

4. Prove that if $\Gamma \nvdash D$, then there is a complete and consistent collection of wffs Σ such that $\Sigma \cup \{\neg D\} \in \Gamma$.

5. Exhibit a strongly complete axiomatization for **D** by using the axiomatization of classical logic in $L(\neg, \wedge)$, p. 84, taking material detachment as the only rule and adding axioms that ensure that D and $\rightarrow$ are evaluated correctly. Comment on why such an axiomatization is not compatible with the motivation of dependence logic. (Compare Exercise 4, p. 124.)

   Such an axiomatization might be considered the wrong approach to **D** from an algebraic viewpoint: the relation $A \sim B$ iff $\vdash_{\mathbf{D}} A \leftrightarrow B$ establishes a congruence relation on Wffs from which we can define algebraic semantics for **D**, as shown in *Epstein, 1987*. However, the relation $A \sim B$ iff $\vdash_{\mathbf{D}} A \supset B \wedge B \supset A$, that is, A is materially equivalent to B, is not a congruence relation on Wffs.

6. (Open) Give a constructive proof of completeness for **D**. Show that there is no constructive proof of strong completeness for **D** by showing that if there were one, then there would be one for classical logic.

7. (Open) Either reduce the axiomatization of **D** by deducing some of the axioms from the remaining ones, or show that the axiomatization is independent

8. (Open) Devise a more natural axiomatization of **D** that does not rely on axioms involving $\supset$. (Suggestion: Devise axioms that allow a proof of: if Γ, $A \vdash B$ and $\Gamma \vdash D(A, B)$, then $\Gamma \vdash A \to B$.)

9. (Open) Because **D** is closed under the rule of substitution, you can prove that the Lewis system **S1** is a subsystem of **D**, reading strict implication as $\to$ (see *Hughes and Cresswell, 1968*, p. 217). Is there a modal logic into which **D** can be translated grammatically (see Chapters VI and X)? More particularly, is there a modal logic **L** and a scheme $\theta(A, B)$ in its language such that $\mathbf{L} \vdash \theta(A, B)$ iff $\mathbf{D} \vdash A \to B$?

10. (Open) Determine whether it is possible to give Kripke-style possible world semantics for **D**. (It is not possible to give finite many-valued semantics for **D**, see Corollary VIII.22.)

10. History

I first presented dependence logic in a preprint in 1978. I was stimulated to do so by conversations I'd had with Niels Egmont Christensen the previous summer. We had been discussing relatedness logic, and I found that I could model his notion of entailment by changing the relationship governing the truth-table for the conditional from nonempty intersection to containment. The wffs under '+' on p. 81 of *Christensen, 1973*, are dependence tautologies, while those under '–' are not, with the exception of $p \to (\neg p \to p)$.

I later realized that I'd been influenced by a brief conversation I'd had with Michael Dunn that summer. He had taken a system which *Parry, 1933,* had devised to axiomatize analytic implication and modified it slightly, presenting semantics for the modification in *Dunn, 1972*. In the last section of that paper Dunn reports that R. K. Meyer noted that the set-assignment semantics of what I have since called 'dependence logic' are complete for Dunn's axiom system. So Dunn's logic and mine are equivalent.

Earlier attempts to axiomatize this logic are apparently faulty. Dunn's axiomatization does not include adjunction as a rule. P. H. Rodenburg has pointed out that my proofs of strong completeness for my earlier axiomatizations (including the one in the first edition of this volume) depend on deriving the associativity of conjunction, $A \wedge (B \wedge C) \to (A \wedge B) \wedge C$ and $(A \wedge B) \wedge C \to A \wedge (B \wedge C)$, which I cannot see how to do in those systems.

Parry, 1971 and *197?,* has detailed the history of this and related syntactical approaches to entailment. Parry's syntactic motivation for his logic provides an interesting comparison with that in Sections A.1 and A.2.

> Our conception of deducibility may be clarified thus: Q is deducible from P if, in any system in which P is asserted, the assertion of Q is justifiable, assuming

> a reasonably complete logic. Now, we ask, if a system of Euclidean geometry contains the assertion that two points determine a straight line, are we justified in asserting in this system: 'Either two points determine a straight line or some mice like cheese'? No, this strange disjunction is not a legitimate assertion in a system of Euclidean geometry, for the simple reason that no such system contains the terms 'mice' or 'cheese', nor can one define by geometric concepts a type of cheese any self-respecting mouse would nibble at. We conclude, then, that it is not true that the statement that either two points determine a straight line or some mice like cheese is deducible from the statement that two points determine a straight line. So the statement that either p or q is not always deducible from the statement that p; putting it the other way around, we find that p does not always entail that p or q.
>
> *Parry, 1971*, p. 5

K. Fine, 1979, has also discussed Parry's system and his intuitions seem similar to mine. *Kielkopf, 1977,* surveys many syntactic and semantic theories of entailment and compares the work of Parry, Dunn, Fine, and others. *Hanson, 1980,* has a stimulating discussion of implication as containment of logical consequences, a topic that I take up in Section F below.

I am grateful to Roger Maddux for conversations and comments that helped me to develop the technical work of Section A.4.

B. Dependence-Style Semantics

The idea that for $A \to B$ to be true the content of A must include the content of B has been put forward for many notions of content other than referential content. And with many of those notions it is classical negation and classical conjunction that are (sometimes implicitly) accepted. So let us call the following *dependence truth-conditions*:

$\neg$ and $\wedge$ are interpreted classically

$\to$ is interpreted by the table for dependent implication:

A	B	$s(A) \supseteq s(B)$	$A \to B$
any	values	fails	F
T	T		T
T	F	holds	F
F	T		T
F	F		T

And let us say that a logic has (or uses) *dependence-style semantics* if it is presented by set-assignment semantics that use dependence truth-conditions. In such a logic we can define $A \vee B$ in any of the three ways I described in Section A.3, p. 168.

Sometimes the idea that the antecedent must contain the consequent requires that containment be strict. Let us call *weak dependence truth-conditions* and *weak dependence-style semantics* those as for dependence style-semantics except that $\supsetneq$ governs the table for the conditional instead of $\supseteq$.

In set-theory there is a notion of duality (similar to the one for classical logic, pp. 56–57). Given any equality in set theory, such as $(\mathsf{A} \cap \mathsf{B}) \cup (\mathsf{A} \cap \mathsf{C}) = \varnothing$, if we interchange the roles of $\cap$ and $\cup$, and those of $\varnothing$ and the universal set (here S), then the resulting equation is called the *dual* of the original one, in this example, $(\mathsf{A} \cup \mathsf{B}) \cap (\mathsf{A} \cup \mathsf{C}) = \mathsf{S}$. Since $\mathsf{A} \supseteq \mathsf{B}$ iff $\mathsf{A} \cap \mathsf{B} = \mathsf{B}$, the dual of which is $\mathsf{A} \cup \mathsf{B} = \mathsf{B}$, we have that the dual of $\mathsf{A} \supseteq \mathsf{B}$ is $\mathsf{A} \subseteq \mathsf{B}$.

Any dependence-style semantics in which the class of set-assignments is defined using only the operations of union, intersection, and complementation has a dual form in which the duals of the conditions defining the set-assignments are used and $\subseteq$ governs the table for the conditional rather than $\supseteq$ (while $\neg$ and $\wedge$ continue to be evaluated classically). Those *dual semantics* give rise to exactly the same logic.

Let us call the following *dual dependence truth-conditions*:

$\neg$ and $\wedge$ are interpreted classically

$\rightarrow$ is interpreted by the table for *dual-dependent implication*:

A	B	$\mathsf{s}(\mathrm{A}) \subseteq \mathsf{s}(\mathrm{B})$	A→B
any	values	fails	F
T	T		T
T	F	holds	F
F	T		T
F	F		T

And call any set-assignment semantics using these truth-conditions *dual dependence-style semantics.*

In the same way we can translate dual dependence-style semantics to dependence-style semantics. If it is clear how to dualize the semantics, I will say that a logic has dependence-style semantics even though I present it using dual dependence-style semantics. For instance, I will say in Chapter VI that many of the modal logics use dependence-style semantics, though they are presented using dual dependence truth-conditions.

Exercises for Section B

1. Formulate the dual semantics for dependence logic and verify that they give rise to exactly the same logic.

2. Prove that given any dependence-style semantics in which the equations defining the class of set-assignments use only union, intersection, and complementation, the dual semantics give rise to the same consequence relation.

3. (Open) Formulate and axiomatize *weak dependence logic*: the logic identical to dependence logic except that $\supsetneq$ governs the table for the conditional instead of $\supseteq$. Give examples and motivation.

4. (Open) Analyze the tautologies and valid deductions of the class of all models that use dependence-style semantics. Can these be axiomatized? Is it reasonable to call these a logic?

C. Dual Dependence Logic, Dual D

Another duality arises from dependence logic if we retain the same class of set-assignments, that is, those satisfying $s(A) = \bigcup\{s(p) : p \text{ appears in } A\}$, and use the dual dependence truth-conditions. This is what I call **dual dependence logic**, **Dual D**.

There is also a syntactic sense in which **D** and **Dual D** are dual. Consider the map from $L(\neg, \rightarrow, \wedge, p_0, p_1, \ldots)$ to itself given by:

$$(p_i)^* = p_i$$
$$(\neg A)^* = \neg(A^*)$$
$$(A \wedge B)^* = A^* \wedge B^*$$
$$(A \rightarrow B)^* = \neg B^* \rightarrow \neg A^*$$

This establishes a translation of **Dual D** into **D**, and also from **D** to **Dual D.** Letting $\Gamma^* = \{B^* : B \in \Gamma\}$, we have:

Theorem 10 $\Gamma \vDash_{\textbf{Dual D}} A$ iff $\Gamma^* \vDash_{\textbf{D}} A^*$

$\Gamma \vDash_{\textbf{D}} A$ iff $\Gamma^* \vDash_{\textbf{Dual D}} A^*$

I leave the proof to you.

We can characterize the dual-dependence relations, that is, those that arise as the relation $\subseteq$ from union set-assignments, using the same method as Lemma 1.

Lemma 11 D is a dual dependence relation iff all of the following hold:

1. D is reflexive
2. D is transitive
3. $\mathsf{D}(A \wedge B, C)$ iff $\mathsf{D}(A, C)$ and $\mathsf{D}(B, C)$
4. $\mathsf{D}(A \rightarrow B, C)$ iff $\mathsf{D}(A, C)$ and $\mathsf{D}(B, C)$
5. $\mathsf{D}(\neg A, A)$
6. $\mathsf{D}(A, \neg A)$

To axiomatize **Dual D**, note that the same formula we used for **D** characterizes the relation governing the truth-table for $\rightarrow$:

$A \rightarrow (B \rightarrow B)$ is true in a model $<\mathsf{v}, \mathsf{s}>$ iff $\mathsf{s}(A) \subseteq \mathsf{s}(B)$

I will use the same abbreviation, 'D(A, B)', for this formula here in order to make it easier to compare the axiomatization for **Dual D** to the one for **D**.

Dual D, Dual Dependence Logic

in $L(\neg, \rightarrow, \wedge)$

$D(A, B) \equiv_{Def} A \rightarrow (B \rightarrow B)$

$A \supset B \equiv_{Def} \neg(A \wedge \neg B)$

axiom schema

1. $(D(A, B) \wedge D(B, C)) \rightarrow (D(A, C) \wedge D(B, C))$
2. $D(A \wedge B, C) \rightarrow D(A, B) \wedge D(A, C)$
3. $D(A, C) \wedge D(B, C) \rightarrow D(A \wedge B, C)$
4. $D(A \wedge B, C) \rightarrow D(A \rightarrow B, C)$
   $D(A \rightarrow B, C) \rightarrow D(A \wedge B, C)$
5. $D(A, \neg A)$
   $D(\neg A, A)$
6. $A \wedge B \rightarrow B \wedge A$
   $A \rightarrow (B \rightarrow (A \wedge B))$
7. $(A \rightarrow B) \rightarrow D(A, B)$
8. $(A \wedge \neg B) \rightarrow \neg(A \rightarrow B)$
9. $(D(A, B) \wedge \neg A) \rightarrow (A \rightarrow B)$
10. $(D(A, B) \wedge B) \rightarrow (A \rightarrow B)$
11. $B \rightarrow (A \supset B)$
12. $A \supset A$
13. $[(A \supset (C \rightarrow D)) \wedge (A \supset C)] \rightarrow (A \supset D \wedge A \supset C)$
14. $(\neg A \supset \neg B) \wedge (\neg A \supset B) \rightarrow [A \wedge D(B, B)]$

rules $\dfrac{A, A \rightarrow B}{B} \qquad \dfrac{A \wedge B}{B}$

I'll write $\vdash_{\mathbf{Dual\ D}}$ for the consequence relation of this system and $\vDash_{\mathbf{Dual\ D}}$ for the semantic consequence relation of the semantics.

In this axiomatization we have a new rule, a "dual" of the rule of adjunction we used for **D**, for now we have $A \rightarrow (B \rightarrow (A \wedge B))$ is a tautology, but $(A \wedge B) \rightarrow B$ is not. And because we must always proceed from less content to more content, we have some odd axioms (1, 13, and 14) in which a conjunct must be added to account for enough content, where the conjunct can be elminated only with this new rule.

I've chosen these axioms to make it as easy as possible for you to prove the strong completeness theorem for dual dependence logic as a modification of the proof for dependence logic (Section A.9).

***Theorem 12 Strong Completeness for* Dual D** $\Gamma \vdash_{\textbf{Dual D}} A$ iff $\Gamma \vDash_{\textbf{Dual D}} A$

Everything we proved about **D** can be duplicated here with the appropriate changes. For instance, **Dual D** is compactly decidable. And $\wedge$ is definable from $\neg$ and $\rightarrow$ since:

$A \wedge B$ is strongly semantically equivalent to

$$\neg(A \rightarrow (B \rightarrow \neg((A \rightarrow B) \rightarrow (A \rightarrow B))))$$

Note that this is the same equivalence as for **S**, p. 114.

Exercises for Section C

1. Formulate a decision procedure for **Dual D**. For each scheme of classical tautologies listed in Chapter II.J.8 (pp. 60–62) decide whether it is a scheme of dual dependence logic tautologies.

2. a. Discuss the examples of Section A.6 in dual dependence logic.
   b. Formalize and discuss the examples of Exercise 3, p. 177, for dual dependence logic.

3. Prove Lemma 11 characterizing dual dependence relations.

4. Use the translation of Theorem 10 to show that in **Dual D** $A \wedge B$ is strongly semantically equivalent to $\neg(A \rightarrow (B \rightarrow \neg((A \rightarrow B) \rightarrow (A \rightarrow B))))$.

5. Show that $\Gamma \vDash_{\textbf{D}} A$ iff $\Gamma^* \vDash_{\textbf{Dual D}} A^*$, and $\Gamma \vDash_{\textbf{Dual D}} A$ iff $\Gamma^* \vDash_{\textbf{D}} A^*$.

6. Derive the following in the axiom system for **Dual D**.
   a. $A \rightarrow A$
   b. $(A \rightarrow B) \rightarrow \neg(A \wedge \neg B)$
   c. $(A \rightarrow B \wedge D(B, A)) \rightarrow (\neg B \rightarrow \neg A)$
   d. $A \wedge B \vDash A$
   e. $A \wedge \neg A \vDash B$

7. Prove the *Strong Completeness Theorem* for **Dual D**.

8. State and prove a normal form theorem for **Dual D**.

9. Use the translation of Theorem to formulate an axiomatization of **Dual D** from the one for **D**.

9. Do Exercises 5, 6, 7, and 9, pp. 181–182, for dual dependence logic.

10. (Open) Walter Carnielli has suggested calling **Dual D** an *external dualization* of **D**. That is, it arises by retaining the same class of set-assignments and dualizing the relations governing the nonclassical truth-tables, here changing $\supseteq$ to $\subseteq$. This is in contrast to an *internal dualization* which arises by retaining the truth-tables but dualizing the conditions establishing the class of set-assignments, here changing from union to intersection set-assignments.

    The internal dualization of **D** is not **D**, for $A \rightarrow (B \rightarrow (A \wedge B))$ is valid in it; nor is it **Dual D**, for taking $\vee$ to be defined classically from $\neg$ and $\wedge$ we have that $[(A \vee B) \rightarrow C] \rightarrow [(A \rightarrow C) \vee (B \rightarrow C)]$ is valid in it.

    Formulate the internal dualization of **D**, axiomatize it, and compare the tautologies to those of both **D** and **Dual D**.

D. A Logic of Equality of Contents, Eq

1. Motivation

In dependence logic if we have a model $<v,s>$ in which $s(p) \not\supseteq s(q)$ and p is false or q is true, then $p \rightarrow q$ is false in the model. But for any proposition r, either true or false, that satisfies $s(r) \supseteq s(q)$, we have $(p \wedge r) \rightarrow q$ is true. For example, we may have the following false:

The moon is made of green cheese $\rightarrow 1 = 0$

While the following is true:

(The moon is made of green cheese $\wedge\ 2+2=4) \rightarrow 1=0$

It's been suggested that such analyses indicate a flaw in dependence logic (and relatedness logic and dual dependence logic). I don't see the problem, for the antecedent does provide the contextual framework for the consequent. Perhaps the latter proposition is unacceptable to some because they want the antecedent to have the *same* referential content as the consequent.

That's precisely what *Sluga, 1986,* suggests as a model of Frege's notion of sense. The logic he appears to be proposing is what I call **Eq**, and differs from dependence logic only in that the relation governing the truth-table for $\rightarrow$ is equality.

2. Set-assignment semantics

We take the formal language to be $L(\neg, \rightarrow, \wedge, p_0, p_1, \ldots)$. **Eq** is the logic given by the class of models $<v,s>$ that use union set-assignments, that is, $s(A) = \bigcup \{ s(p) : \text{p appears in A} \}$, evaluate $\neg$ and $\wedge$ classically, and use the following table for the conditional:

A	B	$\mathsf{s}(A)=\mathsf{s}(B)$	$A\to B$
any	values	fails	F
T	T		T
T	F	holds	F
F	T		T
F	F		T

We assume the *Fully General Abstraction* for these models.

For the semantic consequence relation of **Eq** I'll write $\vDash_{\mathbf{Eq}}$. Note that this logic is compactly decidable by the same methods as for **D** (Section A.5).

As for **D** and **S**, there are various definable choices for $\vee$, which will not affect what follows.

We can view **Eq** as arising from **D**. Consider the following map:

$(p)^* = p$

$(\neg A)^* = \neg(A^*)$

$(A\wedge B)^* = A^* \wedge B^*$

$(A\to B)^* = (A^*\to B^*) \wedge (\neg B^*\to\neg A^*)$

Let $\Gamma^* = \{A^* : A\in\Gamma\}$. I leave to you to prove:

Theorem 13 $\Gamma\vdash_{\mathbf{Eq}} A$ iff $\Gamma^*\vdash_{\mathbf{D}} A^*$

Unfortunately, this translation doesn't give us much grasp on the form of tautologies of **Eq** or how to derive them. For example, imagine trying to prove a simple tautology of **Eq** such as $((A\to B)\wedge\neg B) \to (\neg A\wedge\neg B)$ from the axiomatization of **D** by means of this translation.

We need to axiomatize **Eq** directly. The hard step is to characterize the class of binary relations on wffs that arise from union set-assignments where equality of contents is the condition for A to be related to B. Once we have that, then the method of axiomatization will be the same as for **D**, since we can define:

$E(A,B) \equiv_{\text{Def}} A\to(B\to B)$

For **Eq**-models, $\langle v,\mathsf{s}\rangle\vDash E(A,B)$ iff $\mathsf{s}(A)=\mathsf{s}(B)$.

3. Characterizing Eq-relations

I'll say that E is an *Eq-relation on wffs* if there is a union set-assignment s such that $\mathsf{E}(A,B)$ iff $\mathsf{s}(A)=\mathsf{s}(B)$.

Theorem 14 E is an Eq-relation iff all of the following hold:

1. E is an equivalence relation
2. If $\mathsf{E}(A,B)$, then $\mathsf{E}(A \wedge C, B \wedge C)$
3. $\mathsf{E}(A, \neg A)$
4. $\mathsf{E}(A \wedge B, B \wedge A)$
5. $\mathsf{E}(A \rightarrow B, A \wedge B)$
6. $\mathsf{E}(A \wedge (B \wedge C), (A \wedge B) \wedge C)$
7. If $\mathsf{E}(A, B \wedge C)$ and $\mathsf{E}(B, A \wedge D)$, then $\mathsf{E}(A,B)$
8. $\mathsf{E}(A, A \wedge A)$

Proof: If E is an Eq-relation, then (1)–(8) all hold. For the converse we need more properties of E.

Lemma If E satisfies (1)–(8), then

a. If $\mathsf{E}(A,B)$ and $\mathsf{E}(C,D)$, then $\mathsf{E}(A \wedge C, B \wedge D)$

b. If $\mathsf{E}(A,B)$ and $\mathsf{E}(A,C)$, then $\mathsf{E}(A, B \wedge C)$

c. $\mathsf{E}(B \wedge C, B \wedge B \wedge C)$ for either way of associating

d. $\mathsf{E}(A \rightarrow B, B \rightarrow A)$

e. $\mathsf{E}((A \rightarrow B) \rightarrow C, A \rightarrow (B \rightarrow C))$

f. $\mathsf{E}(A,B)$ iff $\mathsf{E}(A,B^{\dagger})$, where $B^{\dagger} = \bigwedge\{p : p \text{ appears in } B\}$ and the conjuncts appear in the numerical order of their subscripts ($\bigwedge$ indicates the conjunction over all of the elements in the set, associating to the left)

g. Given any scheme $\varphi(A_1, \ldots, A_n, C)$ where $A_1, \ldots, A_n$ are fixed wffs in the language, then: if $\mathsf{E}(A,B)$, then $\mathsf{E}(\varphi(A), \varphi(B))$, where the substitutions are made uniformly for C

Proof: a. If $\mathsf{E}(A,B)$, then by (2), $\mathsf{E}(A \wedge C, B \wedge C)$. If $\mathsf{E}(C,D)$, then by (2) $\mathsf{E}(C \wedge B, D \wedge B)$. By (4), $\mathsf{E}(B \wedge C, C \wedge B)$, so by (1), $\mathsf{E}(A \wedge C, D \wedge B)$. Finally by (2), $\mathsf{E}(D \wedge B, B \wedge D)$, so by (1), $\mathsf{E}(A \wedge C, B \wedge D)$.

Parts (b) and (c) follow from (2), (8), and (1).

d. This follows from (5), (4), and (1).

e. This follows from (4), (5), (6), and (1).

f. I will show that for every A, $\mathsf{E}(A, A^{\dagger})$, from which (f) will follow by (1).

The proof is by induction on the length of A. It is immediate if the length of A is 1. So suppose it's true for all wffs of length $\leq n$ and A has length $n+1$. By (1), $\mathsf{E}(A,A)$. We now have three cases.

Case i. A is $\neg B$. Then $\mathsf{E}(A,B)$ by (3), and $\mathsf{E}(B,B^{\dagger})$ by induction. So $\mathsf{E}(A,B^{\dagger})$ by (1).

Case ii. A is $B \wedge C$. Then by induction, $\mathsf{E}(B,B^{\dagger})$ and $\mathsf{E}(C,C^{\dagger})$. So by part (a), $\mathsf{E}(B \wedge C, B^{\dagger} \wedge C^{\dagger})$. To finish, we need $\mathsf{E}(B^{\dagger} \wedge C^{\dagger}, (B \wedge C)^{\dagger})$, which follows from (4), (6), and (8), using (1) and (2).

Case iii. A is $B \to C$. Then $\mathsf{E}(A, B \wedge C)$ by (5), and we can proceed as in Case ii.

g. Since C is acting as a variable let's replace it by X; then $\varphi(X)$ can be viewed as a propositional function, φ: Wffs $\to$ Wffs. If we treat X as having length 1, then we can assign a length to $\varphi(X)$ in the usual way. To prove (g) we then induct on the length of $\varphi(X)$.

For length 1 it is immediate. Suppose it's true for length n, and $\varphi(X)$ has length $n+1$. Then $\varphi(X)$ is one of i. $\neg\psi(X)$, ii. $\psi(X) \to D$, iii. $D \to \psi(X)$, iv. $\psi(X) \to \gamma(X)$, v. $\psi(X) \wedge D$, vi. $D \wedge \psi(X)$, or vii. $\psi(X) \wedge \gamma(X)$, where D is a fixed wff and ψ, γ are propositional functions. For case (i) the proof is easy by (3). For (ii), if $\mathsf{E}(A,B)$ then by induction, $\mathsf{E}(\psi(A), \psi(B))$, so by (2) and (5) we are done. The other parts follow similarly. ■

We return to the proof of the theorem. Given E satisfying (1)–(8), define:

$\mathsf{D}(A, B)$ iff there is a C such that $\mathsf{E}(A, C)$ and B is a subformula of C

I'll prove that D is a dependence relation. So we can conclude that there is a set-assignment s such that $\mathsf{D}(A,B)$ iff $\mathsf{s}(A) \supseteq \mathsf{s}(B)$. I'll complete the proof by showing that for that s, $\mathsf{E}(A,B)$ iff $\mathsf{s}(A) = \mathsf{s}(B)$.

Lemma D is a dependence relation.

Proof: I'll use the characterization of dependence relations given in Lemma 1. First, D is transitive: suppose $\mathsf{D}(A,B)$ and $\mathsf{D}(B,C)$. Then there are $\varphi(B)$, $\psi(C)$ such that $\mathsf{E}(A, \varphi(B))$ and $\mathsf{E}(B, \psi(C))$. From the latter and part (g) of the previous lemma we get $\mathsf{E}(\varphi(B), \varphi(\psi(C)))$. So $\mathsf{E}(A, \varphi(\psi(C)))$, and so $\mathsf{D}(A,C)$.

D is reflexive as E is.

$\mathsf{D}(A, \neg A)$ and $\mathsf{D}(\neg A, A)$ follow from (3) via (1).

Finally we need to show that $\mathsf{D}(A, B \wedge C)$ iff $\mathsf{D}(A,B)$ and $\mathsf{D}(A,C)$, and then the comparable fact for $B \to C$ will follow similarly. So suppose $\mathsf{D}(A, B \wedge C)$. Then there is a φ such that $\mathsf{D}(A, \varphi(B \wedge C))$. As B is a subformula of $\varphi(B \wedge C)$ and so is C, we have $\mathsf{D}(A,B)$ and $\mathsf{D}(A,C)$. On the other hand, suppose $\mathsf{D}(A,B)$ and $\mathsf{D}(A,C)$. Then there are G, H such that B is a subformula of G, C is a subformula of H, and $\mathsf{E}(A,G)$ and $\mathsf{E}(A,H)$. So by (a) and (8), $\mathsf{E}(A, G \wedge H)$, and so by (f), $\mathsf{E}(A, (G \wedge H)^{\dagger})$. All the propositional variables of both B and C appear in $(G \wedge H)^{\dagger}$. By (4), (6), and (8), we can rearrange and reassociate the conjuncts in $(G \wedge H)^{\dagger}$ to get some K with $(B \wedge C)^{\dagger}$ a subformula of K and $\mathsf{E}(A,K)$. So $\mathsf{D}(A, (B \wedge C)^{\dagger})$, and hence by (f), $\mathsf{D}(A, B \wedge C)$. ■

To complete the proof of the theorem, let s be a union set-assignment such that $\mathsf{D}(A,B)$ iff $\mathsf{s}(A) \supseteq \mathsf{s}(B)$. It remains to show that $\mathsf{E}(A,B)$ iff $\mathsf{s}(A) = \mathsf{s}(B)$.

If $\mathsf{E}(A,B)$, then $\mathsf{D}(A,B)$ and $\mathsf{D}(B,A)$ by the symmetry of E. Hence $\mathsf{s}(A) = \mathsf{s}(B)$.

If $\mathsf{s}(A) = \mathsf{s}(B)$, then $\mathsf{D}(A,B)$ and $\mathsf{D}(B,A)$. Hence there are formulas G and

H such that A is a subformula of G, and B is a subformula of H, and $E(A,H)$ and $E(B,G)$. So by (f), $E(A,H^{\dagger})$ and $E(B,G^{\dagger})$, and as in the last part of the proof of the last lemma above, we can find some P, Q such that $E(A,B^{\dagger} \wedge P)$ and $E(B,A^{\dagger} \wedge Q)$. So by (f), $E(A^{\dagger},B^{\dagger} \wedge Q)$ and $E(B^{\dagger},A^{\dagger} \wedge Q)$, and thus by (7), $E(A^{\dagger},B^{\dagger})$. So by (f) again and (g), we have $E(A,B)$. ■

4. An axiom system for Eq

Eq, a logic of equality of referential contents
in $L(\neg, \rightarrow, \wedge)$

$$E(A, B) \equiv_{Def} A \rightarrow (B \rightarrow B)$$
$$A \supset B \equiv_{Def} \neg(A \wedge \neg B)$$

axiom schema

1. $E(A,A)$
2. $E(A,B) \rightarrow E(B,A)$
3. $(E(A,B) \wedge E(B,C)) \rightarrow (E(A,C) \wedge E(A,B))$
4. $(E(A,B) \wedge E(C,C)) \rightarrow E(A \wedge C, B \wedge C)$
5. $E(A, \neg A)$
6. $E(A \wedge B, B \wedge A)$
7. $E(A \rightarrow B, A \wedge B)$
8. $E(A \wedge (B \wedge C), (A \wedge B) \wedge C)$
9. $(E(A,B \wedge C) \wedge E(B,A \wedge D)) \rightarrow ((E(A,B) \wedge E(C,C)) \wedge E(D,D)$
10. $E(A, A \wedge A)$
11. $(A \wedge B) \rightarrow (B \wedge A)$
12. $(A \rightarrow B) \rightarrow E(A, B)$
13. $(A \wedge \neg B) \rightarrow \neg(A \rightarrow B)$
14. $(E(A, B) \wedge \neg A) \rightarrow (A \rightarrow B)$
15. $(E(A, B) \wedge B) \rightarrow (A \rightarrow B)$
16. $A \supset A$
17. $(E(A, A) \wedge B) \rightarrow (A \supset B)$
18. $[(A \supset (C \rightarrow D)) \wedge (A \supset C)] \rightarrow [(A \supset D) \wedge (A \supset C)]$
19. $(\neg A \supset \neg B) \wedge (\neg A \supset B) \rightarrow (A \wedge E(B, B))$

rules $\dfrac{A,\ A \rightarrow B}{B}$ $\qquad \dfrac{A,\ B}{A \wedge B}$ $\qquad \dfrac{A \wedge B}{B}$

Let $\vdash_{\mathbf{Eq}}$ stand for the consequence relation of this system.

I've chosen these axioms to make it as easy as possible to prove the *Strong Completeness Theorem* for **Eq** by modifying the proof for dependence logic in Section A.9.

***Theorem 15 Strong Completeness for* Eq** $\Gamma \vdash_{\mathbf{Eq}} A$ iff $\Gamma \vDash_{\mathbf{Eq}} A$

The Deduction Theorem fails for **Eq**, but the material implication form of it holds, just as for **D** (Section A.9).

We have that $\mathbf{Eq} \subseteq \mathbf{PC}$ because models in which every wff has the same content, and hence in which the connectives are evaluated classically, are allowed. But *exportation* fails for **Eq**, so the containment is strict, though *importation* and *contraposition* are now tautological.

Can we reduce the list of primitives of the language of **Eq**? As in classical logic, $\neg$ isn't definable from $\wedge$ and $\rightarrow$ since any wff built solely from p_0 using the latter connectives is evaluated as true in every model in which p_0 is assigned T. Also, $\rightarrow$ is not definable from $\neg$ and $\wedge$, since $\rightarrow$ isn't truth-functional. I do not know if $\wedge$ is definable from $\neg$ and $\rightarrow$.

The proof of Theorem 14 characterizing Eq-relations suggests that if quantification over propositions were allowed we could define a connective D(A, B) from $\neg$, $\rightarrow$, $\wedge$ such that in any **Eq**-model, $\langle v, s \rangle \vDash D(A,B)$ iff $s(A) \supseteq s(B)$. But without quantification that's not possible, as the connectives $\neg$, $\rightarrow$, $\wedge$ in **Eq**-models do not distinguish between $s(A) \supsetneq s(B)$ and $s(A) \cap s(B) = \varnothing$.

It's tempting to believe that $\mathbf{Eq} = \mathbf{D} \cap \mathbf{Dual\ D}$. This is so for wffs that involve only the connectives $\neg$ and $\wedge$ or are first degree implications (see the next section). But $\mathbf{Eq} \not\subseteq (\mathbf{D} \cap \mathbf{Dual\ D})$, since $[A \rightarrow (B \rightarrow B)] \rightarrow [B \rightarrow (A \rightarrow A)]$ is an **Eq**-tautology, but is not a tautology of **D** or **Dual D**.

Exercises for Section D

1. State explicitly the *Fully General Abstraction* for **Eq**.
2. Show that *importation* and *transposition* are valid in **Eq**.
3. For each scheme of classical tautologies listed in Chapter II.J.8 (pp. 60–62) decide whether it is a scheme of **Eq** tautologies. How will you interpret $\vee$?
4. Prove the *Strong Completeness Theorem* for **Eq**.
5. Prove for *, p. 189, $\Gamma \vdash_{\mathbf{Eq}} A$ iff $\Gamma^* \vdash_{\mathbf{D}} A^*$, and $\Gamma \vdash_{\mathbf{Eq}} A$ iff $\Gamma^* \vdash_{\mathbf{Dual\ D}} A^*$.
6. Formulate the dual semantics for **Eq**.
7. (Open) Determine whether $\wedge$ is definable from $\neg$ and $\rightarrow$ in **Eq** and whether $\{\neg, \wedge, \rightarrow\}$ is functionally complete.
8. (Open) State and prove a normal form theorem for **Eq**.
9. (Open) Determine whether $\mathbf{Eq} \supset (\mathbf{D} \cap \mathbf{Dual\ D})$.

10. (Open) Either reduce the axiomatization of **Eq** by deducing some of the axioms from the remaining ones, or show that the axiomatization is independent.
11. (Open) Give a a syntactic presentation of **Eq** by analytic tableaux. (Compare *Carnielli, 1987*.)

E. A Syntactic Comparison of D, Dual D, Eq, and S

Some logicians explain the notion of relevance in logic in terms of criteria for variable sharing, for instance, *Anderson and Belnap, 1975*, and *Kielkopf, 1977*. *Fine, 1979,* compares that syntactic view of relevance with a semantic one. For the logics we've studied we have the following criteria, which you can verify using the appropriate completeness theorems. Define $A \rightarrow B$ to be a *first degree implication* or *first degree entailment* (*f.d.e.*) if $\rightarrow$ does not appear in either A or in B.

Theorem 16 **a.** If $\vdash A \rightarrow B$, then for

subject matter relatedness logic, **S**
(*i*) there is at least one propositional variable that appears in both A and B

dependence logic, **D**
(*ii*) every propositional variable that appears in B appears in A

dual dependence logic, **Dual D**
(*iii*) every propositional variable that appears in A appears in B

equality of referential contents logic, **Eq**
(*iv*) the same propositional variables appear in both A and B

b. In **S** if (*i*) fails,
In **D** if (*ii*) fails,
In **Dual D** if (*iii*) fails
In **Eq** if (*iv*) fails
} then if $\vdash C \rightarrow (A \rightarrow B)$, there must be an occurrence of '$\rightarrow$' in C

c. For first degree entailments:

$\vdash A \rightarrow B$ iff $A \rightarrow B$ is a **PC**-tautology and {
for **S**, (*i*) holds
for **D**, (*ii*) holds
for **Dual D**, (*iii*) holds
for **Eq**, (*iv*) holds

For part (b) we need the provisos, since in each of the logics $\vdash A \rightarrow (A \rightarrow A)$.

The equivalences in (c) fail if $A \rightarrow B$ isn't a first degree entailment: consider either *importation* or *exportation* for each logic. However, as each of these logics is contained in **PC**, we have:

if $\vdash A \rightarrow B$, then $A \rightarrow B$ is a **PC**-tautology

Syntactic criteria, however, are too coarse to distinguish logics: the tautologies of nonsymmetric relatedness logic, **R**, satisfy the same criteria as those for **S**.

F. Content as Logical Consequences

Stimulated by the discussion in *Hanson, 1980,* Roger Maddux suggested looking at the content of a proposition as its consequences in classical logic in the framework of dependence-style semantics.

Let's suppose we start with some fixed assignment of propositions to the propositional variables of our formal language. Suppose we interpret $\neg$ and $\wedge$ classically. And suppose we take $A \rightarrow B$ to be tautological iff the **PC**-logical consequences of A contain those of B. Then we'll have **PC** again, as you can show using the *Deduction Theorem* for **PC**.

But this isn't what we want for implication: the truth-values of the propositions matter, too. Yet we don't want to consider truth-values in determining the content of the propositions, for that would trivialize content: every true proposition is a consequence of every proposition, and every false one implies every other in classical logic.

Rather, we begin with an assignment of propositions:

I

$$L(p_0, p_1, \ldots \neg, \rightarrow, \wedge)$$
$$\downarrow$$
$$\{p_0, p_1, \ldots, \text{complex propositions formed from these using } \neg, \rightarrow, \wedge, \vee\}$$

The content of p in this model is then the collection of semi-formal propositions that are **PC**-logical consequences of p. In this way we can apply the logic where we don't already know the truth-values of the propositions.

For $A \rightarrow B$ to be true we should then have that the consequences of A contain those of B, and not both A is true and B is false. This is dependent implication. If we take negation and conjunction to be classical we have a logic.

What happens here if we use the *Fully General Abstraction*? Define a (*fully abstract*) *dependent*-**PC** *model* to be:

II

$$L(\neg, \rightarrow, \wedge, \vee, p_0, p_1, \ldots)$$
$$\downarrow \quad v,\ s, \text{ and truth-tables}$$
$$\{T, F\}$$

Here v is any function, $v: PV \rightarrow \{T, F\}$. There is only one set assignment:

$$s(A) = \{ B : A \vdash_{\mathbf{PC}} B \}$$

And v is extended to all wffs by the dependence truth-conditions: $\neg$ and $\wedge$ are classical, and $\rightarrow$ is evaluated by:

$$v(A \rightarrow B) = T \text{ iff not (both } v(A) = T \text{ and } v(B) = F) \text{ and } s(A) \supseteq s(B)$$

(*Fully Abstract*) *Classically-Dependent Logic*, **DPC**, is the logic presented semantically by these models.

What about the intuition that the logic that takes content to be consequences in classical logic must be classical logic itself? For any wff A that has no occurrences of '$\to$' or is a first degree entailment, A is indeed a dependent-**PC** tautology iff A is a **PC**-tautology. But if p and q are distinct variables, then $q \to (p \to q)$ is a **PC**-tautology but not a dependent-**PC** one: $s(p) \not\supseteq s(q)$, so the consequent is false, yet we may have that q is true. Thus **PC** $\not\subseteq$ **DPC**. More strikingly, **DPC** $\not\subseteq$ **PC** as $\neg(p \to q)$ is a dependent-**PC** tautology if p and q are distinct. For there is only one set assignment for the fully abstract models of type **II**. If we define **DPC** for models of type **I**, however, we could have many different set assignments, one for each realization:

$$s(A) = \{ B: \text{(the realization of A)} \vdash_{\mathbf{PC}} \text{(the realization of B)} \}$$

In that case $\neg(p \to q)$ is not a tautology, for we could have the same proposition assigned to both p and q, so that $s(p)$ would be the same as $s(q)$. Here, then, is a logic in which abstractions do lose some of the original intuition.

Exercises for Section F

1. Show that if we interpret $\neg$ and $\wedge$ classically and take $A \to B$ to be tautological iff the **PC**-logical consequences of A contain those of B, then we have **PC** in $L(\neg, \to, \wedge)$.
2. (Open) Give a decision procedure for **DPC** for models of type **I** and a decision procedure for models of type **II**.
3. (Open) Determine the difference between **DPC** for models of type **I** and models of type **II**.
4. (Open) Axiomatize **DPC** for models of type **I**. Axiomatize **DPC** for models of type **II**.
5. (Open) Maddux has proposed examining the sequence of logics $\mathbf{DPC}_n$ where $\mathbf{DPC}_1 = \mathbf{DPC}$ and $\mathbf{DPC}_{n+1}$ is defined as was **DPC** except that $s(A) = \{B : A \vDash_{\mathbf{DPC}_n} B \}$. Is $\mathbf{DPC}_{n+1} \subseteq \mathbf{DPC}_n$? If so, what's $\bigcap_n \mathbf{DPC}_n$?

VI Modal Logics
– S5, K, T, B, S4, QT, MSI, ML, S4Grz, G, G* –

In reasoning we sometimes consider whether a proposition is not just true or false, but whether it is possible or impossible, necessary on not necessary. Certain *modal* words signal that these aspects of a proposition are important: 'could', 'would', 'must', and others. In this chapter we will look at logics that take account of these modalities, considering semantics based on the idea of a possible world. The set-assignment semantics I will give raise the question whether modal logics are based on a notion of connection of meanings.

Another approach to modeling modalities uses many-valued logics, which I present in Chapter VIII. *Prior, 1955,* discusses that approach and surveys the history of the notions of possibility and necessity, for which you can also consult *Aune, 1968.*

For arguments for and against the (metaphysical) existence of possible worlds see *David Lewis, 1973,* and *Kripke, 1972* , both of which I have drawn on in the following discussions.

There are a number of surveys of modal logics and more advanced mathematical methods for dealing with them: *Chellas, 1980, Hughes and Cresswell, 1984, Segerberg, 1971,* and *Lemmon, 1977.* The last two have historical outlines of the modern development of classical modal logic, as does *Bull and Segerberg, 1984.* I have often followed *Boolos, 1979,* for the technical development here.

A. Implication, Possibility, and Necessity

1. Strict implication vs. material implication

In 1912 C. I. Lewis initiated in modern formal logic a debate that dates back to the Greeks: which implications are sound? The logic of *Principia Mathematica*

(*Whitehead and Russell, 1910–1913*) formalized Philonian implication:

> Philo said that the conditional is true whenever it is not the case that its antecedent is true and its consequent false.
>
> *Sextus Empiricus, Against the Mathematicians,* VIII, 112
> translated in *Mates, 1953*, p. 97

Lewis called this analysis *material implication,* and said that it was inadequate for ordinary reasoning: 'If roses are red, then sugar is sweet' isn't true, for there is nothing in the fact that roses are red that precludes sugar being not sweet. It could have been that roses are red and sugar is not sweet. Just as Chrysippus long ago, he set out what he called *strict implication*:

> The strict implication, P ⥽ Q, means 'It is impossible that P be true and Q false' or 'P is inconsistent with the denial of Q.'
>
> *C. I. Lewis, 1918*, p. 332–333

> And those who introduce connection or coherence say that a conditional holds whenever the denial of its consequent is incompatible with its antecedent.
>
> *Sextus Empiricus, Outlines of Pyrrhonism,* II, 11
> translated in *Mates, 1953*, p. 109

Though Lewis used different symbols, $\supset$ for the material implication and $\rightarrow$ or ⥽ for the strict implication, these are both formalizations of the ordinary language 'if . . . then . . . '. The arguments about which implications are sound are real debates: one single common notion is being analyzed, refined, in different ways. Therefore, the formal language I'll use for modal logics is $L(\neg, \rightarrow, \wedge, p_0, p_1, \ldots)$, at least until Section B.2. The formalization of 'or' is normally given as in classical logic in terms of $\neg$ and $\wedge$.

2. Possible worlds

The basic semantic idea of modal logic is that though conjunction and negation may be evaluated classically, for the conditional we take:

(1) $A \rightarrow B$ is true iff it is not possible that A could be true and B false

Is 'If Caesar crossed the Rubicon, then Shakespeare wrote *Macbeth*' true? By this criterion we want to say 'no', for we believe that Shakespeare could have failed to write *Macbeth*. To analyze an implication we survey the possibilities, the ways things could have been.

We can suggestively say that a way things could have been is a *possible world.* That way of talking comes from Leibniz (see *Mates, 1986*) and was imported into formal semantics for modal logics by *Kripke, 1959.*

A possible world is something we imagine could be the case. It may be a metaphysical "fact" that a certain possible world "exists," but for us to reason together all that is needed is that we agree in general how we will reason about our descriptions of the ways things could be. Then we may withhold judgment and hypothesize that a particular description describes a case that is possible or one that is not possible, and consider the consequences of each.

In particular we must agree on how we are to reason about any one particular description. In any possible world when we invoke (1) we are interested only in the truth-values of the propositions that specify it, so it seems that classical logic is appropriate. Hence:

(2) We agree to use classical logic in reasoning about any particular description of the way things could have been

A convenient abbreviation of (2) is:

In every possible world classical logic holds

(This way of saying things sounds suddenly full of metaphysical import, which is a danger of using the terminology of 'possible worlds'.)

When we describe a way things could have been we usually specify the truth-values of only a few propositions, namely the ones in the implication we're analyzing. However, we may wish to specify infinitely many, for example, all instances of all the physical laws we know. In either case, to employ formal methods we must first be given an assignment of propositions $p_0, p_1, \ldots$ to the propositional variables of the formal language. Then we can formally identify a possible world with an assignment of truth-values to the p_i's. That is, since we're using classical logic within the world, a possible world is a model of classical logic. Looked at syntactically, this is equivalent to saying that relative to the modes of reasoning we've adopted (**PC**) a possible world is a complete and consistent set of sentences.

Now we can be fully precise about what we will take to be a possible world.

Possible Worlds Abstraction Any model of classical logic is a possible world.

With this agreement we have a standard for our imaginings. A possible world is no longer (just) a psychologically acceptable description. Indeed, a description of the way things could be which you offer might be psychologically unacceptable to me, such as conceiving of two lines through a point both parallel to a third. Yet if you can establish that the description you put forward is indeed a model of classical logic, that is a (**PC**-) consistent and complete set of sentences, then by our agreement I must accept that it is a possible world.

Finally, we restrict our attention to just this one new aspect of propositions–possibility–in addition to truth-values and forms:

The Classical Modal Logic Abstraction The only properties of a proposition that matter to (this) logic are its form, its truth-value, and whether it is true or false in any given possible world.

3. Necessity

In our new terminology to say that A is possible is to say that there is some possible world of which A is true. Possibility, then, amounts to a sort of existential quantification over possible worlds.

To say that a proposition is *necessary* is to say that it couldn't be false. That is, it is not possible that it is false: in every possible world, in any way that we could imagine things to be, it's true. So necessity is like a universal quantification over possible worlds. Because of our agreement to use classical logic in every description, we're committed to all classical tautologies being necessary truths. For instance, we may not (or perhaps cannot) imagine that the law of excluded middle fails.

Now we can restate our analysis of implications:

(3) $A \rightarrow B$ is true iff it is not the case that there is some possible world in which A is true and B is false

Abbreviating $\neg(A \wedge \neg B)$ as $A \supset B$, which is material implication, we have:

(4) $A \rightarrow B$ is true iff $A \supset B$ is necessary

4. Different notions of necessity: accessibility relations

The notions of possibility and necessity we've been discussing so far have been determined by requiring that a particular set of sentences, namely the classical logical tautologies, hold in every possible world: the logical laws are necessary. This is *logical necessity,* or more properly, *classical logical necessity.* Other notions of necessity and possibility are important in philosophy and science (see *Aune, 1967)* and we can consider those, too. We won't, however, change our agreement that the classical logical laws are necessary.

Time dependent necessity corresponds approximately to how we reason about future events. At this time I can wonder who will be President of the U.S.A. in 2011, and consider various possibilities. Suppose I assume that it is Ms. Clinton. Then I can consider, relative to that assumption, that is relative to the possible world in which the year 2011 is as I've stipulated, whether Mr. Gore will be President in 2015. All the possible future worlds I imagine will be exactly like the one we live in up to this moment, that is, the true propositions describing what has happened up to that time are the same. All the possible worlds subsequent to the one in which Ms. Clinton is President in 2011 will be exactly like that one up to 2011.

Time dependent necessity or *inevitability at time t* is a notion of necessity *relative to the possible world under consideration.* Right now 'Mr. Gore is President in 2011' is possible. Relative to the world in which Ms. Clinton is President in 2011, call it w, 'Mr. Gore is not President in 2011' is necessary. We might say that a world in which 'Mr. Gore is President in 2011' is true is *accessible* to the present world but not accessible to w. Accessibility is a relation between possible worlds, and for time dependent necessity, world z is accessible relative to world w if w is specified to be at time t and z is exactly like w up to time t. But what does 'exactly like' mean here? To begin with, both worlds must satisfy the classical laws of logic. And in addition any other facts we've specified about w at time t must also hold in world z. We may implicitly assume, say, that world w satisfies the same physical laws as our world, but we must explicitly specify the sentences that are to be true in w at time t if we are to use our formal models. We may also assume that a stipulation of the way things could be in 2011 presupposes a history of how things were before then, but if we do not explicitly stipulate that history we cannot take it into account in our formal reasoning. One thing that must be specified for time dependent analyses is the time at which the world occurs.

What can we say about this accessibility relation for time dependent necessity? Certainly it is reflexive and transitive. It's also anti-symmetric: if z is accessible to w, then z will occur later than w and hence w cannot be exactly like z at that later time unless $w = z$.

For this notion of necessity let's consider how we will evaluate the truth-value of $A \rightarrow B$. Notating 'z is accessible to w' as wRz, we have:

(5) $A \rightarrow B$ is true at world w iff $A \supset B$ is necessary relative to w
iff at every possible world z such that wRz, $A \supset B$ is true

And then:

(6) $A \rightarrow B$ is true iff $A \rightarrow B$ is true at every possible world w

Recall that I said in passing that we might require world z to satisfy the same physical laws as world w. That's *physical necessity* and corresponds roughly to how we might reason in science. I might postulate a physical law A and consider two worlds w and z satisfying all the laws we now know and in w A is true, in z it's false. So w and z are accessible to our present world, yet inaccessible to each other. Or in a science fiction vein, I might consider a world w in which gravitational attraction is proportional to the inverse cubed of the distance between objects. With respect to physical necessity, neither that world nor any world accessible to it is accessible to our present one. For physical necessity a world z is accessible to world w, that is, wRz, if all the physical laws specified to hold in w hold in z. To evaluate an implication with respect to physical necessity we would again use (5) and (6).

Exercises for Section A

1. a. What is Philonian implication?
   b. Compare Philonian implication to the interpretation of the conditional ascribed to Chrysippus.
   c. Compare the strict interpretation of the conditional to the possibility version of semantic consequence, p. 30.
   d. Why should we use $\rightarrow$ for the formalization of strict implication and not a new symbol such as $-\!\!3$?

2. Compare strict implication to Diodorus' analysis of conditionals:

   > But Diodorus says that a conditional is true whenever it neither ever was nor is possible for the antecedent to be true and the consequent false, which is incompatible with Philo's thesis. For, according to Philo, such a conditional as 'If it is day, then I am conversing' is true when it is day and I am conversing, since in that case its antecedent, 'It is day,' is true and its consequent, 'I am conversing,' is true; but according to Diodorus it is false. For it is possible for its antecedent, 'It is day,' to be true and its consequent, 'I am conversing,' to be false at some time, namely, after I have become quiet.
   >
   > *Sextus Empiricus, Against the Mathematicians*, VIII, 113
   > translated in *Mates, 1953*, p. 98

3. a. *C. I. Lewis, 1932,* p. 122, bases a system of modal logic on the definition:

   'p implies q' will be synonymous with 'q is deducible from p'

   Explain why this is not suitable as an interpretation of conditionals because of use-mention confusions. (Hint: How could we interpet an iteration of a conditional?)
   b. Under what conditions would this interpretation be valid in the metalogic? (Hint: When can we identify $\rightarrow$ and $\vdash$?)

4. Argue informally whether each of the following is true or is false according to the strict interpretation of implication.
   a. If roses are red, then roses smell good
   b. If strawberries are red, then some colorblind people cannot see strawberries among their leaves
   c. If $2+2=4$, then $4+4=8$
   d. If roses are red, then roses are red
   e. If Juney is a dog, then dogs bark
   f. If Ralph is a bachelor, then Ralph is a man
   g. If Juney barks and Juney is a dog, then Juney is a dog
   h. If it's not the case that Juney is not a dog, then Juney is a dog
   j. If if Ralph is a dog, then Ralph barks, then if Ralph does not bark,

then Ralph is not a dog

k. If Caesar crossed the Rubicon, then Caesar was assassinated

l. If $2+2=4$, then this sentence is not true

5. a. What is a possible world?
   b. In what sense do possible worlds exist?
   c. If we discover little green men living on mars, would the life there be a possible world different from our own?
   d. What is the *Possible Worlds Abstraction*? *The Classical Modal Logic Abstraction*?

6. Can we have a possible world in which both of 'Ralph is a bachelor' and 'Ralph is married' are true? (See Example 3, p. 173.)

7. a. What is an accessibility relation?
   b. Given a notion of necessity with an associated accessibility relation, what is the interpretation of $\rightarrow$? Of $\neg$ and $\wedge$?
   c. What accessibility relation would be appropriate for logical necessity?

8. a. Choose any science fiction story and decide whether the possible world described in it is accessible to our own.
   b. Give an example of a possible world not accessible to our own under the relation of physical necessity. What makes you think your description is consistent?

9. a. Give a description of the world at the time you're doing this exercise.
   b. Give descriptions of two possible worlds that are not accessible to the world described in (a) and that are not accessible to each other under the relation of time dependent necessity.

B. The General Form of Possible-World Semantics

1. The formal framework

Let us now formalize the notions we've been discussing. Recall that our formal language is $L(\neg, \rightarrow, \wedge, p_0, p_1, \ldots)$. And we set:

$$A \supset B \equiv_{Def} \neg(A \wedge \neg B) \qquad A \vee B \equiv_{Def} \neg(\neg A \wedge \neg B)$$

Suppose we have an assignment of propositions $p_0, p_1, \ldots$ to $p_1, p_2, \ldots$. We said that a possible world is a model of classical logic. So it is determined by the propositions true in it. That is, we can characterize it as a collection of p_i's, namely the ones true in it, using their temporary names $p_1, p_2, \ldots$.

Recall that $PV = \{p_1, p_2, \ldots\}$ and 'Sub' is an abbreviation for 'the subsets of'. We can now say that a *model* is $\langle W, R, e \rangle$ where W is a *nonempty* set whose elements are called *possible worlds*, e is an *evaluation*, $e: W \rightarrow \text{Sub PV}$, and R is a binary relation on W called the *accessibility relation between possible worlds*. We can think of the elements w of W as markers, names of possible worlds. The evaluation e says which propositions (via their temporary names in PV) are true in w, namely $\mathsf{e}(w)$. We inductively define $w \vDash A$, read 'A is true at world w', or 'A is true in world w', or 'w validates A', where $\nvDash$ means '$\vDash$ does not hold' and is read as 'is false at' or 'does not validate':

An inductive definition of $\vDash$

$w \vDash p$ iff $p \in \mathsf{e}(w)$

$w \vDash A \wedge B$ iff $w \vDash A$ and $w \vDash B$

$w \vDash \neg A$ iff $w \nvDash A$

$w \vDash A \to B$ iff for all z such that $w\mathsf{R}\mathsf{z}$, not both $\mathsf{z} \vDash A$ and $\mathsf{z} \nvDash B$

The first three clauses state that a formal possible world is a model of classical logic. An equivalent formulation of the last condition is:

$w \vDash A \to B$ iff for all z such that $w\mathsf{R}\mathsf{z}$, $\mathsf{z} \vDash A \supset B$

We can define $\langle \mathsf{W}, \mathsf{R}, \mathsf{e}, w \rangle$ to be a *model with designated world* w. Then $\langle \mathsf{W}, \mathsf{R}, \mathsf{e}, w \rangle \vDash A$ means $w \vDash A$. This gives an analysis of implication based on necessity and possibility relative to a particular world w. Finally:

A is true in model $\langle \mathsf{W}, \mathsf{R}, \mathsf{e} \rangle$, written $\langle \mathsf{W}, \mathsf{R}, \mathsf{e} \rangle \vDash A$, iff
for all $w \in \mathsf{W}$, $\langle \mathsf{W}, \mathsf{R}, \mathsf{e}, w \rangle \vDash A$

Note: In the literature models are often presented using $\mathsf{e}^*: \mathrm{PV} \to \mathrm{Sub}\, \mathsf{W}$ where the first step of the inductive definition of $\vDash$ is: $w \vDash p$ iff $w \in \mathsf{e}^*(p)$. This is mathematically equivalent to the definition above: given $\mathsf{e}: \mathsf{W} \to \mathrm{Sub\, PV}$, define $\mathsf{e}^*: \mathrm{PV} \to \mathrm{Sub}\, \mathsf{W}$ via $\mathsf{e}^*(p) = \{ w : p \in \mathsf{e}(w) \}$; then $\langle \mathsf{W}, \mathsf{R}, \mathsf{e}^* \rangle \vDash A$ iff $\langle \mathsf{W}, \mathsf{R}, \mathsf{e} \rangle \vDash A$. The reverse translation I'll leave to you. This alternate presentation of models identifies a proposition with the worlds in which it is true; we'll return to it in Section G.

If we are concerned with a particular notion of necessity, we may wish to model it with a particular class of accessibility relations. Call a pair $\langle \mathsf{W}, \mathsf{R} \rangle$ a *frame*, and say that A is *valid in* $\langle \mathsf{W}, \mathsf{R} \rangle$, written $\langle \mathsf{W}, \mathsf{R} \rangle \vDash A$, iff for all e, $\langle \mathsf{W}, \mathsf{R}, \mathsf{e} \rangle \vDash A$. If we specify a class of frames, F, then:

(7) A is *valid* or *tautological* relative to the notion of necessity corresponding to a class of frames F iff for all $\langle \mathsf{W}, \mathsf{R} \rangle \in \mathsf{F}$ and all e, $\langle \mathsf{W}, \mathsf{R}, \mathsf{e} \rangle \vDash A$

We usually say that a frame or model has a property, for example reflexivity, if R does. However, when we say that $\langle \mathsf{W}, \mathsf{R} \rangle$ is *finite* we mean that W is finite. We say that a wff is *valid* or *tautological* if it is valid for the class of all frames.

It is standard to assume a *Fully General Abstraction* for possible-world semantics.

The Fully General Abstraction for (Classical) Possible-World Semantics
Any set W together with any relation R on W and any function $\mathbf{e}: W \to \text{Sub PV}$ comprise a model $\langle W, R, \mathbf{e} \rangle$.

In general it is possible to restrict W to be countable, but there are cases where that would make the mathematics more difficult. However, we may restrict the cardinality of W to be no greater than that of the cardinality of all subsets of the natural numbers.

For a particular notion of necessity we can and do assume a *Fully General Abstraction* that states that any relation R on W that satisfies the appropriate structural conditions for the notion of necessity under consideration is suitable to use in a model. For example, for a logic of time dependent necessity as described in Section A.3 we might take the class of models to be all those $\langle W, R, \mathbf{e} \rangle$ such that R is reflexive, transitive, and anti-symmetric.

Example I'll show that $\neg p \to (p \to q)$ is not valid by exhibiting a model and a world in that model in which it is not true.

Let $W = \{w, z, y\}$ where wRz and zRy, but no other elements are related. Let p be false in z and true in y, and q false in y; we don't need to worry about any other assignments, as we'll see in Lemma 1. We can use a diagram, as below, to present (parts of) a possible-world model: when the accessibility relation between worlds holds we draw an arrow—no arrow means the relation does not hold—and we list below each world the truth-values of various propositions in which we are interested.

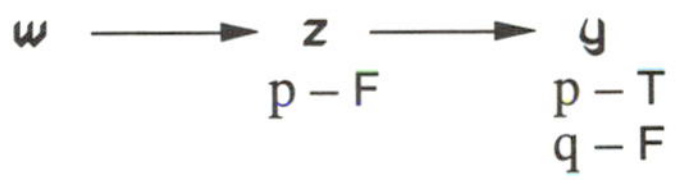

w	→	z	→	y
		p – F		p – T
				q – F

Then $\neg p \to (p \to q)$ is true in w iff $\neg p \supset (p \to q)$ is true in every world related to w. The only world related to w is z, and there p is false and $\neg p$ is true. Thus $p \to q$ must be true in z for the whole wff to be true. But zRy, and in y, $p \supset q$ is false. So $p \to q$ is false in z. Hence $\neg p \to (p \to q)$ is false in w and so is not valid.

I'll leave to you to investigate whether the other "paradoxes" of material and strict implication are tautologies.

Recall that I remarked at the end of Section A.2 that when we describe a way things could have been we usually specify the truth-values of only a few propositions, namely the ones in the implication we're analyzing. That's what we did in the formal possible-world semantics in the example above. I'll now show why that's justified. In doing so, I'll also demonstrate that the only worlds involved in establishing the truth or falsity of a proposition at a world w are those that we can reach in a finite number of steps from that world.

Given $\langle W, R \rangle$, define *the transitive closure of* R to be R^*, the smallest transitive relation that contains R. That is, $w R^* z$ iff there is a sequence $w = y_1, y_2, \ldots, y_n = z$ such that for each i, $y_i R\, y_{i+1}$.

Lemma 1 Let $\langle W, R, e, w \rangle$ be a model with designated world w and $\langle Z, R, e, w \rangle$ that model restricted to $Z = \{w\} \cup \{z : z \in W \text{ and } w R^* z\}$ (that is, in $\langle Z, R, e, w \rangle$ the accessibility relation R and evaluation e are restricted to Z). Then:

a. For all A, $\langle W, R, e, w \rangle \vDash A$ iff $\langle Z, R, e, w \rangle \vDash A$

b. For any e´, if for all p in A and all $z \in Z$, $p \in e'(z)$ iff $p \in e(z)$, then $\langle W, R, e, w \rangle \vDash A$ iff $\langle Z, R, e', w \rangle \vDash A$

Proof: I'll show by induction on the length of A that for all $y \in Z$ and all A, $\langle W, R, e, y \rangle \vDash A$ iff $\langle Z, R, e', y \rangle \vDash A$.

It is immediate if the length of A is 1. Suppose the lemma true for all wffs of length $\leq n$ and that A has length $n+1$. If A is $\neg B$ or $B \wedge C$ the proof is straightforward. So suppose A is $B \rightarrow C$. In that case, $\langle W, R, e, y \rangle \vDash B \rightarrow C$ iff for all $z \in W$ such that yRz, $\langle W, R, e, z \rangle \vDash B \supset C$. If $y \in Z$ and yRz, then $w R^* z$, so $z \in Z$. So we have by induction:

$\langle W, R, e, y \rangle \vDash B \rightarrow C$ iff for all $z \in Z$ such that yRz, $\langle W, R, e, z \rangle \nvDash B$ or $\langle W, R, e, z \rangle \vDash C$

iff for all $z \in Z$ such that yRz, $\langle Z, R, e', z \rangle \nvDash B$ or $\langle Z, R, e', z \rangle \vDash C$

iff for all $z \in Z$ such that yRz, $\langle Z, R, e', y \rangle \vDash B \rightarrow C$ ■

2. Possibility and necessity in the formal language

In the rest of this chapter we will, for the most part, consider logics that correspond to some easily specifiable class of frames. In order to characterize syntactically the tautologies relative to a particular class of frames, it is useful to have an expression in our formal language corresponding to the metalogical notion of necessity, that is, true in all accessible possible worlds. Set:

$$\Box A \equiv_{\text{Def}} \neg A \rightarrow A$$

Then:

(8) $w \vDash \Box A$ iff for all z, if wRz then $z \vDash A$

We usually read $\Box A$ as 'It is necessarily the case that A', or for short, 'Necessarily A', or 'It is necessary that A'. Now we have:

$w \vDash A \rightarrow B$ iff $w \vDash \Box(A \supset B)$

This corresponds to (4) above.

Instead of this definition of $\Box A$, we could use $(A \to A) \to A$. Nothing depends on which of these definitions we choose in what follows: each logic in this chapter contains all instances of $[(A \to A) \to A)] \supset (\neg A \to A)$ and $(\neg A \to A) \supset [(A \to A) \to A]$.

Examples

1. Consider the evaluation of $\Box p \supset p$ using (8). This is true at a world w iff not both $\Box p$ is true and p is false at w. But that can happen:

$$\begin{array}{ccc} w & \longrightarrow & z \\ p - F & & p - T \end{array}$$

Here $\Box p$ is true iff p is true in every world accessible to w, and w need not be accessible to itself. So in this model, $\Box p \supset p$ is false. However, if w is accessible to itself, then $\Box p \supset p$ is true at w. So $\Box p \supset p$ is not valid, though it holds in any model in which R is reflexive, that is, in the class of all reflexive frames.

2. Consider the evaluation of $\Box(p \wedge q) \supset (\Box p \wedge \Box q)$. Suppose $\Box(p \wedge q)$ is true at a world w. Then in every world z related to w, $p \wedge q$ is true. So in every such z, both p is true and q is true. Hence, $\Box p$ is true at w and $\Box q$ is true at w, and so $\Box(p \wedge q) \supset (\Box p \wedge \Box q)$ is true at w. Thus, $\Box(p \wedge q) \supset (\Box p \wedge \Box q)$ is a tautology.

3. The evaluation of $((p \to q) \wedge (q \to r)) \to (p \to r)$ is more complicated, involving iterated conditionals. For $((p \to q) \wedge (q \to r)) \to (p \to r)$ to be true at world w, we must have that for every z such that $w\mathsf{R}z$, if $p \to q$ and $q \to r$ are true at z, then $p \to r$ is true at z. That is if and only if at every world x such that $z\mathsf{R}x$, if $p \supset q$ and $q \supset r$ are true at x, then $p \supset r$ is true at x. But the latter is always the case, since we use classical logic to evaluate those wffs. So, indeed, the wff is a tautology.

$$\begin{array}{ccccc} w & \longrightarrow & z & \longrightarrow & x \\ & & p \to q - T & & p \supset q - T \\ & & q \to r - T & & q \supset r - T \\ & & \text{then} & & \text{then} \\ & & p \to r - T & & p \supset r - T \end{array}$$

Given our definition of $\Box A$ we can define:

$\Diamond A \equiv_{\mathrm{Def}} \neg \Box \neg A$

Then:

$w \vDash \Diamond A$ iff there is some z such that $w\mathsf{R}z$ and $z \vDash A$

Thus $\Diamond A$ true at a world corresponds to A being possible at that world. We read $\Diamond A$ as 'It is possibly the case that A', or, 'Possibly A', or 'It is possible that A'.

Note that in every $\langle W, R, e, w \rangle$:

$w \vDash \Box A$ iff $w \vDash \neg\Diamond\neg A$

This equivalence is the formalization of how we first described what it meant for A to be necessary (p. 200). Alternatively, we could set:

$\Diamond A \equiv_{Def} \neg(A \to \neg A)$

And then:

$\Box A \equiv_{Def} \neg\Diamond\neg A$

Historically, syntactic characterizations of modal logics preceded the semantic analysis of Kripke, and almost all the syntactic characterizations used $\Box$ or $\Diamond$ as a primitive unary connective (see Section E below). But are we justified in regarding 'It is necessary that . . .' and 'It is possible that . . .' as propositional connectives?

Viewing $\Box$ or $\Diamond$ as a connective appears to me to be a use-mention confusion. When we form '$\Box(2+2=4)$' from '$2+2=4$' we are mentioning '$2+2=4$', not using it. The natural reading of '$\Box(2+2=4)$', it seems to me, is 'The proposition '$2+2=4$' is necessary.' What's necessary? Why *that* is necessary, pointing to the proposition '$2+2=4$'. The phrases 'It is necessary that . . .' and 'Necessarily . . .' are indeed used by philosophers, but are best understood, I think, as abbreviations of 'The proposition . . . is necessary.'

What sense, for example, can we make of:

$\Box$ ($\Box$(Ralph is a dog)) $\supset$ $\Box$(Ralph is a dog)

$\Diamond$(Ralph is a dog) $\supset$ $\Diamond$($\Box$($\Diamond$(Ralph is a dog)))

They can be unpacked in terms of the formal semantics, but we have no anterior understanding of them as formalizations of English propositions, because they do not formalize English propositions; they involve metalevel and metalevel again. It's not just that we lose intuition if the proposition becomes too complex, as you might argue for '$\neg\neg\neg\neg\neg\neg\neg\neg\neg\neg\neg$(Ralph is a dog)'. We know that when we use $\neg$, $\to$, $\wedge$, $\vee$ we are forming new propositions by using the propositions to which they are applied. On the face of it, we are not using a proposition but mentioning it when we preface it with 'Necessarily'.

Compare:

Ralph is a pet $\to$ (($\neg$(Ralph meows) $\wedge$ Ralph barks) $\to$ Ralph is a dog)

Understanding $\to$ as a formalization of 'if . . . then . . .' the example makes sense, and we can ask whether it is true or false, whether we should interpret $\to$ as strict or Philonian implication. This is the kind of proposition we have been dealing with in the previous chapters. Modal logics are justified as logics based on particular abstractions of 'if . . . then . . .'.

Nonetheless, some logicians maintain that there is no use-mention confusion involved in taking $\Box$ and $\Diamond$ as connectives. *Haack, 1974,* Chapter 10, gives an introduction to the debates on this question; *Copeland, 1978,* reviews the arguments and shows that if the metalevel distinctions are taken seriously and an infinite hierarchy of languages is produced, then the confusions are comparable to collapsing these languages, which in some sense is technically harmless. Because the use of $\Box$ and $\Diamond$ has become common in the presentations of modal logics, I will discuss the various logics below in the language $L(\neg, \wedge, \Box)$ as well as in $L(\neg, \rightarrow, \wedge)$. In informal abbreviations $\Box$ will bind more strongly than $\wedge$, $\rightarrow$, or $\vee$.

In this chapter you are free to view $\Box$ and $\Diamond$ as sentence connectives if you wish, though I will use them as syntactical abbreviations for metalogical investigations. That is, I use $\Box$ as I did the abbreviation R in relatedness logic or D in dependence logic (see p. 177). In relatedness logic 'R(Ralph is a dog, $2+2=4$)' is true in a model iff 'Ralph is a dog' is related in subject matter to '$2+2=4$'. We are talking about the propositions 'Ralph is a dog' and '$2+2=4$', not using them; the abbreviations allow us to impose restrictions on the semantics via the syntax, but they make no sense as propositions of the semi-formal language. We cannot iterate R: what would we mean by 'R(R($2+2=4$, Ralph is a dog), Juney barks)'? We can unpack this in terms of an examination of the semantics, but not as an intelligible proposition that uses 'Juney barks', and so it is that I view the use of $\Box$.

I will expand on this discussion in Examples 16–18 of Section D below.

Exercises for Section B

1. a. What is a model of classical modal logic?
   b. Why do we identify a possible world with $\mathsf{e}(w)$?
   c. Why do we take W to be nonempty?
   d. Suppose we take $\mathsf{e}': PV \rightarrow \mathsf{W}$. What then, informally, is $\mathsf{e}'(p_i)$?

2. a. Prove: $w \models A \rightarrow B$ iff for all z such that $w\mathsf{R}z$, $z \models A \supset B$.
   b. Prove: $w \models A \rightarrow B$ iff $w \models \neg\Diamond(A \wedge \neg B)$

3. There are six classifications of propositions in terms of necessity and possibility. In the text we discussed possible propositions and necessary propositions. Formalize:
   a. 'Ralph is a dog' is *contingent* (i.e., neither necessary nor not necessary)
   b. 'Ralph is a dog' is *not necessary*
   c. 'Ralph is a dog' is *impossible*
   d. 'Ralph is a dog' is *noncontingent*
   e. 'Ralph is a dog' and 'George is a duck' are *compossible* (i.e., both can be true together)

4. Distinguish the following, giving examples.
   a. A is true at designated world w
   b. A is true in model $\langle W, R, e \rangle$
   c. A is valid in frame $\langle W, R \rangle$
   d. A is valid in a class of frames

5. a. What does it mean to say that a frame is reflexive? finite?
   b. Show that the frame of universal relations characterizes logical necessity.
   c. Show that the frame of equivalence relations characterizes logical necessity.

6. What is the *Fully General Abstraction for Possible-World Semantics*?

7. Determine whether the following are valid for the class of all possible-world models:
   a. $q \rightarrow (p \rightarrow q)$ — a paradox of material implication
   b. $p \rightarrow (q \rightarrow q)$ — a paradox of strict implication
   c. $(p \wedge \neg p) \rightarrow q$ — a paradox of strict implication
   d. $\neg\neg p \rightarrow p$
   e. $p \vee \neg p$
   f. $((p \rightarrow q) \rightarrow p) \rightarrow p$
   g. $(\neg p \rightarrow p) \supset p$
   h. $(\neg p \rightarrow p) \rightarrow p$
   j $(p \rightarrow q) \supset (\neg q \rightarrow \neg p)$
   k. $(p \rightarrow q) \rightarrow (\neg q \rightarrow \neg p)$

8. a. What is the definition of $\Box A$?
   b. Prove: $w \vDash \Box A$ iff for all z, if wRz then $z \vDash A$
   c. What is the definition of $\Diamond A$?
   d. Prove: $w \vDash \Diamond A$ iff for some z, if wRz then $z \vDash A$
   e. Show that for logical necessity, $w \vDash A \rightarrow (A \rightarrow A)$ iff $w \vDash \neg A \rightarrow A$

9. What is the transitive closure of a relation?

10. Determine whether the following are valid for the class of all possible-world models. If one is not valid, try to find a class of frames in which it is valid.
   a. $\Box(p \supset q) \supset (p \rightarrow q)$
   b. $(p \rightarrow q) \supset \Box(p \supset q)$
   c. $\Box(p \supset q) \rightarrow (p \rightarrow q)$
   d. $(p \rightarrow q) \rightarrow \Box(p \supset q)$
   e. $\Box p \supset p$
   f. $p \supset \Box p$
   g. $p \supset \Diamond p$

h. $p \rightarrow \Diamond p$

j. $\Box(\Box p) \supset \Box p$

k. $\Diamond p \supset \Box p$

l. $\Box\Diamond p \supset \Diamond p$

m. $\neg\Box\neg p \supset \Diamond p$

n. $\neg\Diamond\neg p \supset \Box p$

o. $\Diamond(p \wedge q) \supset (\Diamond p \wedge \Diamond q)$

p. $\Diamond p \supset \Box\Diamond p$

11. Argue whether it would be or would not be a use-mention confusion to formalize the following as propositional connectives.

a. 'It is obvious that . . .'

b. 'Obviously . . .'

c. 'It is required that . . .'

d. 'Juney knows that . . .'

e. 'Richard wonders whether . . .'

f. 'Ralph believes that . . .'

g. 'Everyone understands that . . .'

h. 'It is necessary that . . .'

j. 'It is possible that . . .'

k. 'Necessarily . . .'

m. 'Possibly . . .'

n. 'In the future . . .'

o. 'Sometime in the past . . .'

p. 'It is provable that . . .'

C. Semantic Presentations

Most modal logics were originally presented syntactically; only after Kripke's introduction of possible-world semantics were semantic characterizations given. However, I shall present each of the most-studied modal logics semantically first, trying to see what motivation in terms of meanings we can give each. Technical results below that are stated for formulas involving $\Box$ or $\Diamond$ apply to both $L(\neg, \rightarrow, \wedge)$ and to $L(\neg, \wedge, \Box)$.

1. Logical necessity: S5

a. Semantics

Logical necessity and possibility were what we first understood were involved in our analysis of $\rightarrow$: $A \rightarrow B$ is true iff it is not possible that A could be true and B false. We fleshed that out by saying that a possible way things could be is a model of classical logic. So $A \rightarrow B$ is true iff in every classical model, if A is true, then B is true. There was no discussion or need for an accessibility relation. Reverting to that original idea, we could take a model to be $\langle W, e \rangle$ and:

(9)

$w \vDash p$ iff $p \in e(w)$

$w \vDash A \wedge B$ iff $w \vDash A$ and $w \vDash B$

$w \vDash \neg A$ iff $w \nvDash A$

$w \vDash A \rightarrow B$ iff for all z not both $z \vDash A$ and $z \nvDash B$

And then derived we have:

$w \vDash \Box A$ iff for all z, $z \vDash A$

$w \vDash \Diamond A$ iff for some z, $z \vDash A$

Necessity is logical necessity: true in every description of the way things could be.

We can accomodate this understanding within the context of models with accessibility relations by using the one relation that is equivalent to ignoring accessibility: the *universal* relation in which every world is related to every other. That is, for any model $\langle W, R, e \rangle$ in which R is universal, the definition of validity at a world is also (9). We define, therefore:

S5 is the logic of all models $\langle W, R, e \rangle$ where R is universal

b. Semantic consequence

The informal definition of what it means for $A \rightarrow B$ to be true is what we previously called the 'possibility version of semantic consequence' (see p. 30): $A \vDash B$ iff it is impossible for A to be true and B false. The word *entails* is often used for the relation $\vDash$, so that one would say 'A entails B' for '$A \vDash B$'; the relation of semantic consequence is often referred to as *entailment.*

Now comes the confusion: it is obvious that the relation of entailment is not a part of the semi-formal language. When we say 'Juney is a dog $\wedge$ Juney barks *entails* Juney barks' we are mentioning the propositions 'Juney is a dog $\wedge$ Juney barks' and 'Juney barks', not using them. But many logicians working with modal logics wish to read $\rightarrow$ as 'entails', and then claim that 'if . . . then . . .' in modal logics should be understood as '$\supset$'. If that is the case, the whole project of modal logic is built on the most egregious use-mention error, and finding some intelligible understanding of the symbols is hopeless. There cannot be a semi-formal language.

Further, logicians who read $\rightarrow$ in modal logics as 'entails' find no interest in characterizing the relation of semantic consequence in modal logics. After all, they say, that relation is exactly what $\rightarrow$ is supposed to capture.

The only way out of these confusions that I can see is to take modal logic to be, as I originally suggested, a logic just like any other logic we have studied: one that is based on formalizations of 'and', 'not', and 'if . . . then . . .'. Granted that the formalization of 'if . . . then . . .' sounds like the possibility version of semantic consequence. But if that is the case, then it is because we believe that 'if . . . then . . .' functions the same as semantic consequence. We should therefore check whether the notion of 'follows from' in the formal language is the same as the metalinguistic semantic consequence. Define:

$\Gamma \vDash_{\mathbf{S5}} A \equiv_{\mathrm{Def}}$ for every model $\langle W, e \rangle$ of **S5**, if $\langle W, e \rangle \vDash \Gamma$, then $\langle W, e \rangle \vDash A$

Theorem 2 The following are equivalent for all formulas A, B in $L(\neg, \wedge)$:

a. $A \vDash_{\mathbf{PC}} B$

b. $\vDash_{\mathbf{PC}} A \supset B$

c. $\vDash_{\mathbf{S5}} A \supset B$

d. $\vDash_{\mathbf{S5}} A \rightarrow B$

e. $A \vDash_{\mathbf{S5}} B$

Proof: The equivalence of (a) and (b) is the *Semantic Deduction Theorem* of classical logic.

The equivalence of (c) and (d) is proved as follows: Given $\vDash_{\mathbf{S5}} A \supset B$. Then for any $\langle W, e \rangle$, for every $z \in W$, $z \vDash A \supset B$. So for every $w \in W$, $w \vDash \Box(A \supset B)$ and $\vDash_{\mathbf{S5}} A \rightarrow B$. If $\nvDash_{\mathbf{S5}} A \supset B$, then there is some $\langle W, e \rangle$ and some $w \in W$, such that $w \nvDash A \supset B$. So $\langle W, e \rangle \nvDash \Box(A \supset B)$ and $\nvDash_{\mathbf{S5}} A \rightarrow B$.

For the equivalence of (b) and (c), suppose $\vDash_{\mathbf{PC}} A \supset B$. Then given any $\langle W, e \rangle$ and $w \in W$, w is a **PC**-model, so if $w \vDash A$, then $w \vDash B$. Hence $\langle W, e \rangle \vDash A \supset B$. Hence, $\vDash_{\mathbf{S5}} A \supset B$. For the other direction, suppose $\nvDash_{\mathbf{PC}} A \supset B$. Then there is some **PC**-model w such that $w \nvDash A \supset B$. Since A and B are in $L(\neg, \wedge)$, we have $\langle \{w\}, e \rangle \nvDash A \supset B$, where $p \in w$ iff $w(p) = \mathsf{T}$. So $\nvDash_{\mathbf{S5}} A \supset B$.

For the equivalence of (c) and (e), suppose $\vDash_{\mathbf{S5}} A \supset B$. Then given any $\langle W, e \rangle$, if for all $w \in W$, $w \vDash A$, then we must also have that all w, $w \vDash B$; hence if $\langle W, e \rangle \vDash A$, then $\langle W, e \rangle \vDash B$. So $A \vDash_{\mathbf{S5}} B$. In the other direction, suppose $\nvDash_{\mathbf{S5}} A \supset B$. Then for some $\langle W, e \rangle$, some $w \in W$, $w \vDash A$ and $w \nvDash B$. Then $\langle \{w\}, e \rangle \vDash A$ and $\langle \{w\}, e \rangle \nvDash B$, since A, B are in $L(\neg, \wedge)$ and hence the evaluation of A and B at w involves no worlds other than w in W. So $A \nvDash_{\mathbf{S5}} B$. ∎

So, indeed, the formalization of 'if . . . then . . .' in **S5** using logical necessity is exactly the relation of entailment in classical logic. But what about formulas in which $\rightarrow$ (or $\Box$) is iterated?

Theorem 3 For all A in $L(\neg, \rightarrow, \wedge)$ (or $L(\neg, \wedge, \Box)$)

a. $\vDash_{S5} A$ iff $\vDash_{S5} \Box A$

b. If $\vDash_{S5} A \supset B$, then $A \vDash_{S5} B$

c. If $\vDash_{S5} A \rightarrow B$, then $A \vDash_{S5} B$

d. $p \vDash_{S5} \Box p$ but $\nvDash_{S5} p \supset \Box p$

e. $p \vDash_{S5} \neg p \rightarrow p$ but $\nvDash_{S5} p \rightarrow (\neg p \rightarrow p)$

Proof: For (a),

$\vDash_{S5} A$ iff in every model, in every world, A is true
iff in every model, in every world, $\Box A$ is true
iff $\vDash_{S5} A$

b. This is as in the proof of Theorem 2.

c. This is as in the proof of Theorem 2.

d. We have $p \vDash_{S5} \Box p$, since if p is true in $\langle W, e \rangle$, then for every $w \in W$, $w \vDash p$, so for every $w \in W$, $w \vDash \Box p$. And hence $\langle W, e \rangle \vDash \Box p$. But $\nvDash_{S5} p \supset \Box p$, as in:

w	z
p – T	p – F

In this model $w \vDash p$, but $w \nvDash \Box p$, so $w \nvDash p \supset \Box p$.

e. The first part is (d) rewritten according to the definitions. Then since, by (d), $\nvDash_{S5} p \supset \Box p$, we must have by (a), $\nvDash_{S5} \Box(p \supset \Box p)$, which is the second part of (e) by definition. ■

Theorem 4 ***A Semantic Deduction Theorem for* S5**

a. $A \vDash_{S5} B$ iff $\vDash_{S5} \Box A \supset \Box B$

b. $A \vDash_{S5} B$ iff $\vDash_{S5} \Box A \supset B$

Proof: For part (a) we have:

$A \vDash_{S5} B$ iff for all $\langle W, e \rangle$, if all $w \in W$, $w \vDash A$, then all $w \in W$, $w \vDash B$
iff for all $\langle W, e \rangle$, if $w \vDash \Box A$, then $w \vDash \Box B$
iff for all $\langle W, e \rangle$, $\langle W, e \rangle \vDash \Box A \supset \Box B$

For part (b), we have that if $A \vDash_{S5} B$, then $\vDash_{S5} \Box A \supset B$, as for part (a). And if $A \nvDash_{S5} B$, then for some $\langle W, e \rangle$, for all $w \in W$, $w \vDash A$, but for some $z \in W$, $z \nvDash B$. Then $z \vDash \Box A$ but $z \nvDash B$, so $z \nvDash \Box A \supset B$. So $\langle W, e \rangle \nvDash \Box A \supset B$. So $\nvDash_{S5} \Box A \supset B$. ■

Thus, though '$\rightarrow$' does formalize the entailment relation of **PC**, when we consider formulas that use this '$\rightarrow$' but not as the main connective, we have two notions of 'follows from'. We do not have that if $A \vDash_{S5} B$ then $\vDash_{S5} A \rightarrow B$: that would be equivalent to the paradoxes of **PC**, and $\rightarrow$ would then have to be evaluated as in **PC**.

c. *Iterated modalities*

In Section B.2 I discussed the difficulty of taking $\Box$ or $\Diamond$ as primitive: doing so seems to involve use-mention confusions. Iterated modalities are a particular problem, but in **S5** iterated modalities collapse.

Theorem 5

a. The following pairs are equivalent, that is, in every model they are both true or both false:

$\Box\Box A$ and $\Box A$

$\Diamond\Diamond A$ and $\Diamond A$

$\Box\Diamond A$ and $\Diamond A$

$\Diamond\Box A$ and $\Box A$

b. In $L(\neg, \rightarrow, \wedge)$ or $L(\neg, \wedge, \Box)$, every formula of the form $\lambda_1 \cdots \lambda_n A$, where for each i, λ_i is one of $\Box$, $\Diamond$, or $\neg$, is equivalent to one of:
A, $\neg A$, $\Box A$, $\neg\Box A$, $\Diamond A$, $\neg\Diamond A$.

Proof: The proof of (a) I leave as an exercise.

For (b), induct on n. It is true for $n = 1$. Then note that given $\langle W, e\rangle$, if for all $w \in W$, $w \vDash C$ iff $w \vDash D$, then for all w, $w \vDash \neg C$ iff $w \vDash \neg D$, $w \vDash \Box C$ iff $w \vDash \Box D$, $w \vDash \Diamond C$ iff $w \vDash \Diamond D$. So given $\lambda\lambda_1 \cdots \lambda_n A$, this is equivalent to one of: λA, $\lambda\neg A$, $\lambda\Box A$, $\lambda\neg\Box A$, $\lambda\Diamond A$, $\lambda\neg\Diamond A$. Thus, for example, for the last case we have one of: $\neg\neg\Diamond A$, which is equivalent to $\Diamond A$; or $\Box\neg\Diamond A$, which is equivalent to $\neg\Diamond\neg\neg\Diamond A$, which is equivalent to $\neg\Diamond\Diamond A$, which is equivalent to $\neg\Diamond A$; or we have $\Diamond\neg\Diamond A$, which is equivalent to $\neg\Box\neg\neg\Diamond A$, which is equivalent to $\neg\Box\Diamond A$, which is equivalent to $\neg\Diamond A$. The other cases I leave as an exercise. ∎

Nonetheless, there are still formulas that cannot be reduced to ones that use a single modal operator, for example $\Box(p \supset \Box p)$ or $\Box((\Box p \supset p) \supset \Box((\Box p \supset p) \supset \Box p))$, the interpretation of which I shall leave to those who believe it appropriate to take $\Box$ as a primitive connective. Nested uses of $\rightarrow$, however, pose no more problem here than in classical logic: that $p \rightarrow (\neg p \rightarrow p)$ fails in **S5** (Theorem 3.e) is as we expect, since it is a "paradox" of material implication.

d. *Syntactic characterization of the class of universal frames*

In axiomatizing **S5** and other modal logics we shall need formulas whose validity corresponds to evaluating in certain kinds of frames.

We first make some definitions. Given $\langle W, R\rangle$ we say:

R is *reflexive* if for all $w \in W$, wRw

R is *symmetric* if for all $w, z \in W$, if wRz, then zRw

R is *transitive* if for all $w, y, z \in W$, if wRy and yRz, then wRz

R is an *equivalence relation* if R is reflexive, transitive, and symmetric

R is *euclidean* if for all $w, y, z \in W$, if wRy and wRz, then yRz

Lemma 6 R is an equivalence relation iff R is reflexive, transitive, and euclidean.

Proof: Suppose R is reflexive, transitive, and euclidean. Suppose wRz. Then since wRw, we have zRw, so R is symmetric, too.

If R is an equivalence relation, suppose wRy and wRz. Then by symmetry, yRw, and hence by transitivity, yRz, so R is euclidean. ∎

Lemma 7 S5 is the logic of all models $\langle W, R, e \rangle$ such that R is an equivalence relation.

Proof: Let '$\vDash_e$' stand for validity in the class of all equivalence models.

Suppose $\nvDash_{S5} A$. Then for some $\langle W, R, e \rangle$ where R is universal, $\langle W, R, e \rangle \nvDash A$. The universal relation is an equivalence relation. So $\nvDash_e A$.

If $\nvDash_e A$, then for some $\langle W, R, e \rangle$ where R is an equivalence relation, for some $w \in W$, $w \nvDash A$. Fix one such model and world w and consider $Z = \{z \colon wRz\}$. Then R restricted to Z is universal, as I'll let you prove. And by Lemma 1, for any $z \in Z$, $\langle Z, R, e, z \rangle \vDash A$ iff $\langle W, R, e, z \rangle \vDash A$. Hence $\langle Z, R, e \rangle \nvDash A$, and so $\nvDash_{S5} A$.

I'll leave to you to show that $\Gamma \vDash_{S5} A$ iff $\Gamma \vDash_e A$. ∎

Lemma 8 For all $\langle W, R \rangle$,

R is reflexive iff $\langle W, R \rangle \vDash \Box p \supset p$

R is reflexive iff $\langle W, R \rangle \vDash p \supset \Diamond p$

R is symmetric iff $\langle W, R \rangle \vDash p \supset \Box \Diamond p$

R is transitive iff $\langle W, R \rangle \vDash \Box p \supset \Box \Box p$

R is euclidean iff $\langle W, R \rangle \vDash \Diamond p \supset \Box \Diamond p$

Proof: Given $\langle W, R \rangle$. If R is reflexive, then given any e and $w \in W$, if $w \vDash \Box A$, then since wRw, $w \vDash A$. And hence $w \vDash \Box A \supset A$. So $\langle W, R \rangle \vDash \Box A \supset A$. If R is not reflexive, consider some w that is not related to itself. Then take an e such that $p_1 \notin e(w)$ but $p_1 \in e(z)$ for all $z \neq w$. Then $w \vDash \Box p_1$ (even if $W = \{w\}$), but $w \nvDash p_1$. So $\langle W, R \rangle \nvDash \Box p_1 \supset p_1$.

I leave the others as exercises. ∎

Eliminating definitions and simplifying in the language of $L(\neg, \rightarrow, \wedge)$, we have:

Lemma 9 For all $\langle W, R \rangle$,

R is reflexive iff $\langle W, R \rangle \vDash (\neg p \rightarrow p) \supset p$

R is symmetric iff $\langle W, R \rangle \vDash p \supset ((p \rightarrow \neg p) \rightarrow \neg(p \rightarrow \neg p))$

R is transitive iff $\langle W, R \rangle \vDash (\neg p \rightarrow p) \supset (p \rightarrow (\neg p \rightarrow p))$

Proof: The first two come from Lemma 8, using the definitions of $\Box$ and $\Diamond$, and contracting double negations.

For part (c), given $\langle W,R\rangle$ such that R is transitive, take any w and e such that $w\vDash\neg p\rightarrow p$. We need to show that $w\vDash p\rightarrow(\neg p\rightarrow p)$. For any y such that wRy, we must have $y\vDash p$. But then also for any z such that yRz, by transitivity, wRz, so $z\vDash p$, and hence $z\vDash\neg p\supset p$, and so $y\vDash\neg p\rightarrow p$. Thus $y\vDash p\supset(\neg p\rightarrow p)$. And so $w\vDash p\rightarrow(\neg p\rightarrow p)$. Hence $w\vDash(\neg p\rightarrow p)\supset(p\rightarrow(\neg p\rightarrow p))$, and so $\langle W,R\rangle\vDash(\neg p\rightarrow p)\supset(p\rightarrow(\neg p\rightarrow p))$.

If R is not transitive, let w, y, z be such that wRy and yRz, but w is not related to z. Then take any e such that for all x such that wRx, $p\in e(x)$, but $p\notin e(z)$. Then we have:

$$\begin{array}{ccccc} w & \longrightarrow & y & \longrightarrow & z \\ p-T & & p-T & & p-F \end{array}$$

So $w\vDash\neg p\rightarrow p$, $y\vDash p$, but $y\nvDash\neg p\rightarrow p$. So $w\nvDash(\neg p\rightarrow p)\supset(p\rightarrow(\neg p\rightarrow p))$. ■

The following is now a corollary to these lemmas.

Theorem 10 ***a.*** **S5** is the logic of all $\langle W,R\rangle$ in which
$\Box A\supset A$, $A\supset\Box\Diamond A$, $\Box A\supset\Box\Box A$ are valid

b. **S5** is the logic of all $\langle W,R\rangle$ in which
$\Box A\supset A$, $\Box A\supset\Box\Box A$, $\Diamond A\supset\Box\Diamond A$ are valid

c. **S5** is the logic of all $\langle W,R\rangle$ in which
$(\neg p\rightarrow p)\supset p$, $p\supset((p\rightarrow\neg p)\rightarrow\neg(p\rightarrow\neg p))$,
$(\neg p\rightarrow p)\supset(p\rightarrow(\neg p\rightarrow p))$ are valid

e. Some rules valid in S5

Theorem 11 The following rules are valid in **S5**:

a. *Modus ponens* $\dfrac{A,\ A\rightarrow B}{B}$

b. Necessitation $\dfrac{A}{\Box A}$

c. Substitution $\dfrac{A(p)}{\Box A(B)}$ where $\vDash A(p)$ and B is substituted uniformly for p in A

Proof: a. If $\langle W,e\rangle\vDash A$ and $\langle W,e\rangle\vDash A\rightarrow B$, then for every $w\in W$, $w\vDash A$ and $w\vDash A\rightarrow B$. So $w\vDash B$. And hence $\langle W,e\rangle\vDash B$.

b. If $\langle W,e\rangle\vDash A$, then for every $w\in W$, $w\vDash A$. Hence, for every $w\in W$, $w\vDash\Box A$. Hence $\langle W,e\rangle\vDash\Box A$.

c. Suppose $\nvDash_{S5}A(B)$. I will show $\nvDash_{S5}A(p)$. Let $\langle W,e\rangle\nvDash A(B)$. Then for some

$w \in W$, $w \nvDash A(B)$. Define $\langle W, e' \rangle$ by: for all z, for all variables $q \neq p$, $q \in e'(z)$ iff $q \in e'(z)$, and $p \in e'(z)$ iff $z \vDash B$ in $\langle W, e \rangle$. This definition is possible because the only place p can appear in A(B) is in B itself. Now in $\langle W, e' \rangle$, $w \nvDash A(p)$, since the evaluation of A(p) proceeds exactly as the evaluation of A(B) in $\langle W, e, w \rangle$. ■

Note well that the rules of *modus ponens* and necessitation hold on a model-by-model basis, but substitution is valid only for valid formulas.

Exercises for Section C.1

1. a. Why do we call **S5** the logic of logical necessity?
   b. How do we accomodate logical necessity into the framework of necessity and possibility that depend on accessibility relations?
2. Show that if R is universal, then the definition of truth in a model $\langle W, R, e \rangle$ is given as in (9).
3. Explain precisely the relation between the validity of a formula $A \rightarrow B$ in **S5** and the semantic consequence $A \vDash_{PC} B$. Why is '$\rightarrow$' sometimes read as 'entails' in modal logics? How are the paradoxes of material implication involved?
4. Prove Theorem 5.a.
5. Does the elimination of iterated modalities in **S5** fully solve the difficulty of whether using $\Box$ as a primitive involves a use-mention confusion?
6. Given $\langle W, R \rangle$ where R is an equivalence relation and $w \in W$. Set $Z = \{z: wRz\}$. Show that R restricted to Z is universal.
7. Prove that for all $\langle W, R \rangle$:
   R is reflexive iff $\langle W, R \rangle \vDash p \supset \Diamond p$
   R is symmetric iff $\langle W, R \rangle \vDash p \supset \Box\Diamond p$
   R is transitive iff $\langle W, R \rangle \vDash \Box p \supset \Box\Box p$
   R is euclidean iff $\langle W, R \rangle \vDash \Diamond p \supset \Box\Diamond p$
8. Prove that for all $\langle W, R \rangle$,
   R is symmetric iff $\langle W, R \rangle \vDash p \supset ((p \rightarrow \neg p) \rightarrow \neg(p \rightarrow \neg p))$
9. Show that the rule of substitution is not valid if the restriction that it be applied to only valid formulas is lifted. Show that it is also not valid if B may be substituted for some but not necessarily all occurrences of p in A.

2. K—all accessibility relations

The general framework of all possible-worlds semantics allows for all accessibility relations:

K is the logic of all models $\langle W, R, e \rangle$

To my mind, **K** is not a logic: it is not based on some particular notion of necessity or possibility, but encompasses all. It is the formalization of the idea of accessibility without yet choosing one notion of accessibility. It is only by analogy with other systems that it receives the title 'logic'. We say that **K** is the *minimal* logic of possible-world semantics, because any A that is valid in **K** must be valid in all frames. So in studying **K** we are studying all modal logics based on frames.

Let us consider some valid wffs that will be important for later axiomatizations. Recall that '$A \equiv B$' stands for '$(A \supset B) \wedge (B \supset A)$'.

***Theorem 12 The Distribution Axioms are Valid in* K**

a. $\vDash_{\mathbf{K}} \Box(A \supset B) \supset (\Box A \supset \Box B)$

b. $\vDash_{\mathbf{K}} \Box(A \wedge B) \equiv \Box A \wedge \Box B$

c. $\vDash_{\mathbf{K}} (A \rightarrow B) \supset ((\neg A \rightarrow A) \supset (\neg B \rightarrow B))$

d. If $\vDash_{\mathbf{K}} A \equiv B$, then $\vDash_{\mathbf{K}} \Box A \equiv \Box B$

Proof: Part (a) was the Example on p. 646. Part (b) I leave to you. Part (c) is the unabbreviated form of (a) in $L(\neg, \rightarrow, \wedge)$. For part (d), if in every model in every world of that model A and B have the same truth-value, then in every model in every world of that model $\Box A$ and $\Box B$ have the same truth-value. ■

Note that $(\Box A \supset \Box B) \supset \Box(A \supset B)$ is not valid (see Theorem 3.d).

Let's consider what rules hold in **K**.

Theorem 13

a. Material detachment $\dfrac{A,\ A \supset B}{B}$ is valid in **K**.

b. Necessitation $\dfrac{A}{\Box A}$ is valid in **K**.

c. Substitution $\dfrac{A(p)}{\Box A(B)}$ is valid in **K** where $\vDash A(p)$ and B is substituted uniformly for p in A

d. *Modus ponens* $\dfrac{A,\ A \rightarrow B}{B}$ is not valid in **K**.

e. *Modus ponens* is valid in any frame $\langle W, R \rangle$ in which R is reflexive.

Proof: Part (a) I will leave to you. Parts (b) and (c) are done as in the proof of Theorem 11 above.

For part (d), *modus ponens* is not valid in **K** because we have a model:

w →	z (↻)
p – T	p – T
q – F	q – T

Here both $w \vDash p$ and $z \vDash p$, so $\vDash p$ in the model, and $w \vDash p \rightarrow q$ and $z \vDash p \rightarrow q$, since w is not accessible to itself. So $\vDash p \rightarrow q$ in this model, too. But $\nvDash q$.

However, if R is reflexive, *modus ponens* holds. ■

Thus **K** would be an odd logic, indeed, if it were a logic, since *modus ponens* fails. It would be especially awkward to call it a logic for those who read '→' as 'entails'.

Exercises for Section C.2

1. Determine which of the following distribution formulas are valid in **K**.
   a. $\Box(A \wedge B) \equiv \Box A \wedge \Box B$
   b. $\Diamond(A \vee B) \equiv \Diamond A \vee \Diamond B$
   c. $\Diamond(A \wedge B) \equiv \Diamond A \wedge \Diamond B$
   d. $\Box(A \vee B) \equiv \Box A \vee \Box B$
   e. $\Diamond(A \supset B) \supset (\Diamond A \supset \Diamond B)$
   f. $(\Diamond A \supset \Diamond B) \supset \Diamond(A \supset B)$
2. Verify that if $\langle W, R \rangle$ is reflexive, then *modus ponens* is valid in $\langle W, R \rangle$.

3. T, B, and S4

We define:

T is the logic of all models $\langle W, R, e \rangle$ where R is reflexive

B is the logic of all models $\langle W, R, e \rangle$ where R is reflexive and symmetric

S4 is the logic of all models $\langle W, R, e \rangle$ where R is reflexive and transitive

The following characterizations of these logics comes from Lemmas 8 and 9.

Theorem 14 **T** is the logic of all $\langle W, R \rangle$ in which $\Box A \supset A$ is valid

T is the logic of all $\langle W, R \rangle$ in which $(\neg A \rightarrow A) \supset A$ is valid

B is the logic of all $\langle W, R \rangle$ in which $\Box A \supset A$ and $A \supset \Box\Diamond A$ are valid

B is the logic of all $\langle W, R \rangle$ in which $(\neg A \rightarrow A) \supset A$ and $A \supset ((A \rightarrow \neg A) \rightarrow \neg(A \rightarrow \neg A))$ are valid

S4 is the logic of all $\langle W, R \rangle$ in which $\Box A \supset A$ and $\Box A \supset \Box\Box A$ are valid

S4 is the logic of all $\langle W, R \rangle$ in which $(\neg A \rightarrow A) \supset A$ and $(\neg A \rightarrow A) \supset (A \rightarrow (\neg A \rightarrow A))$ are valid

Each of these logics was originally presented syntactically as an extension of **K**, that is, as a logic in which all the formulas of **K** are valid. The first was **S4** in *Gödel, 1933 B*. Gödel's idea was to model the intuitionistic logic of Heyting (see Chapter VII below) by adding to **PC** in $L(\neg, \wedge)$ a unary connective B (we use $\Box$) to be read as *beweisbar* (*it is provable that*). He noted that one could translate from Heyting's logic to what we now call **S4** so that a formula in Heyting's logic was provable iff its translate was provable in **S4**.

The formalism sustains that, as we shall see in Chapter VII. But again we are faced with the question of whether we might actually use **S4** for formalizing reasoning, based as it seems, on confusing metalogic with logic: one of its axioms should be read, on this interpretation, as 'If it is provable that A, then it is provable that it is provable that A.'

The logic **T** was originated by *Feys, 1937* by dropping one of the axioms of Gödel's system (see *Hughes and Cresswell, 1968,* note 20, p. 30).

The formula $A \supset \Box\Diamond A$ was first singled out by *Becker, 1930* because of similarities to reasoning in intuitionistic logic (see *Hughes and Cresswell, 1968,* note 37, p. 58). I do not know who first considered the system **B**, nor if it has ever been proposed as a codification of reasoning.

Each of these systems can be viewed as adopting as axioms formulas that express some idea of necessity and possibility: in **T**, for instance, we have that if A is necessary, then A is true (or if A is true, then it is possible). These give rise to particularly simple restrictions on the accessibility relations.

4. Decidability and the Finite Model Property

We have used general considerations of models and frames to show the validity of certain formulas above. For the logics discussed so far, we actually have decision procedures for all formulas. I will present a method, due to *Lemmon, 1977,* and *Segerberg, 1971,* that relies on our proving that each of the logics above is characterized by finite models.

Given a collection of formulas Γ and a model $\mathsf{M} = \langle \mathsf{W}, \mathsf{R}, \mathsf{e} \rangle$, define a *filtration of* M *with respect to* Γ to be any $\mathsf{M}^* = \langle \mathsf{W}^*, \mathsf{R}^*, \mathsf{e}^* \rangle$ such that:

$x \approx y \equiv_{\text{Def}}$ for all $A \in \Gamma$, $x \vDash A$ iff $y \vDash A$

$x^* = \{y : x \approx y\}$

$\mathsf{W}^* = \{x^* : x \in \mathsf{W}\}$

if $x\mathsf{R}y$, then $x^*\mathsf{R}^*y^*$

if $x^*\mathsf{R}^*y^*$, then if $\Box A \in \Gamma$ and $x \vDash \Box A$, then $y \vDash A$

$\mathsf{e}^*(x^*) = \{p : p \in \Gamma \text{ and } x \vDash p\}$

For any filtration, note that e^* respects the equivalence classes of $\approx$, for if $p \in \mathsf{e}^*(x^*)$, then $p \in \Gamma$ and $x \vDash p$, so if $y \in x^*$, then $y \vDash p$, too.

Given M and Γ, consider the *coarsest* filtration of M with respect to Γ:

$x^* R^* y^*$ iff if $\Box A \in \Gamma$ and $x \vDash \Box A$, then $y \vDash A$

That this is a filtration requires that R^* respects the equivalence classes of $\approx$. Suppose $x^* R^* y^*$ and $w \in x^*$, $z \in y^*$. To show that $w^* R^* z^*$, suppose $\Box A \in \Gamma$ and $w \vDash \Box A$. Then $x \vDash \Box A$. So $y \vDash A$. So $z \vDash A$.

Given a collection of wffs Γ, we say that Γ is *closed under subformulas* if for all $A \in \Gamma$, if B is a subformula of A, then $B \in \Gamma$.

Theorem 15 Given any model $\mathsf{M} = \langle W, R, e \rangle$ and Γ any collection of wffs closed under subformulas. Let $\mathsf{M}^* = \langle W^*, R^*, e^* \rangle$ be a filtration of M with respect to Γ. Then:

a. If Γ is finite, then W^* is finite.

b. For all $A \in \Gamma$, $\mathsf{M} \vDash A$ iff $\mathsf{M}^* \vDash A$.

c. If R is reflexive, then the coarsest filtration is reflexive.

d. If R is reflexive and symmetric, the filtration given by the following is reflexive and symmetric: $x^* R^* y^*$ iff some $w \in x^*$, $z \in y^*$, wRz.

e. If R is reflexive and transitive, the filtration given by the following is reflexive and transitive: $x^* R^* y^*$ iff if $\Box A \in \Gamma$ and $x \vDash \Box A$, then $y \vDash A$ and $y \vDash \Box A$.

f. If R is universal, then R^* is universal.

Proof: For part (a), if $x^* \neq y^*$, there is some $A \in \Gamma$ such that x^* evaluates A differently from y^*. So if Γ is finite, then W^* is finite.

We can show part (b) by showing by induction on the length of A that for all $A \in \Gamma$ and $x \in W$, $\langle W, R, e, x \rangle \vDash A$ iff $\langle W^*, R^*, e^*, x^* \rangle \vDash A$.

If A has length 1, it's true by definition. So suppose it's true for all shorter wffs and A is $B \wedge C$. Then:

$x \vDash B \wedge C$ iff	$x \vDash B$ and $x \vDash C$	
iff	$x^* \vDash B$ and $x^* \vDash C$	by induction, since Γ is closed under subformulas
iff	$x^* \vDash B \wedge C$	

If A is $\neg B$, the proof is similar. If A is $\Box B$, suppose $x^* \vDash \Box B$. If xRy, then $x^* R^* y^*$, so $y^* \vDash B$. Hence by induction as $B \in \Gamma$, $y \vDash B$ and thus $x \vDash \Box B$. In the other direction, suppose $x \vDash \Box B$. If $x^* R^* y^*$, then $y \vDash B$ by the definition of R^*. So by induction $y^* \vDash B$, so $x^* \vDash \Box B$.

I leave parts (c)–(f) as exercises. ∎

Theorem 15 shows that each of **K**, **T**, **B**, **S4**, and **S5** has the *finite model property*: if a wff fails in some model of the logic, then it fails in a finite model.

Corollary 16

$\vDash_{\mathbf{K}} A$	iff	A is valid in all finite frames
$\vDash_{\mathbf{T}} A$	iff	A is valid in all finite reflexive frames
$\vDash_{\mathbf{B}} A$	iff	A is valid in all finite reflexive and symmetric frames
$\vDash_{\mathbf{S4}} A$	iff	A is valid in all finite reflexive and transitive frames
$\vDash_{\mathbf{S5}} A$	iff	A is valid in all finite equivalence frames

Proof: The left to right direction for each is immediate from the definition. For the other direction, suppose $\nvDash_{\mathbf{K}} A$. So there is some model $M \nvDash A$. So by Theorem 15, the coarsest filtration M^* of M with respect to {B: B is a subformula of A} is finite, and $M^* \nvDash A$. The other cases are done similarly using Theorem 15. ■

Now we can show that these systems are decidable using Corollary 16 and the axiomatizations in Section E below.

Corollary 17 For each of **K**, **T**, **B**, **S4**, and **S5** there is an effective procedure to determine whether any given wff is valid.

Proof: Let **L** be any one of these logics. We will see below (Section E) that **L** is axiomatized by a finite number of schema. So we have an effective procedure for listing all theorems of the logic: list all proofs. We can also effectively list all finite frames that satisfy the appropriate conditions for the logics. And we can effectively check whether a given wff is valid in a finite frame because by Lemma 1 we need only consider the propositional variables in that wff.

Now suppose we are given a wff A. To decide if A is a theorem, we dovetail these two listing procedures until we find a proof of A, in which case A is a theorem and hence valid, or we find a finite model of **L** that invalidates A. ■

Exercises for Section C.4

1. a. What is a filtration?
   b. Prove that the coarsest filtration is reflexive if R is reflexive.
   c. If Γ is finite, why is any filtration with respect to Γ finite?
2. Prove Theorem 15.(c)–(f).
3. Prove Theorem 16 for **L = T, B, S4, S5**.
4. Comment on the utility of this method of deciding validity.

D. Examples of Formalization

In this section I present some examples of formalizations of propositions and arguments relative to the assumptions of the previous sections. The format will be the same as in Chapter II.F, and the criteria governing the acceptability of formalizations will be those discussed in Chapter II.E.2.

1. If roses are red, then sugar is sweet

Roses are red $\rightarrow$ sugar is sweet

$p_1 \rightarrow p_2$

Explanation: This is the same formalization we would have given this example before; the difference is our interpretation of '$\rightarrow$'.

In any classical model in which the "obvious" truth-values are assigned, the proposition will be true because 'sugar is sweet' is true. However, in modal logic models the proposition will be false, because we can (**PC**-consistently) imagine that roses could (all) be red while sugar is sour. Or at least we believe we could construct **PC**-consistent models in which 'roses are red' is true and 'sugar is sweet' is false, while most of what we usually believe about "the world" is true.

2. If Ralph is a bachelor, then Ralph is a man

Ralph is a bachelor $\rightarrow$ Ralph is a man

$p_1 \rightarrow p_2$

Explanation: The example is not valid, though some would argue that it is impossible for the antecedent to be true and the consequent false. Our propositional analysis will not justify that view. Not even if we consider the internal structure of propositions (in predicate logic) can we justify the validity of this example (see *Predicate Logic,* Chapter V.K.1). Only by examining the meaning of the nonlogical words involved in the example and adopting assumptions about those meanings could we claim the proposition valid, and to do so would be beyond the scope of logic.

3. If the moon is made of green cheese, then 2 + 2 = 4

The moon is made of green cheese $\rightarrow 2+2=4$

$p_1 \rightarrow p_2$

Explanation: The consequent of this example is usually considered a necessary truth: it would be impossible for $2+2$ not to equal 4. So, it would be argued, the example must be valid. But again we cannot justify the validity of this proposition without recourse to the meaning of the words, incorporating here assumptions about mathematics into the logic. Unless, that is, one is a logicist who argues that mathematics is part of logic.

4. If Juney was a dog, then surely it's possible that Juney was a dog

Juney was a dog $\supset \Diamond$(Juney was a dog)

$p_1 \supset \Diamond p_1$

Explanation: When the words 'possibly' or 'necessarily' or other modals are used with 'if . . . then . . .' it's safe to assume that 'if . . . then . . .' is meant to be read as Philonian (material) implication, not strict implication. Otherwise there would be no need to use these modal words.

The proposition seems evidently valid: what more could ever be required of a proposition to justify that it is possible than that it be true? For if a proposition is true, it's certainly **PC**-consistent to assume it's true (assuming the world as we know it is **PC**-consistent), and hence there is a way for it to be true.

But in some modal systems this example is not valid: the proposition can fail in any modal systems that has models based on a frame that is not reflexive, such as **K**, for the example is (**PC**-) equivalent to $\Box p_1 \supset p_1$ (see Example 1, p. 209). In any such logic *modus ponens* will also fail: B need not follow from A and $A \to B$ (see Theorem 13 above).

It seems to me that it is an integral part of our understanding of the notions of possibility and necessity that the example be true, indeed valid, and that any modal system that does not validate $A \supset \Diamond A$, and with it *modus ponens,* simply does not justify the title 'logic'. Or at least strong arguments based on very different interpretations of the connectives must be put forward, along with examples of reasoning in that system, for us to consider it a logic.

5. If it's possible that Juney was a dog, then Juney was a dog

$\Diamond$(Juney was a dog) $\supset$ Juney was a dog

$\Diamond p_1 \supset p_1$

Explanation: The example is not valid. We can imagine a way for a proposition to be true even if it is false. Thus, '$\Diamond$(Socrates was found innocent by the jury in Athens) $\supset$ Socrates was found innocent by the jury in Athens' is false, since the antecedent, we believe, is true, while the consequent is false.

I keep saying 'we believe', which is not really accurate, since all we are concerned with is exhibiting a **PC**-consistent model. But, it seems to me, we really intend in our analysis here that if everything else were exactly as it had been in Socrates' time, then Socrates still might have been found innocent. That is, for a proposition such as this one, we are not usually concerned with examining its truth or falsity in all possible models of logical necessity, but rather in models of a time-dependent or physical necessity. And then, since we can only with difficulty formalize all our beliefs about the way the world was in Socrates' time, and since it is at best difficult to survey the consistency of those formalized beliefs, it seems apt to me to say, 'the antecedent, we believe, is true.'

One might disagree, however, and claim that $\Diamond p \supset p$ is valid—not for logical necessity, as we have developed it, but perhaps for time-dependent necessity. To adopt that view is to adopt a form of determinism, claiming equivalently that

$p \supset \Box p$ is valid. All modal operators would collapse, for we would have $\Box p \equiv \Diamond p \equiv p$, and we would be left with classical logic, **PC**.

6. It's not possible that Juney was a dog and a cat

$\neg\Diamond$(Juney was a dog $\wedge$ Juney was a cat)

$\neg\Diamond(p_1 \wedge p_2)$

Explanation: This example is similar to Example 2: it is not valid, but we suspect that if we pay attention to the meaning of the words and, here, use some form of physical necessity, then 'Juney was a dog and a cat' would be found to be an impossible proposition, false in every (appropriate) possible world.

7. If it is necessary that Juney is a dog, then it is necessary that it is necessary that Juney is a dog

$\Box$(Juney is a dog) $\supset$ $\Box\Box$(Juney is a dog)

$\Box p_1 \supset \Box\Box p_1$

Explanation: This is valid in the modal logics **S4** and **S5**, but fails in **K**, **T**, and **B**. Is this an instance of an obvious principle of necessity? I don't think we have any intuition in our ordinary language for iterated modals, involving, as they apparently do, use-mention confusions.

8. If this paper is white, it must necessarily be white

This paper is white $\supset$ $\Box$(this paper is white)

$p_1 \supset \Box p_1$

Explanation: The example is clearly invalid; indeed, except for propositions such as '$2 + 2 = 4$', it is hard to imagine a case when it would be true.

Why then does Aristotle think propositions such as this are obviously true (*De Interpretatione,* Chapter 9, 18b)? It is, as *Mates, 1986* (pp. 117–118) calls it, the *fallacy of the slipped modal.* In Greek (according to Mates) and in English, in a modalized conditional it is natural to put the modal operator in the consequent, when what is intended is that the operator takes the entire conditional as scope, and similarly for modals used with other connectives. The example is false, though what is true is the uncontroversial:

$\Box$ (this paper is white $\supset$ this paper is white)

9. If Hoover was elected president, then he must have received the most votes
Hoover was elected president
***Therefore*:**

Hoover must have received the most votes

Hoover was elected president $\supset$ $\Box$(Hoover received the most votes)
Hoover was elected president
Therefore:
$\Box$(Hoover received the most votes)

$$\begin{array}{l} p_1 \supset \Box p_2 \\ \underline{p_1 \qquad\qquad} \\ \Box p_2 \end{array}$$

Explanation: The example is valid. But then we seem to have a paradox: why should it be necessary that Hoover should have received the most votes?

There is no paradox, only the fallacy of the slipped modal. The only plausible motive for accepting the first hypothesis as true is to identify it with '$\Box$(Hoover was elected president $\supset$ Hoover received the most votes)', which for some notions of necessity (legal, time-dependent) could be true. But from '$\Box$(Hoover was elected president $\supset$ Hoover received the most votes)' and 'Hoover was elected president' we cannot conclude '$\Box$(Hoover received the most votes)'.

10. A sea fight must take place tomorrow or not. But it is not necessary that it should take place tomorrow; neither is it necessary that it should not take place. Yet it is necessary that it either should or should not take place tomorrow.

$\Box$(a sea fight takes place tomorrow) $\vee$ $\Box\neg$(a sea fight takes place tomorrow)

$\neg\Box$(a sea fight takes place tomorrow) $\wedge$ $\neg\Box\neg$(a sea fight takes place tomorrow)

$\Box$(a sea fight takes place tomorrow $\vee$ $\neg$(a sea fight takes place tomorrow))

$\Box p_1 \vee \Box\neg p_1$

$\neg\Box p_1 \wedge \neg\Box\neg p_1$

$\Box(p_1 \vee \neg p_1)$

Explanation: This is Aristotle, *De Interpretatione,* 19a, as translated by *McKeon, 1941.* Aristotle finds a paradox here, but what we should find, according to *Mates, 1973,* is a fallacy of the slipped modal. The second and third propositions are plausibly true. But there is no plausibility at all to the first: in colloquial English (and Greek) it sounds right, but that is only by identifying it with the third proposition.

11. It is contingent that US $1 bills are green

$\neg\Box$ (US $1 bills are green) $\wedge$ $\neg\Box\neg$ (US $1 bills are green)

$\neg\Box(p_1) \wedge \neg\Box\neg(p_1)$

Explanation: To say a proposition is contingent is to say that it is not necessary nor is its negation necessary. Or equivalently, both it and its negation are possible. Whether there are any contingent propositions is a question that arises in the same way as whether every true proposition is necessary, as discussed in Example 5.

12. It is possible for Richard L. Epstein to print his own bank notes

$\Diamond$ (Richard L. Epstein prints his own bank notes)

$\Diamond\, p_1$

Explanation: The proposition 'Richard L. Epstein prints his own banknotes' is, we might say, technically possible (logically possible, physically possible), but not legally possible. Legal possibility, however, does not seem to call for an accessibility relation. Rather, to analyze whether a proposition is legally possible we could take models of logical necessity, **S5**, that are restricted to atomic propositions that describe actions. Perhaps time-dependent and physical necessity could be handled without accessibility relations, too.

13. If there were no dogs, then everyone would like cats.

Explanation: This is an example of a *counterfactual*: a conditional asserted on the understanding that the antecedent is false. Classical logic cannot formalize these, for such conditionals would simply be true.

How would we decide whether this conditional is true? We try to imagine a world in which there are no dogs and ask whether everyone would then like cats. This suggests the use of possible worlds, that is, consideration of consistent descriptions of "the way the world could be". But an explanation of subjunctive conditionals in terms of possible worlds will have to differ from our modal logic analyses, since the truth-values of the propositions involved in a subjunctive conditional do not, in general, contribute to the truth-value of the whole. If, that is, we are justified in viewing a subjunctive conditional as formed from other propositions, as, for this example, 'There are no dogs $\rightarrow$ everyone likes cats'.

For subjunctive conditionals *modus ponens* plausibly fails, since the truth of such a conditional depends in some sense on the antecedent being false. Nor is it clear how we might define $\Diamond$ and $\Box$ from $\neg$, $\wedge$, and such an $\rightarrow$, for $\rightarrow$ must be taken as primitive here.

I shall not discuss subjunctive conditionals in this volume. No general agreement on their interpretation is current, though some broad analyses have been proposed, as in, for example, *D. Lewis, 1973,* and *Stalnaker and Thomason, 1970.*

14. It is permissible but not obligatory to kill cats.

Explanation: *Deontic* logic is the study of how to reason with the notions of permission and obligation, in the moral or legal sense. Some logicians have suggested formalizing such notions within the framework of modal logics: they read 'It is obligatory that . . .' as $\Box$, and 'It is permissible that . . .' as $\Diamond$, formalizing the example, for instance, as:

$\Diamond$(one kills cats) $\wedge$ $\neg\Box$(one kills cats)

But on the face of it, this seems wrong: what is permissible is not a proposition, 'one kills a cat', but the doing of an action, 'to kill a cat'. A command or a description of a type of action is not a proposition; obligation and permission are not propositional connectives. Even given a predicate logic analysis, there is no connection between 'Roger kills cats' and 'It is permissible to kill cats', except as we add axioms relating these propositions.

Even granting that a reading as above makes sense, there are difficulties. It may be obligatory not to murder, while it is not true that no one murders. And much of the interest in studying moral questions is in trying to resolve inconsistent moral demands, as in 'It is obligatory not to kill' and 'It is obligatory that every soldier kill his enemy'; in classical modal logic we have $(\Box A \wedge \Box B) \supset \Box(A \wedge B)$, and then a necessary falsehood, from which everything can be deduced. For a survey of deontic logic, see *Åqvist, 1984*.

15. A dog that likes cats is possible

Explanation: There is no need to populate our universe with possible dogs to analyze this proposition. The predicates '— is possible' and '— is necessary' are true only of propositions. Letting **a** stand for the proposition 'There is a dog that likes cats', we can rewrite the example as '**a** is possible'. According to many modal logicians, we should then rewrite the proposition as 'It is possible that there is a dog that likes cats', and formalize that as '$\Diamond$(there is a dog that likes cats)'.

16. Example 2 of Chapter II.F is possible

Explanation: Example 2 of Chaper II.F is 'Ralph is a dog and George is a duck and Howie is a cat'. And that proposition is possible, that is, there is a way in which it could be true, since it once was true.

Why do I not formalize this example as '$\Diamond$ ((Ralph is a dog $\wedge$ George is a duck) $\wedge$ Howie is a cat)'? I think the example is already in it most basic logical form, namely: a predicate applied to an object (see *Predicate Logic*). We have the predicate '— is true', which is true of some propositions and false of all other objects. We have the predicate '— is possible', which is true of some propositions and false of all other objects. We can convert both predicates to apparent connectives: 'It is true that . . .' and 'It is possible that . . .', but we reject the use of the first in that way; why don't we reject the second as well? After all, '— is possible' is, on the motivation given in classical modal logic, '— is possibly true'.

17. Ralph knows that Howie is a cat

Explanation: Modal logic has been used by some to formalize reasoning about knowledge. For each agent, that is person or thing that knows, a distinct modal connective is adopted. For this example, for instance, a formalization might be:

$\Box_{\text{Ralph}}$ (Howie is a cat)

But consider: we say, 'Ralph knows Richard L. Epstein', 'Ralph knows a cat', 'Ralph knows a proposition'. Isn't '. . . knows . . .' a relation between things? The conversion of '. . . knows . . .' from a relation into a connective not only involves the use-mention problems we had with 'It is possible that . . .', but also requires introducing one new connective for each agent. I believe we should name the proposition 'Howie is a cat' as, say, '**a**', and formalize the example as:

Ralph knows **a**

This would allow us to use the full power of first-order logic to axiomatize an analysis of '. . . knows . . .' and clearly establishes that we are discussing a relation between an agent and a proposition, rather than concealing that relation using subscripts (which would be like the aristotelian logicians trying to force reasoning about relations into their system, which was designed to deal only with unary predicates). Modal logic is complicated enough, involving for each model both a notion of truth and a notion of validity, without trying to do a concealed version of first-order logic in it, too.

18. Example 18 is not possible.

Explanation: Suppose Example 18 is not possible. Then since that is exactly what it says, it must be true. And hence it must be possible. A contradiction.

So Example 18 is possible. That's just to say that there is a way in which it could be true. But for it to be true there must be no description of the way things could be in which it is true, a contradiction.

Hence, Example 18 is neither possible nor not possible, neither true nor false. To resolve this paradox we would need to consider all the issues involved in trying to resolve 'This sentence is false'. So why do we believe that the conversion of the predicate '. . . is possible' to a connective 'It is possible that . . .', formalized as $\Diamond$, is any less problematic than converting '. . . is true' to 'It is true that . . .', or involves any less a use-mention confusion?

Exercises for Section D

1. Formalize the following in the format of Section D.
   a. If $2 + 2 = 5$, then the moon is made of green cheese
   b. If Juney is a dog and Juney is not a dog, then the moon is made of green cheese

c. Any wff valid in **K** must be valid in **T**
d. If Sarah is a wife, then Sarah has a husband
e. It's possible for cats to bark
f. It's possible to fly to the moon
g. It's not possible to rob a bank in Cedar City
h. Necessity is the mother of invention
j. If it's necessary that Ralph barks, then Ralph barks
k. It's not necessary that every dog barks
l. It's not necessary to answer every question on the final exam in order to pass
m. There's a possibility that Elvis isn't really dead
n. If Ralph is a dog, then it's absolutely necessary that he's possibly a dog
o. It's possible but not true that Ralph is a dog
p. If Socrates had not drunk Hemlock, he never would have become famous
q. This sentence is possible

E. Syntactic Characterizations of Modal Logics

All the systems presented in this chapter were originally presented syntactically, with the exception of **MSI** in Section H.3. In this section I will give those formulations and prove completeness theorems.

1. The general format

a. Defined connectives

The original formulations of the systems here were given in $L(\neg, \wedge, \Box)$, and I will follow that format first, before considering axiomatizations in $L(\neg, \rightarrow, \wedge)$.

We take the following definitions for both $L(\neg, \wedge, \Box)$ and $L(\neg, \rightarrow, \wedge)$:

$$A \supset B \equiv_{\text{Def}} \neg(A \wedge \neg B)$$

$$A \vee B \equiv_{\text{Def}} \neg(\neg A \wedge \neg B) \qquad A \rightarrow B \equiv_{\text{Def}} \Box(A \supset B)$$

$$A \equiv B \equiv_{\text{Def}} (A \supset B) \wedge (B \supset A) \qquad \Diamond A \equiv_{\text{Def}} \neg\Box\neg A$$

In $L(\neg, \wedge, \Box)$, $A \rightarrow B \equiv_{\text{Def}} \Box(A \supset B)$

In $L(\neg, \rightarrow, \wedge)$, $\Box A \equiv_{\text{Def}} \neg A \rightarrow A$

b. **PC** *in the language of modal logic*

In Chapter II.M.7 I gave an axiomatization of **PC**, classical logic, using only the primitives $\neg$ and $\wedge$. If we allow any formula of $L(\neg, \wedge, \Box)$ to be an instance of A, B, or C in those schema, we then have an *axiomatization of* **PC** *based on* $\neg$ *and* $\wedge$ *for the language of modal logic* (*using* $\Box$). We can then claim that any **PC**-valid wff in that language is provable from the axiomatization via our completeness proof for **PC**, for example $\neg(\Box p_1 \wedge \neg\Box p_1)$.

Similarly, we have an axiomatization of **PC** based on $\neg$ and $\wedge$ for $L(\neg, \rightarrow, \wedge)$.

c. *Normal modal logics*

Every axiomatic system **L** in this chapter will satisfy:

(i) **PC** $\subseteq$ **L**
That is, every theorem of **PC** in the language of modal logic is in **L**

(ii) **L** is closed under the rule of material detachment, $\dfrac{A, A \supset B}{B}$

(iii) **L** contains all instance of the distribution scheme:
$\Box(A \supset B) \supset (\Box A \supset \Box B)$

By (i) and (ii), for each system **L** we will have: if $A \vDash_{\mathbf{PC}} B$, then $\vdash_{\mathbf{L}} A \supset B$.

We call a system a *normal modal logic* if it satisfies (i)–(iii) and:

(iv) **L** is closed under the rule of necessitation $\dfrac{A}{\Box A}$

Let us write '$\vdash$' for derivations in the smallest normal modal logic given by (i), (ii), (iii), and (iv), which we shall shortly see is **K**. The following lemma is a good introduction to the methods of normal modal logics.

Lemma 18 ***a.*** If $\vdash A \equiv B$, then $\vdash \Box A \equiv \Box B$

b. If $\vdash A \equiv B$, then $\vdash \Diamond A \equiv \Diamond B$

c. $\vdash \Box(A \wedge B) \equiv (\Box A \wedge \Box B)$

Proof: a. We have $\mathbf{PC} \vdash (A \equiv B) \supset (A \supset B)$ and hence $\vdash (A \equiv B) \supset (A \supset B)$, so an application of the rule of material detachment gives $\vdash A \supset B$. (In general I will abbreviate an argument like this by saying simply 'by **PC**'.) Now by necessitation, $\vdash \Box(A \supset B)$ and thus using the distribution axioms, $\vdash \Box A \supset \Box B$. Similarly, $\vdash \Box B \supset \Box A$, so by **PC**, $\vdash \Box A \equiv \Box B$.

b. By **PC** we have $\vdash \neg B \supset \neg A$. So by necessitation and using the distribution axioms, $\vdash \Box\neg B \supset \Box\neg A$, and so by **PC**, $\vdash \neg\Box\neg A \supset \neg\Box\neg B$. That is, $\vdash \Diamond A \supset \Diamond B$; and similarly $\vdash \Diamond B \supset \Diamond A$.

c. By **PC** we have $\vdash (A \wedge B) \supset A$, so by necessitation $\vdash \Box((A \wedge B) \supset A)$, and thus using the distribution axioms, $\vdash \Box(A \wedge B) \supset \Box A$. Similarly, $\vdash \Box(A \wedge B) \supset \Box B$. So by **PC**, $\vdash \Box(A \wedge B) \supset (\Box A \wedge \Box B)$.

We also have the distribution axiom $\vdash \Box(B \supset (A \wedge B)) \supset (\Box B \supset \Box(A \wedge B))$. By **PC** we have $\vdash A \supset (B \supset (A \wedge B))$ so using necessitation and distributing we have $\vdash \Box A \supset (\Box(B \supset (A \wedge B)))$, and then by **PC**, $\vdash \Box A \supset (\Box B \supset \Box(A \wedge B))$. So by **PC** (*Importation*), $\vdash (\Box A \wedge \Box B) \supset \Box(A \wedge B)$. ■

2. Axiomatizations and completeness theorems in $L(\neg, \wedge, \Box)$

We define the following axiom systems using schema in $L(\neg, \wedge, \Box)$:

logic	*axioms*	*rules*
K	**PC**	material detachment
	$\Box(A \supset B) \supset (\Box A \supset \Box B)$	necessitation
T	**PC**	material detachment
	$\Box(A \supset B) \supset (\Box A \supset \Box B)$	necessitation
	$\Box A \supset A$	
B	**PC**	material detachment
	$\Box(A \supset B) \supset (\Box A \supset \Box B)$	necessitation
	$\Box A \supset A$	
	$A \supset \Box\Diamond A$	
S4	**PC**	material detachment
	$\Box(A \supset B) \supset (\Box A \supset \Box B)$	necessitation
	$\Box A \supset A$	
	$\Box A \supset \Box\Box A$	
S5	**PC**	material detachment
	$\Box(A \supset B) \supset (\Box A \supset \Box B)$	necessitation
	$\Box A \supset A$	
	$\Box A \supset \Box\Box A$	
	$\Diamond A \supset \Box\Diamond A$	

Another way to describe these systems is to view them as containing the following axioms and *closed under the rules of material detachment and necessitation.*

K is **PC** plus $\Box(A \supset B) \supset (\Box A \supset \Box B)$

T is **K** plus $\Box A \supset A$

B is **T** plus $A \supset \Box\Diamond A$

S4 is **T** plus $\Box A \supset \Box\Box A$

S5 is **S4** plus $\Diamond A \supset \Box\Diamond A$

Our goal now is to prove $\vDash_{\mathbf{L}} A$ iff $\vdash_{\mathbf{L}} A$ for each of these logics, **L**.

Theorem 19 Soundness

If **L** is one of **K**, **T**, **B**, **S4**, or **S5**, then: if $\vdash_{\mathbf{L}} A$, then $\vDash_{\mathbf{L}} A$.

Proof: See Lemma 7 and 8 and Theorems 12, 13, and 14. ■

I'll now show that for each of the five logics **K**, **T**, **B**, **S4**, **S5** there is one particular model, called a canonical model, in which exactly the theorems of that logic hold. The worlds of that model will be complete and consistent sets of wffs, using definitions of those notions which essentially coincide with those for **PC**. This corresponds to the idea that each world is a **PC**-model and that **PC**-models can be identified with complete and consistent sets of wffs.

For the notion of consistency relative to a logic **L** we define:

Γ is **L**-$\wedge$-*inconsistent* if for some $B_1, \ldots, B_n$ in Γ, $\vdash_{\mathbf{L}} \neg(B_1 \wedge \cdots \wedge B_n)$

Γ is **L**-$\wedge$-*consistent* otherwise

We have recourse to this definition in lieu of a deduction theorem here. We can use the ambiguous notation for the conjunction because $\mathbf{PC} \subseteq \mathbf{L}$, so if, for example, $\vdash_{\mathbf{L}} \neg(p_1 \wedge (p_2 \wedge p_3))$, then any result of associating or permuting the p_i's in the formula is also a theorem of **L**. Note that n may be 1.

We define Γ to be *complete* if for all A, at least one of $A, \neg A$ is in Γ.

In an exercise below I ask you to show that these definitions are equivalent to Post-consistency and Post-completeness for theories.

Lemma 20 ***a.*** If Γ is **L**-$\wedge$-consistent, then one of $\Gamma \cup \{\neg A\}$ or $\Gamma \cup \{A\}$ is **L**-$\wedge$-consistent.

b. If Γ is **L**-$\wedge$-consistent and complete, then for each A, exactly one of $A, \neg A$ is in Γ.

c. If Γ is **L**-$\wedge$-consistent and complete, then $\mathbf{L} \subseteq \Gamma$ and Γ is closed under material detachment.

Proof: a. If $\Gamma \cup \{A\}$ is **L**-$\wedge$-inconsistent, then since Γ is **L**-$\wedge$-consistent there must be some $B_1, \ldots, B_n$ such that $\vdash_{\mathbf{L}} \neg(A \wedge B_1 \wedge \cdots \wedge B_n)$. If $\Gamma \cup \{\neg A\}$ is also **L**-$\wedge$-inconsistent, there must be $C_1, \ldots, C_n$ in Γ such that $\vdash_{\mathbf{L}} \neg(\neg A \wedge C_1 \wedge \cdots \wedge C_n)$. Hence by **PC** (compare the remarks in the proof of Lemma 18.a), $\vdash_{\mathbf{L}} \neg(B_1 \wedge \cdots \wedge B_n \wedge C_1 \wedge \cdots \wedge C_n)$, and hence Γ is **L**-$\wedge$-inconsistent, a contradiction. So one of $\Gamma \cup \{\neg A\}$ or $\Gamma \cup \{A\}$ is **L**-$\wedge$-consistent.

I leave part (b) to you.

c. Suppose $\vdash_{\mathbf{L}} A$. Then if $A \notin \Gamma$ we have $\neg A \in \Gamma$. But as $\vdash_{\mathbf{L}} \neg(\neg A)$ (by **PC**) we have a contradiction on the **L**-$\wedge$-consistency of Γ. So $A \in \Gamma$. If A and $A \supset B$ are in Γ, then if $B \notin \Gamma$ we must have $\neg B \in \Gamma$. But $\vdash_{\mathbf{PC}} \neg(A \wedge (A \supset B) \wedge \neg B)$ and that contradicts the **L**-$\wedge$-consistency of Γ; so $B \in \Gamma$. ■

Theorem 21 If Σ is **L**-$\wedge$-consistent, then there is some **L**-$\wedge$-consistent and complete Γ such that $\Sigma \subseteq \Gamma$.

Proof: Let the wffs of the language be ordered as $A_0, A_1, \ldots$. Define:

$$\Gamma_0 = \Sigma$$

$$\Gamma_{n+1} = \begin{cases} \Gamma_n \cup \{A_n\} & \text{if this is } \mathbf{L}\text{-}\wedge\text{-consistent} \\ \Gamma_n \cup \{\neg A_n\} & \text{otherwise} \end{cases}$$

Using the previous lemma, $\Gamma = \bigcup_n \Gamma_n$ is complete and $\mathbf{L}$-$\wedge$-consistent. ■

The *canonical model* for a modal logic $\mathbf{L}$ is $\langle W_{\mathbf{L}}, R_{\mathbf{L}}, e_{\mathbf{L}} \rangle$ where:

$W_{\mathbf{L}} = \{\Gamma : \Gamma$ is a collection of wffs in $L(\neg, \wedge, \Box)$ that is $\mathbf{L}$-$\wedge$-consistent and complete$\}$

$\Gamma R_{\mathbf{L}} \Delta$ iff for all A, if $\Box A \in \Gamma$ then $A \in \Delta$

$e_{\mathbf{L}}(\Gamma) = \{p : p \in \Gamma\}$

Note that $W_{\mathbf{L}}$ is uncountable.

Lemma 22 If $\Gamma \in W_{\mathbf{L}}$ and for all $\Delta \in W_{\mathbf{L}}$ such that $\Gamma R_{\mathbf{L}} \Delta$ we have $B \in \Delta$, then $\Box B \in \Gamma$.

Proof: Suppose $\Gamma \in W_{\mathbf{L}}$ and for all $\Delta \in W_{\mathbf{L}}$ such that $\Gamma R_{\mathbf{L}} \Delta$ we have $B \in \Delta$.

Let $\Sigma = \{A : \Box A \in \Gamma\}$. If $\Sigma \cup \{\neg B\}$ were $\mathbf{L}$-$\wedge$-consistent, then by Theorem 21 there would be some complete and $\mathbf{L}$-$\wedge$-consistent $\Delta \supseteq \Sigma \cup \{\neg B\}$; but $B \notin \Delta$ yet $\Gamma R_{\mathbf{L}} \Delta$, a contradiction. So $\Sigma \cup \{\neg B\}$ is $\mathbf{L}$-$\wedge$-inconsistent.

So for some $B_1, \ldots, B_n$ in Σ, either $\vdash_{\mathbf{L}} \neg(B_1 \wedge \cdots \wedge B_n)$ or $\vdash_{\mathbf{L}} \neg(\neg B \wedge B_1 \wedge \cdots \wedge B_n)$. In either case via **PC**, $\vdash_{\mathbf{L}} \neg B \supset \neg(B_1 \wedge \cdots \wedge B_n)$ and hence $\vdash_{\mathbf{L}} (B_1 \wedge \cdots \wedge B_n) \supset B$. And so by necessitation and the distribution axioms and Theorem 12, $\vdash_{\mathbf{L}} (\Box B_1 \wedge \cdots \wedge \Box B_n) \supset \Box B$, whence by **PC** (*Exportation*) $\vdash_{\mathbf{L}} \Box B_1 \supset (\Box B_2 \supset \cdots \supset (\Box B_n \supset \Box B)) \cdots)$. As each $B_i \in \Sigma$, we have by definition each $\Box B_i \in \Gamma$. So by Lemma 20.c, $\Box B \in \Gamma$. ■

Theorem 23 ***a.*** $\langle W_{\mathbf{L}}, R_{\mathbf{L}}, e_{\mathbf{L}}, \Gamma \rangle \vDash A$ iff $A \in \Gamma$

b. $\vdash_{\mathbf{L}} A$ iff $\langle W_{\mathbf{L}}, R_{\mathbf{L}}, e_{\mathbf{L}} \rangle \vDash A$

Proof: a. Given $\langle W_{\mathbf{L}}, R_{\mathbf{L}}, e_{\mathbf{L}} \rangle$ we proceed for all Γ and A by induction on the length of A. It's true for the propositional variables. Then:

$\Gamma \vDash \neg A$ iff $\Gamma \nvDash A$
iff $A \notin \Gamma$ by induction
iff $\neg A \in \Gamma$ by the completeness of Γ

$\Gamma \vDash A \wedge B$ iff $\Gamma \vDash A$ and $\Gamma \vDash B$
iff $A, B \in \Gamma$ by induction
iff $(A \wedge B) \in \Gamma$ by Lemma 20, as $\mathbf{PC} \subseteq \mathbf{L}$

$\Gamma \models \Box A$ iff for every Δ, if $\Gamma R_L \Delta$ then $\Delta \models A$

iff for every Δ, if $\Gamma R_L \Delta$ then $A \in \Delta$ — by induction

iff $\Box A \in \Gamma$ — by Lemma 22

And thus part (a) is proved.

b. By Lemma 20, for each $\Gamma \in W_L$, $\mathbf{L} \subseteq \Gamma$, so if $\vdash_L A$, then $\langle W_L, R_L, e_L \rangle \models A$. If $\nvdash_L A$, then $\{\neg A\}$ is $\mathbf{L}$-$\wedge$-consistent. By Theorem 21 there is an $\mathbf{L}$-$\wedge$-consistent complete Γ such that $\neg A \in \Gamma$. Hence $\langle W_L, R_L, e_L, \Gamma \rangle \not\models A$, so $\langle W_L, R_L, e_L \rangle \not\models A$. ■

We now need to verify that the accessibility relation of each canonical model has the appropriate properties. We cannot use Lemma 8, since in Theorem 23 we considered only one model from each canonical frame $\langle W_L, R_L \rangle$.

Lemma 24 ***a.*** If $\mathbf{L}$ contains the scheme $\Box A \supset A$, then $\langle W_L, R_L \rangle$ is reflexive.

b. If $\mathbf{L}$ contains the scheme $\Box A \supset \Box\Box A$, then $\langle W_L, R_L \rangle$ is transitive.

c. If $\mathbf{L}$ contains the scheme $A \supset \Box\Diamond A$, then $\langle W_L, R_L \rangle$ is symmetric.

d. If $\mathbf{L}$ contains the scheme $\Diamond A \supset \Box\Diamond A$, then $\langle W_L, R_L \rangle$ is euclidean.

Proof: a. Suppose $\vdash_L \Box A \supset A$. Then for all A, if $\Box A \in \Gamma$, by Lemma 20.c, $A \in \Gamma$, so $\Gamma R_L \Gamma$.

b. If $\Gamma R_L \Delta$ and $\Delta R_L \Sigma$, we need to show that $\Gamma R_L \Sigma$. This is if and only if for all A, if $\Box A \in \Gamma$ then $A \in \Sigma$. But if $\Box A \in \Gamma$ then $\Box\Box A \in \Gamma$ by assumption using Lemma 20, so $\Box A \in \Delta$, so $A \in \Sigma$.

c. Assume $\Gamma R_L \Delta$. To show $\Delta R_L \Gamma$ we need for all B that if $\Box B \in \Delta$ then $B \in \Gamma$. Suppose $\Box B \in \Delta$ and $B \notin \Gamma$. Then $\neg B \in \Gamma$. So by the assumption of this part we get $\Box\Diamond\neg B \in \Gamma$, so $\Diamond\neg B \in \Delta$. That is, $\neg\Box\neg\neg B \in \Delta$, so by Lemma 20.b, $\neg\Box B \in \Delta$, which is a contradiction on the $\mathbf{L}$-$\wedge$-consistency of Δ. Hence $\Delta R_L \Gamma$.

d. Suppose $\Gamma R_L \Delta$ and $\Gamma R_L \Sigma$. We want $\Delta R_L \Sigma$. Suppose not and some $\Box B \in \Delta$, yet $B \notin \Sigma$. Then $\Box B \notin \Gamma$, so $\neg\Box B \in \Gamma$. Thus $\Diamond\neg B \in \Gamma$ by using Lemma 20.b and **PC**. So by the assumption of this part, $\Box\Diamond\neg B \in \Gamma$. Hence $\Diamond\neg B \in \Delta$, so $\neg\Box B \in \Delta$ and $\Box B \notin \Delta$, a contradiction. ■

Theorem 25 ***Completeness of Possible-World Semantics in*** $\mathbf{L}(\neg, \wedge, \Box)$

For each logic **L** of **K**, **T**, **B**, **S4**, and **S5** in $L(\neg, \wedge, \Box)$:

$\vdash_L A$ iff $\models_L A$

Proof: This follows from Theorems 19 and 23 and Lemma 24. ■

3. Axiomatizations and completeness theorems in $L(\neg, \rightarrow, \wedge)$

We define the following axiom systems using schema in $L(\neg, \rightarrow, \wedge)$.

logic	*axioms*	*rules*
K	**PC**	material detachment
	$(\neg(A\supset B)\to(A\supset B)) \equiv A\to B$	necessitation
	$(A\to B) \supset ((\neg A\to A) \supset (\neg B\to B))$	
T	**PC**	material detachment
	$(\neg(A\supset B)\to(A\supset B)) \equiv A\to B$	necessitation
	$(A\to B) \supset ((\neg A\to A) \supset (\neg B\to B))$	
	$(\neg A\to A) \supset A$	
B	**PC**	material detachment
	$(\neg(A\supset B)\to(A\supset B)) \equiv A\to B$	necessitation
	$(A\to B) \supset ((\neg A\to A) \supset (\neg B\to B))$	
	$(\neg A\to A) \supset A$	
	$A \supset ((A\to\neg A) \to \neg(A\to\neg A))$	
S4	**PC**	material detachment
	$(\neg(A\supset B)\to(A\supset B)) \equiv A\to B$	necessitation
	$(A\to B) \supset ((\neg A\to A) \supset (\neg B\to B))$	
	$(\neg A\to A) \supset A$	
	$(\neg A\to A) \supset (A\to(\neg A\to A))$	
S5	**PC**	material detachment
	$(\neg(A\supset B)\to(A\supset B)) \equiv A\to B$	necessitation
	$(A\to B) \supset ((\neg A\to A) \supset (\neg B\to B))$	
	$(\neg A\to A) \supset A$	
	$(\neg A\to A) \supset (A\to(\neg A\to A))$	
	$A \supset ((A\to\neg A) \to \neg(A\to\neg A))$	

By incorporating into the axiomatizations the (defined) equivalence $\Box(A\supset B) \equiv A\to B$, the proofs in Section E.2 carry through as before. Only Lemma 24 requires a new proof for transitivity, but I will leave that to you.

Theorem 26 Completeness of Possible-World Semantics in **L($\neg,\to,\wedge$)**
For each logic **L** of **K**, **T**, **B**, **S4**, and **S5** in L($\neg,\to,\wedge$): $\vdash_{\mathbf{L}} A$ iff $\vDash_{\mathbf{L}} A$

Exercises for Sections E.1–E.3

1. What is a normal modal logic?

2. Define **QT** to be $\mathbf{K} + \Box A \supset A$ closed only under material detachment. Exhibit two theorems of **T** that are not theorems of **QT**. Is every model of **QT** reflexive?

3. a. What is an **L**-$\wedge$-consistent collection of wffs?
   b. Is every **PC**-consistent collection of wffs **L**-$\wedge$-consistent?
   c. Is every **L**-$\wedge$-consistent collection of wffs **PC**-consistent?

4. Prove that if a collection of wffs Γ is **L**-$\wedge$-consistent, then for every A exactly one of A, $\neg A$ is in Γ.

5. Show that for **L** one of **K**, **T**, **B**, **S4**, or **S5**:
   a. Γ is **L**-$\wedge$-consistent iff Γ is Post-consistent
   b. If Γ is a theory, then Γ is complete iff Γ is Post-complete

6. Define the canonical model of a modal logic **L**. Why is the accessibility relation defined as it is?

7. Prove in $L(\neg, \rightarrow, \wedge)$:
   If **L** contains the scheme $(\neg A \rightarrow A) \supset A$,
   then $\langle W_L, R_L \rangle$ is reflexive.
   If **L** contains the scheme $(\neg A \rightarrow A) \supset (A \rightarrow (\neg A \rightarrow A))$,
   then $\langle W_L, R_L \rangle$ is transitive.
   If **L** contains the scheme $A \supset ((A \rightarrow \neg A) \rightarrow \neg(A \rightarrow \neg A))$,
   then $\langle W_L, R_L \rangle$ is symmetric.

8. Given **L** one of **K**, **T**, **B**, **S4**, or **S5** and Γ closed under subformulas, is the coarsest filtration of the canonical model for **L** a model for **L**?

9. (Open) Determine whether the axiom scheme $\Box(A \supset B) \equiv A \rightarrow B$ is superfluous in the axiomatizations in $L(\neg, \rightarrow, \wedge)$.

4. Consequence relations

For normal logics there are two ways we can define a syntactic consequence relation, depending on whether we allow the rule of necessitation to apply to non-theorems.

a. *Without necessitation,* $\vdash_L$

Let **L** be one of the five logics **K**, **T**, **B**, **S4**, or **S5**. For $\Gamma \subseteq$ Wffs, define $\Gamma \vdash_L A$ to mean that there are wffs $B_1, \ldots, B_n = A$ such that each B_i is in **L**, or is in Γ, or is a direct consequence of earlier B_i's by the rule of material detachment. So $\varnothing \vdash_L A$ iff $\mathbf{L} \vdash A$ as defined in the body of the chapter.

This is not our standard definition of proof, for in a derivation we allow B_n to be a theorem and not just an axiom of **L**. Thus, this notion of consequence requires us to interleave two proof procedures, one for theorems of **L** and one for consequences of Γ. We need this because without the rule of necessitation we could not prove the theorems of **L**. The result is that this is the **PC** notion of consequence:

$\Gamma \vdash_{\mathbf{L}} A$ iff $\Gamma \cup \mathbf{L} \vdash_{\mathbf{PC}} A$

Thus the appropriate notions of completeness and consistency are as for **PC**, and the definition of a theory is standard: Γ is a $\vdash_{\mathbf{L}}$- *theory* if $\mathbf{L} \subseteq \Gamma$ and Γ is closed under the proof rule of $\vdash_{\mathbf{L}}$, namely, material detachment. Then the *elements of the canonical model are complete consistent* $\vdash_{\mathbf{L}}$*-theories*: these are the canonical possible worlds.

This notion of syntactic consequence correlates to the idea that only the laws of logic are necessary if true and that $\rightarrow$ should be read as 'entails'.

Theorem 27 ***Strong Completeness with respect to Possible World Models with Designated World***

For each logic **L** listed below we have:

$\Gamma \vdash_{\mathbf{L}} A$ iff for every $\langle W, R, e, w \rangle$ in the class listed, if $w \models \Gamma$, then $w \models A$

K all $\langle W, R, e, w \rangle$

T all reflexive $\langle W, R, e, w \rangle$

B all reflexive and symmetric $\langle W, R, e, w \rangle$

S4 all reflexive and transitive $\langle W, R, e, w \rangle$

S5 all equivalence $\langle W, R, e, w \rangle$

Proof: I'll leave the left to right direction to you.

For the right to left direction, let **L** be one of these logics and suppose that $\Gamma \nvdash_{\mathbf{L}} A$. Then $\Gamma \cup \{\neg A\}$ is **L**-$\wedge$-consistent: if not, then for some $B_1, \ldots, B_n \in \Gamma$ either $\vdash_{\mathbf{L}} \neg(B_1 \wedge \cdots \wedge B_n)$ or $\vdash_{\mathbf{L}} \neg((B_1 \wedge \cdots \wedge B_n) \wedge \neg A)$. In both cases, using the fact that $\mathbf{PC} \subseteq \mathbf{L}$ we get $\vdash_{\mathbf{L}} B_1 \supset (B_2 \supset \cdots \supset (B_n \supset A)) \cdots)$, so $\Gamma \vdash_{\mathbf{L}} A$, a contradiction. Hence there is some complete **L**-$\wedge$-consistent Σ such that $\Gamma \cup \{\neg A\} \subseteq \Sigma$. Choose one.

Define $U = \{ \Delta \in W_{\mathbf{L}} : \neg A \in \Delta \}$. Then $U \neq \varnothing$ as $\Sigma \in U$. So define $M = \langle U, R_{\mathbf{L}}, e_{\mathbf{L}}, \Sigma \rangle$ where $R_{\mathbf{L}}$ and $e_{\mathbf{L}}$ are defined as in the canonical model. Now proceed as in Lemma 22, Theorem 23, and Lemma 24 to prove that M satisfies the appropriate condition stated in the theorem and that for all $\Delta \in U$, $\Delta \models B$ iff $B \in \Delta$. So we have $\Sigma \models \Gamma$ and $\Sigma \nvDash A$. ■

Theorem 28 ***Material Implication Form of the Deduction Theorem***

If **L** is any one of **K**, **T**, **B**, **S4**, or **S5**, then:

$$\Gamma \cup \{A\} \vdash_{\mathbf{L}} B \text{ iff } \Gamma \vdash_{\mathbf{L}} A \supset B$$

$$\Gamma \cup \{A\} \models_{\mathbf{L}} B \text{ iff } \Gamma \models_{\mathbf{L}} A \supset B$$

Proof: The syntactic part comes from our observation above that $\Gamma \vdash_{\mathbf{L}} A$ iff $\Gamma \cup \mathbf{L} \vdash_{\mathbf{PC}} A$; the semantic part is immediate from the definitions. ■

b. With necessitation, $\vdash_{\mathbf{L}}^{\Box}$

Let **L** be any one of the five logics **K**, **T**, **B**, **S4**, or **S5**, and let $\Gamma \subseteq$ Wffs. Define $\Gamma \vdash_{\mathbf{L}}^{\Box} A$ to mean that there are $B_1, \ldots, B_n = A$ such that each B_i is an axiom of **L**, or is in Γ, or is a direct consequence of earlier B_i's by the rule of material detachment or the rule of necessitation. Thus to proceed on the hypothesis of A is also to assume that A is necessary. This is the usual notion of syntactic derivation and here we also have $\varnothing \vdash_{\mathbf{L}}^{\Box} A$ iff $\mathbf{L} \vdash A$.

Theorem 29 Strong Completeness with respect to Frames

For each logic **L** listed below we have:

$\Gamma \vdash_{\mathbf{L}}^{\Box} A$ iff for every frame $\langle W, R \rangle$ in the class listed, for every evaluation e, if $\langle W, R, e \rangle \vDash \Gamma$, then $\langle W, R, e \rangle \vDash A$

K all frames

T all reflexive frames

B all reflexive and symmetric frames

S4 all reflexive and transitive frames

S5 all equivalence frames

The two usual forms of the deduction theorem fail for this notion of syntactic consequence. For each logic, for every A, $A \vdash_{\mathbf{L}}^{\Box} \Box A$, but both $\vdash_{\mathbf{L}} A \supset \Box A$ and $\vdash_{\mathbf{L}} A \rightarrow \Box A$ can fail. However, we do have:

***Theorem 30 Deduction Theorem for* S4 *and* S5**

If **L** is either **S4** or **S5**, then:

a. $\Gamma, A \vdash_{\mathbf{L}}^{\Box} B$ iff $\Gamma \vdash_{\mathbf{L}}^{\Box} \Box A \supset B$

b. $\Gamma, A \vdash_{\mathbf{L}}^{\Box} B$ iff $\Gamma \vdash_{\mathbf{L}}^{\Box} \Box A \supset \Box B$

Proof: For **S5** this follows by the *Semantic Deduction Theorem*, p. 216. I leave the proof for **S4** to you. ∎

Deduction theorems for other modal logics using this notion of syntactic consequence are more complicated. See *Porte, 1982, Surma, 1972,* and *Perzanowski, 1973.*

Exercises for Section E.4

1. Distinguish between $\vdash_{\mathbf{L}}$ and $\vdash_{\mathbf{L}}^{\Box}$

2. a. State the *Deduction Theorem* for $\vdash_{\mathbf{L}}$

   b. Why do the usual forms of the deduction theorem fail for $\vdash_{\mathbf{L}}^{\Box}$?

3. Give an entirely syntactic proof of Theorem 30.
4. (Open?) Establish a syntactic counterpart to the semantic relation:
   if $\langle W,R \rangle \vDash \Gamma$, then $\langle W,R \rangle \vDash A$

F. Quasi-Normal Modal Logics

The discussion of syntactic consequence in the last section suggests the following definition.

A syntactically presented system **L** in $L(\neg, \wedge, \Box)$ is a *quasi-normal* modal logic if:

(i) $\mathbf{K} \subseteq \mathbf{L}$

(ii) **L** is closed under the rule of material detachment

(iii) **L** contains all instance of the distribution scheme
$\Box(A \supset B) \supset (\Box A \supset \Box B)$

A normal modal logic is then a quasi-normal logic that is closed under the rule of necessitation. We need $\mathbf{K} \subseteq \mathbf{L}$ rather than just $\mathbf{PC} \subseteq \mathbf{L}$ in order to ensure that we have possible-world models for quasi-normal logics, since we don't have the rule of necessitation.

For possible-world semantics we have that $\langle W,R,e,w \rangle \vDash \Box A \supset A$ iff wRw. If we require of our class of frames that R be reflexive, that is for all z, zRz, we have **T**. Putting a global condition on $\langle W,R \rangle$ corresponds to closing the logic under the rule of necessitation. However, we may require only a local condition on R by taking a frame with a designated world, $\langle W,R,w \rangle$. Then $\langle W,R,w \rangle \vDash A$ is defined to mean that all e, $\langle W,R,e,w \rangle \vDash A$.

We define:

$\langle W,R,e,w \rangle$ is a *model with designated world* w
if $\langle W,R,e \rangle$ is a model and $w \in W$

And $\langle W,R,w \rangle \vDash A$ iff for every evaluation e on $\langle W,R,w \rangle$, $\langle W,R,e,w \rangle \vDash A$

As an example of a quasi-normal modal logic we define:

QT
in $L(\neg, \wedge, \Box)$
is the closure of $\mathbf{K} \cup \{\Box A \supset A\}$ under the rule of material detachment.

Theorem 31 ***a.*** $\vdash_{\mathbf{QT}} A$ iff for every $\langle W,R,w \rangle$ such that wRw, $\langle W,R,w \rangle \vDash A$

b. $\vdash_{\mathbf{QT}} A$ iff for every finite $\langle W,R,w \rangle$ such that wRw, $\langle W,R,w \rangle \vDash A$

c. $\Gamma \vdash_{\mathbf{QT}} A$ iff for every $\langle W,R,w \rangle$ such that wRw,
if $\langle W,R,w \rangle \vDash \Gamma$, then $\langle W,R,w \rangle \vDash A$

Proof: It's easy to show that if $\vDash_{\mathbf{QT}} A$, then for every $\langle W, R, e, w \rangle$ such that wRw we have $w \vDash A$.

Now suppose $\nvdash_{\mathbf{QT}} A$. First, $\mathbf{QT} \cup \{\neg A\}$ is $\mathbf{QT}$-$\wedge$-consistent and hence is contained in some $\mathbf{QT}$-$\wedge$-consistent and complete set of wffs Σ. Choose such a Σ. Now let $\langle W, R, e \rangle$ be the canonical model for $\mathbf{K}$. Set $M = \langle W, R, e, \Sigma \rangle$. Since $(\Box B \supset B) \in \Sigma$ we have $M \vDash \Box B \supset B$. Hence if $\Box B \in \Sigma$ then $B \in \Sigma$, so $\Sigma R \Sigma$. And since $\neg A \in \Sigma$, $M \nvDash A$.

For part (b) adapt the method of filtrations to models with designated world.

I leave part (c) as an exercise. ■

Is *modus ponens* a valid rule in **QT**? Using the semantic characterization of **QT** from Theorem 31 we have that if wRw in $\langle W, R, w \rangle$ and $\langle W, R, w \rangle \vDash A \rightarrow B$, then $\langle W, R, w \rangle \vDash A \supset B$, so if $\langle W, R, w \rangle \vDash A$, then $\langle W, R, w \rangle \vDash B$.

What then distinguishes **QT** from **T**? I'll let you show that the following is a theorem of **T** but not of **QT**:

$$\Box(\Box p_1 \supset p_1)$$

In accord with the definition of consequence without necessitation, we have that the "laws of logic" (**PC**) are necessary, but new modal principles added to those laws are simply true, not necessary.

Segerberg, 1971, has an exposition of quasi-normal logics; *Blok and Köhler, 1983,* treat them algebraically.

Exercises for Section F

1. Prove Theorem 31.c.
2. Give a syntactic proof that *modus ponens* is valid in **QT**.
3. Define **QS4** and give a completeness theorem for it.
4. Define **QS5** and give a semantic characterization of it. Do iterated modalities collapse in **QS5**? Can **QS5** be viewed as a formalization of logical necessity?

G. Set-Assignment Semantics for Modal Logics

It's often said by modal logicians that the content of a proposition is the possible worlds in which it is true. That suggests that possible worlds exist before we stipulate them in a language. Even were that a metaphysical matter of fact, as *David Lewis, 1973,* Chapter 4, argues, it would be beyond our formal methods: our models allow us to characterize possible worlds only in terms of the sentences true in them. However, given a possible-worlds model we can identify a proposition with the worlds in which it is true, as I pointed out in the Note on p. 206. Set-assignment semantics are based on this idea.

1. Semantics in $L(\neg, \rightarrow, \wedge)$

a. Modal semantics of implication

Given a proposition A, we take a set $s(A)$ for its content, viewing the elements of $s(A)$ as descriptions of the ways things could be if A is true. Because we choose to use classical logic in reasoning about any particular description we must have the following, where $\overline{s(A)}$ denotes the complement of $s(A)$:

$$s(A \wedge B) = s(A) \cap s(B)$$

$$s(\neg A) = \overline{s(A)}$$

Because we want to respect the equivalence of $A \rightarrow B$ and $\Box(A \supset B)$, we want:

$$s(A \rightarrow B) = s(\Box(A \supset B))$$

Any further structural requirements on contents of propositions will depend on what particular notion of necessity we are modeling.

What should the truth-conditions be? We've agreed that $\neg$ and $\wedge$ are to be interpreted classically. How should we analyze $A \rightarrow B$?

Recall the strict analysis of implication:

$A \rightarrow B$ is true iff it's not possible that both A is true and B is false
iff there is no way in which both A is true and B is false
iff $s(A) \cap s(\neg B) = \varnothing$
iff $s(A) \cap \overline{s(B)} = \varnothing$

That is,

(10) $A \rightarrow B$ is true iff $s(A) \subseteq s(B)$

Need we also require that if A is true, then so is B? That seems a *sine qua non* of implication, for without it we cannot use *modus ponens*: from A and $A \rightarrow B$ conclude B. So we take instead of (10):

$A \rightarrow B$ is true iff $s(A) \subseteq s(B)$ and not both A is true and B is false

A *modal semantics for (of) implication* is a set-assignment model or class of models $\langle v, s \rangle$ where:

M1. $s(A \wedge B) = s(A) \cap s(B)$

M2. $s(\neg A) = \overline{s(A)}$

M3. $s(A \rightarrow B) = s(\Box(A \supset B))$

$\neg$ and $\wedge$ are evaluated classically

$\rightarrow$ is evaluated by:

A	B	$s(A) \subseteq s(B)$	$A \to B$
any	values	fails	F
T	T		T
T	F	holds	F
F	T		T
F	F		T

That is, $<v,s>$ satisfies M1–M3 and uses the dual-dependence truth-conditions. We say the model is *finite* if S is finite.

Corresponding to the *Fully General Abstraction* for possible-world semantics, we have:

The Fully General Abstraction for Set-Assignment Semantics for Classical Modal Logics

Any set-assignment s satisfying M1–M3 and valuation v extended to all wffs by the dual dependence truth-conditions comprise a modal semantics for implication.

For particular notions of implication associated with various modal logics we will assume a fully general abstraction that states that any modal semantics of implication satisfying the appropriate structural criteria is a model for that logic.

Here are some observations we will need later.

Lemma 32 If $<v,s>$ is a modal semantics for implication, then:

a. $s(A \supset B) = \overline{s(A)} \cup s(B)$

$s(A \supset B) = \mathsf{S}$ iff $s(A) \subseteq s(B)$

If $\mathbf{PC} \vDash A$, then $v(A) = \mathsf{T}$ and $s(A) = \mathsf{S}$.

b. $v(\Box A) = \mathsf{T}$ iff $v(A) = \mathsf{T}$ and $s(A) = \mathsf{S}$

c. $v((A \to B) \leftrightarrow \Box(A \supset B)) = \mathsf{T}$

d. $v(\Diamond A) = \mathsf{T}$ iff $v(A) = \mathsf{T}$ or $s(A) \neq \varnothing$

e. $v(\Box(B \supset C) \supset (\Box B \supset \Box C)) = \mathsf{T}$

f. $v(\Box A \supset A) = \mathsf{T}$

g. For all A and B, $s(A \to B) \subseteq \overline{s(A)} \cup s(B)$ iff for all A, $s(\Box A) \subseteq s(A)$.

h. If $\mathbf{PC} \vDash A$, then $v(\Box A) = \mathsf{T}$.

Proof: The proofs of these are useful to establish familiarity with these semantics.

The first two parts of part (a) I will leave to you. For the last part of (a), we have that if $\mathbf{PC} \vDash A$, then $v(A) = \mathsf{T}$, since modal semantics of implication evaluate $\neg$

and $\wedge$ as do models of **PC**. To show that if $\mathbf{PC} \vDash A$, then $\mathsf{s}(A) = \mathsf{S}$, consider the strongly complete axiomatization of **PC** in $L(\neg, \wedge)$ given on p. 80. We proceed by induction on the length of a proof of A. If A is an instance of the first axiom, $B \supset (C \supset B)$, then $\mathsf{s}(B \supset (C \supset B)) = \overline{\mathsf{s}(B)} \cup \mathsf{s}(C \supset B) = \overline{\mathsf{s}(B)} \cup (\overline{\mathsf{s}(C)} \cup \mathsf{s}(B)) = \mathsf{S}$. I will leave to you to verify that $\mathsf{s}(A) = \mathsf{S}$ for every instance of every other axiom. Then if A is proved as a consequence of B and $B \supset C$, we must have by induction that $\mathsf{s}(B) = \mathsf{S}$ and $\mathsf{s}(B \supset C) = \mathsf{S} = \overline{\mathsf{s}(B)} \cup \mathsf{s}(C)$, so $\mathsf{s}(C) = \mathsf{S}$.

b. $v(\Box A) = \mathsf{T}$ iff $v(\neg A \rightarrow A) = \mathsf{T}$
iff $\mathsf{s}(\neg A) \subseteq \mathsf{s}(A)$ and $v(\neg A \supset A) = \mathsf{T}$
iff $\overline{\mathsf{s}(A)} \subseteq \mathsf{s}(A)$ and $v(A) = \mathsf{T}$
iff $\mathsf{s}(A) = \mathsf{S}$ and $v(A) = \mathsf{T}$

c. By M3 this reduces to showing that $v(A \rightarrow B) = \mathsf{T}$ iff $v(\Box(A \supset B)) = \mathsf{T}$.
$v(A \rightarrow B) = \mathsf{T}$ iff $v(A \supset B) = \mathsf{T}$ and $\mathsf{s}(A) \subseteq \mathsf{s}(B)$
iff $v(A \supset B) = \mathsf{T}$ and $\mathsf{s}(A \supset B) = \mathsf{S}$ by (a)
iff $v(\Box(A \supset B)) = \mathsf{T}$ by (b)

d. $v(\Diamond A) = \mathsf{T}$ iff $v(\neg\Box\neg A) = \mathsf{T}$
iff $v(\Box\neg A) = \mathsf{F}$
iff $\mathsf{s}(\neg A) \neq \mathsf{S}$ or $v(\neg A) = \mathsf{F}$
iff $\mathsf{s}(A) \neq \varnothing$ or $v(A) = \mathsf{T}$

e. You can establish this by repeated use of (a) and (b).

f. If $v(\Box A) = \mathsf{T}$, then by (b), $v(A) = \mathsf{T}$.

g. If for all A and B we have $\mathsf{s}(A \rightarrow B) \subseteq \overline{\mathsf{s}(A)} \cup \mathsf{s}(B)$, then:

$$\mathsf{s}(\Box A) = \mathsf{s}(\neg A \rightarrow A) \subseteq \overline{\overline{\mathsf{s}(A)}} \cup \mathsf{s}(A) = \mathsf{s}(A)$$

If $\mathsf{s}(\Box A) \subseteq \mathsf{s}(A)$ for all A, then by M3,

$$\mathsf{s}(A \rightarrow B) = \mathsf{s}(\Box(A \supset B)) \subseteq \mathsf{s}(A \supset B) = \overline{\mathsf{s}(A)} \cup \mathsf{s}(B)$$

h. This follows from (a) and (b). ■

Parts (b) and (d) of Lemma 32 show that in any modal semantics of implication the connectives $\Box$ and $\Diamond$ are evaluated by the following tables:

(11)

A	$\mathsf{s}(A) = \mathsf{S}$	$\Box A$
any value	fails	F
T	holds	T
F		F

(12)

A	$\mathsf{s}(A) \neq \varnothing$	$\Diamond A$
T	fails	T
F		F
T	holds	T
F		T

b. Weak modal semantics of implication

There are some modal systems, such as **K**, in which *modus ponens* is not a valid rule.

And $A \supset \Diamond A$ fails. The world at which we are evaluating $A \rightarrow B$ may not be accessible to itself, so the truth-value of $A \rightarrow B$ would not depend on the truth or falsity of A or of B in that world. It is not clear whether such systems are logics, but we may nonetheless give them set-assignment semantics, using (10) instead of the dual-dependence table for $\rightarrow$.

A *weak modal semantics for (of) implication* is a set-assignment model or class of models $<v, s>$ where:

s satisfies M1–M3

$\neg$ and $\wedge$ are evaluated classically

$\rightarrow$ is evaluated by:

(13)

A	B	$s(A) \subseteq s(B)$	$A \rightarrow B$
any	values	fails	F
any	values	holds	T

Aside: This is a wholly intensional reading of $\rightarrow$, as discussed in Chapter IV.G.5. This connective is called *weak implication,* and table (13) is called the *weak table for the conditional.*

I will leave the proof of the following lemma to you.

Lemma 33 If $<v, s>$ is a weak modal semantics, then (a), (c), (e), (g), (h), of Lemma 32 hold and:

$v(\Box A) = T$ iff $s(A) = S$

$v(\Diamond A) = T$ iff $s(A) \neq \varnothing$

So in weak modal semantics of implication we have the following tables:

(14)

A	$s(A) = S$	$\Box A$
any value	fails	F
any value	holds	T

(15)

A	$s(A) \neq \varnothing$	$\Diamond A$
any value	fails	F
any value	holds	T

These, too, are wholly intensional connectives.

2. Semantics in L($\neg, \wedge, \Box$)

If $\neg$, $\wedge$, $\Box$ are taken as primitives, defining $A \rightarrow B$ as $\Box(A \supset B)$, then table (11) replaces the table for dual-dependent implication in the definition of modal semantics for implication. And condition M3 can be deleted since it holds by definition. Lemma 32 continues to hold, except that we replace part (b) by:

$A \to B$ is true iff $s(A) \subseteq s(B)$ and not both A is true and B is false. That is, the dual-dependent table for implication is now derived, rather than taken as primitive.

For weak modal semantics of implication in $L(\neg, \wedge, \Box)$, table (14) replaces table (13); table (13) can then be derived.

Below we shall prove completeness theorems for set-assignment semantics for $L(\neg, \wedge, \Box)$. The proofs apply to $L(\neg, \to, \wedge)$ with minor modifications and the addition, in each case, of a proof that M3 holds.

Exercises for Sections G.1 and G.2

1. a. What is a modal semantics of implication?
   b. Why do we require $v(A) = F$ or $v(B) = T$ for $v(A \to B) = T$?
   c. What modal systems we have studied cannot have modal semantics of implication? (Hint: See Lemma 32.)
2. a. How does a weak modal semantics of implication differ from a modal semantics of implication?
   b. Verify Lemma 33.
3. Prove that in both modal semantics of implication and in weak modal semantics of implication $v(\Box(B \supset C) \supset (\Box B \supset \Box C)) = T$.

3. Connections of meanings in modal logics: the aptness of set-assignment semantics

> No one is likely to deny that the logical impossibility of $(\neg p \wedge q)$ is a *necessary* condition of q's deducibility from p, but it has been suggested that it is not a *sufficient* condition on the ground that a further condition of q's deducibility from p is that there should be some connection of "content" or "meaning" between p and q. It is, however, extremely difficult, if not impossible, to state this additional requirement in precise terms; and to insist on it seems to introduce into an otherwise clear and workable account of deducibility a gratuitously vague element which will make it impossible to determine whether a given formal system is a correct logic of entailment or not.
>
> *Hughes and Cresswell, 1968,* p. 336-337

Thus Hughes and Cresswell defend modal logic, reading what we call $A \to B$ as 'A entails B' or what they take to be equivalent, 'B is deducible from A'. But modal logic does seem based on just such a vague notion of connection of meaning as Hughes and Cresswell wish to exclude from logic, and the set-assignment semantics for implication reflect that.

Modal logicians often say that a proposition is to be identified with the possible worlds in which it is true. To understand a proposition, then, is to be able to conceive of the various possible ways in which it could be true.

> The word 'world' has been used by a number of logicians . . . and seems to be the most convenient one, but perhaps some such phrase as 'conceivable or evisageable state of affairs' would convey the idea better.
>
> *Hughes and Cresswell, 1968*, p. 75

To say that A→B is true is to say that we cannot envisage a state of affairs in which A is true and B is false: the appropriate connection of meaning between A and B obtains. Just because this connection of meaning can be given a rigorous mathematical treatment based on the semantics of classical logic does not mean that in any *application* it is less vague than, say, the referential content of a proposition. Consider how Hughes and Cresswell motivate accessibility relations.

> We can conceive of various worlds which would differ in certain ways from the actual one (a world without telephones, for example). But our ability to do this is at least partly governed by the kind of world we live in: the constitution of the human mind and the human body, the languages which exist or do not exist, and many other things, set certain limits to our powers of conceiving. We could then say that a world, w_1, is accessible to a world, w_2, if w_2 is conceivable by someone living in w_1, and this will make accessibility a relation between worlds as we want it to be.
>
> *Hughes and Cresswell, 1968* , p. 77

The problem, indeed, is determining whether a particular partial description really is an allowed way of imagining, a model of **PC**, as I discussed in Example 5 of Section D, p. 227.

A platonist might object to the way I've presented modal logic, saying that conceiving and imagining have nothing to do with it. Propositions exist as abstract objects, and a possible world is just as real as those. Classical modal logic is the right way to reason about those worlds, those possible states of affairs, which are fixed for all time and independent of us and our language. The difficulty with that view is the same one I have with abstract propositions: we have no direct access to these possible worlds, so how in any application of modal logic are we to proceed? All the arguments concerning connections of meaning and imagining that I've put forward would apply to how a platonist is to use his modal logic. The platonist might counter by saying that it's truth and reality he's studying, not how we deal with it.

Set-assignment semantics for modal logics bring out the structural way in which connections of meanings between propositions function in modal logic. They should not be seen as replacing possible-world semantics, but as bringing out the similarities between modal logic and many other logics, setting all of these within a general semantic framework. Moreover, some modal logics such as **G*** (Section J.3 below) have no possible-world semantics, yet we can characterize them with set-assignment semantics, placing them alongside others in the general framework.

And set-assignment semantics aptly reflect the idea that if when we say that the content of a proposition is the possible worlds in which it is true we also mean to include the actual world, then the content of a proposition incorporates its truth-value. For logics based on that notion formulation (10) and the dual-dependent table for implication should be equivalent, and so they are.

Moreover, it is correct to speak of the truth or falsity of a modal proposition: we are interested in whether 'Roses are red → sugar is sweet' is *true*. And this is a classical conception of truth in which every proposition is true or false but not both. This is something that is obscured by the possible-world semantics but is the basis of the set-assignment ones.

4. S5

S5 is the logic of logical necessity, in which no accessibility relation is needed (see Section C.1 above). This simplicity of the possible-worlds semantics allows us to give simple set-assignment semantics, which can be proved strongly complete by establishing correspondences between possible-worlds models and set-assignment models. We will work in $L(\neg, \wedge, \Box)$.

A modal semantics for implication $<v, s>$ is an **S5**-*model* if it satisfies:

The **S5** *set-assignment conditions*

$s(\Box A) = S$ or $\varnothing$

$s(\Box A) = S$ iff $s(A) = S$

If $s(A) = S$, then $v(A) = T$

These are the first set-assignment semantics we have seen whose structural requirements on set-assignments involve truth-conditions.

Lemma 34 For every possible-worlds model $\langle W, e, w \rangle$, there is an **S5**-model $<v, s>$ such that for all A, $v(A) = T$ iff $w \vDash A$. If $\langle W, e, w \rangle$ is finite, then $<v, s>$ is finite.

Proof: Given $\langle W, e, w \rangle$, set:

$S = W$

$s(A) = \{ z : z \vDash A \}$

$v(p) = T$ iff $w \vDash p$

Extend v to all wffs by the dual-dependence truth-conditions. If $\langle W, e, w \rangle$ is finite, then $<v, s>$ is finite. We must show that $<v, s>$ satisfies M1–M3 and the **S5** set-assignment conditions.

That $s(A \wedge B) = s(A) \cap s(B)$ and $s(\neg A) = \overline{s(A)}$ comes from $\langle W, e \rangle$ being a possible-worlds model. So $<v, s>$ is a modal semantics for implication.

Suppose $s(\Box A) \neq \varnothing$. Then there is some $x \in s(\Box A)$, so $x \vDash \Box A$. So for all z, $z \vDash A$. So for all z, $z \vDash \Box A$. Hence $s(\Box A) = S$.

For the second set-assignment condition:

$$\begin{aligned} \mathsf{s}(\Box A) = \mathsf{S} \ &\text{iff for all } x,\ x \vDash \Box A \\ &\text{iff for all } z,\ z \vDash A \\ &\text{iff } \mathsf{s}(A) = \mathsf{S} \end{aligned}$$

To show that the third condition of **S5**-models holds, we first show by induction on the length of A that $v(A) = \mathsf{T}$ iff $w \vDash A$. It is immediate for the propositional variables. I will leave to you the cases when A is a conjunction or negation. So suppose that A has the form $\Box B$. Then:

$$\begin{aligned} v(\Box B) = \mathsf{T} \ &\text{iff } \mathsf{s}(\Box B) = \mathsf{S} \text{ and } v(B) = \mathsf{T} \\ &\text{iff for all } z,\ z \vDash \Box B \text{ and } w \vDash B \quad \text{by induction} \\ &\text{iff } w \vDash \Box B \end{aligned}$$

Now suppose that $\mathsf{s}(A) = \mathsf{S}$. Then for all z, $z \vDash A$. In particular, $w \vDash A$. Hence, $v(A) = \mathsf{T}$, and $<v,\mathsf{s}>$ is an **S5**-model. ■

Lemma 35 For every **S5**-model $<v,\mathsf{s}>$, there is a possible-worlds model $\langle \mathsf{W}, \mathsf{e}, w \rangle$ such that for all A, $v(A) = \mathsf{T}$ iff $w \vDash A$. If $<v,\mathsf{s}>$ is finite, then $\langle \mathsf{W}, \mathsf{e}, w \rangle$ is finite.

Proof: We will construct the possible-world model by giving an evaluation $\mathsf{e}: PV \to \mathsf{W}$, which is equivalent to a usual presentation (see the Note, p. 206). Let:

$\mathsf{W} = \mathsf{S} \cup \{w\}$ for some object $w \notin \mathsf{S}$

$$\mathsf{e}(p) = \begin{cases} \mathsf{s}(p) \cup \{w\} & \text{if } v(p) = \mathsf{T} \\ \mathsf{s}(p) & \text{if } v(p) = \mathsf{F} \end{cases}$$

If S is finite, then W is finite.

We now show by induction on the length of A, that for all A:

for $x \neq w$, $x \vDash A$ iff $x \in \mathsf{s}(A)$

$w \vDash A$ iff $v(A) = \mathsf{T}$

The only interesting case is when A is of the form $\Box B$. For all x we have:

$$\begin{aligned} x \vDash \Box B \ &\text{iff } w \vDash B \text{ and for all } z \neq w,\ z \vDash B \\ &\text{iff } v(B) = \mathsf{T} \text{ and } \mathsf{s}(B) = \mathsf{S} \quad \text{by induction} \end{aligned}$$

Using this and the **S5** set-assignment condtions, we have:

if $x \neq w$ and $x \vDash \Box B$, then $\mathsf{s}(B) \neq \varnothing$, so $\mathsf{s}(B) = \mathsf{S}$, so $\mathsf{s}(\Box B) = \mathsf{S}$, so $x \in \mathsf{s}(\Box B)$

if $w \vDash \Box B$, then $v(B) = \mathsf{T}$ and $\mathsf{s}(B) = \mathsf{S}$, so $v(\Box B) = \mathsf{T}$

if $v(\Box B) = \mathsf{T}$, then $v(B) = \mathsf{T}$ and $\mathsf{s}(B) = \mathsf{S}$, so $w \vDash \Box B$

We have left to show that if $x \neq w$ and $x \in s(\Box B)$, then $x \vDash \Box B$. But if $x \in s(\Box B)$, then $s(\Box B) = S$, so $s(B) = S$, and hence for all $z \neq w$, $z \in s(B)$. So by induction, for all $z \neq w$, $z \vDash B$. So we only need to show that $w \vDash B$. But if $s(B) = S$, then $v(B) = T$, because $<v, s>$ is an **S5**-model. So for all z, $z \vDash B$, and hence $x \vDash \Box B$. ∎

Using the consequence relation $\vdash_{S5}$ without necessitation (pp. 240–241), we now have:

***Theorem 36 Completeness of Set-Assignment Semantics for* S5**

a. $\vdash_{S5} A$ iff for every **S5**-model $<v, s>$, $v(A) = T$

b. $\vdash_{S5} A$ iff for every finite **S5**-model $<v, s>$, $v(A) = T$

c. $\Gamma \vdash_{S5} A$ iff for every **S5**-model $<v, s>$, if $<v, s> \vDash \Gamma$, then $v(A) = T$

Proof: For parts (a) and (b):

$\vdash_{S5} A$ iff for every (finite) possible-worlds model $\langle W, e, w \rangle$, $w \vDash A$
(Theorem 19, Corollary 17, and the proof of Lemma 7)

iff for every (finite) **S5**-model $<v, s>$, $v(A) = T$
(Lemmas 34 and 35, via the contrapositive of the equivalence)

Part (c) follows as parts (a) and (b) using Theorem 27. ∎

Exercises for Section G.4

1. Given an **S5**-model $<v, s>$, prove directly without using Lemmas 34 and 35:
   a. $s(\Box A) = s(\Box\Box A)$
   b. $s(\Box A) \subseteq s(A)$
   c. $s(\Box(A \supset B)) \subseteq s(\Box A \supset \Box B)$
   d. $s(\Diamond A) = S$ or $\varnothing$
   e. $s(A) \neq \varnothing$ iff $s(\Diamond A) = S$
   f. $s(\Diamond\Diamond A) = s(\Diamond A)$
   g. $s(\Box\Diamond A) = s(\Diamond A)$
2. Use Theorem 19, Corollary 17, and the proof of Lemma 7 to show:
   $\vdash_{S5} A$ iff for every (finite) possible-worlds model $\langle W, e, w \rangle$, $w \vDash A$
3. Formulate conditions for strongly complete set-assignment semantics for **S5** in the language $L(\neg, \rightarrow, \wedge)$.
4. Show that there is a 1-1 onto correspondence between universal possible-worlds models with designated world and **S5**-set-assignment models.
5. (Open) Determine whether $\{\neg, \rightarrow, \wedge\}$ is functionally complete for **S5**. If not, exhibit a collection of connectives that is.

5. S4

in collaboration with **Roger Maddux**

We proceed much as for **S5**, except that here we will prove soundness directly rather than via a correspondence of set-assignment models to possible worlds models using the axiomatization of **S4** from Section E.2, p. 235.

A modal semantics for implication $<v, s>$ is an **S4**-*model* if it satisfies:

The **S4** *set-assignment conditions*

If $s(A) = S$, then $s(\Box A) = S$

$s(\Box(A \supset B)) \subseteq s(\Box A \supset \Box B)$

$s(\Box A) \subseteq s(A)$

$s(\Box A) \subseteq s(\Box\Box A)$

Lemma 37 If $\Gamma \vdash_{S4} A$, then for every **S4**-model $<v, s>$ that validates Γ, $v(A) = T$.

Proof: Let $<v, s>$ be an **S4**-model. To show that every theorem A of **S4** is true in this model we proceed by induction on the length of a proof of A to show $v(A) = T$ and $s(A) = S$.

If A is an axiom of **PC**, then by Lemma 32.a, $v(A) = T$ and $s(A) = S$. Lemma 32.f gives $v(\Box A \supset A) = T$, and by the third condition on **S4** models, $s(\Box A \supset A) = S$ (Lemma 32.a). If $v(\Box A) = T$, then by definition, $s(A) = S$, so by the first condition on **S4** models, $s(\Box A) = S$. Hence $v(\Box\Box A) = T$, and so $v(\Box A \supset \Box\Box A) = T$. By the last condition on **S4** models, $s(\Box A \supset \Box\Box A) = S$.

If A is derived from B and $B \supset C$, then by induction we have $v(B) = T$, $v(B \supset C) = T$, $s(B) = S$, and $s(B \supset C) = S$. Hence $v(A) = T$ and $s(A) = S$.

If A is derived from B by necessitation, we have by induction $v(B) = T$ and $s(B) = S$, so $v(\Box B) = T$, and by the first condition on **S4** models, $s(\Box B) = S$.

If $\Gamma \vdash_{S4} A$ and $<v, s> \vDash \Gamma$, then proceed as above to show $v(A) = T$ (necessitation is not used). ∎

Lemma 38 For every reflexive, transitive possible-worlds model $\langle W, R, e, w \rangle$ there is an **S4**-model $<v, s>$ such that for all A, $v(A) = T$ iff $w \vDash A$. If $\langle W, R, e, w \rangle$ is finite, then $<v, s>$ is finite.

Proof: Let $\langle W, R, e, w \rangle$ be a reflexive, transitive model. Set:

$S = \{ z : wRz \}$

$s(A) = \{ z : wRz \text{ and } z \vDash A \}$

$v(p) = T$ iff $w \vDash p$

By Lemma 1, since R is transitive we need only consider $z \in S$ in evaluating whether, for any A, w validates A. So it is no loss of generality to assume

$W = S$, and then $s(A) =$ the worlds in which A is true.

Extend v to all wffs by the dual-dependence truth-conditions. If $\langle W, R, e, w \rangle$ is finite, $\langle v, s \rangle$ is finite. I will leave to you to show that $\langle v, s \rangle$ satisfies M1 and M2 and hence is a modal semantics of implication. We now must show that the set-assignment satisfies the **S5** conditions.

For the first condition, if $x \in s(\Box A)$, then $x \vDash \Box A$, so by the reflexivity of R, $x \vDash A$, and $x \in s(A)$.

For the second condition, if $x \in s(\Box A)$, then $x \vDash \Box A$, so by the transitivity of R (see Lemma 8, p. 218), $x \vDash \Box\Box A$, and $x \in s(\Box A)$.

We now want to show that for every A, $v(A) = T$ iff $w \vDash A$. We proceed by induction on the length of A. The only hard part is when A is of the form $\Box A$.

If $v(\Box B) = T$, then $v(B) = T$ and $s(B) = S$. So by induction, $w \vDash B$, and for all z such that wRz, $z \vDash B$. Hence $w \vDash \Box B$.

If $v(\Box B) = F$, then $v(B) = F$ or $s(B) \neq S$. So by induction, either $w \nvDash B$ or for some z such that wRz, $z \nvDash B$. So, since R is reflexive, $w \nvDash \Box B$. ■

***Theorem 39 Completeness of Set-Assignment Semantics for* S4**

a. $\vdash_{S4} A$ iff for every **S4**-model $\langle v, s \rangle$, $v(A) = T$

b. $\vdash_{S4} A$ iff for every finite **S4**-model $\langle v, s \rangle$, $v(A) = T$

c. $\Gamma \vdash_{S4} A$ iff for every **S4**-model $\langle v, s \rangle$, if $\langle v, s \rangle \vDash \Gamma$, then $v(A) = T$

Proof: By Lemma 37 if $\vdash_{S4} A$, then for every **S4**-model $\langle v, s \rangle$, $v(A) = T$, and so also for every finite **S4**-model $\langle v, s \rangle$, $v(A) = T$. If $\nvdash_{S4} A$, then there is some reflexive, transitive possible-worlds model $\langle W, R, e, w \rangle$ such that $w \nvDash A$ (Theorem 19). Choose one, which, by Corollary 17, we may assume is finite. By Lemma 38 we then have a finite **S4**-model that does not validate A.

Part (c) uses Theorem 27 and the extension of Lemma 37 to consequences. ■

Exercises for Section G.5

1. Show that every **S5** modal semantics of implication is an **S4** modal semantics of implication. (See Exercise 1, p. 253.)

2. Complete the proof of Lemma 37 to show that if $\langle v, s \rangle$ is an **S4**-model, $\Gamma \vdash_{S4} A$, and $\langle v, s \rangle \vDash \Gamma$, then $v(A) = T$.

3. Show that there is one **S4**-set-assignment model in which exactly the theorems of **S4** are valid. (Hint: See Theorem 23.)

4. Formulate conditions for strongly complete set-assignment semantics for **S5** in the language $L(\neg, \rightarrow, \wedge)$.

5. (Open) Show that the map from the collection of reflexive, transitive possible-worlds models to **S4**-set-assignment models given in Lemma 38 is not onto.

6. Let K be the class of **S4**-set-assignment models that satisfy:

   If $\{y: \text{all A, if } x \in s(\Box A) \text{ then } y \in s(A)\} \subseteq s(B)$, then $x \in s(\Box B)$.

   a. Prove that if $\langle W, R, e, w\rangle$ is a reflexive, transitive possible-worlds model, then there is an **S4**-model $\langle v, s\rangle$ in K such that $v(A) = T$ iff $w \vDash A$.
   (Hint: Check that the model constructed in Lemma 38 is in K.)

   b. Prove that if $\langle v, s\rangle$ is an **S4**-model in K, then there is a reflexive, transitive possible-worlds model $\langle W, R, e, w\rangle$ such that $v(A) = T$ iff $w \vDash A$.
   (Hint: Use the following construction.)

   $W = S \cup \{w\}$ for some object $w \notin S$

   $$e(p) = \begin{cases} s(p) \cup \{w\} & \text{if } v(p) = T \\ s(p) & \text{if } v(p) = F \end{cases}$$

   $$xRy \text{ iff } \begin{cases} \text{for all A, if } x \in s(\Box A) \text{ then } y \in s(A) \\ \text{or} \\ x = w \end{cases}$$

   c. Is K simply presented?

7. Define $\langle W, R, e, w\rangle$ to be a *reduced* **S4**-model if:

   R is reflexive and transitive

   For all z, wRz

   For any $x, y \in W$, there is some B such that either
   $x \vDash B$ and $y \nvDash B$, or $x \nvDash B$ and $y \vDash B$

   a. Show that the class of reduced **S4**-models is strongly complete for **S4**.

   b. Show that the mapping from K to reduced **S4**-models as given in Exercise 5.b is 1-1 and onto.

   c. (Open) Is there a simply presented class of set-assignment models that is strongly complete for **S4** that can be mapped 1-1 onto reduced **S4**-models while preserving truth in a model?

8. Consider the condition on modal semantics of implication:

   P If $v(A) = T$, then $s(A) \neq \varnothing$.

   a. Show that every **S4**-model that arises from a possible-worlds model by the construction of Lemma 38 satisfies condition P.

   Thus we could add P to the list of conditions for **S4**. By adding more conditions we get a better reading of $s(A)$ as the worlds in which A is true. For instance, P says that if A is true, then it is true in some world.

   b. Show that there are **S4** models that do not satisfy P.
   (Hint: Let $S = \{1\}$ and define $s(A) = \{1\}$ iff $\neg p_0 \vdash_{\mathbf{PC}} A^*$, where A^* is A with every occurrence of every p_j replaced by p_0 and $\Box A$ by $\neg A \supset A$ (in the language $L(p_0, p_1, \ldots \neg, \rightarrow, \wedge)$ this would amount to reading $\rightarrow$ as the

PC-connective). In **PC**, $(\Box A)^*$, $(\Diamond A)^*$, $\Diamond A^*$, and $\Box A^*$ are all semantically equivalent to A^*, and for every A, A^* is semantically equivalent to either p_0, or $\neg p_0$, or $p_0 \wedge \neg p_0$, or $\neg(p_0 \wedge \neg p_0)$. Consider $v(p_j) = \mathsf{T}$ for all *i*.)

c. (Open) Prove or disprove that given any **S4**-model $<v, s>$, there is another **S4** modal semantics for implication $<v^*, s^*>$ that satisfies P such that for all A, $v^*(A) = v(A)$.

9. (Open) Give set assignment semantics for **S4** for which $\Gamma \vdash_{\mathbf{L}} \Box A$ iff $\Gamma \vDash A$. (See Section E.4.b.)

6. T

Recall that **T** is the logic of all reflexive possible-worlds models, as axiomatized in Section E.2 (p. 235).

Can we proceed as for **S4**? To respect the rule of necessitation for soundness, it would seem we need that if $s(A) = S$ then $s(\Box A) = S$. But that validates $\Box A \supset \Box\Box A$, which is not a theorem of **T**.

What do we get when we convert a possible-worlds model into a modal semantics of implication? If the accessibility relation is not transitive, then we have two distinct sets: $\{z : wRz\}$ and $\{z : wR^*z\}$, where R^* is the transitive closure of R (see p. 208), where by Lemma 1 we may assume that $W = \{z : wR^*z\}$. We need two content sets for each wff A: $\{z : wRz$ and $z \vDash A\}$ and $\{z : z \vDash A\}$. We can do that by first taking a set C, to correspond to W, and then designating a subset $S \subseteq C$, that will correspond to $\{z : wRz\}$. For each A we assign $t(A) \subseteq C$, corresponding to $\{z : z \vDash A\}$, and then $s(A) = t(A) \cap S$. In this way s will respect necessitation, but only for the theorems of **T**.

We say that s is a **T**-*set-assignment* if there are t, C, and $S \subseteq C$, where S is called the *designated subset,* and:

$t : \text{Wffs} \to \text{Sub } C$

$t(A \wedge B) = t(A) \cap t(B)$

$t(\neg A) = \overline{t(A)}$

$t(A \to B) = t(\Box(A \supset B))$

If $t(A) = C$, then $t(\Box A) = C$

$t(\Box(A \supset B)) \subseteq t(\Box A \supset \Box B)$

$t(\Box A) \subseteq t(A)$

And $s : \text{Wffs} \to \text{Sub } S$ is given by:

$s(A) = t(A) \cap S$

That is, t satisfies the same structural requirements as an **S4** set-assignment, and s is t restricted to the designated subset.

A pair $<v,s>$ is a **T**-*model* if s is a **T**-set-assignment and v is extended to all wffs by the truth-conditions for modal semantics of implication. You can show that every **T**-model is a modal semantics of implication.

Lemma 40 If $\vdash_{\mathbf{T}} A$, then for every **T**-model $<v,s>$, $v(A) = \mathsf{T}$.

Proof: Let $<v,s>$ be a **T**-model with t, C, and S as in the definition above. The proof then follows exactly as for **S4**, Lemma 37, showing by induction on the length of a proof of A that $v(A) = \mathsf{T}$ and $t(A) = C$. Necessitation is respected because if $t(B) = C$, then $s(B) = S$. But if $s(A) = S$, then we may now have $s(\Box A) \neq S$. ■

Lemma 41 For every reflexive possible-worlds model $\langle W, R, e, w \rangle$ there is an **S4**-model $<v,s>$ such that for all A, $v(A) = \mathsf{T}$ iff $w \vDash A$. If $\langle W, R, e, w \rangle$ is finite, then $<v,s>$ is finite.

Proof: Let $\langle W, R, e, w \rangle$ be a reflexive possible-worlds model. Take $C = W$ and $t(A) = \{z : z \vDash A\}$. Take $S = \{z : wRz\}$ and set $s(A) = t(A) \cap S$. Define $v: PV \to \{\mathsf{T}, \mathsf{F}\}$ by $v(p) = \mathsf{T}$ iff $w \vDash p$, extending v to all wffs by the truth-conditions for modal semantics for implication. Then $<v,s>$ is a **T**-model, and the proof that $w \vDash A$ iff $v(A) = \mathsf{T}$ follows as for **S4**. ■

Theorem 42 ***Completeness of Set-Assignment Semantics for* T**

a. $\vdash_{\mathbf{T}} A$ iff for every **T**-model $<v,s>$, $v(A) = \mathsf{T}$

b. $\vdash_{\mathbf{T}} A$ iff for every finite **T**-model $<v,s>$, $v(A) = \mathsf{T}$

c. $\Gamma \vdash_{\mathbf{T}} A$ iff for every **T**-model $<v,s>$, if $<v,s> \vDash \Gamma$, then $v(A) = \mathsf{T}$

Proof: Proceed as in the proof for **S4**, Theorem 39. ■

The class of **T**-models given here is not simply presented by the criteria of Chapter IV.D. By complicating the truth-conditions for the models rather than the set-assignments, we can give a class of models that is simply presented and that is strongly complete for **T**. Consider models $<v,s>$ such that $s: \text{Wffs} \to S$ is an **S4** set-assignment and $T \subseteq S$. Extend v to all wffs by the classical evaluation of $\neg$ and $\wedge$ and: $v(\Box A) = \mathsf{T}$ iff $v(A) = \mathsf{T}$ and $T \subseteq s(A)$ (in $L(\neg, \to, \wedge)$ this is replaced by: $v(A \to B) = \mathsf{T}$ iff $(s(A) \cap T) \subseteq (s(B) \cap T)$ and ($v(A) = \mathsf{F}$ or $v(B) = \mathsf{T}$).) Then, as you can show, these semantics are simply presented and are strongly complete for **T**.

7. B

Recall that **B** is the logic of all reflexive, symmetric possible-worlds models, which is axiomatized in Section E.2, p. 235.

To give set-assignment semantics for **B** we use designated subsets as for **T** and mix conditions for v and s, as for **S5**. A pair $<v,s>$ is a **B**-*model* if there are t, C and $S \subseteq C$ such that $t: \text{Wffs} \to \text{Sub } C$ and:

$$t(A \wedge B) = t(A) \cap t(B)$$
$$t(\neg A) = \overline{t(A)}$$
$$t(A \rightarrow B) = t(\Box(A \supset B))$$
If $t(A) = C$, then $t(\Box A) = C$
$$t(\Box(A \supset B)) \subseteq t(\Box A \supset \Box B)$$
$$t(\Box A) \subseteq t(A)$$
$$t(A) \subseteq t(\Box\Diamond A)$$
And $s: \text{Wffs} \rightarrow \text{Sub } S$ given by $s(A) = t(A) \cap S$
satisfies: if $v(A) = T$, then $s(\Diamond A) = S$
And v is extended to all wffs by the truth-conditions for modal semantics of implication.

That is, $\langle v, s \rangle$ is a **T**-model that also satisfies: $t(A) \subseteq t(\Box\Diamond A)$ and if $v(A) = T$ then $s(\Diamond A) = S$. Here, too, every **B**-model is a modal semantics of implication.

Theorem 43 ***Completeness of Set-Assignment Semantics for* B**

a. $\vdash_{\mathbf{B}} A$ iff for every **B**-model $\langle v, s \rangle$, $v(A) = T$

b. $\vdash_{\mathbf{B}} A$ iff for every finite **B**-model $\langle v, s \rangle$, $v(A) = T$

c. $\Gamma \vdash_{\mathbf{B}} A$ iff for every **B**-model $\langle v, s \rangle$, if $\langle v, s \rangle \models \Gamma$, then $v(A) = T$

Proof: Proceed as for **T**. But here we need to show that for every **B**-model $\langle v, s \rangle$, $v(A \supset \Box\Diamond A) = T$. Suppose $v(A) = T$. Then $v(\Diamond A) = T$ and by the conditions on set-assignments for these models, $s(\Diamond A) = S$. Hence $v(\Box\Diamond A) = T$.

The rest of the proof is as for **T**. The only new point to verify is that the model constructed as in Lemma 41 from a symmetric, reflexive possible-worlds model satisfies: $t(A) \subseteq t(\Box\Diamond A)$ and if $v(A) = T$ then $s(\Diamond A) = S$. The first is easy because the possible-worlds model is symmetric. The second part is proved after first showing that $v(A) = T$ iff $w \models A$. So if $v(A) = T$, then $w \models A$. Hence for any z for which wRz we have zRw, so $z \models \Diamond A$. Hence $s(\Diamond A) = S$. ■

Exercises for Sections G.6 and G. 7

1. Show that every **T**-model is a modal semantics of implication.
2. Show that the set-assignment model derived from the possible-worlds model in Lemma 41 is a **T**-model.
3. Show that the set-assignment models for **T** described at the end of Section E.6 are simply presented and strongly complete for **T**.
4. Show that the set-assignment model derived from the possible-worlds model in the version of Lemma 41 for **B** is a **B**-model.
5. Give simply presented set-assignment semantics for **B** in the same way as for **T**.

H. The Smallest Logics Characterized by Various Semantics

What logics are characterized by the various semantics considered in this chapter?

1. K

The system **K** was introduced in Section C.2 as the logic of all possible-world models. As I remarked there, it is not clear that **K** is in any sense a logic: it formalizes the idea of possible-world semantics with accessibility relations without specifying any particular view among those. Moreover, as noted in Theorem 13, *modus ponens* is not valid in **K**, so **K** cannot be considered a formalization of entailment. Nonetheless, let us see if we can produce set-assignment semantics for **K**. We will have recourse to the axiomatization of **K** of Section E.2:

K *in* $L(\neg, \wedge, \Box)$

PC axioms

$\Box(A \supset B) \supset (\Box A \supset \Box B)$

rules material detachment
necessitation

Since *modus ponens* is not valid in **K**, set-assignment semantics for **K** cannot be modal semantics of implication, which use the dual-dependence truth-table for $\rightarrow$. We will use instead weak modal semantics for **K**, coupled with a designated subset approach as for **T**.

We say that s is a **K**-*set-assignment* if there are t, C, and $\mathsf{S} \subseteq \mathsf{C}$ such that

$\mathsf{t}: \text{Wffs} \rightarrow \text{Sub } \mathsf{C}$

$\mathsf{t}(A \wedge B) = \mathsf{t}(A) \cap \mathsf{t}(B)$

$\mathsf{t}(\neg A) = \overline{\mathsf{t}(A)}$

$\mathsf{t}(A \rightarrow B) = \mathsf{t}(\Box(A \supset B))$

If $\mathsf{t}(A) = \mathsf{C}$, then $\mathsf{t}(\Box A) = \mathsf{C}$

$\mathsf{t}(\Box(A \supset B)) \subseteq \mathsf{t}(\Box A \supset \Box B)$

And $\mathsf{s}: \text{Wffs} \rightarrow \text{Sub } \mathsf{S}$ is given by:

$\mathsf{s}(A) = \mathsf{t}(A) \cap \mathsf{S}$

These are the conditions for **T** with $\mathsf{t}(\Box A) \subseteq \mathsf{t}(A)$ deleted.

A pair $<\mathsf{v}, \mathsf{s}>$ is a **K**-*model* if s is a **K**-set-assignment and v uses the truth-conditions of *weak modal semantics of implication.*

You can show that every **K**-model is a weak modal semantics of implication.

Theorem 44 $\vdash_{\mathbf{K}} A$ iff for every **K**-model $<\mathsf{v}, \mathsf{s}>$, $\mathsf{v}(A) = \mathsf{T}$ and $\mathsf{s}(A) = \mathsf{S}$.

The proof is as for **T** using Lemma 33.

Hence, **K** *is the smallest logic that has both possible-world semantics and weak modal semantics.*

2. QT and quasi-normal logics

What is the smallest logic that has both possible-world semantics and modal semantics of implication? By Theorem 23 it must extend **K** and be closed under material detachment. By Lemma 32.f it must contain the scheme $\Box A \supset A$. But it need not be closed under necessitation. This is exactly the logic **QT** we examined in Section F.

Define $<v,s>$ to be a **QT**-*model* if s is a **K**-set-assignment and v uses the truth-conditions for *modal semantics for implication.*

Note that every **QT**-model is a modal semantics for implication.

Theorem 45 $\vdash_{QT} A$ iff for every **QT**-model $<v,s>$, $v(A) = T$.

The proof is as for **T**. We cannot claim that if $\vdash_{QT} A$ then for every **QT**-model $<v,s>$, $s(A) = S$. Consider the model $\langle W, R, e, w\rangle$ for **QT** where $W = \{w, z\}$, $w \vDash p$, $z \nvDash p$ and wRw, wRz, zRw, but not zRz. Then $z \nvDash \Box p \supset p$. So the model $<v,s>$ derived from this possible-world model as in the proof of Theorem 41 will not satisfy $s(\Box p \supset p) = S$.

Possible-world semantics and set-assignment semantics for other quasi-normal logics can be given along the lines of the ones we've given for **QT**.

3. The logic characterized by modal semantics of implication

We define:

MSI is the logic of all modal semantics of implication in $L(\neg, \rightarrow, \Box)$

Let $\mathbf{PC}^{\Box}$ be the closure under the rule of necessitation of **PC** in the language $L(\neg, \wedge, \Box)$. Since **MSI** is closed under both *modus ponens* and material detachment, considering Lemma 33 I conjecture that $\mathbf{MSI}^{\Box}$ is the same as:

ML
in $L(\neg, \wedge, \Box)$

$\mathbf{PC}^{\Box}$

$\Box(A \supset B) \supset (\Box A \supset \Box B)$

$\Box A \supset A$

rule material detachment

Note that $\mathbf{ML} \subseteq \mathbf{QT}$ and $\mathbf{ML} \subseteq \mathbf{MSI}$. Also, $\vdash_{ML} (A \wedge (A \rightarrow B)) \supset B$: $\vdash_{ML} \Box(A \supset B) \supset (A \supset B)$, and then $\vdash_{ML} (A \wedge \Box(A \supset B)) \supset B$, since $\mathbf{PC} \subseteq \mathbf{ML}$.

Thus **ML** is closed under *modus ponens.* I do not know if we get the same logic if we replace the scheme $\Box A \supset A$ by the rule of *modus ponens.*

The natural language for **MSI** and **ML** is $L(\neg, \rightarrow, \wedge)$, and we could take **MSI** to be the logic of all modal semantics of implication in $L(\neg, \rightarrow, \wedge)$. Then we define:

ML

in $L(\neg, \rightarrow, \wedge)$

$\Box A \equiv_{\text{Def}} \neg A \rightarrow A \qquad A \leftrightarrow B \equiv_{\text{Def}} (A \rightarrow B) \wedge (B \rightarrow A)$

PC$^{\Box}$

$\Box(A \supset B) \supset (\Box A \supset \Box B)$

$(A \wedge (A \rightarrow B)) \supset B$

$\Box(A \supset B) \leftrightarrow (A \rightarrow B)$

rule material detachment

Once more **ML** $\subseteq$ **MSI**. And we can prove that $\vdash_{\mathbf{ML}} \Box A \supset A$: we have $\vdash_{\mathbf{ML}} (\neg A \wedge (\neg A \rightarrow A)) \supset A$, so by **PC**, $\vdash_{\mathbf{ML}} (A \wedge (\neg A \rightarrow A)) \supset A$, and then again by **PC** $\vdash_{\mathbf{ML}} (\neg A \rightarrow A) \supset A$; that is, $\vdash_{\mathbf{ML}} \Box A \supset A$.

Each classical modal logic we have seen can be presented in either $L(\neg, \rightarrow, \wedge)$ or $L(\neg, \wedge, \Box)$. In what sense are the two presentations the same logic?

Given one of the classical modal logics, **L**, in $L(\neg, \rightarrow, \wedge)$, we can translate to $L(\neg, \wedge, \Box)$ by translating $\neg$ and $\wedge$ homophonically and setting:

$(A \rightarrow B)^* = \Box(A^* \supset B^*)$

In translating from $L(\neg, \wedge, \Box)$ to $L(\neg, \rightarrow, \wedge)$, we translate $\neg$ and $\wedge$ homophonically, and set:

$(\Box A)^\dagger = \neg(A^\dagger) \rightarrow A^\dagger$

For **MSI** we have only one notion of consequence, semantic, and:

In $L(\neg, \rightarrow, \wedge)$, $\Gamma \vDash_{\mathbf{MSI}} A$ iff $\Gamma^* \vDash_{\mathbf{MSI}} A^*$

In $L(\neg, \wedge, \Box)$, $\Gamma \vDash_{\mathbf{MSI}} A$ iff $\Gamma^\dagger \vdash_{\mathbf{MSI}} A^\dagger$

For the other logics we have two notions of syntactic consequence, and these translations preserve both, as you can show using the completeness theorems:

Theorem 46 In $L(\neg, \rightarrow, \wedge)$, $\Gamma \vdash_{\mathbf{L}} A$ iff $\Gamma^* \vdash_{\mathbf{L}} A^*$

$\Gamma \vdash_{\mathbf{L}}{}^{\Box} A$ iff $\Gamma^* \vdash_{\mathbf{L}}{}^{\Box} A^*$

In $L(\neg, \wedge, \Box)$, $\Gamma \vdash_{\mathbf{L}} A$ iff $\Gamma^\dagger \vdash_{\mathbf{L}} A^\dagger$

$\Gamma \vdash_{\mathbf{L}}{}^{\Box} A$ iff $\Gamma^\dagger \vdash_{\mathbf{L}}{}^{\Box} A^\dagger$

Exercises for Section H

1. What is the smallest logic that has both modal semantics of implication and possible-world semantics?
2. Show that every **K**-model is a weak modal semantics of implication.
3. a. Give set-assignment semantics for **QS4** (See Exercise 4, p. 244).
   b. Give set-assignment semantics for **QS5**.
4. Give simply presented set-assignment semantics for **QT**.
5. Prove that the maps * and † are translations (Theorem 46).
6. (Open) Prove or disprove that **ML** axiomatizes **MSI**.
7. (Open) Prove or disprove that if the scheme $\Box A \supset A$ is deleted from **ML** and the rule of *modus ponens* is added, we get the same collection of valid wffs and the same consequence relation.
8. (Open) Give complete semantics for $\mathbf{PC}^{\Box}$.
9. (Open) Determine whether $\{\neg, \rightarrow, \wedge\}$ is functionally complete for **MSI**. If not, exhibit a collection of connectives that is.
10. (Open) Determine whether the class of finite models for **MSI** yields the same consequence relation as the class of all models.

J. Modal Logics Modeling Notions of Provability

1. '□' read as 'It is provable that'

> Deducibility and provability are strange notions, and different though their properties may be from those of implication and necessity, the symbolism of modal logic turns out to be exceedingly useful notation for representing the forms of sentences of formal theories that have to do with the notions of deducibility, provability, and consistency, and the techniques devised to study systems of modal logic disclose facts about these notions that are of great interest.
>
> *Boolos, 1979*, p.4

A survey of the relation of modal logics to notions of provability in arithmetic can be found in *Boolos, 1980 B*. In this section I'll give a brief synopsis of the main connections, presupposing some familiarity with classical first-order logic, and then turn to semantic analyses of the modal systems involved.

Let **PA** denote *Peano Arithmetic,* the first-order theory of arithmetic with induction (see, for example, *Epstein and Carnielli,* Chapter 23). We may Gödel number the formulas of the language, denoting by $[\![A]\!]$ the Gödel number of A. Then we may define in the language of **PA** a predicate *Bew* (for the German *beweisbar,* meaning *provable*) which corresponds to provability under that Gödel numbering. That is, for any natural number *m*:

$$\text{Bew}(m) \text{ holds iff } m = [\![A]\!] \text{ and } \mathbf{PA} \vdash A$$

A *realization* is a map $\varphi: PV \rightarrow$ sentences of the language of **PA**. The *provability translation* of modal sentences *under realization* φ is the map from the sentences of the language of modal logic to those of the language of **PA** defined inductively by:

$$p^{\varphi} = \varphi(p)$$

$$(\neg A)^{\varphi} = \neg(A^{\varphi})$$

$$(A \wedge B)^{\varphi} = A^{\varphi} \wedge B^{\varphi}$$

$$(\Box A)^{\varphi} = \text{Bew}([\![A^{\varphi}]\!])$$

That is, we read $\Box A$ as 'it is provable that'. If we take $L(\neg, \rightarrow, \wedge)$ to be the language of modal logic, then the last part of the translation is replaced by:

$$(A \rightarrow B)^{\varphi} = \text{Bew}([\![A^{\varphi} \supset B^{\varphi}]\!])$$

Two modal systems are closely connected to such translations. The first is **G**, the closure of $\mathbf{K} \cup \{\Box(\Box A \supset A) \supset \Box A\}$ under material detachment and necessitation. The other, called **G***, is the closure of $\mathbf{G} \cup \{\Box A \supset A\}$ under the rule of material detachment.

Thus **G*** is to **G** as **QT** is to **K**. We'll find that the set-assignment semantics of **G*** are related to those of **G** just as those of **QT** are to **K**: they use the same set-assignments, while the former are modal semantics of implication and the latter are weak modal semantics.

The connection of **G** and **G*** to provability in arithmetic is established by the following two theorems of Solovay, which can be found in *Boolos, 1979.*

Theorem 47 $\mathbf{G} \vdash A$ iff every provability translation of A is a theorem of **PA**. That is, for all φ, $\mathbf{PA} \vdash A^{\varphi}$.

If $\mathbf{G} \vdash A$, then $\mathbf{G} \vdash \Box A$, as **G** is closed under the rule of necessitation. Hence **G** can be viewed as *the sentences that express provable principles of provability.*

Theorem 48 $\mathbf{G}^* \vdash A$ iff every provability translation of A is true in the standard model of **PA**. That is, if **N** represents the natural numbers, then for all φ, $\mathbf{N} \models A^{\varphi}$.

Hence **G*** can be viewed as *the sentences that express true principles of provability.*

The *provability and truth translation* of modal sentences under realization φ is the map defined by:

$p^\dagger = \varphi(p) \wedge \mathrm{Bew}(\llbracket \varphi(p) \rrbracket)$

$(\neg A)^\dagger = \neg A^\dagger$

$(A \wedge B)^\dagger = A^\dagger \wedge B^\dagger$

$(\Box A)^\dagger = \mathrm{Bew}(\llbracket A^\dagger \rrbracket) \wedge A^\dagger$

We can construe this as reading $\Box A$ as 'it is provable and true that'.

Similarly we can define a map # as we defined $\dagger$ except, taking $p^\# = \varphi(p)$.

The modal system connected to these translations is **S4Grz**, the closure under material detachment and necessitation of **S4** $\cup \{ (\Box(\Box(A \supset \Box A) \supset A)) \supset A \}$.

Theorem 49 *a.* $\vdash_{\mathbf{S4Grz}} A$ iff every provability-and-truth translation of A is a theorem of **PA**

b. $\vdash_{\mathbf{S4Grz}} A$ iff for every map #, $\mathbf{PA} \vdash A^\#$

A proof of (a) can be found in *Goldblatt, 1978,* and of (b) in *Boolos, 1979* and *1980 B.* Thus **S4Grz** can be viewed as *the sentences that express provable principles of provability-and-truth.* In Chapter VII.C we'll see that the intuitionist propositional logic can be construed as expressing correct principles of provability and truth similar to these.

Let us turn to semantics for these logics.

2. S4Grz

S4Grz

in $L(\neg, \wedge, \Box)$

PC

$\Box(A \supset B) \supset (\Box A \supset \Box B)$

$\Box A \supset A$

$\Box A \supset \Box\Box A$

$(\Box(\Box(A \supset \Box A) \supset A)) \supset A$

rules material detachment
necessitation

Note that the new scheme is: $((A \rightarrow \Box A) \rightarrow A) \supset A$. The initials *Grz* are given to this system because of the work of *Grzegorczyk, 1967.*

We say that R is *anti-symmetric* if $y R z$ and $z R y$ together imply that $y = z$. We say that a frame $\langle W, R \rangle$ is a *finite weak partial order* if W is finite and R is reflexive, transitive, and anti-symmetric.

Theorem 50 $\vdash_{\mathbf{S4Grz}} A$ iff for every finite weak partial order $\langle W, R \rangle$, $\langle W, R \rangle \models A$

For a proof of Theorem 50 see *Segerberg, 1971,* pp. 96–103, or *Boolos, 1979,* Chapter 13 (note that both Segerberg and Boolos use '$\rightarrow$' for what I call '$\supset$'). As Segerberg notes, 'finite' is essential here.

For the set-assignment semantics, we say that an **S4**-model $<v, s>$ is an **S4Grz**-*model* if it is finite and satisfies:

$$s(\Box(\Box(A \supset \Box A) \supset A)) \subseteq s(A)$$

$$\text{If } v(A) = F, \text{ then } s(\Box(A \supset \Box A)) \nsubseteq s(A)$$

The first condition ensures that the models respect necessitation for consequences of the new axiom scheme; the second condition ensures that every instance of the new scheme is true.

Theorem 51 $\vdash_{S4Grz} A$ iff for every **S4Grz** -model $<v, s>$, $v(A) = T$

Proof: It is straightforward to show that these models are sound for **S4Grz** by the same method as for **S4** (Lemma 37, p. 254).

To show that the semantics are complete, the proof is as for **S4** except that we now have to show that the model $<v, s>$ we construct for Lemma 38 satisfies the two new conditions. The first I will leave to you. For the second, suppose $v(A) = F$. Then $w \nvDash A$. We have:

$$\langle W, R, e, w \rangle \vDash (\Box(\Box(A \supset \Box A) \supset A)) \supset A$$

So we must have $w \nvDash \Box(\Box(A \supset \Box A) \supset A)$. Hence there is a z such that wRz and $z \nvDash \Box(A \supset \Box A) \supset A$. This can only be if $z \nvDash A$ and $z \vDash \Box(A \supset \Box A)$. Thus $z \in s(\Box(A \supset \Box A))$ and $z \notin s(A)$. ■

3. G

G
in $L(\neg, \wedge, \Box)$

PC
$\Box(A \supset B) \supset (\Box A \supset \Box B)$
$\Box(\Box A \supset A) \supset \Box A$

rules material detachment
necessitation

A relation is *anti-reflexive* if no element is related to itself. A frame $\langle W, R \rangle$ is a *finite strict partial order* if W is finite and R is transitive, anti-reflexive, and anti-symmetric. *Boolos, 1979* proves the following.

Theorem 52 $\vdash_G A$ iff for every finite strict partial order $\langle W, R \rangle$, $\langle W, R \rangle \vDash A$

We define a **G**-*set-assignment* to be one that satisfies:

$s(A \wedge B) = s(A) \cap s(B)$

$s(\neg A) = \overline{s(A)}$

If $s(A) = S$, then $s(\Box A) = S$

$s(\Box(A \supset B)) \subseteq s(\Box A \supset \Box B)$

$s(\Box(\Box A \supset A)) \subseteq s(\Box A)$

$s(\Box A) \subseteq s(A)$ iff $s(A) = S$

We say that $<v,s>$ is a **G**-*model* if it is a weak modal semantics of implication in which s is a **G**-set-assignment.

Note that in **G**-models the truth-value assignment v is independent of s.

Lemma 53 If $\vdash_G A$, then for every **G**-model $<v,s>$, $v(A) = T$.

Proof: The proof follows as for Lemma 37, except that we use Lemma 33 for weak modal semantics. We need to show that $v(\Box(\Box A \supset A) \supset \Box A) = T$. If $v(\Box(\Box A \supset A)) = T$, then $s(\Box A \supset A) = S$. Hence $s(\Box A) \subseteq s(A)$. So by the last condition for **G**-set-assignments, $s(A) = S$, and then since we are using weak modal semantics of implication, $v(\Box A) = T$. ■

Lemma 54 If $\nvdash_G A$, then for some finite **G**-model $<v,s>$, $v(E) = F$.

Proof: If $\nvdash_G A$, then there is some $\langle W, R, e, w \rangle$ such that $w \nvDash A$ and $\langle W, R \rangle$ is a finite strict partial order. Choose one and set $S = \{z : wRz\}$ and $v(p) = T$ iff $w \vDash p$. Extend v to all wffs by the truth-conditions of weak modal semantics.

I will leave to show that $<v,s>$ satisfies the first five conditions for **G**-set-assignments, where we need that R is transitive to show that it satisfies the third (see Lemma 38). So we need to show that $<v,s>$ satisfies the last condition.

Suppose $s(A) \neq S$. Then for some x, $x \nvDash A$. If $x \vDash \Box A$, we are done. If not, then for some y, xRy and $y \nvDash A$. Again, if $y \vDash \Box A$, we are done. If not, we can continue; as W is finite this process must terminate in an end point, that is, a z with no world related to it, such that $z \nvDash A$. But since z is an end point, it vacuously validates $\Box A$. Hence $s(\Box A) \nsubseteq s(A)$.

$$\begin{array}{ccccccc} w & \longrightarrow & x & \longrightarrow & y & \dashrightarrow & z \\ & & A \text{ is } F & & A \text{ is } F & & A \text{ is } F \\ & & & & \Box A \text{ is } F & & \end{array}$$

Finally we show that $v(A) = T$ iff $w \vDash A$ by induction on the length of A. The only hard case is if A is $\Box B$. Then $v(\Box B) = F$ iff $s(B) \neq S$, which is iff there is some z such that wRz and $z \nvDash B$. And that is if and only if $w \nvDash \Box B$. ■

Combining Lemmas 53 and 54 we have:

Theorem 55 $\vdash_{\mathbf{G}} A$ iff for every **G**-model $<\nu, s>$, $\nu(A) = T$
iff for every finite **G**-model $<\nu, s>$, $\nu(A) = T$

We make one last observation about **G** before we turn to **G***.

Lemma 56 $\nvdash_{\mathbf{G}} \Box A \supset A$

Proof: Suppose to the contrary that $\vdash_{\mathbf{G}} \Box A \supset A$. Then for any A, by necessitation, $\vdash_{\mathbf{G}} \Box(\Box A \supset A)$. Hence via the second axiom scheme, $\vdash_{\mathbf{G}} \Box A$, and so $\vdash_{\mathbf{G}} A$. That is, every wff is a theorem of **G**, which means **G** is inconsistent. But that is a contradiction, since by the completeness theorems we can show that $p_1 \wedge \neg p_1$ is not a theorem of **G**. ■

4. G*

G*
in $L(\neg, \wedge, \Box)$
is the closure of $\mathbf{G} \cup \{\Box A \supset A\}$ under the rule of material detachment

By Lemma 56, $\mathbf{G}^* \neq \mathbf{G}$. Since **G*** is consistent (Theorem 57 below), the same argument shows that **G*** is not closed under necessitation, that is, it is not normal. Thus **G*** is quasi-normal and is to **G** as **QT** is to **K**.

No characterization of **G*** in terms of a class of possible-world models is known. We can't modify the possible-world semantics given for **G** in the last section by adding the requirement that each model be stipulated with respect to a designated world w for which $w R w$, for such a model may no longer validate **G**. For example, if $w \nvDash p$ yet for all other $z \in W$, $z \vDash p$, then $w \nvDash \Box(\Box p \supset p) \supset \Box p$.

Boolos, 1980 A, has given semantics for **G*** in terms of a notion of a formula being "eventually true" at a world in a possible-world model. Let $\langle W, R, e \rangle$ be a possible-world model. For this model we have already defined $w \vDash B$ for every $w \in W$. We now define by induction on j and the length of B for every natural number j, $w \vDash^{(j)} B$.

$w \vDash^{(j)} p$ iff $w \vDash p$

$w \vDash^{(j)} \neg B$ iff $w \nvDash^{(j)} B$

$w \vDash^{(j)} B \wedge C$ iff $w \vDash^{(j)} B$ and $w \vDash^{(j)} C$

$w \vDash^{(j)} \Box B$ iff $w \vDash \Box B$ and for all $k < j$, $w \vDash^{(k)} B$

Then define:

$w \vDash^{*} A$ iff for some i, for all $j \geq i$, $w \vDash^{(j)} A$

We read $w \vDash^* A$ as A *is eventually true at* w, or w *eventually validates* A.

Finally define: $\langle W, R, e\rangle \vDash^* A$ iff for all $w \in W$, $w \vDash^* A$. And $\langle W, R\rangle \vDash^* A$ iff for all e, $\langle W, R, e\rangle \vDash^* A$. *Boolos, 1980,* proves:

Theorem 57 $\vdash_{\mathbf{G}^*} A$ iff for every finite strict partial order $\langle W, R\rangle$, $\langle W, R\rangle \vDash^* A$

Before turning to set-assignment semantics for **G*** we need some observations.

Lemma 58 *a.* $w \vDash^{(0)} B$ iff $w \vDash B$

b. $w \vDash^{(j+1)} \Box B$ iff $w \vDash^{(j)} \Box B$ and $w \vDash^{(j)} B$

c. If $w \vDash^* \Box B$, then for all j, $w \vDash^{(j)} B$

Proof: I'll do (c) and leave the rest to you. If $w \vDash^* \Box B$, then for any j there is some $k > j$ such that $w \vDash^{(k)} \Box B$. Hence by definition, for all $i < k$, $w \vDash^{(i)} B$, so $w \vDash^{(j)} B$. ■

To give set-assignment semantics for **G*** we use the same set-assignments as for **G**, but replace the weak modal truth-conditions by those for modal semantics for implication, just as we did in modifying the semantics of **K** to get those for **QT**.

We say that $<v, s>$ is a **G***-*model* if s is a **G**-set-assignment and $<v, s>$ is a modal semantics of implication.

It's straightforward to show that every **G***-model validates **G*** by recalling that every modal semantics for implication validates $\Box A \supset A$ (Lemma 32.f).

Lemma 59 If $\nvdash_{\mathbf{G}^*} A$, then for some **G***-model $<v, s>$, $v(E) = F$.

Proof: If $\nvdash_{\mathbf{G}^*} A$, then by Theorem 57 there is some $\langle W, R, e, w\rangle$ such that $\langle W, R\rangle$ is a finite strict partial order and $\langle W, R, e, w\rangle \nvDash A$. Choose one.

As R is transitive we may assume without loss of generality that $W - \{w\} = \{z : wRz\}$. We may also assume, by relabeling if necessary, that no natural number is in W.

Define:

$$S = (W - \{w\}) \cup \{j : j \geq 0\}$$

$$s(A) = \{z : z \neq w \text{ and } z \vDash A\} \cup \{j : w \vDash^{(j)} A\}$$

Take $v(p) = T$ iff $w \vDash p$, and extend v to all wffs by the truth-conditions for modal semantics of implication. Note that we use '$z \vDash$' and not '$z \vDash^*$' in the definition of $s(A)$.

We now show that $v(A) = T$ iff $w \vDash^* A$, and that $<v, s>$ is a **G***-model.

The proof that $v(A) = T$ iff $w \vDash^* E$ is by induction on the length of A. The only difficult part is when A is of the form $\Box B$. If $v(\Box B) = T$, then $v(B) = T$ and $s(B) = S$. So for every z such that wRz, $z \vDash B$ and for all j, $w \vDash^{(j)} A$. Hence by

Lemma 58.b, $w \vDash^* \Box B$.

If $v(\Box B) = F$, then $v(B) = F$ or $s(B) \neq S$. If the latter, then for some z, wRz and $z \nvDash B$, so $w \nvDash \Box B$, so $w \nvDash^* \Box B$. If the former, then by induction $w \nvDash^* B$, so for some j, $w \nvDash^{(j)} B$. By Lemma 58.c, $w \nvDash^* \Box B$.

We now proceed to show that $<v,s>$ is a **G***-model. I'll leave to you to show that the first two conditions for **G**-set-assignments hold.

To show that if $s(A) = S$ then $s(\Box A) = S$, suppose $s(A) = S$. Then for all z such that wRz, $z \vDash A$ and for all j, $w \vDash^{(j)} A$. Hence $w \vDash \Box A$. And so by induction using Lemma 58, for all j, $w \vDash^{(j)} \Box A$. And also for all z, $z \vDash \Box A$. So $s(\Box A) = S$.

To show that $s(\Box(A \supset B)) \subseteq s(\Box A \supset \Box B)$ and $s(\Box(\Box A \supset A)) \subseteq s(\Box A)$, we'll prove that if $\vdash_{\mathbf{G}} A$, then $s(A) = S$. If $\vdash_{\mathbf{G}} A$, then $(W - \{w\}) \subseteq s(A)$. To show that every j is in $s(A)$ we induct first on j and then on the length of a proof of A. It is immediate for $j = 0$. Suppose it's true for j. I'll leave to you that $w \vDash^{(j+1)} \Box(B \supset C) \supset (\Box B \supset \Box C)$. For an axiom of the form $\Box(\Box B \supset B) \supset \Box B$, suppose $w \vDash^{(j+1)} \Box(\Box B \supset B)$. Then $w \vDash^{(j)} \Box B \supset B$ and by Lemma 58.b, $w \vDash^{(j)} \Box(\Box B \supset B)$, hence by induction (on this scheme) $w \vDash^{(j)} \Box B$ and also $w \vDash^{(j)} B$. Thus $w \vDash^{(j+1)} \Box B$ by Lemma 58.c. It's now straightforward to complete the proof for consequences of the axioms.

Finally we establish that $s(\Box A) \subseteq s(A)$ iff $s(A) = S$. Suppose $s(A) \neq S$. If there is some z such that wRz and $z \nvDash A$, then proceed as in the proof of Lemma 53 to get that $s(\Box A) \nsubseteq s(A)$. Otherwise, suppose that for all z such that wRz, $z \vDash A$, so that $w \vDash \Box A$. Let j be minimal such that $j \notin s(A)$, that is, $w \nvDash^{(j)} A$. Then for all $k < j$, $w \vDash^{(k)} A$. Hence $w \vDash^{(j)} \Box A$, so $j \in s(\Box A)$. ■

Theorem 60 **G***$\vdash$A iff for every **G***-model $<v,s>$, $v(E) = T$

VII Intuitionism – Int and J –

Intuitionism is an approach to doing mathematics constructively, without infinitistic assumptions. In this chapter we will look at Heyting's formalization of intuitionist reasoning and Kripke semantics for Heyting's formalization. In doing so we will see for the first time a nonclassical table for negation.

A. Intuitionism and Logic

Toward the end of the nineteenth century mathematicians began to use completed infinite totalities in mathematical constructions and proofs. The use of such collections was justified on a platonist conception of mathematics, sometimes related to a formalist reduction of mathematics to logic.

> In particular, in introducing new numbers, mathematics is only obliged to give definitions of them, by which such a definiteness and, circumstances permitting, such a relation to the older numbers are conferred upon them that in given cases they can definitely be distinguished from one another. As soon as a number satisfies all these conditions, it can and must be regarded as existent and real in mathematics.
>
> *Cantor, 1883,* p. 182

But there were mathematicians who objected to using nonconstructive proofs or definitions. *Brouwer, 1907* and *1908,* went further, however, and argued that the classical laws of logic that are valid in finite domains do not necessarily apply to (potentially) infinite collections. In the following years he and his colleagues developed a distinct program of mathematics, now called *intuitionism.*

The intuitionist believes that mathematics, that is the doing of mathematics, is prior to logic. Formal logic may be interesting and useful, but it cannot be the basis of mathematics.

> And in the construction of [all mathematical sets of units which are entitled to that name] neither the ordinary language nor any symbolic language can have any other rôle than that of serving as a nonmathematical auxiliary, to assist the mathematical memory or to enable different individuals to build up the same set.
>
> *Brouwer, 1912,* p. 81

The fundamental concepts of mathematics can be built by us independently of sense experience:

> This neo-intuitionism considers the falling apart of moments of life into qualitatively different parts, to be reunited only while remaining separated by time, as the fundamental phenomenon of the human intellect, passing by abstracting from its emotional content into the fundamental phenomenon of mathematical thinking, the intuition of the bare two-oneness. . . . Finally this basal intuition of mathematics, in which the connected and the separate, the continuous and discrete are united, gives rise immediately to the intuition of the

> linear continuum, i.e., of the "between," which is not exhaustible by the interposition of new units and which therefore can never be thought of as a mere collection of units.
>
> *Brouwer, 1912,* p. 80

It is on the question of how to reason about the infinite, or the potentially infinite that the intuitionists disagree with the classical mathematician, for in the realm of the finite they concur that classical logic is appropriate. Let us look at some examples.

1. Given any natural number, there is a prime number greater than it

Explanation: The intuitionist and the platonist agree that this proposition is true. We have a method that given any natural number can produce a prime number greater than that one. For if n is given, then either $(1 \cdot 2 \cdot 3 \cdots (n\text{-}1) \cdot n) + 1$ is prime, or if not, then since no number less than n divides this number, it is divisible by some other number that is prime, and that number must be between n and $(1 \cdot 2 \cdot 3 \cdots (n\text{-}1) \cdot n) + 1$. So given n, check each number successively between n and $(1 \cdot 2 \cdot 3 \cdots (n\text{-}1) \cdot n) + 1$ until a prime is found.

2. There is a prime pair greater than $10^{8489728}$ or there is not a prime pair greater than $10^{8489728}$

Explanation: A prime pair is a pair of primes that differ by 2, for example, 3 and 5, or 11 and 13. No one knows whether there is a largest prime pair.

On the platonist conception of mathematics the example is true. It is an informal instance of the law of excluded middle: $A \vee \neg A$. So in classical logic this is not only true, but valid.

The intuitionist, however, argues that the law of excluded middle is not valid, and that in particular this example is not to be accepted. We know of many prime pairs, but no one has ever exhibited a prime pair greater than $10^{8489728}$, nor has anyone demonstrated that there is no such pair. It does no good to say, 'Well, just start checking,' for that process would be endless if there is no prime pair greater than $10^{8489728}$.

> The solution is to abandon the principle of bivalence, and suppose our statements to be true just in case we have established that they are, i.e., if mathematical statements are in question, when we at least have an effective method of obtaining a proof of them.
>
> *Dummett, 1977,* p. 375

Brouwer takes as a basic insight that the validity of $A \vee \neg A$ is identified with the principle that every mathematical problem is solvable (see *Brouwer, 1928,* pp. 41–42, translated in *Bochenski, 1970,* p. 295).

3. The square root of 2 is not rational

Explanation: The intuitionist and the classical mathematician give the same proof that $\sqrt{2}$ is not rational, that is, $\sqrt{2}$ cannot be expressed as the quotient of two integers, p/q. For suppose it were of the form p/q, and suppose further that p/q is in lowest terms, that is, no number other than 1 divides both p and q. Then $p = \sqrt{2}\,q$, and so $p^2 = 2 \cdot q^2$. Hence, p^2 is even and p must be even, too, say $p = 2 \cdot r$. And q must be odd, since p/q is in lowest terms. Yet we have $(2 \cdot r)^2 = 2 \cdot q^2$, so $2 \cdot r^2 = q^2$, and we can conclude that q is even, a contradiction. So the assumption that the square root of 2 is rational has led us to a contradiction. Hence 'The square root of 2 is not rational' is true.

4. There are irrational numbers a and b such that a^b is rational

Explanation: The classical proof that there are two numbers a and b that are not rational such that a^b is rational begins by noting that $\sqrt{2}$ is irrational, as we proved above. Now consider $\sqrt{2}^{\sqrt{2}}$. If this is rational, we are done. Otherwise, both $\sqrt{2}^{\sqrt{2}}$ and $\sqrt{2}$ are irrational, and $(\sqrt{2}^{\sqrt{2}})^{\sqrt{2}} = \sqrt{2}^2 = 2$.

But the intuitionist argues that this is no proof at all, for we have not exhibited the required pair of irrationals. Which pair is it? $\sqrt{2}$ and $\sqrt{2}$? Or $\sqrt{2}^{\sqrt{2}}$ and $\sqrt{2}$? To prove A or B we must either prove A or prove B.

5. Consider the decimal $b = .b_1 b_2 \ldots b_n \ldots$, where:

$$b_n = \begin{cases} 3 & \text{if no string of 7 consecutive 7's appears before the } n^{\text{th}} \text{ decimal place in the expansion of } \pi \\ 0 & \text{otherwise} \end{cases}$$

Then b is rational.

Explanation: The platonist proves that this example is true: if b were not rational, then it could not be a finite string of 3's, $.33\ldots3$. But in that case it would have to be 1/3, a rational, which is a contradiction.

But the intuitionist says that the platonist has only drawn a contradiction from the assumption that b is not rational; that is, the platonist has proved 'It is not the case that it is not the case that b is rational'. But he has not proved that b is rational, for he has not exhibited two integers of which it is the quotient. And no method is now known to compute numbers p and q such that $b = p/q$. So 'b is rational' is unproved.

Thus the intuitionist rejects that $\neg\neg A \rightarrow A$ is valid. In the realm of the infinite or potentially infinite we cannot conclude A from $\neg\neg A$.

It is clear that the intuitionist must understand the connectives $\neg$, $\wedge$, $\vee$, $\rightarrow$ differently from the classical logician. Dummett explains them in terms of proofs:

A proof of $A \wedge B$ is anything that is a proof of A and of B.

A proof of $A \vee B$ is anything that is a proof of A or of B.

A proof of $A \rightarrow B$ is a construction of which we can recognize that, applied to any proof of A, it yields a proof of B.

A proof of $\neg A$ is usually characterized as a construction of which we can recognize that, applied to any proof of A, it will yield a proof of a contradiction.

Dummett, 1977, pp. 12–13

In the explication of '$\neg$' Dummett warns that 'a contradiction' must not be understood to be some statement of the form $B \wedge \neg B$, lest the characterization be circular. Rather, 'a contradiction' is intended to mean some particular statement, such as '$0 = 1$'; or negation is assumed to be clear when applied to arithmetic equations involving no variables, such as '$47 \times 23 = 1286$', and a contradiction is then understood as $(A \rightarrow B) \wedge \neg B$ where B is such an equation.

With this background let's turn to how intuitionistic reasoning has been formalized.

Exercises for Section A

1. Explain why the intuitionist rejects $A \vee \neg A$ as valid. Why can Brouwer claim that the validity of $A \vee \neg A$ is equivalent to the principle that every mathematical problem can be solved? Why does the intuitionist reject $\neg\neg A \rightarrow A$ as valid?
2. Show that one of De Morgan's Laws $\neg(A \wedge B) \rightarrow (\neg A \vee \neg B)$ is not valid on a constructive interpretation of the logical connectives.
3. In doing real analysis constructively, the intuitionists use the notion of a *free-choice sequence* in place of an arbitrary sequence of real numbers. At every stage of writing down the sequence a free choice may be made. For example, time is divided into discrete stages, and at any moment n we can tell whether, say, we have a proof of Fermat's Last Theorem. So we may define a real number $c = (c_n)$ by taking:

$$c_{2n} = \begin{cases} 1 & \text{if there are } u, v, t, w \text{ such that } 3 \leq w \leq n+2 \\ & \text{and } 0 < u, v, t \leq n \text{ such that } u^w + v^w = t^w \\ 0 & \text{otherwise} \end{cases}$$

$$c_{2n+1} = \begin{cases} 1 & \text{if a proof of Fermat's Last Theorem has been} \\ & \text{obtained by stage } n \\ 0 & \text{otherwise} \end{cases}$$

This is an example of a choice sequence for which there is no determinate procedure for calculating c_n, because 'proof' is not to be understood as a proof in some formal system, but any arbitrary correct proof.

Give a constructive proof that $c \neq 0$ and that $\neg(c < 0)$. Show that we cannot conclude, at present, that $c > 0$.

4. Consult a textbook on real analysis to find a proof by contradiction that claims to prove the existence of a mathematical object but that does not exhibit (constructively) the object.

B. Heyting's Formalization of Intuitionism

> Intuitionistic mathematics is an activity of thought, and every language—even the formalistic—is for it only a means of communication. It is impossible in principle to establish a system of formulae that would have the same value as intuitionistic mathematics, since it is impossible to reduce the possibilities of thought to a finite number of rules that thought can previously lay down. The endeavour to reproduce the most important parts of mathematics in a language of formulae is justified exclusively by the great conciseness and definiteness of this last as compared with customary languages, properties which fit it to facilitate penetration of the intuitionistic concepts and their application in research. . . . The relationship between this [formal] system and mathematics is this, that on a determinate interpretation of the constants and under certain restrictions on substitution for variables every formula expresses a correct mathematical proposition. (E.g. in the propositional calculus the variables must be replaced only by senseful [sinnerfülte] mathematical sentences.)
>
> *Heyting, 1930,* translated in *Bochenski,* pp. 293–294

Heyting presented his formal system syntactically as a collection of theorems. It is now standard to refer to it as the *intuitionist propositional calculus.*

1. Heyting's axiom system Int

Int *in* $L(\neg, \to, \wedge, \vee)$

I. $A \to (A \wedge A)$

II. $(A \wedge B) \to (B \wedge A)$

III. $(A \to B) \to ((A \wedge C) \to (B \wedge C))$

IV. $((A \to B) \wedge (B \to C)) \to (A \to C)$

V. $A \to (B \to A)$

VI. $(A \wedge (A \to B)) \to B$

VII. $A \to (A \vee B)$

VIII. $(A \vee B) \to (B \vee A)$

IX. $((A \to C) \wedge (B \to C)) \to ((A \vee B) \to C)$

X. $\neg A \to (A \to B)$

XI. $((A \to B) \wedge (A \to \neg B)) \to \neg A$

rules *modus ponens* adjunction

$$\frac{A,\ A \rightarrow B}{B} \qquad \frac{A,\ B}{A \wedge B}$$

Heyting used the rule of substitution rather than schema.

Until Section E, whenever I write $\vdash$ I mean $\vdash_{\textbf{Int}}$.

What is the nature of a project to give formal semantics to this system? By Heyting's quote, with which apparently all intuitionists agree, we cannot hope to fully capture or accurately represent the intuitionists' notion of meaning with formal semantics for a formal language. At best formal semantics can give us a *projective knowledge* of how intuitionists reason: we can gain enough insight to be able to reason in agreement with them if we wish. But the same could be said for any logic. What distinguishes the intuitionists is the degree to which they claim the precedence of intuition over logical systems and the extent to which they feel their notions have been misunderstood by classically trained mathematicians and logicians.

2. Kripke semantics for Int

I'll present formal semantics for **Int** along the lines of *Kripke, 1965.* See *Troelstra and van Dalen, 1988,* for a survey of formal semantics for **Int**, and *Dummett, 1977,* pp. 213–214, for an historical account.

We say that $\langle \mathsf{W}, \mathsf{R}, \mathsf{e} \rangle$ is a (Kripke) *model* if W is a nonempty set, R is a *reflexive, transitive* relation on W, and $\mathsf{e}: \text{PV} \rightarrow \text{Sub } \mathsf{W}$. We call e an *evaluation* and say that the model is *finite* if W is finite. The pair $\langle \mathsf{W}, \mathsf{R} \rangle$ is a *frame*.

We define a relation $\models$, read as 'validates', between elements w of W and wffs, where $\not\models$ means '$\models$ does not hold':

1. $w \models p$ iff for all z such that $w\mathsf{R}z$, $z \in \mathsf{e}(p)$
2. $w \models A \wedge B$ iff $w \models A$ and $w \models B$
3. $w \models A \vee B$ iff $w \models A$ or $w \models B$
4. $w \models \neg A$ iff for all z such that $w\mathsf{R}z$, $z \not\models A$
5. $w \models A \rightarrow B$ iff for all z such that $w\mathsf{R}z$, $z \not\models A$ or $z \models B$

Then $\langle \mathsf{W}, \mathsf{R}, \mathsf{e} \rangle \models A$ iff for all $w \in \mathsf{W}$, $w \models A$.

Here and throughout this chapter *except in* Section B.4 *we do not necessarily assume that* PV *and* Wffs *are completed infinite totalities*. Assignments or evaluations such as e above can be understood as meaning that we have a method such that given any variable p we can produce a subset of W.

This is the presentation given by *Fitting, 1969. Dummett, 1977,* gives an equivalent formulation by requiring that for each p, $\mathsf{e}(p)$ is *closed under* R, or for short is R*-closed*: if $w \in \mathsf{e}(p)$ and $w\mathsf{R}z$, then $z \in \mathsf{e}(p)$. He can then replace condition (1) by: $w \models p$ iff $w \in \mathsf{e}(p)$. I ask you to show in an exercise that that formulation is equivalent to the one given above.

***Lemma 1 Soundness of the Kripke semantics for* Int**

a. For any $\langle \mathsf{W}, \mathsf{R}, \mathsf{e} \rangle$ and $w \in \mathsf{W}$, $w \vDash \mathrm{A}$ iff for all z such that $w\mathsf{R}z$, $z \vDash \mathrm{A}$.

b. If $\vdash_{\mathbf{Int}} \mathrm{A}$, then A is valid in every Kripke model.

c. If $\Gamma \vdash_{\mathbf{Int}} \mathrm{A}$, then every Kripke model that validates Γ also validates A.

I leave the proof of this lemma as an exercise: part (a) is proved by induction on the length of wffs, part (b) by induction on the length of proofs. Part (c) follows by proving that the collection of wffs validated at any w is closed under deduction.

Dummett further classifies a subcollection of frames as *Kripke trees.* For the purposes of this chapter I will take these to be $\langle \mathsf{W}, \mathsf{R} \rangle$ such that $\langle \mathsf{W}, \mathsf{R} \rangle$ is a weak partial order (i.e., R is reflexive, transitive and anti-symmetric: if $x\mathsf{R}y$ and $y\mathsf{R}x$, then $x = y$) and there is an *initial point* $w \in \mathsf{W}$ that has no predecessor under R (i.e., for no $z \neq w$ do we have $z\mathsf{R}w$ and w is related to all elements of W). Whenever I refer to a model $\langle \mathsf{W}, \mathsf{R}, \mathsf{e}, w \rangle$ as a *Kripke tree* I will mean that w is the initial point.

How are these semantics supposed to reflect the intuitionists' understanding of logic, particularly Dummett's reading of the connectives? Let's first quote Fitting.

> W is intended to be a collection of . . . states of knowledge. Thus a particular w in W may be considered as a collection of (physical) facts known at a particular time. The relation R represents (possible) time succession. That is, given two states of knowledge w and z in W, to say $w\mathsf{R}z$ is to say: if we now know w, it is possible that later we will know z. Finally, to say that $w \vDash \mathrm{A}$ is to say: knowing w, we know A, or: from the collection of facts w, we may deduce the truth of A.
>
> Under this interpretation condition [4 above] for example, may be interpreted as follows: from the facts w we may conclude $\neg\mathrm{A}$ if and only if from no possible additional facts can we conclude A. . . . [Lemma 1.a is interpreted as:] If from a certain amount of information we can deduce A, given additional information, we still can deduce A, or if at some time we know A is true, at any later time we still know A is true.
>
> *Fitting, 1969,* p. 21

Dummett refers to the points in W as 'states of information' and says that p *is true at* w iff $w \in \mathsf{e}(\mathrm{p})$ (recall that he requires $\mathsf{e}(\mathrm{p})$ to be an R-closed set). He then says:

> Given any set of formulas, the sentence-letters occurring in them represent unanalysed constituent statements: we are considering states of information only in so far as they bear on the verification of these constituent statements. A state of information consists in a knowledge of two things: which of the constituent statements have been verified; and what future states of information are possible. That the constituent statement represented by a sentence letter p has been verified in the state of information represented by a point w is itself represented by the fact that $w \in \mathsf{e}(\mathrm{p})$. That the state of information represented by w may subsequently be improved upon by achieving the state represented by

a point z is represented by the fact that wRz. Note that there is no assumption that, at any point, we shall actually every [sic] acquire more information.

The requirement that $e(p)$ be an [R-closed set] . . . corresponds intuitively to the assumption that, once a constituent statement has been verified, it remains verified; i.e., that we do not forget what we have verified.

Dummett, 1977, p. 182

Let us see how these semantics can be used.

Example $A \vee \neg A$ is not valid.

Consider the model:

$z \notin e(p_1)$ ← $w \in e(p_1)$ → $y \in e(p_1)$

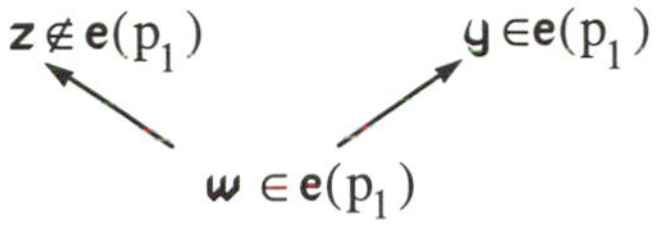

Since wRy and wRz, we have $w \nvDash p_1$ and $w \nvDash \neg p_1$. So $w \nvDash p_1 \vee \neg p_1$.

Example $\neg\neg A \rightarrow A$ is not valid.

Consider the model:

$w \notin e(p_1)$ ⟶ $z \notin e(p_1)$ ⟶ $y \in e(p_1)$

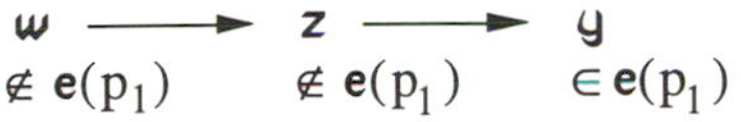

Since wRy and zRy and $y \vDash p_1$, we have that $w \vDash \neg\neg p_1$. But $w \nvDash p_1$. Hence $w \nvDash \neg\neg p_1 \rightarrow p_1$. ■

Using classical reasoning we can see how for these semantics $A \vee B$ is valid iff A is valid or B is valid (Example 4 above).

Theorem 2 $A \vee B$ is valid in every finite Kripke tree iff A is valid in every finite Kripke tree or B is valid in every finite Kripke tree

Proof: The right to left direction is immediate.

I will give an intuitionistically acceptable proof that if both A and B are not valid then $A \vee B$ is not valid. Classically, that is equivalent to what we want; it is more difficult to establish this intuitionistically.

If both A and B are not valid, then there is some $\langle W_1, R_1, e_1, w_1 \rangle$ such that $w_1 \nvDash A$, and $\langle W_2, R_2, e_2, w_2 \rangle$ such that $w_2 \nvDash B$. Define $\langle W, R, e, z \rangle$, where $z \notin W_1 \cup W_2$, by:

$W = W_1 \cup W_2$

xRy iff $(x = z)$ or $(x, y \in W_1$ and $xR_1y)$ or $(x, y \in W_2$ and $xR_2y)$

$e(p) = \{z\} \cup e_1(p) \cup e_2(p)$ for all p

Then $\langle W, R, e, z \rangle$ is a finite Kripke tree as you can check. And $z \nvDash A$, since zRw_1 and $w_1 \nvDash A$; and $z \nvDash B$, since zRw_2 and $w_2 \nvDash B$. Hence $z \nvDash A \vee B$. ■

We now want to show that these semantics characterize the logic of Heyting.

Exercises for Section B

1. Show that Dummett's formulation of semantics for **Int** (p. 277) is equivalent to the one given in the text.
2. Prove Lemma 1.
3. Show that both $A \vee \neg A$ and $\neg\neg A \to A$ fail in the model:

   $w \longrightarrow y$

   $\notin \mathbf{e}(p_1) \qquad \in \mathbf{e}(p_1)$

4. Determine whether the following are valid.
   a. $(\neg A \to A) \to A$
   b. $\neg(A \wedge B) \to \neg A \vee \neg B$
   c. $((A \vee B) \wedge \neg A) \to B$
   d. $\neg A \to (A \to B)$
   e. $(A \to B) \to (\neg B \to \neg A)$
   f. $(\neg B \to \neg A) \to (A \to B)$
5. Prove:
   a. $\vDash (A \vee \neg A) \to (\neg\neg A \to A)$
   b. $\vDash (\neg\neg A \to A) \to (A \vee \neg A)$
   c. $\vDash [(A \to B) \to ((\neg A \to B) \to B)] \to (B \vee \neg B)$
   d. $\vDash \neg\neg(A \wedge B) \to \neg\neg A \wedge \neg\neg B$
   e. $\vDash \neg\neg(A \to B) \to (\neg\neg A \to \neg\neg B)$
6. Show that each instance of each axiom scheme of Heyting's system is valid in every finite Kripke tree.
7. Determine which axioms of the axiomatization of **PC** in $L(\neg, \to, \wedge, \vee)$, pp. 82–83, are valid in every finite Kripke tree.
8. Determine which axioms of the axiomatization of the positive fragment of **PC** in $L(\to, \wedge, \vee)$, pp. 87–88, are valid in every finite Kripke tree.
9. Give a constructive proof of Theorem 2.

C. Completeness of Kripke Semantics for Int

In the first section I will give demonstrations in **Int** leading to the *Syntactic Deduction Theorem*. In the second section I will give a classical proof of the *Strong Completeness Theorem* for Kripke semantics for **Int**. Then I will discuss other proofs of completeness of these semantics and an alternate axiomatization of **Int**.

1. Some syntactic derivations and the *Deduction Theorem*

The following syntactic derivations are used to prove the *Syntactic Deduction Theorem* for **Int** and are needed in the completeness proof of the next section. Nowhere in this section do we use axiom scheme X.

Lemma 3 *a.* If $\Gamma \vdash A \rightarrow B$ and $\Gamma \vdash B \rightarrow C$, then $\Gamma \vdash A \rightarrow C$.

b. $\vdash A \wedge B \rightarrow A$

c. $\vdash A \wedge B \rightarrow B$

d. $\vdash A \rightarrow A$

e. $\vdash B \rightarrow (A \vee B)$

f. If $\Gamma \vdash A \rightarrow B$ and $\Gamma \vdash A \rightarrow (B \rightarrow C)$, then $\Gamma \vdash A \rightarrow C$.

g. If $\Gamma \vdash A \rightarrow B$ and $\Gamma \vdash A \rightarrow C$, then $\Gamma \vdash A \rightarrow (B \wedge C)$.

Proof:

a. i.	$A \rightarrow B$	premise
ii.	$B \rightarrow C$	premise
iii.	$(A \rightarrow B) \wedge (B \rightarrow C)$	rule of adjunction on (i) and (ii)
iv.	$((A \rightarrow B) \wedge (B \rightarrow C)) \rightarrow (A \rightarrow C))$	axiom IV
v.	$A \rightarrow C$	*modus ponens* on (iii) and (iv)
b. i.	$A \rightarrow (B \rightarrow A)$	axiom V
ii.	$(A \rightarrow (B \rightarrow A)) \rightarrow (A \wedge B \rightarrow ((B \rightarrow A) \wedge B)$	axiom III
iii.	$(A \wedge B) \rightarrow ((B \rightarrow A) \wedge B)$	*modus ponens* on (i) and (ii)
iv.	$((B \rightarrow A) \wedge B) \rightarrow (B \wedge (B \rightarrow A))$	axiom II
v.	$(B \wedge (B \rightarrow A)) \rightarrow A$	axiom VI
vi.	$(A \wedge B) \rightarrow A$	by (iii), (iv) and (v) using part (a)
c. i.	$(A \wedge B) \rightarrow (B \wedge A)$	axiom III
ii.	$(B \wedge A) \rightarrow B$	by part (b)
iii.	$(A \wedge B) \rightarrow B$	by (i) and (ii) using part (a)
d. i.	$A \rightarrow (A \wedge A)$	axiom I
ii.	$(A \wedge A) \rightarrow A$	by part (b)
iii.	$A \rightarrow A$	by (i) and (ii) using part (a)
e. i.	$B \rightarrow (B \vee A)$	axiom VII
ii.	$(B \vee A) \rightarrow (A \vee B)$	axiom VIII
iii.	$B \rightarrow (A \vee B)$	by (i) and (ii) using part (a)
f. i.	$A \rightarrow (B \rightarrow C)$	premise
ii.	$(A \rightarrow (B \rightarrow C)) \rightarrow ((A \wedge B) \rightarrow ((B \rightarrow C) \wedge B))$	axiom III

iii.	$(A \wedge B) \to ((B \to C) \wedge B)$	*modus ponens* on (i) and (ii)
iv.	$((B \to C) \wedge B) \to (B \wedge (B \to C))$	axiom II
v.	$(B \wedge (B \to C)) \to C$	axiom VI
vi.	$(A \wedge B) \to C$	by (iii), (iv) and (v), using part (a)
vii.	$(B \wedge A) \to (A \wedge B)$	axiom II
viii.	$(B \wedge A) \to C$	by (vi) and (vii) using part (a)
ix.	$(A \to B) \to ((A \wedge A) \to (B \wedge A))$	axiom III
x.	$A \to B$	premise
xi.	$(A \wedge A) \to (B \wedge A)$	*modus ponens* on (ix) and (x)
xii.	$A \to (A \wedge A)$	axiom I
xiii.	$A \to (B \wedge A)$	by (xi) and (xii) using part (a)
xiv.	$A \to C$	by (xiii) and (viii) using part (a)

g. i.	$(A \to B) \to (A \wedge C \to B \wedge C)$	axiom III
ii.	$A \to B$	premise
iii.	$(A \wedge C) \to (B \wedge C)$	*modus ponens* on (i) and (ii)
iv.	$(A \to C) \to ((A \wedge A) \to (A \wedge C))$	axiom III
v.	$(A \to C)$	premise
vi.	$(A \wedge A) \to (A \wedge C)$	*modus ponens* on (iv) and (v)
vii.	$A \to (A \wedge A)$	axiom I
viii.	$A \to (A \wedge C)$	by (vi) and (vii) using part (a)
ix.	$A \to (B \wedge C)$	by (viii) and (iii) using part (a) ■

***Theorem 4 The Syntactic Deduction Theorem for* Int**

$\Gamma \cup \{A\} \vdash_{\mathbf{Int}} B$ iff $\Gamma \vdash_{\mathbf{Int}} A \to B$

Proof: Let $A_0, A_1, \ldots, A_n = B$ be a proof of B from $\Gamma \cup \{A\}$. I will show by induction that for all $i \leq n$, $\Gamma \vdash A \to A_i$.

Either A_0 is an axiom, or $A_0 \in \Gamma$, or A_0 is A. For the first two we have the result by using axiom V. If A_0 is A, then we are done by Lemma 7.d.

Suppose $\Gamma \vdash A \to A_i$ for all $i < k$. Then if A_k is an axiom, is in Γ, or is A, we are done as before. Otherwise there are $i, j < k$ such that A_k is obtained from A_i and A_j by one of the rules. In that case we are done by Lemma 7.f, g. ■

Set $\neg^1 A \equiv_{\text{Def}} \neg A$, and for $n \geq 1$, $\neg^{n+1} A \equiv_{\text{Def}} \neg(\neg^n A)$. And set $A \leftrightarrow B \equiv_{\text{Def}} (A \to B) \wedge (B \to A)$.

Corollary 5 ***a.*** $\vdash_{\mathbf{Int}} A \to \neg\neg A$

b. $\vdash_{\mathbf{Int}} \neg\neg\neg A \to \neg A$

c. $\vdash_{\mathbf{Int}} \neg A \to \neg\neg\neg A$

d. For $n \geq 1$, $\vdash_{\mathbf{Int}} \neg^{2n+1} A \leftrightarrow \neg A$

e. For $n \geq 1$, $\vdash_{\mathbf{Int}} \neg^{2n+2} A \leftrightarrow \neg\neg A$

Proof: a. $\vdash_{\mathbf{Int}} \neg A \rightarrow \neg A$ by Lemma 3.d. And $A \vdash_{\mathbf{Int}} \neg A \rightarrow A$ by axiom V. Hence by the rule of adjunction, $A \vdash_{\mathbf{Int}} (\neg A \rightarrow A) \wedge (\neg A \rightarrow \neg A)$. So by axiom XI and Lemma 3.a, $A \vdash_{\mathbf{Int}} \neg\neg A$. Hence by the *Deduction Theorem,* $\vdash_{\mathbf{Int}} A \rightarrow \neg\neg A$. Part (c) is an instance of part (a).

b. By part (a), $\neg\neg\neg A \vdash_{\mathbf{Int}} A \rightarrow \neg\neg A$. By axiom V and the *Deduction Theorem,* $\neg\neg\neg A \vdash_{\mathbf{Int}} A \rightarrow \neg\neg\neg A$. So by adjunction, $\neg\neg\neg A \vdash_{\mathbf{Int}} (A \rightarrow \neg\neg A) \wedge (A \rightarrow \neg\neg\neg A)$. Hence by axiom XI, $\neg\neg\neg A \vdash_{\mathbf{Int}} \neg A$. So $\vdash_{\mathbf{Int}} \neg\neg\neg A \rightarrow \neg A$.

Parts (d) and (e) are proved by induction on *n*. ■

2. Completeness theorems for Int

I will prove in this section that the class of finite Kripke trees is strongly complete for **Int**. We make the following definitions:

Σ is a *theory* if $\Sigma \supseteq$ **Int** and Σ is closed under *modus ponens* and adjunction

Σ is *consistent* if for some A, $\Sigma \nvdash A$ (Post-consistency)

Σ is *full* if Σ is a consistent theory such that for every A and B,
if $(A \vee B) \in \Sigma$, then $A \in \Sigma$ or $B \in \Sigma$

Lemma 6 *a.* Σ is consistent iff for every A, $\Sigma \nvdash A$ or $\Sigma \nvdash \neg A$

b. If Σ is full, then:

i. $\Sigma \vdash A$ iff $A \in \Sigma$

ii. $A \wedge B \in \Sigma$ iff $A \in \Sigma$ and $B \in \Sigma$

iii. $A \vee B \in \Sigma$ iff $A \in \Sigma$ or $B \in \Sigma$

iv. $\Sigma \vdash A \vee B$ iff $\Sigma \vdash A$ or $\Sigma \vdash B$

Proof: For part (a), if $\Sigma \vdash B$ and $\Sigma \vdash \neg B$, then by axiom X for every A, $\Sigma \vdash A$. The converse is immediate.

For part (b) (i), note that $\Sigma \supseteq$ **Int** and is closed under the rules.

Part (ii) follows from part (a) and Lemma 3.

For part (iii), if $A \vee B \in \Sigma$, then by definition $A \in \Sigma$ or $B \in \Sigma$. If $A \in \Sigma$, then by axiom VII and part (i), $A \wedge B \in \Sigma$. If $B \in \Sigma$, then by Lemma 3 and part (i), $A \vee B \in \Sigma$.

Part (iv) follows from parts (i) and (iii). ■

The first use of infinitistic intuitionistically unacceptable reasoning in this section occurs in the proof of the next lemma.

Lemma 7 If $\Gamma \nvdash_{\mathbf{Int}} E$, then there is some full $\Sigma \supseteq \Gamma$ such that $E \notin \Sigma$.

Proof: Let $B_1, B_2, \ldots$ be a listing of all wffs. Define:

$$\Sigma_0 = \Gamma$$

$$\Sigma_{n+1} = \begin{cases} \Sigma_n \cup \{B_n\} & \text{if } \Sigma_n \nvdash B_n \to E \\ \Sigma_n & \text{otherwise} \end{cases}$$

$$\Sigma = \bigcup_n \Sigma_n$$

By Lemma 3.d, $\vdash E \to E$, so for all n, $\Sigma_n \vdash E \to E$, and hence $E \notin \Sigma$. I'll show by induction that for all n, $\Sigma_n \nvdash E$. It's true for $n = 0$. Suppose it's true for all $i \leq n$. If $\Sigma_{n+1} \vdash E$, then by induction we must have $\Sigma_{n+1} = \Sigma_n \cup \{B_n\}$, but then by the *Deduction Theorem,* $\Sigma_n \vdash B_n \to E$, a contradiction. Hence $\Sigma \nvdash E$, so Σ is consistent. It remains to show that Σ is full.

To show that Σ is a theory, suppose that $\Sigma \vdash A$. Were $A \notin \Sigma$, then by construction $\Sigma \vdash A \to E$, so we would have $\Sigma \vdash E$ which is a contradiction. Hence $A \in \Sigma$.

Suppose $A \vee B \in \Sigma$. If $A \notin \Sigma$ and $B \notin \Sigma$, then by the construction $\Sigma \vdash A \to E$ and $\Sigma \vdash B \to E$. But then by axiom IX, $\Sigma \vdash A \vee B \to E$, a contradiction. So $A \in \Sigma$ or $B \in \Sigma$. Hence Σ is full. ■

Lemma 8 If Γ is full, then for all E:

a. $\Gamma \vdash E$ iff for every $\Sigma \supseteq \Gamma$ which is full, $E \in \Sigma$

b. $\Gamma \vdash \neg E$ iff for every $\Sigma \supseteq \Gamma$ which is full, $E \notin \Sigma$

Proof: a. If $\Gamma \vdash E$, then for all full $\Sigma \supseteq \Gamma$, $\Sigma \vdash E$. So by Lemma 6.b, $E \in \Sigma$. The converse is immediate.

b. Suppose $\Gamma \vdash \neg E$. Then for all full $\Sigma \supseteq \Gamma$, $\Sigma \vdash \neg E$, so by the consistency of Σ, $E \notin \Sigma$.

Now suppose that for all full $\Sigma \supseteq \Gamma$, $E \notin \Sigma$. So $\Gamma \cup \{E\}$ is inconsistent, for were it not then taking $\Gamma \cup \{E\}$ for Γ in Lemma 7 we would have a contradiction. Hence $\Gamma \cup \{E\} \vdash \neg E$. So by the *Deduction Theorem,* $\Gamma \vdash E \to \neg E$. By Lemma 3.d, $\Gamma \vdash E \to E$. Using Axiom XI, we have $\Gamma \vdash \neg E$. ■

Define the *canonical model* for **Int** to be:

$\langle \mathsf{W}_{\mathbf{Int}}, \subseteq, \mathsf{e}_{\mathbf{Int}} \rangle$, where:

$$\mathsf{W}_{\mathbf{Int}} = \{\Gamma : \Gamma \text{ is full}\}$$

$$\mathsf{e}_{\mathbf{Int}}(\Gamma) = \{p : p \in \Gamma\}$$

I leave to you to show that the canonical model is reflexive, transitive, and anti-symmetric. It is also uncountably infinite.

Lemma 9 ***a.*** $\langle \mathsf{W}_{\mathbf{Int}}, \subseteq, \mathsf{e}_{\mathbf{Int}}, \Gamma \rangle \vDash A$ iff $A \in \Gamma$

b. $\langle \mathsf{W}_{\mathbf{Int}}, \subseteq, \mathsf{e}_{\mathbf{Int}} \rangle \vDash A$ iff $\vdash_{\mathbf{Int}} A$

Proof: a. By induction on the length of A. It is immediate if the length of A is 1. Suppose A has length greater than 1 and it is true for all wffs shorter than A and all Γ. We have the following cases, depending on the form of A:

$\Gamma \vDash B \wedge C$	iff $\Gamma \vDash B$ and $\Gamma \vDash C$	
	iff $B \in \Gamma$ and $C \in \Gamma$	by induction
	iff $B \wedge C \in \Gamma$	by Lemma 6
$\Gamma \vDash B \vee C$	iff $\Gamma \vDash B$ or $\Gamma \vDash C$	
	iff $B \in \Gamma$ or $C \in \Gamma$	by induction
	iff $B \vee C \in \Gamma$	because Γ is full and by Lemma 6
$\Gamma \vDash \neg B$	iff for all $\Sigma \supseteq \Gamma$, $\Sigma \nvDash B$	
	iff for all $\Sigma \supseteq \Gamma$, $B \notin \Sigma$	by induction
	iff $\Gamma \vdash \neg B$	by Lemma 8
	iff $\neg B \in \Gamma$	by Lemma 6

If A has the form $B \rightarrow C$, first note:

$\Gamma \vDash B \rightarrow C$	iff for all $\Sigma \supseteq \Gamma$, $\Sigma \nvDash B$ or $\Sigma \vDash C$
	iff for all $\Sigma \supseteq \Gamma$, $B \notin \Sigma$ or $C \in \Sigma$

Now suppose $\Gamma \vDash B \rightarrow C$. I'll show that $\Gamma \cup \{B\} \vdash C$ from which, by the *Deduction Theorem,* $\Gamma \vdash B \rightarrow C$, and hence by Lemma 6, $B \rightarrow C \in \Gamma$. If $\Gamma \cup \{B\}$ is inconsistent, we are done. So suppose $\Gamma \cup \{B\}$ is consistent. If $\Gamma \cup \{B\} \nvdash C$, then by Lemma 7, for some full $\Sigma \supseteq \Gamma \cup \{B\}$, $C \notin \Sigma$, which contradicts that $\Gamma \vDash B \rightarrow C$.

Now suppose $B \rightarrow C \in \Gamma$. If $\Sigma \supseteq \Gamma$ is full and $B \in \Sigma$, then $C \in \Sigma$ as Σ is closed under deduction. So $\Gamma \vDash B \rightarrow C$.

b. This follows by Lemma 7. ■

Note: An *endpoint* of a Kripke model is a z such that for no y do we have zRy. Any endpoint validates **PC**, since the evaluation at the endpoint proceeds as in a classical model taking $\mathsf{v}(p) = \mathsf{T}$ iff $\mathsf{z} \in \mathsf{e}(p)$. In particular, if Γ is an endpoint of the canonical model, then $\Gamma \vDash$ **PC**, so **PC** $\subset \Gamma$. As **PC** $\vdash A \vee \neg A$, for every A, $A \in \Gamma$ or $\neg A \in \Gamma$. That is, the endpoints of the canonical model are **PC**-complete and consistent sets of wffs.

We say that Γ is *closed under subformulas* if given any $A \in \Gamma$, if B is a subformula of A, then $B \in \Gamma$.

Lemma 10 Given any Kripke model $\langle \mathsf{W}, \mathsf{R}, \mathsf{e} \rangle$ and Γ a finite collection of wffs closed under subformulas, there is a finite anti-symmetric Kripke model $\langle \mathsf{W}^*, \mathsf{R}^*, \mathsf{e}^* \rangle$ such that for every $A \in \Gamma$, $\langle \mathsf{W}, \mathsf{R}, \mathsf{e} \rangle \vDash A$ iff $\langle \mathsf{W}^*, \mathsf{R}^*, \mathsf{e}^* \rangle \vDash A$.

Proof: The proof mimics that of Theorem VI.15.

Define an equivalence relation on W:

$x \approx y$ iff for all $A \in \Gamma$, $x \vDash A$ iff $y \vDash A$

Denote by x^* the equivalence class of x. Define:

$\mathsf{W}^* = \{ x^* : x \in \mathsf{W} \}$

$x^* \mathsf{R}^* y^*$ iff for all $A \in \Gamma$, if $x \vDash A$ then $y \vDash A$

$\mathsf{e}(x^*) = \{ p : p \in \Gamma$ and $x \vDash p \}$

Because of the definition of $\approx$, R^* and e^* are well-defined, R^* is reflexive, transitive, and anti-symmetric, and W^* is finite. Moreover by Lemma 1.a, if $x\mathsf{R}y$ then $x^*\mathsf{R}^*y^*$.

In order to show that for all $A \in \Gamma$, all $x \in \mathsf{W}$, $\langle \mathsf{W}, \mathsf{R}, \mathsf{e}, x \rangle \vDash A$ iff $\langle \mathsf{W}^*, \mathsf{R}^*, \mathsf{e}^*, x^* \rangle \vDash A$, we proceed by induction on the length of A. The only interesting cases are if A is of the form $B \rightarrow C$ or $\neg B$.

If A is $B \rightarrow C$, suppose $x \vDash B \rightarrow C$. By way of contradiction suppose that there is some y^* such that $x^*\mathsf{R}^*y^*$ and $y^* \vDash B$ yet $y^* \nvDash C$. Then by induction, since $B, C \in \Gamma$, $y \vDash B$ and $y \nvDash C$. But then $y \nvDash B \rightarrow C$, contradicting $x^*\mathsf{R}^*y^*$. So $x^* \vDash B \rightarrow C$.

Now suppose $x^* \vDash B \rightarrow C$. If $x\mathsf{R}y$ and $y \vDash B$, then $x^*\mathsf{R}^*y^*$ and, by induction, $y^* \vDash B$. But then $y^* \vDash C$, so by induction $y \vDash C$. Hence $x \vDash B \rightarrow C$.

If A is $\neg B$, suppose $x \vDash \neg B$ and $x^*\mathsf{R}^*y^*$. If $y^* \vDash B$, then by induction $y \vDash B$, contradicting $x^*\mathsf{R}^*y^*$. So $y^* \nvDash B$, and $x^* \vDash \neg B$.

If $x^* \vDash \neg B$ and $x\mathsf{R}y$, then $x^*\mathsf{R}^*y^*$ and $y^* \nvDash B$. So by induction $y \nvDash B$. Hence $x \vDash \neg B$. ■

***Theorem 11 Completeness of the Kripke Semantics for* Int**

a. $\Gamma \vdash_{\mathbf{Int}} A$ iff every Kripke model that validates Γ also validates A

b. $\Gamma \vdash_{\mathbf{Int}} A$ iff every Kripke tree that validates Γ also validates A

c. For finite Γ, $\Gamma \vdash_{\mathbf{Int}} A$ iff every finite Kripke tree that validates Γ also validates A

d. **Int** is decidable

Proof: a. From left to right is Lemma 1.c.

If $\Gamma \nvdash_{\mathbf{Int}} A$, then consider the submodel of the canonical model, $\langle \mathsf{W}_\Gamma, \subseteq, \mathsf{e}_{\mathbf{Int}} \rangle$ where $\mathsf{W}_\Gamma = \{ \Sigma : \Sigma$ is full and $\Sigma \supseteq \Gamma \}$. Since $\Gamma \nvdash A$, by Lemma 7, $\mathsf{W}_\Gamma \neq \varnothing$; and for all B, $\langle \mathsf{W}_\Gamma, \subseteq, \mathsf{e}_{\mathbf{Int}} \rangle \vDash B$ iff $B \in \Gamma$. Hence $\langle \mathsf{W}_\Gamma, \subseteq, \mathsf{e}_{\mathbf{Int}} \rangle \vDash \Gamma$ and $\langle \mathsf{W}_\Gamma, \subseteq, \mathsf{e}_{\mathbf{Int}} \rangle \nvDash A$.

b. This follows by part (a), using the methods of Lemma VI.1, p. 208, to cull a Kripke tree from the model $\langle \mathsf{W}_\Gamma, \subseteq, \mathsf{e}_{\mathbf{Int}} \rangle$.

c. This follows from part (b) by Lemma 10.

d. Given a wff we have the following procedure: we constructively list every finite Kripke model and test in it whether the wff fails, and at the same time we list out all theorems using the constructive proof process. If a wff is not a tautology, we will find a model in which it fails. If it is a tautology, we will find a proof of it. Since, classically at least, every wff is either a tautology or not, we have a decision procedure. ■

Exercises for Sections C.1 and C.2

1. Prove parts (d) and (e) of Corollary 5.
2. What does it mean to say that a collection of wffs is *full*?
3. Prove the canonical model for **Int** is reflexive, transitive, and anti-symmetric.
4. Prove Lemma 9.b.
5. Axiomatize the fragment of **Int** in $L(\rightarrow, \wedge, \vee)$ and establish a completeness theorem for the Kripke models.
6. Prove that substitution is a derived rule for **Int**.

3. On completeness proofs for Int, and an alternate axiomatization

We proved in the last section:

1. $\Gamma \vdash_{\mathbf{Int}} A$ iff every Kripke model that validates Γ also validates A
2. $\Gamma \vdash_{\mathbf{Int}} A$ iff every Kripke tree that validates Γ also validates A
3. For finite Γ, $\Gamma \vdash_{\mathbf{Int}} A$ iff every finite Kripke tree that validates Γ also validates A

Our proofs used classical logic and infinitistic, nonconstructive methods. *Dummet, 1977, gives an intuitionistically acceptable proof* of the finite strong completeness proof for finite Kripke trees (3), though he uses a different axiomatization of **Int**. He exhibits a method such that given any finite Γ and any A such that $\Gamma \nvdash_{\mathbf{Int}} A$, he can produce a finite Kripke tree that validates Γ and invalidates A. Therefore, intuitionistic logic can be used for its own metalogic.

Thus if we confine our attention to finite collections of propositions or wffs, finite models suffice for analyzing our logic: about these the intuitionist agrees we can reason classically. And since only finite models are involved we can, if we wish, understand the *Fully General Abstraction* for these semantics in intuitionistic terms.

Note that if we allow Γ to be infinite and $\Gamma \nvdash A$, then it is not clear how to proceed intuitionistically to produce a model of Γ that invalidates A: we may not be able to survey all of Γ at once. But even if we reason classically, the reduction from the class of all Kripke trees to the class of finite Kripke trees seems to require that we restrict ourselves to finite Γ.

Here is the axiomatization of *Dummett, 1977,* p.126:

Int
in $L(\neg, \rightarrow, \wedge, \vee)$

axiom schema

1. $A \rightarrow (B \rightarrow A)$
2. $A \rightarrow (B \rightarrow (A \wedge B))$
3. $(A \wedge B) \rightarrow A$
4. $(A \wedge B) \rightarrow B$
5. $A \rightarrow (A \vee B)$
6. $B \rightarrow (A \vee B)$
7. $(A \vee B) \rightarrow ((A \rightarrow C) \rightarrow ((B \rightarrow C) \rightarrow C))$
8. $(A \rightarrow B) \rightarrow ((A \rightarrow (B \rightarrow C)) \rightarrow (A \rightarrow C))$
9. $(A \rightarrow B) \rightarrow ((A \rightarrow \neg B) \rightarrow \neg A)$
10. $A \rightarrow (\neg A \rightarrow B)$

rule $\dfrac{A,\ A \rightarrow B}{B}$

Dummett's proof that his system is characterized by finite Kripke trees is enough to establish via Theorem 11 that his system is equivalent to Heyting's. To establish the equivalence of Dummett's system with Heyting's in an intuitionistically acceptable manner, first note that Dummett's system contains **Int** via Lemma 1.b. To show the containment in the other direction you can derive in **Int** each (instance of each) scheme of Dummett using Lemma 3, the *Deduction Theorem,* and the observation that $\{A, B\} \vdash_{\mathbf{Int}} C$ iff $A \wedge B \vdash_{\mathbf{Int}} C$.

Two other intuitionistically acceptable proofs of completeness of intuitionistic logic, as extended to contain quantifiers, can be found in *Veldman, 1976,* and *De Swart, 1976.*

Exercises for Section C.3

1. Establish that Dummett's axiom system is equivalent to Heyting's without invoking Theorem 9.

D. Translations and Comparisons with Classical Logic

1. Translations of Int into modal logic and classical arithmetic

If you are familiar with possible-world semantics for modal logics (Chapter VI), you may have noticed that the semantics for **Int** are similar to those for **S4** and **S4Grz**. Using Theorem 11 of the previous section we can interpret **Int** in **S4** and **S4Grz**. Consider the following map from $L(\neg, \rightarrow, \wedge, \vee)$ to $L(\neg, \wedge, \Box)$:

$$p^* = \Box p$$
$$(A \wedge B)^* = A^* \wedge B^*$$
$$(A \vee B)^* = A^* \vee B^*$$
$$(A \rightarrow B)^* = \Box(A^* \supset B^*)$$
$$(\neg A)^* = \Box\neg(A^*)$$

Set $\Gamma^* = \{A^*: A \in \Gamma\}$. Note that $(A \to B)^* = A^* \to B^*$.

Theorem 12 ***a.*** $\Gamma \vdash_{\mathbf{Int}} A$ iff $\Gamma^* \vdash_{\mathbf{S4}} A^*$

b. $\vdash_{\mathbf{Int}} A$ iff $\vdash_{\mathbf{S4Grz}} A^*$

Proof: Any Kripke model for **Int** can be viewed as a Kripke model for modal logic via the Note on p. 206. And $\langle W, R, e, w \rangle \vDash A$ in the intuitionist semantics iff $\langle W, R, e, w \rangle \vDash A^*$ in the modal semantics. Part (a) is then a consequence of Theorem 11 and Theorem VI.29. Part (b) follows by Theorem 11 and the characterization of **S4Grz** quoted in Chapter VI.J.2, Theorem VI.49. ∎

Gödel, 1933 B, was the first to interpret **Int** in **S4**, long before formal semantics had been given for either logic. His translation was suggested by reading '$\Box$' as 'it is provable that':

$$\begin{aligned} p' &= p \\ (A \wedge B)' &= \Box A' \wedge \Box B' \\ (A \vee B)' &= \Box A' \vee \Box B' \\ (A \to B)' &= \Box A' \supset \Box B' \\ (\neg A)' &= \neg \Box A' \end{aligned}$$

Alternatively, we may take $(\neg A)' = \Box \neg \Box A'$.

We can also interpret **Int** in terms of provability in classical arithmetic via the translations of Chapter VI.J.1. We can compose the map $*$ above with any provability-and-truth translation $\#$ (p. 265) by defining $(A \vee B)^\# = A^\# \vee B^\#$, thus obtaining a map of the language $L(\neg, \to, \wedge, \vee)$ of **Int** to that of Peano Arithmetic, **PA** :

$$\begin{aligned} p_i^+ &= \alpha_i \wedge \text{Bew}([\![\alpha_i]\!]) \quad \text{for some sentence } \alpha_i \text{ of the language of } \mathbf{PA} \\ (A \wedge B)^+ &= A^+ \wedge B^+ \\ (A \vee B)^+ &= A^+ \vee B^+ \\ (A \to B)^+ &= (A^+ \supset B^+) \wedge \text{Bew}([\![A^+ \supset B^+]\!]) \\ (\neg A)^+ &= \neg(A^+) \wedge \text{Bew}([\![\neg A^+]\!]) \end{aligned}$$

Combining Theorem 11 with Theorem VI.48 we have:

Theorem 13 $\mathbf{Int} \vdash A$ iff for every translation $^+$ as above, A^+ is a theorem of Peano Arithmetic

An intuitionist implication in arithmetic asserts the truth and provability of a material implication; an intuitionist negation asserts that the sentence is false and provably so (compare the first quote by Dummett in Section A, p. 273).

By a series of observations about the logics **G*** and **S4Grz**, *Goldblatt, 1978,* invokes Theorem VI.47 to prove:

Theorem 14 **Int**$\vdash$A iff for every translation $^+$ as above, A^+ is true of the natural numbers

2. Translations of classical logic into Int

The relation of **Int** to **PC** is complicated. To begin with, **Int** $\subset$ **PC**, as every axiom of **Int** is a classical tautology and the rules are classically valid. All the intuitionist's principles are acceptable to the classical logician.

And if we consider only the positive part of **PC**, the classical logician's reasoning is acceptable to an intuitionist, too:

Theorem 15 In $L(\rightarrow, \wedge, \vee)$, $\Gamma \vDash_{\mathbf{PC}} A$ iff $\Gamma \vDash_{\mathbf{Int}} A$

Proof: From right to left is because **Int** $\subset$ **PC**. From left to right, consider the axiomatization of the fragment of **PC** in $L(\rightarrow, \wedge, \vee)$, Theorem II.27, p. 88. The rule is valid in **Int**, and every axiom (scheme) is valid in **Int**. For example, axiom scheme 2 is $(A \rightarrow (B \rightarrow C)) \rightarrow ((A \rightarrow B) \rightarrow (A \rightarrow C))$. By Lemma 3, $\{A \rightarrow (B \rightarrow C), (A \rightarrow B)\} \vDash_{\mathbf{Int}} (A \rightarrow C)$. So by the *Deduction Theorem,* the wff is valid. I will leave the others to you. ■

Note that the proof that the axiomatization of **PC** in $L(\rightarrow, \wedge, \vee)$ is complete is established using nonconstructive methods.

What if we include negation? As *Kolmogorov, 1925* said, for every "truth" A of classical logic (arithmetic), there is a "pseudo-truth" $\neg\neg A$, such that $\vdash_{\mathbf{PC}} A$ iff $\vdash_{\mathbf{Int}} \neg\neg A$. Let $\neg\neg\Gamma = \{\neg\neg A : A \in \Gamma\}$. The following is due to *Glivenko, 1929.*

Theorem 16 In $L(\neg, \rightarrow, \wedge, \vee)$:

a. $\Gamma \vdash_{\mathbf{PC}} A$ iff $\neg\neg\Gamma \vdash_{\mathbf{Int}} \neg\neg A$

b. $\vdash_{\mathbf{PC}} \neg A$ iff $\vdash_{\mathbf{Int}} \neg A$

Proof: a. From right to left, we have **Int** $\subset$ **PC**. *Modus ponens* is a rule of **PC**, and adjunction is a valid rule in **PC**. So if $\neg\neg\Gamma \vdash_{\mathbf{Int}} \neg\neg A$, then $\neg\neg\Gamma \vdash_{\mathbf{PC}} \neg\neg A$. Since for all B, $\vdash_{\mathbf{PC}} \neg\neg B \leftrightarrow B$, we have $\Gamma \vdash_{\mathbf{PC}} A$.

From left to right, I will give two quite different proofs, the first semantic, the second (essentially) syntactic.

First proof: We first observe that for any B and finite Kripke tree $\langle W, R, e\rangle$:

$\langle W, R, e\rangle \vDash \neg\neg B$ iff for all w, $w \vDash \neg\neg B$

iff for all w and all z, if wRz, then $z \nvDash \neg B$

iff for all w and all z for which wRz, there is some x such that zRx and $x \vDash B$

For any endpoint x of $\langle W, R, e\rangle$ (i.e., for no $z \neq x$ do we have xRz) we have that if $\vdash_{\mathbf{PC}} B$, then $x \vDash B$ because the evaluation at an endpoint proceeds as in the classical model taking $v(p) = T$ iff $x \in e(p)$.

Suppose we have $\vdash_{\mathbf{PC}} A$. Then for any point w of any finite Kripke tree model and any z such that wRz, there is some endpoint x such that zRx. Since $x \vDash A$ we have, as observed above, $w \vDash \neg\neg A$. So $\vdash_{\mathbf{Int}} \neg\neg A$.

Suppose now that $\Gamma \vdash_{\mathbf{PC}} A$. Since the syntactic consequence relation is compact we may assume that Γ is finite. At any endpoint t of a finite Kripke tree model $\langle W, R, e\rangle$, if $t \vDash \Gamma$ then $t \vDash A$. So for any point z in the model, if $z \vDash \neg\neg\Gamma$ then for every y such that zRy there is an x such that yRx and $x \vDash \Gamma$. For such an x there is some endpoint t such that xRt, and by Lemma 1.a, $t \vDash \Gamma$. Hence $t \vDash A$, so $z \vDash \neg\neg A$. Hence by Theorem 2, $\neg\neg\Gamma \vdash_{\mathbf{Int}} \neg\neg A$.

Second proof: I will list the wffs we need to show are theorems of **Int** in order to prove that if $\vdash_{\mathbf{PC}} A$, then $\neg\neg\Gamma \vdash_{\mathbf{Int}} \neg\neg A$.

1. Every instance of the axiom schema for **PC** given in Chaper II.N.6, except $(A \to B) \to ((\neg A \to B) \to B)$, is a theorem of **Int**: axiom 1 is an axiom of **Int**, and axioms 2, 4–10 do not involve negation, and hence are theorems of **Int** by Theorem 15 above.
2. $\neg\neg[(A \to B) \to ((\neg A \to B) \to B)]$
3. $\neg\neg(A \to B) \to (\neg\neg A \to \neg\neg B)$

By Corollary 5, if $\vdash_{\mathbf{Int}} A$, then $\vdash_{\mathbf{Int}} \neg\neg A$. So by (1) and (2) we have for every **PC** axiom A, $\vdash_{\mathbf{Int}} \neg\neg A$.

Now we induct on the length of a proof of A in **PC** to show if $\vdash_{\mathbf{PC}} A$, then $\vdash_{\mathbf{Int}} \neg\neg A$. If the length is 1, then either A is an axiom of **PC** or else $A \in \Gamma$, and so we are done as above. So suppose we have shown it for any wff with proof of length less than or equal to n steps, and $A_1, \ldots, A_n, A_{n+1}$ is a proof of A from Γ in **PC**. The last step must be an application of *modus ponens* on $A_i \to A_{n+1} = A_j$, where $i, j \leq n$. By induction $\neg\neg\Gamma \vdash_{\mathbf{Int}} \neg\neg A_i$ and $\neg\neg\Gamma \vdash_{\mathbf{Int}} \neg\neg(A_i \to A_{n+1})$. By (3), $\vdash_{\mathbf{Int}} \neg\neg(A_i \to A_{n+1}) \to (\neg\neg A_i \to \neg\neg A_{n+1})$, so using *modus ponens* twice we have $\neg\neg\Gamma \vdash_{\mathbf{Int}} \neg\neg A_{n+1}$.

b. From right to left is because $\mathbf{Int} \subset \mathbf{PC}$. So suppose $\vdash_{\mathbf{PC}} \neg A$. Then by part (a), $\vdash_{\mathbf{Int}} \neg\neg\neg A$, so by Corollary 5.b, $\vdash_{\mathbf{Int}} \neg A$. ■

Gödel, 1933 A, capitalized on this double negation translation to show that in one sense **Int** can be viewed as an extension of **PC:**

Corollary 17 If A is a wff of $L(\neg, \wedge)$, then $\vdash_{\mathbf{PC}} A$ iff $\vdash_{\mathbf{Int}} A$.

Proof: From right to left is because $\mathbf{Int} \subset \mathbf{PC}$.

Suppose $\vdash_{\mathbf{PC}} A$. If A is a negation, then we are done by the previous theorem. So suppose A is not a negation. Then A must be of the form $B_1 \wedge \ldots \wedge B_n$, $n \geq 1$, where each B_i is not a conjunction. Therefore, each B_i must be either a negation or a propositional variable. Since $\vdash_{\mathbf{PC}} A$ we have $\vdash_{\mathbf{PC}} B_i$ for each *i*. But no variable is a theorem of **PC**. Thus A has the form $\neg C_1 \wedge \ldots \wedge \neg C_n$ where for each *i*, $\vdash_{\mathbf{PC}} \neg C_i$. But then by the previous theorem we have $\vdash_{\mathbf{Int}} \neg C_i$. So by the rule of adjunction, $\vdash_{\mathbf{Int}} A$. ■

Corollary 18 Not both $\rightarrow$ and $\vee$ can be defined from $\neg$ and $\wedge$ in **Int**.

Proof: If both could be defined, then we'd have **Int** = **PC** in $L(\neg, \rightarrow, \wedge, \vee)$. ■

Actually, the four connectives of **Int** are all independent: no one of them can be defined in terms of the other three. I present part of a proof of that by *McKinsey, 1939* in Chapter VIII.G, and outline an alternate proof in Exercise 7 below.

Though we can view the collection of tautologies of **Int** as extending those of **PC**, we cannot view **Int** as a logic extending **PC**, for in $L(\neg, \wedge)$:

$$\neg\neg p \vdash_{\mathbf{PC}} p \quad \text{but} \quad \neg\neg p \nvdash_{\mathbf{Int}} p$$

Corollary 17 suggests that by using the definition of $\rightarrow$ and $\vee$ in **PC** we can define **PC** as a collection of theorems within **Int**, as first suggested by Gödel and established by *Łukasiewicz, 1952*. Define the translation from $L(\neg, \rightarrow, \wedge, \vee)$ to itself:

$$\begin{aligned} p^{\dagger} &= p \\ (A \wedge B)^{\dagger} &= A^{\dagger} \wedge B^{\dagger} \\ (\neg A)^{\dagger} &= \neg(A^{\dagger}) \\ (A \vee B)^{\dagger} &= \neg(\neg A^{\dagger} \wedge \neg B^{\dagger}) \\ (A \rightarrow B)^{\dagger} &= \neg(A^{\dagger} \wedge \neg B^{\dagger}) \end{aligned}$$

Corollary 19 $\vdash_{\mathbf{PC}} A$ iff $\vdash_{\mathbf{Int}} A^{\dagger}$

But again, we have $\neg\neg p \vdash_{\mathbf{PC}} p$, and as $(\neg\neg p)^{\dagger} = \neg\neg p$, $(\neg\neg p)^{\dagger} \nvdash_{\mathbf{Int}} p$; the translation does not preserve consequences. However, the following translation from $L(\neg, \rightarrow, \wedge, \vee)$ to itself, due to *Gentzen, 1936,* does preserve consequences.

$$\begin{aligned} (p)^{\circ} &= \neg\neg p \\ (A \wedge B)^{\circ} &= A^{\circ} \wedge B^{\circ} \\ (A \rightarrow B)^{\circ} &= A^{\circ} \rightarrow B^{\circ} \\ (\neg A)^{\circ} &= \neg(A^{\circ}) \\ (A \vee B)^{\circ} &= \neg(\neg A^{\circ} \wedge \neg B^{\circ}) \end{aligned}$$

Theorem 20 $\Gamma \vdash_{\mathbf{PC}} A$ iff $\Gamma^{\circ} \vdash_{\mathbf{Int}} A^{\circ}$

Proof: Gentzen's proof was entirely syntactic, whereas I will use the completeness theorem.

First suppose that $\Gamma^\circ \vdash_{\mathbf{Int}} A^\circ$. Note that $\mathbf{PC} \vdash A \leftrightarrow A^\circ$. So if $\Gamma^\circ \vdash_{\mathbf{Int}} A^\circ$, then as $\mathbf{Int} \subset \mathbf{PC}$, $\Gamma^\circ \vdash_{\mathbf{PC}} A^\circ$. So $\Gamma \vdash_{\mathbf{PC}} A$.

For the other direction we first need:

Lemma $\vdash_{\mathbf{Int}} \neg\neg(A^\circ) \rightarrow A^\circ$

Proof: We proceed by induction on the length of A. If A is a variable, p, then by Corollary 5.e, $\vdash_{\mathbf{Int}} \neg\neg(\neg\neg p) \rightarrow \neg\neg p$. So suppose A has length greater than 1 and the lemma is true for all shorter wffs.

If A is $\neg B$, then we are done by Corollary 5.

If A is $B \wedge C$, then we have by induction $\vdash_{\mathbf{Int}} \neg\neg B^\circ \rightarrow B^\circ$ and $\vdash_{\mathbf{Int}} \neg\neg C^\circ \rightarrow C^\circ$. The lemma then follows by $\vdash_{\mathbf{Int}} \neg\neg(A \wedge B) \rightarrow (\neg\neg A \wedge \neg\neg B)$, which I leave as an exercise.

If A is $B \vee C$, then since $(B \vee C)^\circ = \neg(\neg B^\circ \wedge \neg C^\circ)$ we are done as above.

Finally, we have the case where A is $B \rightarrow C$. We have $\vdash_{\mathbf{Int}} \neg\neg(D \rightarrow E) \rightarrow (\neg\neg D \rightarrow \neg\neg E)$ (Exercise 8), and hence $\vdash_{\mathbf{Int}} \neg\neg(B^\circ \rightarrow C^\circ) \rightarrow (\neg\neg B^\circ \rightarrow \neg\neg C^\circ)$. By Corollary 5.a, $\vdash_{\mathbf{Int}} B^\circ \rightarrow \neg\neg B^\circ$, and by induction we have $\vdash_{\mathbf{Int}} \neg\neg C^\circ \rightarrow C^\circ$. So by Lemma 3.a, $\vdash_{\mathbf{Int}} \neg\neg(B^\circ \rightarrow C^\circ) \rightarrow (B^\circ \rightarrow C^\circ)$. This ends the proof of the lemma.

To return to the proof of the theorem, suppose $\Gamma \vdash_{\mathbf{PC}} A$. Then for some $A_1, \ldots, A_n \in \Gamma$, $\{A_1, \ldots, A_n\} \vdash_{\mathbf{PC}} A$. Hence $\{A_1^\circ, \ldots, A_n^\circ\} \vdash_{\mathbf{PC}} A^\circ$, so $(A_1^\circ \wedge \cdots \wedge A_n^\circ) \vdash_{\mathbf{PC}} A^\circ$. Thus by Theorem 18, $\neg\neg(A_1^\circ \wedge \cdots \wedge A_n^\circ) \vdash_{\mathbf{Int}} \neg\neg A^\circ$. By the lemma and Corollary 4.e we have $(A_1^\circ \wedge \cdots \wedge A_n^\circ) \vdash_{\mathbf{Int}} A^\circ$, hence by Lemma 3, $\{A_1^\circ, \ldots, A_n^\circ\} \vdash_{\mathbf{Int}} A^\circ$. That is, $\Gamma^\circ \vdash_{\mathbf{Int}} A^\circ$. ■

Theorem 20 can be extended to classical and intuitionist arithmetic, too. In the language of arithmetic, extend the translation to include:

$$(\exists x A^\circ) = \neg \forall x \neg A^\circ$$

(See *Kleene, 1952*, §81.) Thus, even from the intuitionist's viewpoint, no inconsistency can arise in classical arithmetic if there is not already an inconsistency in intuitionist arithmetic. But from the intuitionist's viewpoint, that is not the problem; the problem is that the classical logician proves false (meaningless) theorems, especially of the form $\neg \forall x \neg A(x) \rightarrow \exists x A(x)$. For despite the invocation of this translation by *Gödel, 1933 A,* to assert that intuitionist arithmetic is only apparently narrower than classical arithmetic, there is a real difference from the intuitionist's perspective. The connectives (and $\exists$) mean something different to the intuitionist, and translations of classical logic into intuitionistic logic do not preserve meaning, as I discuss in Section E.2.b below and in Chapter X.B.4.

3. Axiomatizations of classical logic relative to Int

Theorem 21 In $L(\neg, \rightarrow, \wedge, \vee)$:

a. **PC** = the closure under *modus ponens* of **Int** $\cup \{(A \rightarrow B) \rightarrow ((\neg A \rightarrow B) \rightarrow B)\}$

b. **PC** = the closure under *modus ponens* of **Int** $\cup \{\neg\neg A \rightarrow A\}$

c. **PC** = the closure under *modus ponens* of **Int** $\cup \{A \vee \neg A\}$

Proof: a. Consider the axiomatization of **PC** in $L(\neg, \rightarrow, \wedge, \vee)$ given in Chapter II.M.6, p. 82. Every schema there other than $(A \rightarrow B) \rightarrow ((\neg A \rightarrow B) \rightarrow B)$ is valid in every finite Kripke tree, and hence every instance of any of those schema is a theorem of **Int**. Thus if we add $(A \rightarrow B) \rightarrow ((\neg A \rightarrow B) \rightarrow B)$ to **Int**, we can prove every theorem of **PC**.

b. **Int** $\cup \{\neg\neg A \rightarrow A\} \subseteq$ **PC**, which is closed under *modus ponens.* It remains to show that every theorem of **PC** can be derived from **Int** $\cup \{\neg\neg A \rightarrow A\}$. Take $\vdash$ to mean 'derivable from using only *modus ponens*.' Suppose **PC**$\vdash A$. Then by Theorem 16, $\neg\neg$**Int**$\vdash \neg\neg A$. So $\neg\neg$**Int** $\cup \{\neg\neg A \rightarrow A\} \vdash A$. Finally $\neg\neg$**Int** $\subseteq$ **Int** since by Corollary 5.a, **Int**$\vdash A \rightarrow \neg\neg A$.

b. This follows from part (b) as $\vdash_{\mathbf{Int}} (A \vee \neg A) \rightarrow (\neg\neg A \rightarrow A)$ (Exercise 5.a of Section B or Exercise 8 below). ■

Exercises for Section D

1. Prove that $\langle W, R, e, w\rangle \vDash A$ in the intuitionist semantics iff $\langle W, R, e, w\rangle \vDash A^*$ in the modal semantics.

2. For each of the translations of **Int** into **S4** in Section 1, give the translation of:
   a. $A \vee \neg A$
   b. $\neg\neg A \rightarrow A$
   c. $(A \rightarrow B) \rightarrow ((\neg A \rightarrow B) \rightarrow B)$
   d. $(\neg A \rightarrow A) \rightarrow A$
   e. $(A \rightarrow B) \rightarrow (\neg B \rightarrow \neg A)$

3. Prove that consequences are preserved for either version of Gödel's translation of **Int** into **S4** given in Section 1: $\Gamma \vDash A$ iff $\Gamma' \vDash A'$.

4. For each of the translations of classical logic into **Int** given in Section 2, give the translation of:
   a. $A \vee \neg A$
   b. $\neg\neg A \rightarrow A$
   c. $(A \rightarrow B) \rightarrow ((\neg A \rightarrow B) \rightarrow B)$

5. Prove Corollary 19, $\vdash_{\mathbf{PC}} A$ iff $\vdash_{\mathbf{Int}} A^{\dagger}$.

6. Prove syntactically:
   a. $\vdash_{\mathbf{Int}} \neg\neg(A \rightarrow B) \rightarrow (\neg\neg A \rightarrow \neg\neg B)$
   b. $\vdash_{\mathbf{Int}} \neg\neg(A \wedge B) \rightarrow (\neg\neg A \wedge \neg\neg B)$

c. $\vdash_{\mathbf{Int}} \neg\neg[(A \rightarrow B) \rightarrow ((\neg A \rightarrow B) \rightarrow B)]$

d. $\vdash_{\mathbf{Int}} (A \vee \neg A) \rightarrow (\neg\neg A \rightarrow A)$

7. Show that:

   PC = the closure under *modus ponens* of $\mathbf{Int} \cup \{(\neg A \rightarrow A) \rightarrow A\}$

8. Prove Theorem 21.a using Hilbert's axiomatization of **PC** (Exercise 7, p. 85).

9. a. Give a proof from $\mathbf{Int} \cup \{(A \rightarrow B) \rightarrow ((\neg A \rightarrow B) \rightarrow B)\}$ of the following, using the rule of *modus ponens* only: $\neg\neg A \rightarrow A$, $A \vee \neg A$

   b. Give a proof of $(A \rightarrow B) \rightarrow ((\neg A \rightarrow B) \rightarrow B)$ from $\mathbf{Int} \cup \{A \vee \neg A\}$ using *modus ponens* only.

10. What is wrong with the following argument: since the evaluation of the connectives $\rightarrow$, $\wedge$, $\vee$ are the same in Kripke models of **S4** and **Int**, the positive fragment of those logics is the same?

11. Fill in the following outline to show that no one of the connectives of **Int** can be defined from the other three:

    a. $\neg$ is independent, via Theorem 15

    b. $\vee$ is independent via Theorem 20

    c. $\rightarrow$ is independent via Corollary 19. Or show that if A has no occurrence of $\rightarrow$, then for any p, q, if $\vdash (p \rightarrow q) \rightarrow A$, then $\vdash (p \rightarrow \neg\neg q) \rightarrow A$. So we would get $\vdash (p \rightarrow q) \rightarrow (p \rightarrow \neg\neg q)$ if $\rightarrow$ were definable.

    d. $\wedge$ is independent by considering a model:

    $x \in e(p)$ → $z \in e(p), e(q)$

    $w \in e(q)$ → $z \in e(p), e(q)$

    Every A in which $\wedge$ does not appear is either false or is forced at x or w.

E. Set-assignment semantics for Int

Dummett is explicit in attributing content to propositions:

> If we take it as a primary function of a sentence to convey information, then it is natural to view a grasp of the meaning of a sentence as consisting in an awareness of its *content*; and this amounts to knowing the conditions under which an assertion made by it is correct.
>
> *Dummett, 1977,* p. 363

In terms of the formal Kripke semantics and their interpretation given above, this would amount to identifying a proposition with those elements of W, that is those states of knowledge or information, in which it is valid. Symbolically, $s(A) = \{ z : z \vDash A \}$. This is what we did for **S4** (Section VI.G.5), and the translation of intuitionistic logic into **S4** (Theorem 12) suggests we try it here.

The basis of that approach is to convert a model $\langle W, R, e, w\rangle$ to a model $<v,s>$ such that $s(A) = \{ z : z \vDash A \}$ and $v(A) = T$ iff $w \vDash A$. But here we have a complication: $w \nvDash A$ does not imply $w \vDash \neg A$. We may have both $w \nvDash A$ and $w \nvDash \neg A$, corresponding to the intuition that from the information of w we may not be able to deduce either A or $\neg A$. Thus we may not set $v(\neg A) = T$ iff $v(A) = F$. Rather we need to take into account the content of A.

1. The semantics

A set-assignment model $<v,s>$ uses *intuitionist truth-conditions* if:

$\wedge$ and $\vee$ are evaluated classically,

$\rightarrow$ is evaluated by the dual dependence table (as for **S4**):

A	B	$s(A) \subseteq s(B)$	$A \rightarrow B$
any	values	fails	F
T	T		T
T	F	holds	F
F	T		T
F	F		T

and $\neg$ is evaluated by the table for *intuitionist negation*:

A	$s(A) = \varnothing$	$\neg A$
any value	fails	F
T	holds	F
F		T

That is, $v(\neg A) = T$ iff $v(A) = F$ and $s(A) = \varnothing$.

Then $<v,s>$ is an **Int**-*model* if it uses intuitionist truth-conditions and satisfies:

Int 1. $s(A \wedge B) = s(A) \cap s(B)$

Int 2. $s(A \vee B) = s(A) \cup s(B)$

Int 3. $s(\neg A) \cup s(B) \subseteq s(A \rightarrow B)$

Int 4. $s(A) \cap s(A \rightarrow B) \subseteq s(B)$

Int 5. $s(A \rightarrow B) \subseteq s((A \wedge C) \rightarrow (B \wedge C))$

Int 6. $s(A \rightarrow B) \cap s(B \rightarrow C) \subseteq s(A \rightarrow C)$

Int 7. $s(A \rightarrow C) \cap s(B \rightarrow C) = s((A \vee B) \rightarrow C)$

Int 8. $s(A \rightarrow B) \cap s(A \rightarrow \neg B) = s(\neg A)$

Int 9. If $v(A) = T$, then $s(A) = S$

Int 10. $s(A \wedge \neg A) = \varnothing$

We say that the model is *finite* if S is finite.

We define $\Gamma \vDash_{\mathbf{Int}} A$ for these semantics in the usual way: for every **Int**-model $<v,s>$ that validates Γ, $v(A) = T$.

Example $\neg(A \wedge (A \rightarrow \neg A))$ is true in every **Int**-model $<v,s>$

We need to show that $s(A \wedge (A \rightarrow \neg A)) = \varnothing$ and $v(A \wedge (A \rightarrow \neg A)) = F$. For the first, by Int 1, $s(A \wedge (A \rightarrow \neg A)) = s(A) \cap s(A \rightarrow \neg A)$, so by Int 4, $s(A \wedge (A \rightarrow \neg A)) \subseteq s(\neg A)$. Hence by Int 10, $s(A \wedge (A \rightarrow \neg A)) = \varnothing$. If $v(A) = F$, then $v(A \wedge (A \rightarrow \neg A)) = F$. If $v(A) = T$, then $v(\neg A) = F$, so $v(A \rightarrow \neg A) = F$ and $v(A \wedge (A \rightarrow \neg A)) = F$. Hence $v(\neg(A \wedge (A \rightarrow \neg A))) = T$.

Lemma 22 If $\Gamma \vdash_{\mathbf{Int}} A$, then $\Gamma \vDash_{\mathbf{Int}} A$.

Proof: It's enough to show that the axioms are valid in **Int**-models and that the rules preserve validity. I will simply note which conditions are involved in the verification of the axioms:

I and II follow from Int 1

III follows from Int 5 and Int 1

IV follows from Int 6 and Int 1

V follows from Int 3 and Int 9

VI follows from Int 1 and Int 4

VII and VIII follow from Int 2

IX follows from Int 7, Int 1, and Int 2

X follows from Int 3, since if $v(\neg A) = T$ we must have $s(A) = \varnothing$

XI follows from Int 8 and Int 10 ∎

Lemma 23 Given any Kripke tree $\langle W, R, e, w \rangle$ there is an **Int**-model $<v,s>$ such that $v(A) = T$ iff $w \vDash A$ and $s(A) = \{ z : z \vDash A \} - \{w\}$.

Proof: We need to show that the pair $<v,s>$ described in the lemma is an **Int**-model.

I'll leave to you to show that Int 1–Int 8 and Int 10 are satisfied. To establish Int 9 we can use induction on the length of A. If A is a propositional variable, then the result follows by Lemma 1. For conjunctions and disjunctions it is easy. If A is $\neg B$ and $v(\neg B) = T$, then $v(B) = F$ and $s(B) = \varnothing$, so $s(\neg B) = \mathsf{S}$. If A is $B \rightarrow C$ and $v(B \rightarrow C) = T$, then $s(B) \subseteq s(C)$ so $s(B \rightarrow C) = \mathsf{S}$.

I leave as an exercise to establish by induction on the length of wffs that every wff is evaluated by the intuitionist truth-conditions. ∎

Using Lemmas 23, 24, and Theorem 11 we have:

***Theorem 24 Strong Completeness of the Set-Assignment Semantics for* Int**

$\Gamma \vdash_{\mathbf{Int}} A$ iff $\Gamma \models_{\mathbf{Int}} A$

For finite Γ the class of finite **Int**-models is strongly complete.

The only place where nonconstructive reasoning might have entered into the proof of Theorem 24 is in claiming the existence of the Kripke tree and evaluation used in the proof of Lemma 23. But Dummett explicitly constructs such a model in his proof of Theorem 11.c (see Section C.3 above), so we can claim that Theorem 24 is intuitionistically acceptable so long as we confine ourselves to finite Γ.

2. Observations and refinements of the set-assignment semantics

a. If we wished to give a reading of $\mathsf{s}(A)$ as the constructive mathematical content of A, as I'd once hoped to do, dependent implication rather than dual dependent implication would be appropriate: $A \rightarrow B$ is true iff the constructive mathematical content of A contains that of B, and not both A is true and B is false. That has a nice sound to it, and it's not hard to modify the semantics here to give ones based on dependent implication. For instance, condition Int 1 would read $\mathsf{s}(A \wedge B) = \mathsf{s}(A) \cup \mathsf{s}(B)$, the constructive mathematical content of $A \wedge B$ is the constructive mathematical content of A plus the constructive mathematical content of B. Surprisingly, though, I can find in the literature no explication of the constructive mathematical content of a proposition, nothing that would tell me how the content of $\forall x(x+2=2+x)$ differs from that of $\forall x \forall y\,(x+y=y+x)$ in such a way that I could see how the contents affect proofs or derivations: we can derive the former from the latter, but does that mean that the latter has *more* constructive content?

b. Is classical negation definable from the intuitionist connectives? That is, is there some scheme N(A) built from $\neg, \rightarrow, \wedge, \vee$ that is evaluated by the following table in every **Int**-model $<\mathsf{v},\mathsf{s}>$?

A	N(A)
T	F
F	T

If there is, then I believe it could be shown that $\{\neg, \rightarrow, \wedge, \vee\}$ is functionally complete.

This question is closely related to whether there is a semantically faithful translation of **PC** into **Int** (see Chapter X.B.4) and whether **Int** is narrower than **PC** (see the comment following Theorem 20, p. 293 above). By each criteria of Chapter IV.F.2, the semantics I have given for **Int** are incompatible with those for **PC**.

c. If for every every state of information and every A, either A or $\neg$A is derivable, then we have classical logic (Theorem 22). In terms of the set-assignments, if we add to the conditions on an **Int**-model $s(\neg A) = \overline{s(A)}$, then we also have by Int 3 and Int 4, $s(A \to B) = \overline{s(A)} \cup s(B)$. Thus the set-assignments would be restricted to boolean algebras of sets, for which we have $\mathbf{PC} \vdash A$ iff $s(A) = S$ (see *Rasiowa, 1974*). Since $s(A) = \varnothing$ would imply $s(\neg A) = S$, for such models we would have $v(A) = T$ iff $s(A) = S$, which is iff $\mathbf{PC} \vdash A$.

d. From Corollary 4 we can picture how set-assignments operate on negations, noting that in some models we may have $s(A) = s(\neg\neg A)$.

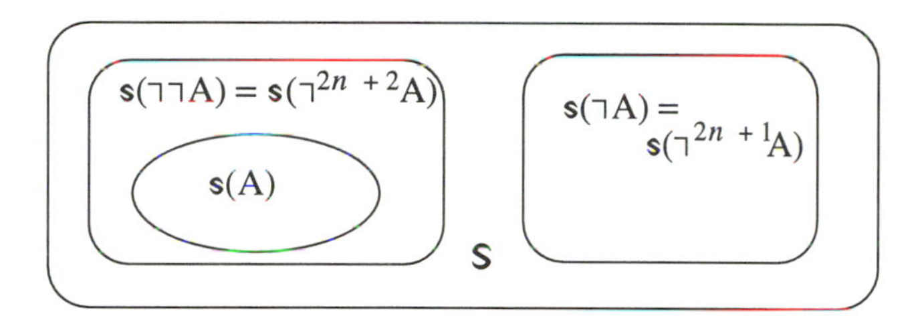

e. Do set-assignment semantics for **Int** allow an inductive definition of truth?

Suppose we know $v(A)$, $s(A)$, and $v(B)$, $s(B)$. Then $v(A \wedge B)$, $v(A \vee B)$ can be calculated directly. We can also calculate $v(A \to B)$, and if that is true, we know that we must assign $s(A \to B) = S$. But if it is false, then the conditions that $s(A \to B)$ must satisfy are global, for example, Int 5. Similarly, we may calculate $v(\neg A)$, and if it is true, then we must assign $s(\neg A) = S$. If $v(A) = T$, then by Int 10 we must assign $s(\neg A) = \varnothing$. However, if $v(A) = v(\neg A) = F$, then the conditions that $s(\neg A)$ must satisfy are global. We need to know the content of these wffs in order to evaluate wffs of which they are constituent. So to the extent that we can assign $s(A \to B)$ and $s(\neg A)$ when these formulas are false, we have an inductive definition of truth.

> The principal reason for suspecting these explanations [of the logical constants] of incoherence is their apparently highly impredicative character: if we know which constructions are proofs of the atomic statements of any first-order theory, then the explanations of the logical constants, taken together, determine which constructions are proofs of any of the statements of that theory; yet the explanations require us, in determining whether or not a construction is a proof of a conditional or of a negation, to consider its effect when applied to an arbitrary proof of the antecedent or of the negated statement, so that we must, in some sense, be able to survey or grasp some totality of constructions which will include all possible proofs of a given statement. The question is whether such a set of explanations can be acquitted of the charge of vicious circularity.
>
> *Dummett, 1977,* p. 390

f. Does every set-assignment model for **Int** arise from a Kripke model by the construction of Lemma 23? I suspect not. However, we can pick out those that do with the following condition:

Int K. If $\bigcap\{\mathsf{s}(C) : x \in \mathsf{s}(C)\} \subseteq \overline{\mathsf{s}(A)} \cup \mathsf{s}(B)$, then $x \in \mathsf{s}(A \to B)$

Let $\mathcal{I}$ be the class of set-assignment models that use the intuitionist truth-conditions and satisfy Int 1, Int 2, Int 4, Int 8, Int 10, and Int K.

Theorem 25 *a.* If $<\mathsf{v},\mathsf{s}>$ is an **Int**-model that arises from a Kripke tree by the construction of Lemma 23, then $<\mathsf{v},\mathsf{s}>$ satisfies Int K.

b. Given any $<\mathsf{v},\mathsf{s}> \in \mathcal{I}$, there is a Kripke model $\langle W, R, e, w\rangle$ such that $\mathsf{v}(A) = \mathsf{T}$ iff $w \vDash A$ and $\mathsf{s}(A) = \{z : z \vDash A\} - \{w\}$.

c. $\mathcal{I}$ is strongly complete for **Int**, and the class of finite models in $\mathcal{I}$ is finitely strongly complete.

Proof: a. Suppose that $<\mathsf{v},\mathsf{s}>$ arises from $\langle W, R, e, w\rangle$. And suppose $\bigcap\{\mathsf{s}(C) : x \in \mathsf{s}(C)\} \subseteq \overline{\mathsf{s}(A)} \cup \mathsf{s}(B)$. Then if xRy, by Lemma 1.a, $y \in \bigcap\{\mathsf{s}(C) : x \in \mathsf{s}(C)\}$. So if xRy, then $y \in \overline{\mathsf{s}(A)}$ or $y \in \mathsf{s}(B)$, so $y \nvDash A$ or $y \vDash B$. Hence $x \vDash A \to B$. So $x \in \mathsf{s}(A \to B)$.

b. Given $<\mathsf{v},\mathsf{s}>$ define:

$W = S \cup \{w\}$ for some object $w \notin S$

$$e(p) = \begin{cases} \mathsf{s}(p) \cup \{w\} & \text{if } \mathsf{v}(p) = \mathsf{T} \\ \mathsf{s}(p) & \text{if } \mathsf{v}(p) = \mathsf{F} \end{cases}$$

xRy iff $y \in \bigcap\{\mathsf{s}(C) : x \in \mathsf{s}(C)\}$ or $x = w$

Note that if $x \neq w$, xRy iff for all C, if $x \in \mathsf{s}(C)$ then $y \in \mathsf{s}(C)$. So R is reflexive and transitive.

I'll now show by induction on the length of A that for $x \neq w$, $x \in \mathsf{s}(A)$ iff $x \vDash A$. It's easy to check for propositional variables. Suppose now that it's true for all wffs shorter than A. If A is a conjunction or disjunction, the proof is immediate from Int 1 and Int 2.

So suppose that A is $\neg B$. If $x \in \mathsf{s}(\neg B)$, then for all y such that xRy, $y \in \mathsf{s}(\neg B)$ and hence by Int 10 and Int 1, $y \notin \mathsf{s}(B)$. So by induction, $y \nvDash B$. So $x \vDash \neg B$.

If $x \vDash \neg B$, then for all y such that xRy, $y \vDash \neg B$. Hence for all y such that xRy, $y \notin \mathsf{s}(B)$ by induction. Thus $\bigcap\{\mathsf{s}(C) : x \in \mathsf{s}(C)\} \subseteq \overline{\mathsf{s}(B)} \cup \mathsf{s}(\neg B)$. Hence by Int K, $x \in \mathsf{s}(B \to \neg B)$. But $x \in \mathsf{s}(B \to B)$ since $B \to B$ is a tautology. So by Int 8, $x \in \mathsf{s}(\neg B)$.

If A is $B \to C$, suppose first that $x \in \mathsf{s}(B \to C)$. If xRy, then $y \in \mathsf{s}(B \to C)$. So if $y \vDash B$, then by induction $y \in \mathsf{s}(B)$. So by Int 4, $y \in \mathsf{s}(C)$, and thus $y \vDash C$. Hence $x \vDash B \to C$.

If $x \vDash B \to C$, suppose $y \in \bigcap \{ s(C) : x \in s(C) \}$. Then xRy, so $y \vDash B \to C$. Hence either $y \nvDash B$ or $y \vDash C$, and so by induction, $y \notin s(B)$, or $y \in s(C)$. Thus by Int K, $x \in s(B \to C)$.

Now I'll prove by induction on the length of A that $v(A) = T$ iff $w \vDash A$. Recall that by Int 9, if $v(A) = T$, then $s(A) = S$. The only interesting cases in the proof are when A is a negation or conditional.

If A is $\neg B$,

$$\begin{aligned} v(\neg B) = T \ \text{ iff } & \ v(B) = F \text{ and } s(B) = \varnothing \\ \text{iff } & \ w \nvDash B \text{ and for all } y \text{ such that } wRy,\ y \notin s(B) \\ \text{iff } & \ w \nvDash B \text{ and for all } y \text{ such that } wRy,\ y \nvDash B \\ \text{iff } & \ w \vDash \neg B \end{aligned}$$

If A is $B \to C$,

$$\begin{aligned} v(B \to C) = T \ \text{ iff } & \ (v(B) = F \text{ or } v(C) = T) \text{ and } s(B) \subseteq s(C) \\ \text{iff } & \ (w \nvDash B \text{ or } w \vDash C) \text{ and for all } y \text{ such that } wRy, \\ & \ y \notin s(B) \text{ or } y \in s(C) \\ \text{iff } & \ (w \nvDash B \text{ or } w \vDash C) \text{ and for all } y \text{ such that } wRy,\ y \nvDash B \text{ or } y \vDash C \\ \text{iff } & \ w \vDash B \to C \end{aligned}$$

c. This follows by Theorems 2 and 3 using the 1–1 correspondence between Kripke models and models in $\mathcal{I}$ established in Lemma 23 and parts (a) and (b). ■

Note that $\mathcal{I}$ is not simply presented (Chapter IV.D, pp. 141–142).

g. The conditions on **Int**-models may be considerably weakened and still yield a class that is complete for **Int**. First, we can use the minimal intuitionist truth-table for $\neg$ from the next section: $v(\neg A) = T$ iff $v(A) = F$ and $s(A) \subseteq s(\neg A)$. In the presence of Int 10, this table is equivalent to the one already given. However, Int 10 is not needed and may be deleted: to verify that axioms X and XI are valid see the proof of Lemma 17 below. At least I don't believe that Int 10 is a consequence of Int 1–Int 9; Int 3 and Int 4 together yield only $s(A) \cap s(\neg A) \subseteq s(B)$ for all B. We could apparently have some subset $Q \subseteq S$ that is contained in $s(B)$ for every B: Q would be the conditions justifying the assertion of any contradiction. That reading is apt for the minimal calculus of the next section. However Int 10 would follow from the assumption that there is even one proposition with no content.

Int 4 can be changed to: $s(A \to B) \subseteq \overline{s(A)} \cup s(B)$, though it's not clear to me that this is a weaker condition.

Finally, in Int 7 and Int 8 '=' can be changed to '$\subseteq$'.

Exercises for Sections E.1 and E.2

1. Determine directly whether the following are valid in every set-assignment model for **Int**:

a. $\neg A \to (A \to B)$ d. $((A \lor B) \land \neg A) \to B$ g. $(\neg A \to A) \to A$
b. $A \to \neg\neg A$ e. $((A \to B) \to A) \to A$
c. $A \to A \lor B$ f. $((A \to B) \to (A \to \neg B)) \to \neg$

2. Prove Lemma 22.
3. For the pair $<v,s>$ given in Lemma 23 show that Int 1–Int 8 and Int 10 are satisfied and every wff is evaluated by the intuitionist truth-conditions.
4. (Open) Is there a simply presented collection of set-assignment models for **Int** for which we can establish a 1-1 correspondence between Kripke models and set-assignment models as in Lemma 25? (The correspondence is important for showing that the translation of **Int** into **S4** is semantically faithful, Theorem X.15.)
5. (Open) Is the connective ~ interpreted by $v(\sim A) = T$ iff $s(A) \subseteq s(\sim A)$ definable in **Int**?

3. Bivalence in intuitionism: the aptness of set-assignment semantics

Dummett explicitly says that the intuitionist abandons the principle of bivalence for truth. Yet it is not so easy for someone speaking our language to fully escape the Yes–No dichotomy that we all practice and impose on experience. Given any proposition and any particular state of information it is either correct or incorrect to assert A, there being no third way. And given any proposition A and any finite collection of states of information ordered under time, either it is always correct to assert A or it is not. *Tertium non datur.*

> It is evident that it is fundamental to the notion of an assertion that it be capable of being either correct or incorrect; and therefore, in so far as assertion is taken to be the primary mode of employment of sentences, it is fundamental to our whole understanding of language that sentences are capable of being true or false, where a sentence is true if an assertion could be correctly made by uttering it, and false if such an assertion would be incorrect.
>
> *Dummett, 1973,* p. 371

From the viewpoint of the general framework for semantics for propositional logics proposed in Chapter IV, the intuitionist reasons analogously to a modal logician, or a many-valued, or a relevance logician. His notion of the truth of a complex proposition is based on two aspects of its constituent propositions: truth-value and some epistemological mathematical content. The intuitionist disagrees with the classical logician not on the truth-values of the atomic arithmetic formulas, but on the use of the connectives, particularly $\neg$ and $\to$, which take into account both aspects of the constituent propositions. From this point of view it seems perfectly apt to read T and F as 'true' and 'false' in the tables for **Int**.

We may take Dummett's interpretation of the content of a proposition as the conditions under which it is correct to assert A. Here we must understand s(A) to be a collection of various conditions under each of which it is correct to assert A. So, for example, Int 4 can be read as 'Any condition that justifies my asserting A and that justifies my asserting $A \to B$ also justifies my asserting B'. Then $v(A) = T$ in a model means that it is always correct to assert A, and in that case $v(\neg A) = F$: we cannot always correctly assert $\neg A$. If, however, $v(A) = F$ we must ask whether there's any state of knowledge or information in our finite model that can verify A, that is, whether $s(A) = \varnothing$. If $s(A) = \varnothing$, then it's always correct to assert $\neg A$ and hence $v(\neg A) = T$. If $s(A) \neq \varnothing$, then sometimes it's correct to assert A; so we can't always assert $\neg A$, and thus $v(\neg A) = F$. Granted this does impose some global (platonic?) point of view on whether we can *ever* verify A. However, we can confine ourselves to finite models: it's no more platonic than the reading Dummett gives in his intuitionistically valid completeness proof of Theorem 2.

You might argue that we should use new symbols here, say C and I for 'correct to assert' and 'incorrect to assert' rather than T and F. But that would obscure the similarity of the duality that the intuitionist imposes on propositions with that which the classical logician uses. Both use a proposition to deduce further ones just in case it is true. And the formal logic of each is designed to capture those schema that are invariably true and that can be used to deduce true propositions for any assignment of propositions. The situation is the same as when I chose to use '$\to$' to represent whatever notion of one proposition following from another that a logic proposes: I believe the underlying similarities represent some shared background assumptions and my notation represents that.

Dummett, 1977, and also *McCarty, 1983,* discuss the aptness of various other semantics for modeling the intuitionist's point of view, such as Beth trees and De Swart models (see also *De Swart, 1977*), for which comparable readings of v and s can be given. To the extent that any of these semantics capture the basis of intuitionism, so will ours. I will return to this point in discussing translations between logics in Chapter X.B.7.

Dummett, 1973, apparently rejects the kind of reading for the semantics of intuitionistic logic that I give. But his comments on theories of meaning sound much like the motivation for the general framework of Chapter IV.

> A theory of meaning, at least of the kind with which we are most familiar, seizes upon some one general feature of sentences . . . as central: the notion of the content of an individual sentence is then to be explained in terms of this central feature. . . . The justification for thus selecting some one single feature of sentences as central—as being that in which their individual meanings consist—is that it is hoped that every other feature of the use of sentences can be derived, in a uniform manner, from this central one.
>
> *Dummett, 1973,* pp. 222–223

I agree with Dummett: many logicians make the unreasonable claim that all features of use that are of significance to logic can be reduced to the one upon which their logic is based. If some feature, such as relevance, cannot be derived then it's argued that it must not be significant, or not logical.

But Dummett himself wishes us to believe that intuitionism manages to capture all features of the use of a sentence; "meaning is use" where:

> The "use" of a sentence is not, in this sense, a *single* feature; the slogan simply restricts the *kind* of feature that may legitimately be appealed to as constituting or determining meaning. . . . It is the multiplicity of the different features of the use of sentences, and the consequent legitimacy of the demand, given a molecular view of language, for harmony between them, that makes it possible to criticise existing practice, to call in question uses that are actually made of sentences of the language.
>
> *Dummett, 1973,* p. 223

This argues for one overarching semantic theory that encompasses all or at least many of these features of sentences that are important to reasoning. Dummett argues that intuitionism does so, not only for mathematical statements, but generally for natural language (*Dummett, 1977*, Chapter 7.1). But how can that be? The intuitionists' notion of content seems to me only one among many, hardly able to model such features of use as, say, relevance (axiom scheme V is a standard fallacy of relevance). And even if it did model those in some general fashion, it doesn't tell us how to reason in accord with any particular one. It seems to me that Dummett has fallen into the same error as the logicians he has criticized: he has taken one notion of content to be central, claiming that the numerous features of sentences that are significant to reasoning are thus taken into account.

F. The Minimal Calculus J

1. The minimal calculus

In 1936 Johansson commented on Heyting's formal axioms for intuitionism.

> Among the logical axioms that Heyting set up for the derivation of the formal laws of intuitionistic logic there are two at which one starts:
>
> $\vdash B \to (A \to B)$
>
> $\vdash \neg A \to (A \to B)$
>
> The sense of these axioms is naturally only that the relation of implication in calculus has a different meaning than in ordinary speech. One can write $A \to B$ in the following three cases:
>
> 1. If B is recognized as a logical conclusion of A.
> 2. If B is recognized to be true.
> 3. If A is recognized to be false.

> One can easily become reconciled with the second case; however, the third case means an easily overlooked extension of the meaning of the conclusion. It is worth the effort to see if this can be avoided.
>
> *Johansson, 1936,* p. 119

Johansson then proposed a reduced *minimal calculus* of intuitionism in which (3) need not be accepted.

J
in $L(\neg, \rightarrow, \wedge, \vee)$
as for **Int** except delete axiom scheme X. $\neg A \rightarrow (A \rightarrow B)$

Johansson's axiomatization was in a different language. He notes that in line with an analysis given by *Kolmogorov, 1932,* one may introduce a propositional constant $\perp$ and define $\neg A$ as $A \rightarrow \perp$. He explains this constant:

> The interpretation of $\perp$ as an undefined basic statement is related to the 'problem theoretical' meaning of the intuitionistic logic offered by Kolmogoroff. Namely, $\neg A$ refers to (using Kolmogroff) the task 'assuming that the solution of A is given, to find a contradiction,' and that agrees with the definition of $\neg A$ as $A \rightarrow \perp$ if one interprets $\perp$ as the task 'to obtain a contradiction.' This task is not defined; though it is an implicit assumption with Kolmogoroff that $\perp \rightarrow B$ is valid , i.e., that the following problem has been solved: 'Assuming that a contradiction has been obtained, to solve an arbitrary problem. ' If we leave out this assumption and thus obtain a sharper $\rightarrow$, then we obtain a 'problem theoretical' meaning for the minimal calculus.
>
> *Johansson, 1936,* p. 131

2. Kripke-style semantics

I will present the semantics for **J** of *Fitting, 1969,* which are a modification of the Kripke semantics for **Int**. I will modify those to apply to $L(\neg, \rightarrow, \wedge, \vee)$ here. In Section 3 below I present an alternate axiomatization of **J** using $\perp$ as primitive. At the end of Section 4 I will comment on the aptness of these semantics.

A *model for* **J** is $\langle \mathsf{W}, \mathsf{R}, \mathsf{Q}, \mathsf{e} \rangle$ where:

$\langle \mathsf{W}, \mathsf{R} \rangle$ is transitive and reflexive

$\mathsf{Q} \subseteq \mathsf{W}$ is R-closed (i.e., if $w \in \mathsf{Q}$ and $w\mathsf{R}z$, then $z \in \mathsf{Q}$)

$\mathsf{e} : PV \rightarrow \text{Sub}\,\mathsf{W}$

Q is to be thought of as those states of information that are inconsistent.

Validity in such a model is defined as in the Kripke semantics for **Int** with the exception of the evaluation of negations:

$w \vDash \neg A$ iff for all z such that $w\mathsf{R}z$, $z \nvDash A$ or $z \in \mathsf{Q}$

Example $\neg A \to (A \to B)$ is not valid.

Consider the model with only one world w such that $w \vDash p_1$, $w \nvDash p_2$, and $w \in \mathcal{Q}$. Then $w \vDash \neg p_1$ and $w \nvDash p_1 \to p_2$. So $w \nvDash \neg p_1 \to (p_1 \to p_2)$.

To prove a completeness theorem for **J**, first note that all the derivations in Section C.1 hold for **J**, as pointed out on p. 281.

Lemma 26 $B, \neg B \vdash_{\mathbf{J}} \neg A$

Proof:

$\vdash B \to (A \to B)$	axiom V
$\vdash \neg B \to (A \to \neg B)$	axiom V
$B, \neg B \vdash (A \to B) \wedge (A \to \neg B)$	adjunction
$B, \neg B \vdash \neg A$	axiom XI and Lemma 3.a

■

Now we proceed as in Section C.2, using the same definitions, noting that Lemma 6.b, Lemma 7, and Lemma 8.a hold for **J**.

The *Canonical Model for* **J** is defined as:

$\langle \mathcal{W}_{\mathbf{J}}, \subseteq, \mathcal{Q}_{\mathbf{J}}, \mathbf{e}_{\mathbf{J}} \rangle$, where

$$\mathcal{W}_{\mathbf{J}} = \{\Gamma : \Gamma \text{ is full}\}$$

$$\mathcal{Q}_{\mathbf{J}} = \{\Gamma : \text{for some A, both A and } \neg A \text{ are in } \Gamma\}$$

$$\mathbf{e}_{\mathbf{J}}(\Gamma) = \{p : p \in \Gamma\}$$

I leave to you to show that the canonical model is reflexive, transitive, and anti-symmetric. It is also uncountably infinite.

Lemma 27 $\Gamma \vdash \neg E$ iff for every $\Sigma \supseteq \Gamma$ which is full, $E \notin \Sigma$ or $\Sigma \in \mathcal{Q}$

Proof: Suppose $\Gamma \vdash \neg E$. Then for all full $\Sigma \supseteq \Gamma$, $\Sigma \vdash \neg E$, and hence $\neg E \in \Sigma$. So either $E \notin \Sigma$, or if $E \in \Sigma$, then $\Sigma \in \mathcal{Q}$.

Now suppose that for every $\Sigma \supseteq \Gamma$ which is full, $E \notin \Sigma$ or $\Sigma \in \mathcal{Q}$. Consider $\Gamma \cup \{E\}$. If this is inconsistent or in $\mathcal{Q}$, by Lemma 26 we are done. But this must be the case, since otherwise by (the new version of) Lemma 7, there is some $\Sigma \supseteq \Gamma \cup \{E\}$ that is full and, because $\Sigma \nvdash \neg E$, is not in $\mathcal{Q}$. ■

Lemma 28 ***a.*** $\langle \mathcal{W}_{\mathbf{J}}, \subseteq, \mathcal{Q}_{\mathbf{J}}, \mathbf{e}_{\mathbf{J}}, \Gamma \rangle \vDash A$ iff $A \in \Gamma$

b. $\langle \mathcal{W}_{\mathbf{J}}, \subseteq, \mathcal{Q}_{\mathbf{J}}, \mathbf{e}_{\mathbf{J}} \rangle \vDash A$ iff $\vdash_{\mathbf{J}} A$

Proof: As for Lemma 9, except for the inductive stage on negation in part (a):

$\Gamma \vDash \neg B$ iff for all $\Sigma \supseteq \Gamma$, $\Sigma \nvDash B$ or $\Sigma \in \mathcal{Q}_{\mathbf{J}}$

iff for all $\Sigma \supseteq \Gamma$, $B \notin \Sigma$ by induction or $\Sigma \in \mathcal{Q}_{\mathbf{J}}$

iff $\Gamma \vdash \neg B$ by Lemma 27

iff $\neg B \in \Gamma$ by Lemma 8.a ■

I will leave to you to modify Lemma 10 to show that every Fitting model for a finite collection of formulas closed under subformulas can be filtered to a finite model. So we can prove, as for **Int**:

***Theorem 29 Completeness of the Fitting Semantics for* Int**

a. $\Gamma \vdash_{\mathbf{J}} A$ iff every Fitting model that validates Γ also validates A

b. For finite Γ, $\Gamma \vdash_{\mathbf{J}} A$ iff every finite anti-symmetric Fitting model that validates Γ also validates A

c. **J** is decidable

As long as we ignore negation, then **PC**, **Int**, and **J** all agree:

Corollary 30 In $L(\rightarrow, \wedge, \vee)$,

$\Gamma \vDash_{\mathbf{PC}} A$ iff $\Gamma \vDash_{\mathbf{Int}} A$

iff $\Gamma \vDash_{\mathbf{J}} A$

Proof: As for Theorem 15, noting that $\rightarrow$, $\wedge$, $\vee$ are evaluated the same in Fitting and Kripke models. ■

3. An alternate axiomatization

Here is an axiomatization of **J** in the language $L(\bot, \rightarrow, \wedge, \vee)$, where $\bot$ is a propositional constant. It is due to *Segerberg, 1968,* p. 30.

J

in $L(\bot, \rightarrow, \wedge, \vee)$

$\neg A \equiv_{\text{Def}} A \rightarrow \bot$

1. $(A \wedge B) \rightarrow A$
2. $(A \wedge B) \rightarrow B$
3. $A \rightarrow (A \vee B)$
4. $B \rightarrow (A \vee B)$
5. $(A \rightarrow C) \rightarrow ((B \rightarrow C) \rightarrow ((A \vee B) \rightarrow C))$
6. $(A \rightarrow B) \rightarrow ((A \rightarrow C) \rightarrow (A \rightarrow (B \wedge C))$
7. $(A \rightarrow (B \rightarrow C)) \rightarrow ((A \rightarrow B) \rightarrow (A \rightarrow C))$
8. $A \rightarrow (B \rightarrow A)$

rule $\dfrac{A,\ A \rightarrow B}{B}$

Segerberg shows that the closure of $\mathbf{J} \cup \{\bot \to A\}$ under *modus ponens* is strongly complete for the class of Kripke trees, where $\bot$ is evaluated as invalid at all elements of the tree. Hence we have the following axiomatization of **Int**.

Int
in $L(\to, \wedge, \vee, \bot)$
is the closure under *modus ponens* of $\mathbf{J} \cup \{\bot \to A\}$

4. Kolmogorov's axiomatization of intuitionistic reasoning in L(¬, →)

Kolmogorov, 1925, was the first to axiomatize intuitionistic reasoning, though his paper was apparently not known to logicians outside of Russia until much later. Kolmogorov began by considering the axiomatization of classical propositional logic in $L(\neg, \to)$ by *Hilbert, 1922* (see Exercise 7, p. 85 above). The axiom $(A \to B) \to ((\neg A \to B) \to B)$ had to be deleted, for from it we could deduce $\neg\neg A \to A$ (see Theorem 21). Of the other axiom of negation, he said:

> The axiom $[A \to (\neg A \to B)]$ does not have and cannot have any intuitive foundation since it asserts something about the consequences of something impossible: we have to accept B if the true judgment A is regarded as false. . . . Thus, [this axiom] cannot be an axiom of the intuitionistic logic of judgments.
>
> *Kolmogorov, 1925,* p. 421

Instead, he took the positive part of Hilbert's axiomatization and added the *principle of contradiction*: $(A \to B) \to ((A \to \neg B) \to \neg A)$.

Kolmogorov's axiomatization of intuitionistic reasoning, Kol
in $L(\neg, \to)$

$$A \to (B \to A)$$

$$(A \to (A \to B)) \to (A \to B)$$

$$(A \to (B \to C)) \to (B \to (A \to C))$$

$$(B \to C) \to ((A \to B) \to (A \to C))$$

$$(A \to B) \to ((A \to \neg B) \to \neg A)$$

rule $\dfrac{A,\ A \to B}{B}$

Kolmogorov proves:

If $PV(A) = \{q_1, \ldots, q_n\}$ and $\vdash_{\mathbf{PC}} A$, then

$$\{\neg\neg q_1 \to q_1, \ldots, \neg\neg q_n \to q_n\} \vdash_{\mathbf{Kol}} A$$

> The precise boundary of the domain in which the special logic of judgments [classical logic] is applicable has thus been found; the domain coincides with the domain in which the formula of double negation [$\neg\neg A \to A$] is applicable.
>
> *Kolmogorov, 1925,* p. 427

From this he concludes that, if we take A^* to be A with every propositional variable p replaced by $\neg\neg p$ (compare Corollary 5.d, e):

If $\vdash_{\mathbf{PC}} A$, then $\vdash_{\mathbf{Kol}} A^*$

However, as I ask you to show in Exercise 9 below, the double negation translation of **PC** into **Int** does not work for **J**. Kolmogorov took instead the following translation:

$$\begin{aligned} \nu(p) &= \neg\neg p \\ \nu(\neg A) &= \neg\neg\nu(A) \\ \nu(A \to B) &= \neg\neg(\nu(A) \to \nu(B)) \end{aligned}$$

That is, $\nu(A)$ is A with every subformula preceded by $\neg\neg$. Letting $\nu(\Gamma) = \{\nu(A): A \in \Gamma\}$, Kolmogorov proves:

$\Gamma \vdash_{\mathbf{PC}} A$ iff $\nu(\Gamma) \vdash_{\mathbf{Kol}} \nu(A)$

Though the Fitting semantics work for **J**, they do not reflect the semantic intuition behind the logic. Neither Kolmogorov nor Johansson, nor any intuitionist talks of "inconsistent worlds". Kolmogorov and Johansson excised the principle $\neg A \to (A \to B)$ from the minimal logic so that one could have a theory in which both A and $\neg A$ appear, not because both are true or equally likely, but because for neither do we have, now or in the foreseeable future, a proof. To my knowledge, it is an open problem to give sensible semantics for **J**.

Exercises for Sections F.1–F.4

1. Why do Johansson and Kolmogorov reject the principle of negation, $\neg A \to (A \to B)$, that Heyting accepted?
2. Show that there are A, and B such that $A, \neg A \nvDash_{\mathbf{J}} B$.
3. Determine whether each of the following is valid in **J**. If it is not valid, exhibit a model in which it fails.

   a. $(A \wedge \neg A) \to B$

   b. $(A \to B) \to ((\neg A \to B) \to B)$

   c. $(\neg A \to A) \to A$

   d. $(A \to B) \to (\neg B \to \neg A)$

   e. $(A \vee \neg A) \to (\neg\neg A \to A)$

   f. $\neg\neg[(A \to B) \to ((\neg A \to B) \to B)]$

4. (Open?) The presence or absence of what formulas in a theory ensure that it is Post-consistent for **J**?
5. Show that in the canonical model for **J**, $\mathbf{Q}_{\mathbf{J}} = \{\Gamma : \text{for every } A, \neg A \in \Gamma\}$.
6. Prove Lemma 10 for Fitting models for **J**.
7. Explain the differences in motivation, syntax, and semantics between formalizing intuitionistic reasoning in $L(\neg, \rightarrow, \wedge, \vee)$ and in $L(\bot, \rightarrow, \wedge, \vee)$.
8. Prove a strong completeness theorem for Segerberg's axiomatization of **J**.
9. Establish a translation between **J** formulated in the language of $L(\neg, \rightarrow, \wedge, \vee)$ and **J** in the language $L(\rightarrow, \wedge, \vee, \bot)$.
10. Show $\nvdash_{\mathbf{J}} \neg\neg(p \rightarrow (\neg p \rightarrow q))$. (Hint: Put an endpoint in $\mathbf{Q}$.)
11. a. (*Prawitz and Malmnas, 1965*) Define in $L(\rightarrow, \wedge, \vee, \bot)$:
    $A^{\dagger}$ is A with every subformula B replaced by replaced by $(B \vee \bot)$.
    Prove: $\Gamma \vdash_{\mathbf{Int}} A$ iff $\Gamma^{\dagger} \vdash_{\mathbf{J}} A^{\dagger}$.
    b. Define in $L(\rightarrow, \wedge, \vee, \bot)$: A^* is A with every p_i replaced by $(p_i \vee \bot)$.
    Show that $\Gamma \vdash_{\mathbf{Int}} A$ iff $\Gamma^* \vdash_{\mathbf{J}} A^*$. (See *Leivant, 1985*, p. 686.)
12. Extend the mapping v to $L(\neg, \rightarrow, \wedge, \vee)$ and prove that it is a translation: $\Gamma \vdash_{\mathbf{PC}} A$ iff $v(\Gamma) \vdash_{\mathbf{Kol}} v(A)$.
13. (Open?) Establish a translation of **J** into a modal logic.
14. (Open?) Prove that Kolmogorov's axiomatization is strongly complete for Fitting models restricted to $L(\neg, \rightarrow)$.
15. (Open) Is there a translation of **J** into **Int**?
16. (Open?) Give semantics for **J** that reflect the semantics assumptions of Kolmogorov and Johansson, as quoted in the text.

5. Set-assignment semantics

To give set-assignment semantics for **J** we modify those for **Int**. Negation cannot be evaluated by the general table for negation (Chapter IV.B.1), since that would give $A, \neg A \vDash B$. Instead, we use a wholly intensional negation.

A set-assignment model $\langle \upsilon, \mathsf{s} \rangle$ uses *minimal intuitionist truth-conditions* if:

$\wedge$ and $\vee$ are evaluated classically

$\rightarrow$ is evaluated by the dual dependence table as for **Int**:

$$\upsilon(A \rightarrow B) = \mathsf{T} \text{ iff } \mathsf{s}(A) \subseteq \mathsf{s}(B) \text{ and not both } \upsilon(A) = \mathsf{T} \text{ and } \upsilon(B) = \mathsf{F}$$

and $\neg$ is evaluated by the *minimal intuitionist negation*:

$$\upsilon(\neg A) = \mathsf{T} \text{ iff } \mathsf{s}(A) \subseteq \mathsf{s}(\neg A)$$

Then $<\upsilon,s>$ is a **J**-*model* if it uses the minimal intuitionist truth-conditions and satisfies Int 1, Int 2, Int 4–7, Int 9, and:

Int 3′. $s(B) \subseteq s(A \to B)$

Int 8′. $s(A \to B) \cap s(A \to \neg B) \subseteq s(\neg A)$

Lemma 31 If $\Gamma \vdash_J A$, then for every **J**-model $<\upsilon,s>$ that validates Γ, $\upsilon(A) = T$.

Proof: The verification is as in the proof of Lemma 22 except for axiom schema V and XI. By Int 3′, axiom scheme V is valid.

For axiom scheme XI, suppose $\upsilon(A \to B) = T$ and $\upsilon(A \to \neg B) = T$. Then $s(A) \subseteq s(B) \cap s(\neg B)$. By Int 3′, $s(B) \subseteq s(A \to B)$ and $s(\neg B) \subseteq s(A \to \neg B)$. By Int 8 we then have $s(B) \cap s(\neg B) \subseteq s(\neg A)$. Hence $s(A) \subseteq s(\neg A)$. Thus, using the minimal intuitionist negation, $\upsilon(\neg A) = T$. ■

Lemma 32 If $\Gamma \nvdash_J E$, then there is some **J**-model $<\upsilon,s>$, such that $<\upsilon,s> \vDash \Gamma$ and $\upsilon(E) = F$. If Γ is finite, then we may take $<\upsilon,s>$ to be finite.

Proof: Given $\Gamma \nvdash_J E$, by Theorem 29 there is some Fitting model $\langle W, R, Q, e \rangle \vDash \Gamma$ and $w \nvDash E$. If Γ is finite, then we may assume W is finite, too. Define $<\upsilon,s>$ by:

$\upsilon(A) = T$ iff $w \vDash A$

$s(A) = \{z : wRz \text{ and } z \vDash A\}$

I leave to you to show that $<\upsilon,s>$ is a **J**-model, and in particular that it uses the minimal intuitionist truth-conditions.

Then $<\upsilon,s> \vDash \Gamma$ and $\upsilon(E) = F$. ■

We now conclude:

Theorem 33 ***Strong Completeness of the Set-Assignment Semantics for* J**

$\Gamma \vdash_J A$ iff $\Gamma \vDash_J A$

For finite Γ the class of finite **J**-models is strongly complete.

We can give almost the same reading to these semantics as to those for **Int**. The only difference is that now we admit that the information available to us at some particular time may be inconsistent, leading us to correctly assert, relative to that bad information, both some proposition A and its negation. But there may be nothing in that faulty information which would lead us to assert some other proposition B. So $\varnothing \neq s(A) \cap s(\neg A)$ and $s(A) \cap s(\neg A)$ may not be contained in $s(B)$; hence $(A \wedge \neg A) \to B$ could fail.

Are these semantics more acceptable than the ones of Fitting? Since we may still have a model in which both A and $\neg A$ are true, they do not reflect the assumptions of Johansson and Kolmogorov. Negation is taken as wholly intensional, but it could even be taken as truth-default (see Chapter IX.G).

Note the similarity of the explanation of these set-assignment semantics to the motivation for dependence logic, Chapter V. But for dependence logic $B \to (A \to B)$ fails. Fifty years after Johansson wrote his paper one still starts at that axiom scheme. In Kolmogorov's terms it would read, 'Given a solution to B, convert it to a method for *converting* a solution of A to one for B'. Perhaps this is acceptable if intuitionistic logic is to be applied only to mathematics. But Dummett argues that the justification of a semantics is in its extension to a theory of meaning for natural language. Outside the domain of mathematics it would seem unreasonable to assume that the "trivial" conversion of ignoring A and taking the proffered solution to B is not tantamount to "changing the subject," and hence unacceptable. Can semantics be given for a dependent intuitionistic calculus: **Int** with both $B \to (A \to B)$ and $\neg A \to (A \to B)$ deleted?

Exercises for Section F.5

1. Show that the following table cannot be used for negation in **J**.

A	$s(A) \subseteq s(\neg A)$	$\neg A$
any value	fails	F
T	holds	F
F		T

2. Verify that the set-assignment semantics for **J** are sound.
3. Show that the <v,s> defined in Lemma 32 is a **J**-model.
4. (Open) Can a connective be defined in **J** that has the same set-assignment table as negation in **Int**? (See Exercise 13, p. 310.)
5. (Open) Determine whether $\{\neg, \to, \wedge, \vee\}$ is functionally complete for **J**. If not, exhibit a collection of connectives that is.
6. (Open) Give strongly complete semantics for a dependent minimal intuitionistic logic, as suggested in the last paragraph of this chapter.

VIII Many-Valued Logics – L_3, L_n, $L_{\aleph}$, K_3, G_3, G_n, $G_{\aleph}$, S5 –

Some believe that more than two truth-values are needed in reasoning. I will discuss the history of that idea and then explain how additional truth-values can be accounted for as paying attention to different aspects of propositions. Then I will present several examples of logics using more than two truth-values, as well as applications of those systems to analyses of other logics we have studied. I also include for reference a general definition of many-valued logics.

General sources on many-valued logics are *Rescher, 1968* and *1969, Wójcicki, 1988,* and, for the philosophical issues, *Haack, 1974.* A major bibliography is *Wolf, 1977.*

A. How Many Truth-Values?

1. History

The applicability of the classical dichotomy of true-false has been questioned since antiquity. In the beginning of the modern development of formal logic, De Morgan considered using more than two values for his calculus.

> But we should be led to extend our formal system if we considered propositions under three points of view, as true, false, or inapplicable. We may confine ourselves to single alternatives either by introducing not-true (including both false and inapplicable) as the recognized contrary of true; or else by confining our results to universes in which there is always applicability, so that true or false holds in every case. The latter hypothesis will best suit my present purpose.
>
> *De Morgan, 1847,* p.149

In the early twentieth century tables for the connectives using three or more values were used by several logicians to establish the independence of particular axioms in formal systems, an example of which I give in Section G. *Łukasiewicz* (pronounced 'Woo-kah-sheay-vitch') *and Tarski, 1930,* p.43 (footnote 5), give a short history of that early work.

The experience of working with those tables suggested to Łukasiewicz that a three-valued formal system would be appropriate to reason with future contingent propositions, such as 'There will be a sea battle tomorrow', which, he argued, are neither true nor false but rather possible. The first many-valued system proposed as a logic, that is as a formal system for reasoning, was set out in *Łukasiewicz, 1920* (Section C.1 below). In that system a third formal value is introduced into the truth-tables not as an additional truth-value, but as a marker to indicate that a proposition to which it is assigned has no truth-value.

At the same time *Post, 1921,* formulated a class of many-valued systems as generalizations of the 2-valued classical calculus. These are called 'logics' only by analogy with other logical systems, for to my knowledge no one has proposed any of them as a formalization of reasoning with propositions. Many-valued systems are often interesting to mathematicians as a clearly demarcated area of finite combinatorics that can be developed in analogy with formal logics. In Section B I'll present the general definition of a many-valued system in order to have a uniform terminology; with that we can show in Section E that the logics we've already studied cannot be characterized as finite-valued systems. The rest of the chapter will be concerned with many-valued systems that are either proposed as logics, that is as descriptive or prescriptive models of reasoning, or that were devised to reveal facets of other logics we have encountered in the previous chapters. An example of the latter are the systems that Gödel devised to investigate intuitionistic logic (Section F).

Kleene, 1952, proposed a 3-valued system suited to reasoning with propositions, such as undecidable arithmetical statements, that are true or false, but which of these alternatives we do not or cannot know. I present his logic in Section D. Other systems have been developed based on a similar motivation that there are degrees of truth or falsity roughly corresponding to degrees of certainty.

Another motive for many-valued systems has been to deal with paradoxical or inconsistent sentences. For instance, *Moh Shaw-Kwei, 1954,* proposed reserving the third value of Łukasiewicz's system for sentences such as 'This sentence is false'. He, and later *Kripke, 1975,* who instead used Kleene's logic, viewed such sentences as neither true nor false. On the other hand *Yablo, 1985,* argues that we should think of such sentences as both true *and* false and indicates how many-valued systems can be appropriate for reasoning on that basis. In the next chapter I'll present a many-valued logic along those lines.

2. Hypothetical reasoning and aspects of propositions

In Chapter I I argued that in most applications of logic we reason *on the hypothesis* that this or that proposition is true or is false, since we cannot make our communications precise enough to be unequivocal nor, in general, can we know with certainty the truth-value of the propositions we deal with. Of course we may have doubts about the hypothesis. It seems to me that many-valued logics factor those doubts into the logic by ascribing them to the content of the proposition. For any proposition p:

I can assent to p or *I can not assent to p*

There is no third choice. I can be in doubt about whether I should assent or not. But that doubt is not a third choice; it is doubt about *whether* to make the one choice or the other. If the doubt predominates, so that I don't assent, then that is the choice: I do not assent. Or I may assent with doubt about the wisdom or propriety doing so: but I have assented. Every many-valued logic recognizes this division of choices by partitioning the *n* values ascribed to propositions into two classes: designated and undesignated.

$$\left.\begin{matrix}1\\ \vdots \\ m\end{matrix}\right\} \text{ assent — the } \textit{designated values} \qquad \left.\begin{matrix}m+1\\ \vdots \\ n\end{matrix}\right\} \text{ do not assent — the } \textit{undesignated values}$$

A many-valued logic provides us with a calculus to enable us to know whether and how much to doubt a complex proposition from knowledge of the content of (doubt about) its constituents, and how to reason accordingly. But there is no need for a new notion of proposition: a proposition is a written or uttered declarative sentence with which we agree to proceed hypothetically as being either true or false, sometimes ascribing additional content to it.

Smiley has given a similar analysis:

> The way to defend [the method of designating truth-values] is to read 'true' for 'designated'. The method of defining logical consequence then needs no justification, for it now reads as saying that a proposition follows from others if and only if it is true whenever they are all true. What does need explaining is how there can be more than two truth-values. The answer is that propositions can be classified in other ways than as true or untrue, and by combining such a classification with the true/untrue one we in effect subdivide the true and untrue propositions into a larger number of types. For example, given any property φ of propositions, there are prima facie four possible types of proposition: true and φ, true and not φ, untrue and φ, untrue and not φ. If φ is unrelated to truth, like 'obscene' or 'having to do with geometry', all four types can exist and we get four truth-values, two being designated and two undesignated. If φ has any bearing on truth some of the types may be ruled out; e.g., if φ is (perhaps) 'about the future' or 'meaningless', the type 'true and φ' will be empty, leaving three truth-values of which just one is designated. One cannot foretell how the connectives will behave with respect to this or that classification of propositions, but to the extent that the types of compound propositions turn out to be functions of the types of their constituents, so we shall get a many-valued logic.
>
> *Smiley, 1976,* pp. 86–87

B. A General Definition of Many-Valued Semantics

The definitions I will present here give a uniform terminology for many-valued logics and are necessary when we investigate whether the logics we have already studied have finite-valued semantics. They are quite abstract, and as such have been studied extensively, for instance by *Wójcicki, 1988*, and *Carnielli, 1987 B*. In the following sections I will repeat these definitions for specific logics, and you may wish to look at those examples first, after which the definitions of this section will seem natural.

We take as our language $L(\neg, \rightarrow, \wedge, \vee, p_0, p_1, \ldots)$. I will leave to you to modify the definitions to apply to languages with other connectives.

M is a *matrix* if

$$\mathsf{M} = \langle \mathsf{U}, \mathsf{D}, \neg_{\mathsf{M}}, \rightarrow_{\mathsf{M}}, \wedge_{\mathsf{M}}, \vee_{\mathsf{M}} \rangle$$

where:

U is a set with $\mathsf{D} \subseteq \mathsf{U}$

$\rightarrow_{\mathsf{M}}, \wedge_{\mathsf{M}}, \vee_{\mathsf{M}}$ are binary operations on U

$\neg_{\mathsf{M}}$ is a unary operation on U

We call D the set of *designated*, or *distinguished elements* of M. The operations are called the *(truth-) tables for* $\neg, \rightarrow, \wedge, \vee$. M is called *finite-valued, infinite-valued,* or *n-valued* according to whether U is finite, infinite, or has *n* elements.

An *evaluation* e *with respect to* M is a function $\mathsf{e}: PV \rightarrow \mathsf{U}$ that is extended inductively to all wffs by:

$$\mathsf{e}(\neg A) = \neg_{\mathsf{M}} \mathsf{e}(A)$$

$$\mathsf{e}(A \rightarrow B) = \rightarrow_{\mathsf{M}}(\mathsf{e}(A), \mathsf{e}(B))$$

$$\mathsf{e}(A \wedge B) = \wedge_{\mathsf{M}}(\mathsf{e}(A), \mathsf{e}(B))$$

$$\mathsf{e}(A \vee B) = \vee_{\mathsf{M}}(\mathsf{e}(A), \mathsf{e}(B))$$

If we consider U to be a collection of truth-values, then this definition says that every evaluation is truth-functional.

We say that e *validates* A, written $\mathsf{e} \models A$, if $\mathsf{e}(A) \in \mathsf{D}$. We write $\mathsf{e} \models \Gamma$ if for all $A \in \Gamma$, $\mathsf{e} \models A$. When more than one matrix is under consideration we write $\models_{\mathsf{M}}$ for $\models$.

A wff A is *valid* (*with respect to* M), written $\models_{\mathsf{M}} A$ or simply $\models A$, if for every evaluation e, $\mathsf{e} \models A$. That is, every evaluation of A takes a designated value. We say that M is a *characteristic matrix for a set of wffs* Γ if $\Gamma = \{A : \models_{\mathsf{M}} A\}$. In that case we say that Γ *has many-valued semantics*, or that Γ *is a many-valued logic*. The term *many-valued system* may sometimes be used to refer to either such a logic or the semantics of it.

Every many-valued matrix in this chapter will have only one designated value. In Chapter IX we will look at a logic characterized by a 3-valued matrix with two designated values.

We can define the notion of *semantic consequence* (*with respect to* M) in the usual way: $\Gamma \models A$ iff for every evaluation e, if $\mathsf{e} \models \Gamma$ then $\mathsf{e} \models A$. That is, a proposition follows from others if it takes a designated value whenever they do. For some many-valued logics other notions of semantic consequence have been suggested: see, for example, *Smiley, 1976,* or *Wójcicki, 1988.*

It is useful to be able to give finite-valued semantics for a logic, for then there is a simple decision method for validity like the one for the two-valued classical logic (Chapter II.J.7). But some logics that are decidable, such as **Int** and **S5**, cannot be given finite-valued semantics (see Section E). In contrast, *any* set of wffs Γ that is closed under the rule of substitution has infinite-valued semantics: take the matrix M to be $\langle \text{Wffs}, \Gamma, \neg, \rightarrow, \wedge, \vee \rangle$, the designated values are the wffs of Γ, and the operations are the connectives; an evaluation of A is then a substitution instance of it. But if Γ is first presented as a syntactic or semantic consequence relation, then the semantic consequence relation for this matrix need not coincide with that (see *Dummet, 1977,* Chapter 5.1).

C. The Łukasiewicz Logics

Łukasiewicz argued that a two-valued logic is incompatible with the view that some propositions about the future are not predetermined.

> I can assume without contradiction that my presence in Warsaw at a certain moment of next year, e.g., at noon on 21 December, is at the present time determined neither positively nor negatively. Hence it is *possible,* but not *necessary*, that I shall be present in Warsaw at the given time. On this assumption the proposition 'I shall be in Warsaw at noon on 21 December of next year', can at the present time be neither true nor false. For if it were true now, my future presence in Warsaw would have to be necessary, which is contradictory to the assumption. If it were false now, on the other hand, my future presence in Warsaw would have to be impossible, which is also contradictory to the assumption. Therefore, the proposition considered is at the moment *neither true nor false* and must possess a third value, different from '0' or falsity and '1' or truth. This value we can designate by '$\frac{1}{2}$'. It represents 'the possible', and joins 'the true' and 'the false' as a third value.
>
> *Łukasiewicz, 1930,* pp. 165–166

Łukasiewicz saw his work partly as a formalization of Aristotle's views concerning future contingent propositions and the notions of necessity and possibility (see in particular *Łukasiewicz, 1922* and *1930*; in the latter he has a history of the principle of *tertium non datur*). Whether his interpretation of Aristotle is correct, and if correct reasonable, has been discussed by a number of authors. *Haack, 1974,* Chapter 3, argues that his program was misconceived on the basis of a modal fallacy; in doing so she reviews the literature on the subject. *Prior, 1955,* pp. 230– 250, gives a sustained defense and explication of Łukasiewicz's work, relating it to Aristotle's and Ockham's views. Without judging the matter beyond what I have already said in Section A, I will try to present Łukasiewicz's views by following in large part Prior's presentation.

1. The 3-valued logic $\mathbf{L_3}$

a. The truth-tables and their interpretation

This explanation will, for the most part, follow *Prior, 1955* and *1967.*

A proposition may take one of three values under an interpretation: 1, $\frac{1}{2}$, or 0 , where we are to understand:

$\mathbf{e}(A) = 1$ means that A is determinately true

$\mathbf{e}(A) = 0$ means that A is determinately false

$\mathbf{e}(A) = \frac{1}{2}$ means that A is neither determinately true nor false; in this case we say that A is possible or *neuter*

Future contingent propositions are the ones that are to take value $\frac{1}{2}$, reflecting Łukasiewicz's rejection of determinism.

The matrix for Łukasiewicz's 3-valued logic is given by the following tables:

$A \wedge B$	B: 1	$\frac{1}{2}$	0
A 1	1	$\frac{1}{2}$	0
$\frac{1}{2}$	$\frac{1}{2}$	$\frac{1}{2}$	0
0	0	0	0

$A \vee B$	B: 1	$\frac{1}{2}$	0
A 1	1	1	1
$\frac{1}{2}$	1	$\frac{1}{2}$	$\frac{1}{2}$
0	1	$\frac{1}{2}$	0

A	$\neg A$
1	0
$\frac{1}{2}$	$\frac{1}{2}$
0	1

$A \to B$	B: 1	$\frac{1}{2}$	0
A 1	1	$\frac{1}{2}$	0
$\frac{1}{2}$	1	1	$\frac{1}{2}$
0	1	1	1

Define $A \leftrightarrow B \equiv_{\text{Def}} (A \to B) \wedge (B \to A)$.

An $\mathbf{L_3}$*-evaluation* is a map $\mathbf{e}: \text{PV} \to \{0, \frac{1}{2}, 1\}$ that is extended to all wffs of $L(\neg, \to, \wedge, \vee, p_0, p_1, \ldots)$ by these tables. The *sole designated value* is 1, so that $\mathbf{e} \models A$ means $\mathbf{e}(A) = 1$; then A *is valid*, written $\models A$, means that $\mathbf{e}(A) = 1$ for all $\mathbf{L_3}$-evaluations. And $\Gamma \models_{\mathbf{L_3}} A$ means that for every $\mathbf{L_3}$-evaluation $\mathbf{e}$, if $\mathbf{e}(B) = 1$ for every B in Γ, then $\mathbf{e}(A) = 1$. The collection of valid wffs we call $\mathbf{L_3}$, or the $\mathbf{L_3}$*-tautologies.*

Three important rules are valid for $\mathbf{L_3}$: *modus ponens* (as you can check), substitution of logical equivalents, and substitution. We have for any $\mathbf{e}$:

$\mathbf{e}(A \to B) = 1$ iff $\mathbf{e}(A) \leq \mathbf{e}(B)$

$\mathbf{e}(A \leftrightarrow B) = 1$ iff $\mathbf{e}(A) \leq \mathbf{e}(B)$ and $\mathbf{e}(B) \leq \mathbf{e}(A)$ iff $\mathbf{e}(A) = \mathbf{e}(B)$

Hence, the rule of substitution is valid: if C(B) is the result of replacing some but not necessarily all occurrences of A in C with B,

$$\frac{A \leftrightarrow B}{C(A) \leftrightarrow C(B)}$$

I will let you show that the rule of substitution is valid, too, where A(B) is A(p) with every occurrence of p in A replaced by B:

$$\frac{\models A(p)}{\models A(B)}$$

Both $A \vee \neg A$ and $\neg(A \wedge \neg A)$ fail to be $\mathbf{L_3}$-tautologies in accord with Łukasiewicz's rejection of bivalence: they take the value $\frac{1}{2}$ if A has value $\frac{1}{2}$. We have instead a principle of trivalence. To express that, we first pick out a formula that identifies a proposition as having value $\frac{1}{2}$:

$$IA \equiv_{Def} A \leftrightarrow \neg A$$

The letter 'I' is for 'indeterminate'. The table for IA is:

A	IA
1	0
$\frac{1}{2}$	1
0	0

With the aid of IA we can express what I call the *law of excluded fourth*:

$$A \vee \neg A \vee IA$$

This is an $\mathbf{L_3}$-tautology, a *principle of trivalence* for $\mathbf{L_3}$.

Moh Shaw-Kwei, 1954, suggested that a paradoxical sentence is one that is equivalent to its negation, and in $\mathbf{L_3}$ a paradoxical sentence should be assigned $\frac{1}{2}$. Then IA could be read as ''A' is paradoxical'. However, it is not enough to invoke 3 values for propositions in order to resolve the liar paradox, 'This sentence is false': Łukasiewicz's system spawns the equally problematic *strengthened liar paradox,* 'This sentence is false or paradoxical'.

Łukasiewicz's original motivation was to give a logic suitable for reasoning about possibilities and necessities. He considered the following function to be appropriate to formalize possibility:

A	$\Diamond$A
1	1
$\frac{1}{2}$	1
0	0

A proposition is possible if it is true or neuter, but not possible if it is false. Tarski noted that this table is definable in $\mathbf{L_3}$:

$\Diamond A$ has the same table as $\neg A \to A$

Note that this is the definition we used in Chapter VI for necessity.

It is harmless to understand $\Diamond A$ as an abbreviation useful in our metalogical investigations; but if we view $\Diamond$ as a connective we are under an obligation to explain why this is not a use-mention confusion, as discussed in Chapter VI.B.2.

Given a possibility operator, we define a necessity operator:

$\Box A \equiv_{Def} \neg \Diamond \neg A$

Its table is:

A	$\Box$A
1	1
$\frac{1}{2}$	0
0	0

Prior, 1955, p. 421, argues that this connective serves to model Aristotle's notion of necessity, and he quotes Aristotle:

> Once it is, that which is is-necessarily, and once it is not, that which is not necessarily-is-not.

Correspondingly we have that $A \to (A \to \Box A)$ is an $\mathbf{L_3}$-tautology, though $A \to \Box A$ can fail for neuter A. And on pp. 248–249, Prior points out that $\Box A$ should not be interpreted as 'It is logically necessary that A', as for **S5**. On that reading we should have '$\Box$(Socrates is dead $\to$ Socrates is dead)' is true on the basis of its form, whereas '$\Box$(Socrates is dead)' would not be true. Yet on Łukasiewicz's interpretation of necessity they would both be true since 'Socrates is dead' is true.

> And should the view that there are "neuter" propositions be accepted, not only the law of excluded middle but the whole structure of two-valued logic can be preserved by the understanding that the "propositions" substitutable for its variables are only those referring to matters of present, past or otherwise [e.g. timeless] determinate fact.
>
> *Prior, 1955*, p. 250

For such propositions $\Box A$ and $\Diamond A$ are both equivalent to A.

However, with this definition of possibility, if $\mathbf{e}(p) = \frac{1}{2}$, then $\mathbf{e}(\Diamond(p \wedge \neg p)) = \frac{1}{2}$. That is, if p is possible, then both p and $\neg p$ are together possible, an anomaly at best, though *Łukasiewicz, 1953* defended his interpretation against this. The problem arises from the validity of the general principle $(\Diamond A \wedge \Diamond B) \to \Diamond(A \wedge B)$.

Is there a *Deduction Theorem* for $\mathbf{L_3}$? We cannot identify $\to$ and $\vdash$: $A \wedge \neg A \vDash_{\mathbf{L_3}} \neg(A \to B)$, but $\nvDash_{\mathbf{L_3}} (A \wedge \neg A) \to \neg(A \to B)$. Define:

$A \rightarrow_3 B \equiv_{Def} A \rightarrow (A \rightarrow B)$

Its table is:

$A \rightarrow_3 B$		B 1	$\frac{1}{2}$	0
A	1	1	$\frac{1}{2}$	0
	$\frac{1}{2}$	1	1	1
	0	1	1	1

So $\mathbf{e}(A \rightarrow B) \neq 1$ iff $\mathbf{e}(A) = 1$ and $\mathbf{e}(B) \neq 1$. Thus we have:

Theorem 1 A Semantic Deduction Theorem for $\mathbf{L_3}$

$\Gamma \cup \{A\} \vDash_{\mathbf{L_3}} B$ iff $\Gamma \vDash_{\mathbf{L_3}} A \rightarrow_3 B$

The table for $\rightarrow_3$ was first given by *Monteiro, 1967,* who defined the connective as $\Diamond \neg A \vee B$. He noted that we could take $\neg$, $\wedge$, $\rightarrow_3$ as primitives, defining $A \rightarrow B$ as $(A \rightarrow_3 B) \wedge (\neg B \rightarrow_3 \neg A)$. In the form given above, however, the connective generalizes to Łukasiewicz's *n*-valued logics (below) by iterating '$A \rightarrow$'. *Wójcicki, 1988,* counting the number of arrows, calls this '$\rightarrow_2$'.

Though not every classical tautology is valid in $\mathbf{L_3}$, we do have the converse, $\mathbf{L_3} \subset \mathbf{PC}$: the $\mathbf{L_3}$-tables restricted to the values 0 and 1 are the **PC**-tables reading T for 1, and F for 0. We cannot retrieve **PC** by identifying 1 and $\frac{1}{2}$ with T, though, for the **PC**-tautology $\neg(A \rightarrow \neg A) \vee \neg(\neg A \rightarrow A)$ takes value 0 when A has value $\frac{1}{2}$.

But we can interpret **PC** in $\mathbf{L_3}$ by means of a translation, based on the ideas of Tokarz (see *Wójcicki, 1988,* pp. 71–72). Consider the map * from $L(\neg, \rightarrow)$ to itself given by:

$$(p)^* = p$$

$$(A \rightarrow B)^* = A^* \rightarrow_3 B^*$$

$$(\neg A)^* = A^* \rightarrow_3 \neg(A^* \rightarrow A^*)$$

And $\Gamma^* = \{A^* : A \in \Gamma\}$.

Theorem 2 The map * is a translation of **PC** into $\mathbf{L_3}$: $\Gamma \vDash_{\mathbf{PC}} A$ iff $\Gamma^* \vDash_{\mathbf{L_3}} A^*$

Proof: If $\mathbf{e}$ is an $\mathbf{L_3}$-evaluation and υ a 2-valued model of **PC** such that $\mathbf{e}(p) = 1$ iff $\upsilon(p) = \mathsf{T}$, then for every B, $\mathbf{e}(B^*) = 1$ iff $\upsilon(B) = \mathsf{T}$, as you can check by induction on the length of B. Hence $\Gamma \vDash_{\mathbf{PC}} A$ iff $\Gamma^* \vDash_{\mathbf{L_3}} A^*$. ∎

We may reduce the primitives of the language of $\mathbf{L_3}$ to just $\neg$ and $\rightarrow$:

$A \vee B$ has the same table as $(A \rightarrow B) \rightarrow B$

$A \wedge B$ has the same table as $\neg(\neg A \vee \neg B)$

However, $\{\neg, \rightarrow, \wedge, \vee\}$ is not functionally complete, in the sense that there is a function of the three values that cannot be given by any scheme. Consider a connective T, the 'Słupecki operator', that is evaluated as $\frac{1}{2}$ regardless of the value of A. TA is not definable from $\neg$, $\rightarrow$, $\wedge$, $\vee$: if it were, then, since $\neg$TA and TA have the same value, $\neg \text{TA} \rightarrow \text{TA}$ would be an $\mathbf{L_3}$-tautology, whereas no wff of the form $\neg \text{B} \rightarrow \text{B}$ is a **PC**-tautology, and $\mathbf{L_3} \subset \mathbf{PC}$.

Exercises for Section C.1.a

1. a. Show that the following are not valid in $\mathbf{L_3}$:

   a. $\text{A} \vee \neg \text{A}$ b. $\neg(\text{A} \wedge \neg \text{A})$ c. $(\text{A} \wedge \neg \text{A}) \rightarrow \neg(\text{A} \rightarrow \text{B})$

2. Determine which of the following are valid in $\mathbf{L_3}$:

   a. $\neg\neg \text{A} \rightarrow \text{A}$
   b. $(\text{A} \rightarrow \text{B}) \rightarrow (\neg \text{B} \rightarrow \neg \text{A})$
   c. $(\text{A} \rightarrow (\text{B} \rightarrow \text{C})) \rightarrow ((\text{A} \wedge \text{B}) \rightarrow \text{C})$
   d. $\text{B} \rightarrow (\text{A} \rightarrow \text{B})$
   e. $(\text{A} \rightarrow \text{B}) \rightarrow ((\neg \text{A} \rightarrow \text{B}) \rightarrow \text{B})$
   f. $(\text{A} \rightarrow \text{B}) \rightarrow ((\text{A} \rightarrow \neg \text{B}) \rightarrow \neg \text{A})$
   g. $(\text{A} \rightarrow \text{B}) \rightarrow \neg(\text{A} \wedge \neg \text{B})$
   h. $(\text{A} \rightarrow \text{B}) \rightarrow (\neg \text{A} \vee \text{B})$
   j. $((\text{A} \rightarrow \text{B}) \rightarrow \text{A}) \rightarrow \text{A}$
   k. $(\text{A} \rightarrow (\text{B} \rightarrow \text{C})) \rightarrow ((\text{A} \rightarrow \text{B}) \rightarrow \text{A} \rightarrow \text{C})$

4. Show that $\text{A} \vee \neg \text{A} \vee \text{IA}$ is a tautology of $\mathbf{L_3}$.

5. Show that in $\mathbf{L_3}$ 'This sentence is false or paradoxical' cannot be given a truth-value. Explain whether in your reasoning about this sentence you are using classical logic or $\mathbf{L_3}$.

6. a. Show that $\Diamond \text{A}$ has the same table as $\neg \text{A} \rightarrow \text{A}$.
   b. Show that $\text{A} \rightarrow (\text{A} \rightarrow \Box \text{A})$ is a tautology.
   c. Show that $(\Diamond \text{A} \wedge \Diamond \text{B}) \rightarrow \Diamond(\text{A} \wedge \text{B})$ is a tautology.

7. a. Prove the *Semantic Deduction Theorem* for $\mathbf{L_3}$.
   b. Show that $\Diamond \neg \text{A} \vee \text{B}$ has the same table as $\text{A} \rightarrow_3 \text{B}$.
   c. Show that $\text{A} \rightarrow \text{B}$ has the same table as $(\text{A} \rightarrow_3 \text{B}) \wedge (\neg \text{B} \rightarrow_3 \neg \text{A})$.

8. Why would it be inappropriate to devise an $\mathbf{L_3}$-modal logic in which truth at a possible-world is evaluated by the $\mathbf{L_3}$-tables?

9. a. Prove that $\mathbf{L_3} \subset \mathbf{PC}$.
   b. Prove that if 1 and $\frac{1}{2}$ are both designated then $\neg(\text{A} \rightarrow \neg \text{A}) \vee \neg(\neg \text{A} \rightarrow \text{A})$ is not valid.
   c. Prove that $\mathbf{L_3}$ in $\text{L}(\rightarrow, \wedge, \vee)$ is not the same as **PC** in $\text{L}(\rightarrow, \wedge, \vee)$.

10. Show that no wff of the form $\neg B \rightarrow B$ is a **PC**-tautology.

11. Formalize in $\mathbf{L_3}$ and comment on the following modal principles that *Łukasiewicz , 1930,* discusses:
   a. If it is not possible that p, then not p
   b. Whatever is, when it is, is necessary
   c. If it is supposed that not-p, then it is (on this supposition) not possible that p
   d. For some p: it is possible that p and it is possible that not-p

12. a. Formalize in $\mathbf{L_3}$ the examples in Chapter VI.D and comment on them.
   b. Formalize Examples 2, 4, 5, 6, and 7 of Chapter V.A.7.

13. Which of the following are valid in $\mathbf{L_3}$, where $\supset$ is defined from $\neg$ and $\wedge$?
   a. $\Box A \supset A$ d. $\Diamond A \supset \Box\Diamond A$
   b. $\Diamond A \supset A$ e. $\Box A \supset \Box\Box A$
   c. $A \supset \Box A$ f. $A \supset \Box\Diamond A$

14. As for Exercise 14 replacing '$\supset$' by '$\rightarrow$'.

15. a. Compare $\mathbf{L_3}$ to the modal logics **K**, **T**, **B**, **S4**, and **S5**.
   b. Show that there is no translation of any of **K**, **T**, **B**, **S4**, or **S5** into $\mathbf{L_3}$. (Hint: Use Corollary 20, below.)
   c. (Open) Is there a translation of $\mathbf{L_3}$ into any of the logics **K**, **T**, **B**, **S4**, or **S5**?

16. Show that the rule of substitution is valid.

17. Show that the following rules of substitution are valid, where C(B) is the result of replacing some but not necessarily all occurrences of A in C with B.

   a. $$\frac{(A \leftrightarrow B) \wedge (IA \leftrightarrow IB)}{C(A) \leftrightarrow C(B)}$$ b. $$\frac{(A \leftrightarrow B) \wedge (IA \leftrightarrow IB)}{(C(A) \leftrightarrow C(B)) \wedge (IC(A) \leftrightarrow IC(B))}$$

18. For the translation * of Theorem 2, give the translation of:
   a. $A \vee \neg A$ c. $A \leftrightarrow \neg A$ e. $(A \rightarrow B) \rightarrow (\neg B \rightarrow \neg A)$
   b. $\neg(A \wedge \neg A)$ d. $(\neg A \rightarrow A) \rightarrow A$

19. a. Exhibit a collection of connectives that are functionally complete for $\mathbf{L_3}$.
   b. Prove a normal form theorem for $\mathbf{L_3}$.

b. A finite axiomatization of $\mathbf{L_3}$

We have two routes to axiomatizing $\mathbf{L_3}$: we can use nonconstructive infinitistic methods to give a strong completeness proof similar to the one for classical logic, using appropriate notions of completeness and consistency; or we can give a constructive proof of finite strong completeness by reducing tautologies to normal forms. In this section I will do the first, emphasizing the similarities to the methods for classical logic. I will note the first use of each axiom in the proof. *Avron, 1991,* has also given an axiomatization using similar methods.

In the next section I present another axiomatization arrived at constructively.

L_3 *in* $L(\neg, \rightarrow)$

$A \rightarrow_3 B \equiv_{Def} A \rightarrow (A \rightarrow B)$ $\qquad$ $IA \equiv_{Def} A \leftrightarrow \neg A$

1. $B \rightarrow_3 (A \rightarrow B)$
2. $(A \rightarrow_3 (B \rightarrow C)) \rightarrow ((A \rightarrow_3 B) \rightarrow (A \rightarrow_3 C))$
3. $\neg A \rightarrow (A \rightarrow B)$
4. $A \rightarrow_3 (IA \rightarrow B)$
5. $\neg A \rightarrow_3 (IA \rightarrow B)$
6. $(\neg A \rightarrow_3 A) \rightarrow ((IA \rightarrow_3 A) \rightarrow A)$
7. $\neg\neg A \rightarrow A$
8. $A \rightarrow \neg\neg A$
9. $A \rightarrow (\neg B \rightarrow \neg(A \rightarrow B))$
10. $A \rightarrow_3 (IB \rightarrow I(A \rightarrow B))$
11. $IA \rightarrow (\neg B \rightarrow I(A \rightarrow B))$
12. $IA \rightarrow (IB \rightarrow (A \rightarrow B))$

rule $\dfrac{A,\ A \rightarrow B}{B}$

I will write $\vdash$ instead of $\vdash_{L_3}$ throughout this section for the consequence relation of this axiom system. I leave to you to check that the system is sound.

Lemma 3 ***a.*** $\vdash (A \rightarrow B) \rightarrow (A \rightarrow_3 B)$

b. $\vdash A \rightarrow_3 A$

c. $\{A,\ A \rightarrow_3 B\} \vdash B$

Proof: a. **Axiom 1** yields $(A \rightarrow B) \rightarrow_3 (A \rightarrow (A \rightarrow B))$, which by definition is (a).

b. $A \rightarrow_3 (A \rightarrow A)$ — axiom 1

$A \rightarrow_3 ((A \rightarrow A) \rightarrow A)$ — axiom 1

$[A \rightarrow_3 ((A \rightarrow A) \rightarrow A)] \rightarrow [(A \rightarrow_3 (A \rightarrow A)) \rightarrow (A \rightarrow_3 A)]$ — **axiom 2**

$A \rightarrow_3 A$ — *modus ponens* twice

c. Apply *modus ponens* twice. ■

Lemma 4 ***A Syntactic Deduction Theorem for*** L_3 $\quad \Gamma, A \vdash B$ iff $\Gamma \vdash A \rightarrow_3 B$

Proof: The proof is as for classical logic, Theorem II.8, using **axioms 1** and **2** and Lemma 3. ■

The definition of a complete consistent theory should correspond to the set of sentences true in a model. Given any $\mathbf{L_3}$-evaluation exactly one of $\mathbf{e}(A)$, $\mathbf{e}(\neg A)$, $\mathbf{e}(IA)$ has value 1, and the one that does determines the value of $\mathbf{e}(A)$. So define:

Γ is *consistent* (*relative to* $\mathbf{L_3}$) if *at most one of* $A, \neg A, IA$ is a consequence of Γ

Γ is *complete* (*relative to* $\mathbf{L_3}$) if for every A *at least one of* $A, \neg A, IA$ is in Γ

As usual, $\text{Th}(\Gamma) \equiv_{\text{Def}} \{A: \Gamma \vdash A\}$ and Γ is *theory* if $\Gamma = \text{Th}(\Gamma)$.

Lemma 5 ***a.*** Γ is consistent iff for some B, $\Gamma \nvdash B$ (Post-consistency)

b. If Γ is complete and consistent, then Γ is a theory.

c. If Γ is consistent and $\Gamma \nvdash A$, then $\Gamma \cup \{\neg A\}$ or $\Gamma \cup \{IA\}$ is consistent.

d. If Γ is consistent, then one of $\Gamma \cup \{A\}$, $\Gamma \cup \{\neg A\}$, $\Gamma \cup \{IA\}$ is consistent.

Proof: a. From left to right is immediate. So suppose Γ is inconsistent. Then by **axioms 3–5** for any B, $\Gamma \vdash B$.

b. Suppose Γ is complete and consistent and $\Gamma \vdash A$. If $A \notin \Gamma$, then one of $\neg A$, IA is in Γ. But then by axioms 3 and 4, $\Gamma \vdash B$ for every B, a contradiction on the consistency of Γ by part (a). So $A \in \Gamma$.

c. Suppose Γ is consistent and $\Gamma \nvdash A$. Suppose further that both $\Gamma \cup \{\neg A\}$ and $\Gamma \cup \{IA\}$ are inconsistent. Then by part (a), $\Gamma \cup \{\neg A\} \vdash A$ and $\Gamma \cup \{IA\} \vdash A$. So by the *Syntactic Deduction Theorem,* $\Gamma \vdash \neg A \rightarrow_3 A$ and $\Gamma \vdash IA \rightarrow_3 A$. So by **axiom 6**, $\Gamma \vdash A$, a contradiction. So $\Gamma \cup \{\neg A\}$ or $\Gamma \cup \{IA\}$ is consistent.

d. If Γ is consistent, suppose both $\Gamma \cup \{\neg A\}$ and $\Gamma \cup \{IA\}$ are inconsistent. Then as in (c), $\Gamma \vdash A$. So $\Gamma \cup \{A\}$ is consistent. ■

Lemma 6 The following are equivalent:

a. Γ is complete and consistent

b. There is some $\mathbf{L_3}$-evaluation $\mathbf{e}$ such that

$$\mathbf{e}(A) = 1 \quad \text{iff} \quad A \in \Gamma$$
$$\mathbf{e}(A) = \tfrac{1}{2} \quad \text{iff} \quad IA \in \Gamma$$
$$\mathbf{e}(A) = 0 \quad \text{iff} \quad \neg A \in \Gamma$$

c. There is some $\mathbf{L_3}$-evaluation $\mathbf{e}$ such that $\Gamma = \{A: \mathbf{e} \vDash A\}$

Proof: I will prove that (a) implies (b) and leave to you that (b) implies (a) and the equivalence of parts (c) and (b).

If Γ is complete and consistent, then the $\mathbf{e}$ given in the lemma is well-defined.

It remains to show that **e** is an $\mathbf{L_3}$-evaluation. By Lemma 5, Γ is a theory.

For negation,

$\mathbf{e}(\neg A) = 1$ iff $\neg A \in \Gamma$ iff $\mathbf{e}(A) = 0$

$\mathbf{e}(\neg A) = 0$ iff $\neg(\neg A) \in \Gamma$ iff (by **axioms 7** and **8**) $A \in \Gamma$ iff $\mathbf{e}(A) = 1$

So by process of elimination, $\mathbf{e}(\neg A) = \frac{1}{2}$ iff $\mathbf{e}(A) = \frac{1}{2}$.

For the conditional we have many cases:

If $\mathbf{e}(B) = 1$, then $B \in \Gamma$, so by **axiom 1**, $(A \rightarrow B) \in \Gamma$, so $\mathbf{e}(A \rightarrow B) = 1$.

If $\mathbf{e}(A) = 0$, then $\neg A \in \Gamma$, so by **axiom 3**, $(A \rightarrow B) \in \Gamma$, so $\mathbf{e}(A \rightarrow B) = 1$.

If $\mathbf{e}(A) = 1$ and $\mathbf{e}(B) = 0$, then $A, \neg B \in \Gamma$, so by **axiom 9**, $\neg(A \rightarrow B) \in \Gamma$, and hence $\mathbf{e}(A \rightarrow B) = 0$.

If $\mathbf{e}(A) = 1$ and $\mathbf{e}(B) = \frac{1}{2}$, then $A, IB \in \Gamma$, so by **axiom 10**, $I(A \rightarrow B) \in \Gamma$, and hence $\mathbf{e}(A \rightarrow B) = \frac{1}{2}$.

If $\mathbf{e}(A) = \frac{1}{2}$ and $\mathbf{e}(B) = 0$, then $IA, \neg B \in \Gamma$, so by **axiom 11**, $I(A \rightarrow B) \in \Gamma$, and hence $\mathbf{e}(A \rightarrow B) = \frac{1}{2}$.

If $\mathbf{e}(A) = \frac{1}{2}$ and $\mathbf{e}(B) = \frac{1}{2}$, then $IA, IB \in \Gamma$, so by **axiom 12**, $(A \rightarrow B) \in \Gamma$, and hence $\mathbf{e}(A \rightarrow B) = 1$.

Hence **e** evaluates all wffs correctly. ■

Lemma 7 If $\Gamma \nvdash A$, then there is a complete and consistent theory Σ such that $A \notin \Sigma$ and $\Gamma \subseteq \Sigma$.

Proof: Let $B_0, B_1, \ldots$ be a listing of all wffs. Define:

$$\Sigma_0 = \begin{cases} \Gamma \cup \{\neg A\} & \text{if that is consistent} \\ \Gamma \cup \{IA\} & \text{otherwise} \end{cases}$$

and

$$\Sigma_{n+1} = \begin{cases} \Sigma_n \cup \{IB_n\} & \text{if that is consistent; if not, then} \\ \Sigma_n \cup \{\neg B_n\} & \text{if that is consistent; if not, then} \\ \Sigma_n \cup \{B_n\} & \end{cases}$$

By Lemma 5, each Σ_n is consistent. Hence $\Sigma = \bigcup_n \Sigma_n$ is consistent, and by the choice of Σ_0, $A \notin \Sigma$. By construction Σ is complete. So by Lemma 5, Σ is a theory. ■

It is now routine to prove strong completeness using Lemmas 6 and 7.

Theorem 8 Strong Completeness for $\mathbf{L_3}$ $\Gamma \vdash_{\mathbf{L_3}} A$ iff $\Gamma \vDash_{\mathbf{L_3}} A$

We have given a strongly complete axiomatization of the finite matrix of $\mathbf{L_3}$ by representing the values of the matrix with wffs. I will use the same method again in Chapter IX.E.1, using there collections of wffs to represent the values of the matrix.

This method has been generalized by *Carnielli, 1987 B.*

However, the method is not universally applicable. There is a finite matrix for which there is no finite axiomatization, that is, no axiomatization using only a finite number of schema and rules. See *Wojtylak, 1984.*

Exercises for Section C.1.b

1. a. Show that the axiom system of this section is sound for $\mathbf{L_3}$.
   b. (Open) Which, if any, subscript 3's can be deleted from the axioms to give a complete axiomatization of $\mathbf{L_3}$?
2. Verify the proof of the *Syntactic Deduction Theorem.*
3. a. Which axioms are used to show the equivalence of consistency with Post-consistency?
   b. Without using the completeness of the axiomatization, show that if Γ is a theory, then Γ is complete iff Γ is Post-complete, i.e., for any $A \notin \Gamma$, for any B, $\Gamma \cup \{A\} \vdash B$.
4. Prove that the following are equivalent:
   i. There is some $\mathbf{L_3}$-evaluation **e** such that
   $$\mathbf{e}(A) = 1 \text{ iff } A \in \Gamma, \quad \mathbf{e}(A) = \tfrac{1}{2} \text{ iff } \mathrm{I}A \in \Gamma, \quad \mathbf{e}(A) = 0 \text{ iff } \neg A \in \Gamma$$
   ii. There is some $\mathbf{L_3}$-evaluation **e** such that $\Gamma = \{A : \mathbf{e} \vDash A\}$.
5. Prove a *Syntactic Deduction Theorem for Finite Consequences*:
   $\{B_1, \ldots, B_n\} \vdash A$ iff $\vdash B_1 \rightarrow_3 (B_2 \rightarrow_3 (\ldots (B_n \rightarrow_3 A)) \ldots)$
6. Axiomatize $\mathbf{L_3}$ in $L(\neg, \rightarrow, \wedge, \vee)$.
7. (Open?) Axiomatize the positive fragment of $\mathbf{L_3}$ in $L(\rightarrow, \wedge, \vee)$.
8. (Open) Reduce the axiom system of this section by eliminating schema that are dependent

c. *Wajsberg's axiomatization of* $\mathbf{L_3}$

The following axiomatization of $\mathbf{L_3}$ is due to *Wajsberg, 1931.*

$\mathbf{L_3}$ *in* $L(\neg, \rightarrow)$

$WL_3 1.$ $A \rightarrow (B \rightarrow A)$

$WL_3 2.$ $(A \rightarrow B) \rightarrow ((B \rightarrow C) \rightarrow (A \rightarrow C))$

$WL_3 3.$ $(\neg A \rightarrow \neg B) \rightarrow (B \rightarrow A)$

$WL_3 4.$ $((A \rightarrow \neg A) \rightarrow A) \rightarrow A)$

rule $\dfrac{A,\ A \rightarrow B}{B}$

Write $\Gamma \vdash_{\mathbf{WL_3}} A$ if there is a proof of A from Γ in this system.

Axiom $L_3 4$ is a weak form of the law of excluded middle. It is definitionally equivalent to $(A \rightarrow \neg A) \vee A$, which in **PC** is provably equivalent to $A \vee \neg A$.

Theorem 9 $\vdash_{\mathbf{WL_3}} A$ iff A is an $\mathbf{L_3}$-tautology.

Wajsberg, 1931, gives a constructive proof of Theorem 9. His proof is a complex combinatorial argument reducing tautologies to certain normal forms. It is not possible to prove the strong completeness of the axiomatization by such finitistic means (cf. Chapter II.M.1).

However, using the results of the last section, we can prove that Wajsberg's axiomatization is strongly complete, too. For if $\Gamma \vDash A$, then $\Gamma \vdash_{\mathbf{L_3}} A$ in the first axiomatization, so there are $B_1, \ldots, B_n$ such that $\{B_1, \ldots, B_n\} \vdash_{\mathbf{L_3}} A$. So by the *Syntactic Deduction Theorem* (and Exercise 5 above)

$$\vdash_{\mathbf{L_3}} B_1 \rightarrow_3 (B_2 \rightarrow_3 (\ldots (B_n \rightarrow_3 A)) \ldots)$$

Hence,

$$\vDash_{\mathbf{L_3}} B_1 \rightarrow_3 (B_2 \rightarrow_3 (\ldots (B_n \rightarrow_3 A)) \ldots)$$

And by Theorem 9,

$$\vdash_{\mathbf{WL_3}} B_1 \rightarrow_3 (B_2 \rightarrow_3 (\ldots (B_n \rightarrow_3 A)) \ldots)$$

So $\{B_1, \ldots, B_n\} \vdash_{\mathbf{WL_3}} A$, and so $\Gamma \vdash_{\mathbf{WL_3}} A$.

d. Set-assignment semantics for $\mathbf{L_3}$

We can give set-assignment semantics for $\mathbf{L_3}$ that imitate the 3-valued tables by allowing only three choices for content sets. These simple semantics use the intuitionist truth-tables, which generalize to Łukasiewicz's infinite-valued logic in Section C.2.c. But we can also give richer semantics that allow greater variation in content, and those might be interpreted to apply to paradoxical sentences.

We say that $\langle v, s \rangle$ is an $\mathbf{L_3}$-*model* for $L(\neg, \rightarrow)$ if $\neg$ and $\rightarrow$ are evaluated by the intuitionist tables:

$v(\neg A) = \mathsf{T}$ iff $s(A) = \varnothing$ and $v(A) = \mathsf{F}$

$v(A \rightarrow B) = \mathsf{T}$ iff $s(A) \subseteq s(B)$ and (not both $v(A) = \mathsf{T}$ and $v(B) = \mathsf{F}$)

and s satisfies:

1. $$s(\neg A) = \begin{cases} \overline{s(A)} & \text{if } s(A) = \varnothing \text{ or } s(A) = S \\ s(A) & \text{otherwise} \end{cases}$$

2. $$s(A \rightarrow B) = \begin{cases} S & \text{if } s(A) \subseteq s(B) \\ s(B) & \text{if } s(B) \subset s(A) \text{ and } s(A) = S \\ s(A) & \text{if } s(B) \subset s(A) \text{ and } s(A) \neq S \end{cases}$$

3. $\mathsf{v}(p) = \mathsf{T}$ iff $\mathsf{s}(p) = \mathsf{S}$

4. If both $\varnothing \subset \mathsf{s}(A) \subset \mathsf{S}$ and $\varnothing \subset \mathsf{s}(B) \subset \mathsf{S}$, then $\mathsf{s}(A) = \mathsf{s}(B)$.

Note that condition 3 allows an inductive definition of truth from the assignment of contents and truth-values to the atomic propositions. Condition 4 ensures that there are at most 3 possibilities for content sets: $\varnothing$, S, and some U such that $\varnothing \subset \mathsf{U} \subset \mathsf{S}$.

The proof of the following is straightforward, though lengthy, and I will leave it to you.

Lemma 10

a. $\mathsf{v}(A) = \mathsf{T}$ iff $\mathsf{s}(A) = \mathsf{S}$

b. $\mathsf{v}(A \vee B) = \mathsf{T}$ iff $\mathsf{v}(A) = \mathsf{T}$ or $\mathsf{v}(B) = \mathsf{T}$

c. $\mathsf{s}(A \vee B) = \mathsf{s}(A) \cup \mathsf{s}(B)$

d. $\mathsf{v}(A \wedge B) = \mathsf{T}$ iff $\mathsf{v}(A) = \mathsf{T}$ and $\mathsf{v}(B) = \mathsf{T}$

e. $\mathsf{s}(A \wedge B) = \mathsf{s}(A) \cap \mathsf{s}(B)$

f. $\mathsf{s}(A \rightarrow B) = \mathsf{s}(\neg A) \cup \mathsf{s}(B)$

g. Given an $\mathbf{L_3}$-evaluation $\mathbf{e}$: Wffs $\rightarrow \{0, \frac{1}{2}, 1\}$, define

$\mathsf{v}(p) = \mathsf{T}$ iff $\mathbf{e}(p) = 1$

$\mathsf{s}(A) = \{x : x \in [0,1] \text{ and } x < \mathbf{e}(A)\}$

where $[0,1]$ is the collection of real numbers between 0 and 1. Extend v to all wffs by the intuitionist tables for $\neg$ and $\rightarrow$. Then $<\mathsf{v},\mathsf{s}>$ is an $\mathbf{L_3}$-model and $\mathsf{v}(A) = \mathsf{T}$ iff $\mathbf{e}(A) = 1$.

h. Given an $\mathbf{L_3}$-model $<\mathsf{v},\mathsf{s}>$, define $\mathbf{e}$: Wffs $\rightarrow \{0, \frac{1}{2}, 1\}$ via

$$\mathbf{e}(A) = \begin{cases} 1 & \text{if } \mathsf{s}(A) = \mathsf{S} \\ \frac{1}{2} & \text{if } \varnothing \subset \mathsf{s}(A) \subset \mathsf{S} \\ 0 & \text{if } \mathsf{s}(A) = \varnothing \end{cases}$$

Then $\mathbf{e}$ is an $\mathbf{L_3}$-evaluation and $\mathbf{e}(A) = 1$ iff $\mathsf{v}(A) = \mathsf{T}$.

Note that $\mathbf{L_3}$-models thus evaluate (the defined) $\wedge$, $\vee$ by the intuitionist truth-conditions, too.

From parts (g) and (h) we have that the consequence relation for the set-assignment semantics is the same as for the $\mathbf{L_3}$ matrix, which by Theorem 8 coincides with the syntactic consequence relation.

Theorem 11 Strong Completeness of the Set-Assignment Semantics

$\Gamma \vDash_{\mathbf{L_3}} A$ iff every set-assignment $\mathbf{L_3}$-model that validates Γ also validates A iff $\Gamma \vdash_{\mathbf{L_3}} A$

We can also give strongly complete semantics for $\mathbf{L_3}$ that are not limited to only three content sets in each model. Define $<\upsilon, s>$ to be a *rich* $\mathbf{L_3}$*-model* if:

$\upsilon(\neg A) = T$ iff $s(A) = \varnothing$ and $\upsilon(A) = F$ [as before]

$$\upsilon(A \to B) = T \text{ iff } (s(A) \subseteq s(B) \text{ or both } \varnothing \subset s(A) \subset S \text{ and } \varnothing \subset s(B) \subset S) \text{ and (not both } \upsilon(A) = T \text{ and } \upsilon(B) = F)$$

s satisfies conditions (1) and (3) as before, and:

$$2.\ s(A \to B) = \begin{cases} S & \text{if } s(A) \subseteq s(B) \text{ or (both } \varnothing \subset s(A) \subset S \text{ and } \varnothing \subset s(B) \subset S) \\ s(B) & \text{if } s(B) \subset s(A) \text{ and } s(A) = S \\ s(A) & \text{if } s(B) \subset s(A) \text{ and } s(A) \neq S \end{cases}$$

Parts (a), (g), and (h) of Lemma 10 can be proved for rich models, from which follows the strong completeness of these semantics for $\mathbf{L_3}$. Note that every $\mathbf{L_3}$-model is a rich $\mathbf{L_3}$-model.

In these semantics all content sets other than $\varnothing$ and S play the same role in determining the truth-value of complex wffs. Nonetheless, the intermediate content sets need not all be the same. Dummett says of many-valued logics:

> On one intuitive interpretation of 'true', 'is true' can then be taken to mean 'has a designated value' and 'is false' to mean 'has an undesignated value'. The different individual designated values are then to be taken not as degrees of truth, but, rather, as corresponding to different ways in which a sentence might be true. We cannot determine the truth or falsity of a complex sentence just from the truth or falsity of its constituents; to do this we must know the particular ways in which they are true or false.
>
> *Dummett, 1977*, p. 166

Consider then a proposition A such that $\varnothing \subset s(A) \subset S$. It has the same content as its negation. So on Dummett's reading of content the ways in which A could be true are the same as those in which $\neg A$ could be true. This would be appropriate for paradoxical sentences not all of which need have the same content. Different circumstances might distinguish different paradoxical sentences, though we may choose to ignore the distinctions in calculating truth-values.

Exercises for Section C.1.d

1. Prove Lemma 10.
2. Distinguish between $\mathbf{L_3}$-models and rich $\mathbf{L_3}$-models.
3. Prove Lemma 10 for rich models of $\mathbf{L_3}$.
4. Give set-assignment semantics for $\mathbf{L_3}$ in $L(\neg, \to, \wedge, \vee)$.

2. The logics $\mathbf{L}_n$ and $\mathbf{L}_\aleph$

a. Generalizing the 3-valued tables

Łukasiewicz and Tarski, 1930, generalized the 3-valued matrix for $\mathbf{L_3}$ by allowing evaluations to take any value in [0,1], the real numbers between 0 and 1.

An **L**-*evaluation* is a map $\mathbf{e}: \text{PV} \to [0,1]$ that is extended to all wffs of $L(\neg, \to, p_0, p_1, \ldots)$ by the following tables:

$$\mathbf{e}(\neg A) = 1 - \mathbf{e}(A)$$

$$\mathbf{e}(A \to B) = \begin{cases} 1 & \text{if } \mathbf{e}(A) \leq \mathbf{e}(B) \\ (1 - \mathbf{e}(A)) + \mathbf{e}(B) & \text{if } \mathbf{e}(B) < \mathbf{e}(A) \end{cases}$$

As before we define:

$$A \vee B \equiv_{\text{Def}} (A \to B) \to B$$

$$A \wedge B \equiv_{\text{Def}} \neg(\neg A \vee \neg B)$$

$$A \leftrightarrow B \equiv_{\text{Def}} (A \to B) \wedge (B \to A)$$

These have the following tables:

$$\mathbf{e}(A \vee B) = \max(\mathbf{e}(A), \mathbf{e}(B))$$

$$\mathbf{e}(A \wedge B) = \min(\mathbf{e}(A), \mathbf{e}(B))$$

$$\mathbf{e}(A \leftrightarrow B) = \begin{cases} 1 & \text{if } \mathbf{e}(A) = \mathbf{e}(B) \\ (1 - \mathbf{e}(A)) + \mathbf{e}(B) & \text{if } \mathbf{e}(A) > \mathbf{e}(B) \\ (1 - \mathbf{e}(B)) + \mathbf{e}(A) & \text{if } \mathbf{e}(B) > \mathbf{e}(A) \end{cases}$$

Note that the function for disjunction is associative: the way that disjuncts are associated does not affect the evaluation of an alternation.

We define for $n \geq 2$:

$$\mathbf{L}_n = \{ A : \mathbf{e}(A) = 1 \text{ for every } \mathbf{L}\text{-evaluation } \mathbf{e}: \text{PV} \to \{ \tfrac{m}{n-1} : 0 \leq m \leq n-1 \} \}$$

$$\mathbf{L}_{\aleph_0} = \{ A : \mathbf{e}(A) = 1 \text{ for every } \mathbf{L}\text{-evaluation } \mathbf{e} \text{ that takes rational values in } [0,1] \}$$

$$\mathbf{L}_\aleph = \{ A : \mathbf{e}(A) = 1 \text{ for every } \mathbf{L}\text{-evaluation } \mathbf{e} \}$$

Note that when we restrict the values that **e** may take on PV in these definitions, then the extension of **e** to all wffs obeys the same restriction.

Theorem 12 *a.* $\mathbf{L}_2 = \mathbf{PC}$

b. $\mathbf{L}_n \neq \mathbf{L}_{n+1}$

c. $\mathbf{L}_n \supset \mathbf{L}_{\aleph_0}$

d. $\mathbf{L}_{\aleph_0} = \mathbf{L}_\aleph$

e. $\mathbf{L}_{m+1} \subseteq \mathbf{L}_{n+1}$ iff n divides m

Proof: Part (a) follows by reading T for 1, F for 0.

For parts (b) and (c), define the sequence of wffs:

$$D_n \equiv_{\text{Def}} \bigvee_{1 \le i \ne k < n+1} (p_i \leftrightarrow p_k)$$

That is, D_n is the disjunction of the indexed wffs associating to the left. For example:

$$D_2 \equiv_{\text{Def}} ((p_1 \leftrightarrow p_2) \vee (p_1 \leftrightarrow p_3)) \vee (p_2 \leftrightarrow p_3)$$

Then, as I will let you prove, for $m \ge n$, $D_m \in \mathbf{L}_n$; for $m < n$, $D_m \notin \mathbf{L}_n$; and for all n, $D_n \notin \mathbf{L}_{\aleph_0}$. But $\mathbf{L}_n \supseteq \mathbf{L}_{\aleph_0}$, since the matrix for $\mathbf{L}_n$ is a submatrix of the one for $\mathbf{L}_{\aleph_0}$.

Part (d) follows from Theorem 13 below.

Part (e) is a combinatorial argument that I outline in Exercise 9 below. ■

In developing these systems Łukasiewicz said,

> It was clear to me from the outset that among all the many-valued systems only two can claim any philosophical significance: the three-valued and the infinite-valued ones. For if values other than '0' and '1' are interpreted as 'the possible', only two cases can reasonably be distinguished: either one assumes that there are no variations in degree of the possible and consequently one arrives at the three-valued system; or one assumes the opposite, in which case it would be most natural to suppose (as in the theory of probabilities) that there are infinitely many degrees of possibility, which leads to the infinite-valued propositional calculus. I believe the latter system is preferable to all others.
>
> *Łukasiewicz, 1930,* p. 173

Later, *Łukasiewicz, 1953,* argued that $\mathbf{L}_4$ could be interpreted as a reconstruction of Aristotle's modal notions. But I will look only at the infinite-valued logic here. See *Wójcicki, 1988,* Chapter 4.3, for more information about these logics.

b. An axiom system for $\mathbf{L}_\aleph$

$\mathbf{L}_\aleph$ *in* $L(\neg, \rightarrow)$

$WL_3 1.$	$A \rightarrow (B \rightarrow A)$	as for $\mathbf{WL_3}$
$WL_3 2.$	$(A \rightarrow B) \rightarrow ((B \rightarrow C) \rightarrow (A \rightarrow C))$	
$WL_3 3.$	$(\neg A \rightarrow \neg B) \rightarrow (B \rightarrow A)$	
$WL_\aleph.$	$(A \rightarrow B) \vee (B \rightarrow A)$	

rule $\dfrac{A,\ A \rightarrow B}{B}$

I denote the consequence relation of this system $\vdash_{\mathbf{L}_\aleph}$.

Theorem 13 $\vdash_{\mathbf{L}_{\aleph}} A$ iff $\vDash_{\mathbf{L}_{\aleph_0}} A$ iff $\vDash_{\mathbf{L}_{\aleph}} A$

Turquette, 1959, gives the history and references for the proof of the first part of this theorem. The second equivalence follows because every axiom and hence every consequence of the system is an $\mathbf{L}_{\aleph}$-tautology, so $\mathbf{L}_{\aleph_0} \subseteq \mathbf{L}_{\aleph}$; and $\mathbf{L}_{\aleph} \subseteq \mathbf{L}_{\aleph_0}$, since the matrix for $\mathbf{L}_{\aleph_0}$ is part of the matrix for $\mathbf{L}_{\aleph}$.

c. Set-assignment semantics for $\mathbf{L}_{\aleph}$

It might seem obvious that set-assignment semantics for $\mathbf{L}_{\aleph}$ could be obtained by modifying those for $\mathbf{L}_3$ to require the content sets to be linearly ordered under inclusion. But those models do not validate $\mathrm{WL}_3 3$. It would seem that we need to explicitly postulate some measure or topological structure on the collection of content sets to mock the operations of + and − on [0,1].

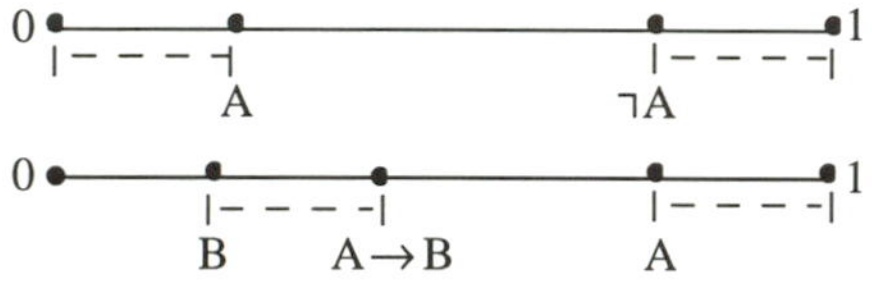

0 |- - - -| A ¬A |- - - -| 1

0 B |- - - -| A→B A |- - - -| 1

But the solution is simpler: just postulate enough structure on the collection $\{s(A): A \text{ is a wff}\}$ to validate $\mathbf{L}_{\aleph}$.

We say that $\langle v, s \rangle$ is an $\mathbf{L}_{\aleph}$*-model* for $L(\neg, \rightarrow)$ if $\neg$ and $\rightarrow$ are evaluated by the intuitionist truth-conditions (as for $\mathbf{L}_3$):

$v(\neg A) = T$ iff $s(A) = \varnothing$ and $v(A) = F$

$v(A \rightarrow B) = T$ iff $s(A) \subseteq s(B)$ and (not both $v(A) = T$ and $v(B) = F$)

and s satisfies:

L1. $s(A \rightarrow B) = S$ iff $s(A) \subseteq s(B)$

L2. $s(\neg A) = S$ iff $s(A) = \varnothing$

L3. $s(B) \subseteq s(A \rightarrow B)$

L4. $s(A \rightarrow B) \subseteq s((B \rightarrow C) \rightarrow (A \rightarrow C))$

L5. If $s(A) \subseteq s(B)$, then $s(B \rightarrow C) \subseteq s(A \rightarrow C)$.

L6. $s(\neg A \rightarrow \neg B) \subseteq s(B \rightarrow A)$

L7. $s(\neg B) \subseteq s(\neg A)$ iff $s(A) \subseteq s(B)$

L8. $s(A) \subseteq s(B)$ or $s(B) \subseteq s(A)$

L9. $v(p) = T$ iff $s(p) = S$

and for the defined connectives:

L10. $s(A \vee B) = s(A) \cup s(B)$

L11. $s(A \wedge B) = s(A) \cap s(B)$

I will leave to you to prove the following Lemma.

Lemma 14 For every $\mathbf{L}_{\aleph}$-model $<\upsilon, \mathsf{s}>$:

a. $\upsilon(A) = \mathsf{T}$ iff $\mathsf{s}(A) = \mathsf{S}$

b. $\upsilon(A \vee B) = \mathsf{T}$ iff $\upsilon(A) = \mathsf{T}$ or $\upsilon(B) = \mathsf{T}$

c. $\upsilon(A \wedge B) = \mathsf{T}$ iff $\upsilon(A) = \mathsf{T}$ and $\upsilon(B) = \mathsf{T}$

Thus $\mathbf{L}_{\aleph}$-models use the intuitionist truth-tables for all four connectives.

The algebra of sets $\{\mathsf{s}(A)\colon A \text{ is a wff}\}$ inherits enough structure to prove that the semantic consequence relation of these set-assignment models is the same as the syntactic and semantic consequence relation of $\mathbf{L}_{\aleph}$.

Theorem 15 $\vdash_{\mathbf{L}_{\aleph}} A$ iff for every $\mathbf{L}_{\aleph}$-model $<\upsilon, \mathsf{s}>$, $\upsilon(A) = \mathsf{T}$

Proof: You can check that the semantics are sound.

If $\nvdash_{\mathbf{L}_{\aleph}} A$, then there is some $\mathbf{L}$-evaluation e such that $\mathsf{e}(A) \neq 1$. Define $<\upsilon, \mathsf{s}>$ via $\mathsf{S} = (0,1]$, $\mathsf{s}(A) = (0, \mathsf{e}(A)]$, and $\upsilon(p) = \mathsf{T}$ iff $\mathsf{e}(p) = 1$, where $(0,0]$ is understood to mean $\varnothing$. Then $<\upsilon, \mathsf{s}>$ is an $\mathbf{L}_{\aleph}$-model, and by Lemma 14.a, $\upsilon(A) = \mathsf{F}$. ■

Exercises for Section C.2

1. Show that for every $\mathbf{L}$-evaluation e, $\mathsf{e}(A \vee B) = \max(\mathsf{e}(A), \mathsf{e}(B))$, and $\mathsf{e}(A \wedge B) = \min(\mathsf{e}(A), \mathsf{e}(B))$.
2. Show that $\mathsf{e}(A \leftrightarrow B) = 1$ iff $\mathsf{e}(A) = \mathsf{e}(B)$.
3. Prove (using the definitions in the proof of Lemma 12):
   a. For $m \geq n$, $D_m \in \mathbf{L}_n$
   b. For $m < n$, $D_m \notin \mathbf{L}_n$
   c. For all n, $D_n \notin \mathbf{L}_{\aleph_0}$
4. Prove $\mathbf{L}_{\aleph} \subseteq \mathbf{L}_{\aleph_0}$.
5. Show that the axiom system $\mathbf{L}_{\aleph}$ is sound.
6. Show that the set-assignment semantics for $\mathbf{L}_3$ with condition (4) replaced by the requirement that the content sets be linearly ordered under inclusion does not validate $WL_3 3$.
7. (Open) Reduce the list of conditions that s must satisfy for $\mathbf{L}_{\aleph}$-models.
8. Prove Lemma 14.
9. Fill in the details of the following outline that $\mathbf{L}_{m+1} \subseteq \mathbf{L}_{n+1}$ iff n divides m.
   a. If $m < n$, use Exercise 3.b to show $\mathbf{L}_{m+1} \nsubseteq \mathbf{L}_{n+1}$.

b. If $m = qn$ the $\mathbf{L}_{n+1}$ - matrix is a submatrix of $\mathbf{L}_{m+1}$, so $\mathbf{L}_{m+1} \subseteq \mathbf{L}_{n+1}$.

c. If $n < m$, but n does not divide m, consider the formulas H_k for $k \geq 1$, defined by: $H_1(A) = \neg A$, $H_{k+1}(A) = A \rightarrow H_k(A)$. In any $\mathbf{L}_s$, $e(H_k(A)) = (k+1)(1 - e(A))$ if this is < 1, otherwise 1. In $\mathbf{L}_{m+1}$, $e(H_{n-2}(p) \leftrightarrow p) = 1$ iff $\frac{n-1}{n} = e(p)$, so for any e, $e(H_{n-2}(p) \leftrightarrow p) \neq 1$, since n does not divide m. So in $\mathbf{L}_{m+1}$, for any e, $e(H_m(H_{n-2}(p) \leftrightarrow p))$ $= 1$. But in $\mathbf{L}_{n+1}$, $e(H_m(H_{n-2}(p) \leftrightarrow p)) = 0$. So $H_m(H_{n-2}(p) \leftrightarrow p)$ is valid in $\mathbf{L}_{m+1}$ but not in $\mathbf{L}_{n+1}$.

D. Kleene's 3-Valued Logic

1. The truth-tables

Kleene, 1952, pp. 332–340, proposed a 3-valued logic as a way to reason with arithmetical propositions whose truth-value we either do not or cannot know. Three "values" are postulated for propositions: T for 'true', F for 'false', and U for 'undefined' or 'unknown'. The value U is not to be considered a third truth-value. Rather, in accord with Kleene's platonist view of propositions, it marks our ignorance of the actual truth-value of the proposition.

> Here 'unknown' is a category into which we can regard any proposition as falling whose value we either do not know or choose for the moment to disregard; and it does not then exclude the other two possibilities 'true' and 'false'. . . .
>
> The strong 3-valued logic can be applied to completely defined predicates $Q(x)$ and $R(x)$ from which composite predicates are formed using $\neg$, $\vee$, $\wedge$, $\rightarrow$, $\leftrightarrow$ in the usual 2-valued meanings, thus. Suppose that there are fixed algorithms which decide the truth or falsity of $Q(x)$ and of $R(x)$, each on a subset of the natural numbers (as occurs, e.g., after completing the definition of any two partial recursive predicates classically). Let T, F, U mean 'decidable by the algorithms (i.e., by use of only such information about $Q(x)$ and $R(x)$ as can be obtained by the algorithms) to be true', 'decidable by the algorithms to be false', 'undecidable by the algorithms whether true or false'. [Or] assume a fixed state of knowledge about $Q(x)$ and $R(x)$ (as occurs, e.g., after pursuing algorithms for each of them up to a given stage). Let T, F, U mean 'known to be true', 'known to be false', 'unknown whether true or false'.
>
> *Kleene, 1952,* pp. 335–336

Formally, we begin by taking as our language $L(\neg, \rightarrow, \wedge, \vee, p_0, p_1, \ldots)$. Kleene gives the following tables for what he calls the *strong connectives* :

A ∧ B		B: T	U	F
A	T	T	U	F
	U	U	U	F
	F	F	F	F

A ∨ B		B: T	U	F
A	T	T	T	T
	U	T	U	U
	F	T	U	F

A → B		B: T	U	F
A	T	T	U	F
	U	T	U	U
	F	T	T	T

A	¬A
T	F
U	U
F	T

A $\mathbf{K_3}$-*evaluation* is a map $\mathsf{e}\colon \mathrm{PV} \to \{\mathsf{T}, \mathsf{F}, \mathsf{U}\}$ that is extended to all wffs by these tables. The *sole designated value* is T.

If we read 1 for T, 0 for F, and $\frac{1}{2}$ for U, these tables agree with those for Łukasiewicz's 3-valued logic $\mathbf{L_3}$ with one exception: if $\mathsf{e}(A) = \mathsf{e}(B) = \mathsf{U}$, then Łukasiewicz assigns $\mathsf{e}(A \to B) = \mathsf{T}$, whereas Kleene assigns value U. So with Kleene's tables we assign U to any complex proposition built from propositions all of which take value U. *Kripke, 1975,* finds this apt to deal with paradoxical sentences such as 'What I am now saying is false', and uses $\mathbf{K_3}$-evaluations to give a theory of truth for a first-order language that contains its own truth predicate (see *Epstein, 1992,* for a description and critique of Kripke's analysis). But now there are no tautologies.

Lemma 16 ***a.*** There is no wff A that takes value T for all $\mathbf{K_3}$-evaluations.

b. If both T and U are taken as designated values for these tables, then the set of tautologies is **PC**. That is, $\vDash_{\mathbf{PC}} A$ iff for every $\mathbf{K_3}$-evaluation e, $\mathsf{e}(A) = \mathsf{T}$ or $\mathsf{e}(A) = \mathsf{U}$.

c. The consequence relation using both T and U as designated values does not coincide with $\vDash_{\mathbf{PC}}$.

Proof: a. If for all $p \in \mathrm{PV}(A)$, $\mathsf{e}(p) = \mathsf{U}$, then $\mathsf{e}(A) = \mathsf{U}$.

b. I'll show that $\nvDash_{\mathbf{PC}} A$ iff there is some $\mathbf{K_3}$-evaluation e such that $\mathsf{e}(A) = \mathsf{F}$.

Every 2-valued **PC**-model is a $\mathbf{K_3}$-evaluation, so if $\nvDash_{\mathbf{PC}} A$, then there is a $\mathbf{K_3}$-evaluation e such that $\mathsf{e}(A) = \mathsf{F}$. If there is some $\mathbf{K_3}$-evaluation e such that $\mathsf{e}(A) = \mathsf{F}$, then define a **PC**-model υ by $\upsilon(p) = \mathsf{T}$ iff $\mathsf{e}(A) = \mathsf{T}$ or $\mathsf{e}(A) = \mathsf{U}$. You can then prove by induction on the length of a wff that if $\mathsf{e}(B) = \mathsf{T}$ then $\upsilon(B) = \mathsf{T}$, and if $\mathsf{e}(B) = \mathsf{F}$ then $\upsilon(B) = \mathsf{F}$. So $\upsilon(A) = \mathsf{F}$ and $\nvDash_{\mathbf{PC}} A$.

c. If both T and U are designated, then $\{A, A \to B\} \nvDash B$, for we may have $\mathsf{e}(A) = \mathsf{U}$ and $\mathsf{e}(B) = \mathsf{F}$. ■

Since there are no tautologies for this logic, we must understand the matrix as a semantic presentation of a logic in terms of a consequence relation only:

$\Gamma \vDash_{\mathbf{K_3}} A$ iff for every $\mathbf{K_3}$-evaluation e, if $e \vDash \Gamma$ then $e(A) = \mathsf{T}$

This relation is not empty: for instance, $A \rightarrow B \vDash_{\mathbf{K_3}} \neg B \rightarrow \neg A$, and $\neg(A \wedge \neg A) \vDash_{\mathbf{K_3}} A \vee \neg A$.

The absence of tautologies makes the usual Hilbert-style syntactic system inappropriate. *Cleave, 1974,* has given a natural deduction style definition of syntactic consequence that coincides with $\vDash_{\mathbf{K_3}}$.

2. Set-assignment semantics

The goal here is to give set-assignment semantics whose consequence relation coincides with $\vDash_{\mathbf{K_3}}$.

A pair $<v,s>$ is a $\mathbf{K_3}$-*model* for $L(\neg, \rightarrow, \wedge, \vee, p_0, p_1, \ldots)$ if the truth-tables for $\wedge$ and $\vee$ are classical, and

$$v(A \rightarrow B) = \mathsf{T} \text{ iff } v(\neg A) = \mathsf{T} \text{ or } v(B) = \mathsf{T}$$

$$v(\neg A) = \mathsf{T} \text{ iff } v(A) = \mathsf{F} \text{ and } s(A) = \varnothing$$

and s satisfies:

K1. $s(A) \subseteq s(B)$ or $s(B) \subseteq s(A)$

K2. $s(A \rightarrow B) = s(\neg A) \cup s(B)$

K3. $s(\neg A) = \begin{cases} s(A) & \text{if } \varnothing \subset s(A) \subset \mathsf{S} \\ \overline{s(A)} & \text{otherwise} \end{cases}$

K4. $s(A \wedge B) = s(A) \cap s(B)$

K5. $s(A \vee B) = s(A) \cup s(B)$

K6. $v(p) = \mathsf{T}$ iff $s(p) = \mathsf{S}$

The difference between these semantics and the semantics for **PC** lies solely in the table and set-assignments for negation. Note that we do not require the content sets to be chosen from $\varnothing \subset \mathsf{U} \subset \mathsf{S}$ as we originally did for $\mathbf{L_3}$.

Theorem 17

a. In every $\mathbf{K_3}$-model $<v,s>$, $v(A) = \mathsf{T}$ iff $s(A) = \mathsf{S}$

b. Given a $\mathbf{K_3}$-evaluation e, if we define $<v,s>$ by:

$$v(p) = \mathsf{T} \text{ iff } e(p) = \mathsf{T} \qquad \mathsf{S} = \{1,2\}$$

$$s(A) = \begin{cases} \mathsf{S} & \text{if } e(A) = \mathsf{T} \\ \{1\} & \text{if } e(A) = \mathsf{U} \\ \varnothing & \text{if } e(A) = \mathsf{F} \end{cases}$$

then $<v,s>$ is a $\mathbf{K_3}$-model and $v(A) = \mathsf{T}$ iff $e(A) = \mathsf{T}$.

c. Given a $\mathbf{K_3}$-model $<\upsilon,s>$, if we set

$$e(A) = \begin{cases} T & \text{if } s(A) = S \\ U & \text{if } \varnothing \subset s(A) \subset S \\ F & \text{if } s(A) = \varnothing \end{cases}$$

then **e** is a $\mathbf{K_3}$-evaluation, and $e(A) = T$ iff $\upsilon(A) = T$.

d. $\Gamma \vDash_{\mathbf{K_3}} A$ iff for every $\mathbf{K_3}$-model $<\upsilon,s>$,
if for every $B \in \Gamma$, $\upsilon(B) = T$, then $\upsilon(A) = T$

The proofs are straightforward though long, and I will leave them to you.

Our $\mathbf{K_3}$-models reflect that what is important about a wff is whether it takes an extreme value (T, F; or $s(A) = \varnothing, S$). We don't even need to require that the content sets be linearly ordered by inclusion, for we can prove Theorem 19 if we replace K1 and K2 by:

$$\text{K7.} \quad s(A \to B) = \begin{cases} S & \text{if } s(\neg A) = S \text{ or } s(B) = S \\ \varnothing & \text{if } s(A) = S \text{ and } s(B) = \varnothing \\ s(B) & \text{otherwise} \end{cases}$$

Exercises for Section D

1. a. Prove:

   $A \to B \vDash_{\mathbf{K_3}} \neg B \to \neg A$

   $\neg(A \wedge \neg A) \vDash_{\mathbf{K_3}} A \vee \neg A$

   b. For every **PC**-tautology of the form $A \to B$ in Chapter II.J.8, pp. 60–62, determine whether $A \vDash_{\mathbf{K_3}} B$.

2. Why is a Hilbert-style syntactic system inappropriate for $\mathbf{K_3}$?

3. Prove Theorem 17.

E. Logics Having No Finite-Valued Semantics

1. General criteria

The following theorem is due to *Gödel, 1932.*

Theorem 18 There is no finite-valued matrix that characterizes either **Int** or **J**.

Proof: For $n \geq 2$ set:

$$D_n \equiv_{\text{Def}} \bigvee_{1 \leq i \neq k < n+1} (p_i \leftrightarrow p_k)$$

That is, D_n is the disjunction of the indexed wffs associating to the left.
For example:

$D_2 \equiv_{Def} ((p_1 \leftrightarrow p_2) \vee (p_1 \leftrightarrow p_3)) \vee (p_2 \leftrightarrow p_3)$

For each n, **Int** $\nvdash D_n$, as you can check using the Kripke semantics for **Int** from Chapter VII.B.2. Now assume to the contrary that M is an n-valued matrix characteristic for **Int**. I will show that $M \vDash D_n$, which is a contradiction.

For any evaluation **e** with respect to M, there must be some i, k with $1 \leq i < k \leq n+1$, such that $e(p_i) = e(p_k)$. Choose such a pair, and then $e \vDash D_n$ iff $e \vDash D_n(p_i/p_k)$, where the latter formula is D_n with p_k substituted for p_i throughout. But then one of the disjunctions in the alternation $D_n(p_i/p_k)$ is $(p_k \leftrightarrow p_k)$, so that **Int** $\vdash D_n(p_i/p_k)$. As M is characteristic for **Int** we must have that $e \vDash D_n(p_i/p_k)$, hence $e \vDash D_n$. Since **e** was an arbitrary evaluation, $M \vDash D_n$.

The same proof applies to **J**, noting that if **Int** $\nvdash D_n$, then **J** $\nvdash D_n$ by Corollary VII.30. ■

A straightforward generalization of the proof of Theorem 18 establishes the following, as suggested by Walter Carnielli.

Theorem 19 Let **L** be a logic presented semantically. Suppose there is a sequence of formulas E_n for $n \geq 2$ in the language of **L** such that:

a. The propositional variables appearing in E_n are exactly $p_1, \ldots p_n, p_{n+1}$

b. For all n, **L** $\nvDash E_n$

c. For all $i, k \leq n+1$, if $i \neq k$ then **L** $\vDash E_n(p_i/p_k)$

Then there is no finite-valued matrix that is characteristic for **L**.

Theorem 19 applies also to logics that are presented syntactically by replacing $\vDash$ with $\vdash$ everywhere.

Corollary 20 None of the following logics has finite-valued semantics:

R, **S**, **D**, **Dual D**, **Eq** (Chapters III and V)

The modal logics **K**, **T**, **B**, **QT**, **S4**, **S5**, **S4Grz**, **MSI**, **G**, **G*** (Chapter VI)

Proof: Take E_n in Theorem 19 to be D_n from the proof of Theorem 20, using $\vee$ defined from $\neg$ and $\wedge$. ■

These logics, then, recognize infinitely many possibilities for propositions.

2. Infinite-valued semantics for the modal logic S5

Though none of the classical modal logics of Chapter VI can be characterized by a finite-valued matrix, there is a simple and intuitive characterization of **S5** as an infinite-valued logic, as shown by *Carnap, 1954.*

Let's assume we're given a Kripke model of **S5** that contains a countable number of possible worlds. Each proposition can be identified with the worlds in which it is true. So we can represent a proposition as an infinite sequence of 0's and 1's:

$$\mathbf{e}(A) = (a_1, \ldots, a_n, \ldots)$$

where for each i, a_i is 1 if A is true in the ith world, 0 if A is false in the ith world. The evaluation of the connectives is:

$$\begin{aligned}\mathbf{e}(\neg A) &= (1, 1, 1, \ldots, 1, \ldots) - \mathbf{e}(A)\\ &= \text{the result of interchanging 0's and 1's in } \mathbf{e}(A)\end{aligned}$$

$$\begin{aligned}\mathbf{e}(A \wedge B) &= \mathbf{e}(A) \times \mathbf{e}(B) \quad \text{(the coordinatewise product)}\\ &= \text{the result of taking } \min(\mathbf{e}(A), \mathbf{e}(B)) \text{ at each coordinate}\end{aligned}$$

Recall that the necessity of **S5** models is logical necessity: every possible world is conceivable in every other. So the evaluation of the necessity operator is:

$$\mathbf{e}(\Box A) = \begin{cases} \text{all 1's} & \text{if } \mathbf{e}(A) \text{ is all 1's} \\ \text{all 0's} & \text{otherwise} \end{cases}$$

Recalling the definition $A \rightarrow B \equiv_{\text{Def}} \Box(\neg(A \wedge \neg B))$, the evaluation of conditionals is:

$$\mathbf{e}(A \rightarrow B) = \begin{cases} \text{all 1's if there is no index for which} \\ \qquad \text{the entry in } \mathbf{e}(A) \text{ is 1 and in } \mathbf{e}(B) \text{ is 0} \\ \text{all 0's otherwise} \end{cases}$$

The only designated value is the sequence consisting entirely of 1's.

We can use Theorem VI.25 to establish that this class of evaluations characterizes **S5** and that the semantic consequence relation of this matrix coincides with the consequence relation without necessitation for **S5** (Theorem VI.27). We only need to note that for any A, any finite Kripke model in which A is not valid can be extended to an infinite one in which A is not valid, which I'll leave to you.

By Theorem VI.25, **S5** is also characterized by the class of finite equivalence Kripke models, so we could use only arbitrarily long finite sequences instead of infinite ones.

F. The Systems $\mathbf{G}_n$ and $\mathbf{G}_{\aleph}$

In the proof of Theorem 18 we used Kripke semantics to show $\mathbf{Int} \nvdash D_n$. *Gödel, 1932,* devised a sequence of finite-valued matrices each of which validates **Int** but in which the D_n's successively failed. I'll present those here.

A **G**-*evaluation* is a map $\mathbf{e}: \mathrm{PV} \to [0,1]$ that is extended to all wffs of $L(\neg, \to, \wedge, \vee, p_0, p_1, \ldots)$ by the following tables:

$$\mathbf{e}(\neg A) = \begin{cases} 1 & \text{if } \mathbf{e}(A) = 0 \\ 0 & \text{if } \mathbf{e}(A) \neq 0 \end{cases}$$

$$\mathbf{e}(A \to B) = \begin{cases} 1 & \text{if } \mathbf{e}(A) \leq \mathbf{e}(B) \\ \mathbf{e}(B) & \text{otherwise} \end{cases}$$

$$\mathbf{e}(A \wedge B) = \min(\mathbf{e}(A), \mathbf{e}(B))$$

$$\mathbf{e}(A \vee B) = \max(\mathbf{e}(A), \mathbf{e}(B))$$

Defining $A \leftrightarrow B \equiv_{\mathrm{Def}} (A \to B) \wedge (B \to A)$, we have:

$$\mathbf{e}(A \leftrightarrow B) = \begin{cases} 1 & \text{if } \mathbf{e}(A) = \mathbf{e}(B) \\ \min(\mathbf{e}(A), \mathbf{e}(B)) & \text{otherwise} \end{cases}$$

For $n \geq 2$ define:

$\mathbf{G}_n = \{ A : \mathbf{e}(A) = 1$ for every **G**-evaluation $\mathbf{e}: \mathrm{PV} \to \{ \frac{m}{n-1} : 0 \leq m \leq n-1 \} \}$

$\mathbf{G}_{\aleph_0} = \{ A : \mathbf{e}(A) = 1$ for every **G**-evaluation **e** that takes rational values in $[0,1] \}$

$\mathbf{G}_{\aleph} = \{ A : \mathbf{e}(A) = 1$ for every **G**-evaluation **e** $\}$

The tables for $\mathbf{G}_3$ are:

A	¬A
1	0
$\frac{1}{2}$	0
0	1

A → B		B: 1	$\frac{1}{2}$	0
A	1	1	$\frac{1}{2}$	0
	$\frac{1}{2}$	1	1	0
	0	1	1	1

and (as for $\mathbf{L}_3$)

A ∧ B		B: 1	$\frac{1}{2}$	0
A	1	1	$\frac{1}{2}$	0
	$\frac{1}{2}$	$\frac{1}{2}$	$\frac{1}{2}$	0
	0	0	0	0

A ∨ B		B: 1	$\frac{1}{2}$	0
A	1	1	1	1
	$\frac{1}{2}$	1	$\frac{1}{2}$	$\frac{1}{2}$
	0	1	$\frac{1}{2}$	0

These were originally devised by *Heyting, 1930*, p. 56, to show that $\mathbf{Int} \nvdash \neg\neg A \to A$.

The proof of the following is as for Theorem 12. In particular, part (e) follows from Theorem 23 below.

Theorem 21 *a.* $\mathbf{G}_2 = \mathbf{PC}$

b. $\mathbf{G}_n \supseteq \mathbf{G}_{\aleph_0} \supseteq \mathbf{G}_\aleph$

c. $\mathbf{G}_n \neq \mathbf{G}_{n+1}$

d. $\mathbf{G}_n \neq \mathbf{G}_{\aleph_0}$

e. $\mathbf{G}_{\aleph_0} = \mathbf{G}_\aleph$

Theorem 22 $\mathbf{G}_n \supset \mathbf{G}_\aleph \supset \mathbf{Int} \supset \mathbf{J}$

Proof: Theorem 21 says that $\mathbf{G}_n \supset \mathbf{G}_\aleph$. We have that $\mathbf{Int} \supset \mathbf{J}$ because the axioms of the latter are among the former, whereas $\mathbf{J} \nvdash \neg A \rightarrow (A \rightarrow B)$, which is an axiom of **Int** (see p. 306). That $\mathbf{G}_\aleph \supseteq \mathbf{Int}$ follows by checking that every axiom of **Int** receives value 1 under every **G**-evaluation, and if $\mathsf{e}(A) = 1$ and $\mathsf{e}(A \rightarrow B) = 1$, then $\mathsf{e}(B) = 1$, and if $\mathsf{e}(A) = 1$ and $\mathsf{e}(B) = 1$, then $\mathsf{e}(A \wedge B) = 1$.

Finally, $\mathbf{G}_\aleph \neq \mathbf{Int}$ because $[(A \rightarrow B) \vee (B \rightarrow A)] \in \mathbf{G}_\aleph$, but that wff is not an **Int**-tautology, as you can check using the Kripke semantics for **Int**. ■

Dummett, 1959, has shown how to characterize $\mathbf{G}_{\aleph_0}$ syntactically:

$\mathbf{G}_{\aleph_0}$

in $L(\neg, \rightarrow, \wedge, \vee)$

All the schema and rules of **Int** (p. 276 or p. 288) plus:

$(A \rightarrow B) \vee (B \rightarrow A)$

Theorem 23 $\vdash_{\mathbf{G}_{\aleph_0}} A$ iff for every G-evaluation e that takes only rational values, $\mathsf{e}(A) = 1$

This axiom system is also complete for $\mathbf{G}_\aleph$, as every theorem A of this system is a $\mathbf{G}_\aleph$-tautology and $\mathbf{G}_\aleph \subseteq \mathbf{G}_{\aleph_0}$.

There are two ways in which we can give set-assignment semantics for $\mathbf{G}_\aleph$. In the first we say that $\langle \upsilon, \mathsf{s} \rangle$ is a $\mathbf{G}_\aleph$-*model* if υ uses the intuitionist truth-conditions, that is $\wedge$ and $\vee$ are classical, and:

$\upsilon(A \rightarrow B) = \mathsf{T}$ iff $\mathsf{s}(A) \subseteq \mathsf{s}(B)$ and (not both $\upsilon(A) = \mathsf{T}$ and $\upsilon(B) = \mathsf{F}$)

$\upsilon(\neg A) = \mathsf{T}$ iff $\upsilon(A) = \mathsf{F}$ and $\mathsf{s}(A) = \varnothing$

and s satisfies:

G1. $\mathsf{s}(\neg A) = \begin{cases} \mathsf{S} & \text{if } \mathsf{s}(A) = \varnothing \\ \varnothing & \text{otherwise} \end{cases}$

G2. $\mathsf{s}(A \rightarrow B) = \begin{cases} \mathsf{S} & \text{if } \mathsf{s}(A) \subseteq \mathsf{s}(B) \\ \mathsf{s}(A) \cap \mathsf{s}(B) & \text{otherwise} \end{cases}$

G3. $s(A \wedge B) = s(A) \cap s(B)$

G4. $s(A \vee B) = s(A) \cup s(B)$

G5. $v(p) = T$ iff $s(p) = S$

G6. $s(A) \subseteq s(B)$ or $s(B) \subseteq s(A)$

The proof that the consequence relation for these semantics is the same as for the matrix for $\mathbf{G}_{\aleph}$ follows as in Lemma 10.

Suppose we replace G6 by:

There is some U such that $s : \text{Wffs} \to \{\varnothing, U, S\}$

Then the consequence relation of that subclass of models coincides with the consequence relation for the matrix of $\mathbf{G}_3$.

Alternatively, we can add G6 to the list of conditions for an **Int**-model $\langle v, s \rangle$ (see Chapter VII.E.1, pp. 296–297). We can then establish that the resulting models are complete for $\mathbf{G}_{\aleph}$ by using Theorem 23, first proving a version of Lemma 10. In those set-assignment models we also have

$$s(\neg A) = \begin{cases} S & \text{if } s(A) = \varnothing \\ \varnothing & \text{otherwise} \end{cases}$$

Exercises for Section F

1. Prove Theorem 21.
2. Fill in the sketch in the proof of Theorem 22 that $\mathbf{G}_{\aleph} \supseteq \mathbf{Int}$.
3. Prove $\mathbf{Int} \nvdash (A \to B) \vee (B \to A)$
4. Prove that for all n, $\mathbf{G}_n \supseteq \mathbf{G}_{n+1}$.
5. (Open) Establish whether the two set-assignment semantics for $\mathbf{G}_{\aleph}$ yield the same class of models

G. A Method for Proving Axiom Systems Independent

Given an axiom system, we say that an *axiom is independent* of the others if it is not a consequence of them. If the axiomatization is by schema we say that an *axiom scheme is independent* of the others if some instance of it is independent of all instances of the other schema. An axiom (scheme) that is a consequence of the others in a system is superfluous, though of course it may serve to shorten proofs.

One way to show that an axiom (scheme) A is independent is to exhibit a finite matrix M that validates all the axioms (schema) except for A, and validates the conclusion of a rule if it validates the hypotheses. In that case all consequences of the other axioms (schema) will be validated by M. So A cannot be a consequence of the others. By using a finite matrix, validity is (easily) decidable.

As an example, George Hughes has provided the following proof that the axiomatization of **PC** that I gave in Chapter II.L.1 is independent.

PC *in* $L(\neg, \rightarrow)$

1. $\neg A \rightarrow (A \rightarrow B)$
2. $B \rightarrow (A \rightarrow B)$
3. $(A \rightarrow B) \rightarrow ((\neg A \rightarrow B) \rightarrow B)$
4. $(A \rightarrow (B \rightarrow C)) \rightarrow ((A \rightarrow B) \rightarrow (A \rightarrow C))$

rule $\dfrac{A,\ A \rightarrow B}{B}$

For each of the matrices below the only designated value is 1. And for each if $\mathbf{e}(A) = 1$ and $\mathbf{e}(A \rightarrow B) = 1$, then $\mathbf{e}(B) = 1$, so that deductions preserve the designated value. In each case we will use the following table for $\neg$:

A	¬A
1	4
2	3
3	2
4	1

To show scheme 1 is independent we use the following table for $\rightarrow$:

		B			
A → B		1	2	3	4
A	1	1	2	3	4
	2	1	1	3	4
	3	1	2	1	4
	4	1	1	1	1

This table and the table for $\neg$ above validate axiom schema (2), (3), and (4): the proof is tedious but mechanical. However, the tables invalidate axiom scheme 1: $\neg 2 \rightarrow (2 \rightarrow 4) = 4$.

To show scheme (2) is independent we use the following table for $\rightarrow$, which together with the table for $\neg$ above validates schema (1), (3), and (4):

		B			
A → B		1	2	3	4
A	1	1	3	1	3
	2	1	1	3	3
	3	1	1	1	1
	4	1	1	1	1

Axiom scheme (2), however, is invalid: $2 \to (1 \to 2) = 3$.

To show that scheme (3) is independent we use the following table for $\to$, which together with the table for $\neg$ above validates schema (1), (2), and (4):

		B			
	A → B	1	2	3	4
A	1	1	2	3	4
	2	1	1	1	1
	3	1	2	1	2
	4	1	1	1	1

But axiom scheme (3) is invalid: $(3 \to 3) \to ((\neg 3 \to 3) \to 3) = 3$.

To show scheme (4) is independent we use the following table for $\to$, which together with the table for $\neg$ above validates schema (1), (2), and (3):

		B			
	A → B	1	2	3	4
A	1	1	2	3	4
	2	1	1	3	1
	3	1	2	1	2
	4	1	1	1	1

But axiom scheme (4) is invalid : $(2 \to (4 \to 3)) \to ((2 \to 4) \to (2 \to 3)) = 3$.

Note that each tables validates $A \to A$, too. So each scheme is independent of the others plus $A \to A$.

Unfortunately, this method of proving independence need not not work for every finite axiom system. In Exercise 3 below I outline the proof of *Gödel, 1933 C,* that there is a finite axiom system and one schema that is independent of the system but cannot be shown to be independent by any finite-valued matrix.

McKinsey, 1939, used finite matrices in a similar way to prove that the four connectives $\{\neg, \to, \wedge, \vee\}$ are independent in **Int**. As an example I'll show that $\neg$ cannot be defined from $\{\to, \wedge, \vee\}$. Suppose it could. Then there is some scheme S(A) in which $\neg$ does not appear such that $\mathbf{Int} \vdash S(A) \leftrightarrow \neg A$. Since the $\mathbf{G_3}$-tables validate **Int** we would therefore have that for every **G**-evaluation taking values in $\{0, \frac{1}{2}, 1\}$, $\mathbf{e}(S(A)) = \mathbf{e}(\neg A)$. But if $\mathbf{e}(p) = 1$ for all p, then $\mathbf{e}(S(A)) = 1$; whereas $\mathbf{e}(\neg A) = 0$. Therefore, there is no such S(A).

Exercises for Section G

1. Show that if we define an axiom scheme independent of others if all instances of it are independent of all instances of all others, then axiom scheme 1 is not independent of the other schema. (Hint: Consider $\neg A \to (A \to (C \to C))$.)

2. a. Show that every scheme of the axiomatization of **PC** is independent of the other schema plus $A \to A$. (Hint: Show that $A \to A$ is validated by the tables for $\to$.)

   b. Show $A \to A$ is independent of schemas (1), (3), and (4). (Hint: Modify the table that shows scheme (2) independent to read: $(1 \to 3) = 3$ and $(2 \to 2) = 3$.)

3. (*Gödel, 1933 C*)

   a. Show that the following axiom system is contained in **PC**.

      1. $A \to A$
      2. $(A \to \neg\neg B) \to (A \to B)$
      3. $(\neg\neg A \to \neg\neg B) \to (A \to B)$

      rule $\dfrac{A,\ A \to B}{B}$

   b. Show that $A \to \neg\neg A$ is independent of the axiom system in (a).
   (Hint: Consider the infinite matrix whose elements are the natural numbers and $\mathbf{e}(\neg A) = \mathbf{e}(A) + 1$, $\mathbf{e}(A \to B) = 0$ if $\mathbf{e}(A) \geq \mathbf{e}(B)$, and $\mathbf{e}(A \to B) = 1$ if $\mathbf{e}(A) < \mathbf{e}(B)$, with designated element 0.)

   b. Show that any finite matrix that validates the axiom system in (a) (respecting the rule, too) also validates $A \to \neg\neg A$. (Hint: Given such a matrix, there must be some $k < n$ such that $\mathbf{e}(\neg^{2k} A) = \mathbf{e}(\neg^{2n} A)$; then $\mathbf{e}(\neg^{2k} A \to \neg^{2n} A)$ is designated by (1). So by (3) $\mathbf{e}((\neg^{2k} A \to \neg^{2n} A) \to (\neg^{2k-2} A \to \neg^{2n-2} A))$ is designated, too. Use that the matrix respects *modus ponens* to obtain that $\mathbf{e}(A \to \neg^{2(n-k)} A)$ is designated, too. Then apply (2) for $(n - k - 1)$ times to show that $\mathbf{e}(A \to \neg\neg A)$ is designated.)

IX A Paraconsistent Logic – J3 –

in collaboration with **Itala M. L. D'Ottaviano**

A paraconsistent logic is one in which a nontrivial theory may include both a proposition and its negation. The study of one particular paraconsistent logic, $\mathbf{J_3}$, will illuminate the notions of consistency and completeness and call into question the adequacy of the general framework of set-assignment semantics.

I will first introduce the general notion of a paraconsistent logic. After presenting the semantics for the 3-valued paraconsistent logic $\mathbf{J_3}$, I will axiomatize it, and in doing so will suggest that paraconsistent logics are inconsistent only with respect to classical semantics, not with respect to their own formal or informal semantic notions. An analysis of set-assignment semantics for $\mathbf{J_3}$ will highlight the way in which the general framework for semantics of Chapter IV uses falsity as a default truth-value.

A. Paraconsistent Logics

Bivalence in logic, as we have studied it so far, has been reflected in the semantics for negation: not both A and $\neg$A can be true. From A and $\neg$A together we can conclude anything.

The only exception to this we have seen is the intuitionist minimal calculus, **J**, Chapter VII.F. Bivalence, according to intuitionists, is a platonist view that all propositions are decidable. Still, the intuitionism of Brouwer and Heyting did accept that from a false proposition one could deduce anything, $\neg A \rightarrow (A \rightarrow B)$, so that, as argued in Chapter VII.E.3, a bivalent set-assignment semantics could be given. But Kolmogorov and Johansson excised that principle from the minimal logic, so that one could have a theory in which both A and $\neg$A appeared, not because both were true or equally likely, but because for neither did we have a proof. The set-assignment semantics for **J** took negation, then, as a purely intensional connective: the truth-value of a negation depends wholly on the content of the proposition being negated, not in any way on its truth-value.

But others say that inconsistencies, situations in which it is correct to assert both a proposition and its negation, arise naturally in the course of life, are unavoidable, and indeed are fruitful. Hegelians, those who study paradoxes, modelers of reasoning in everyday contexts who must reckon, as time goes on, that some of our beliefs are contradicted, and even the poet Whitman have embraced contradictions:

> Do I contradict myself?
> Very well then . . . I contradict myself;
> I am large . . . I contain multitudes.

Jaśkowski, 1948, proposed constructing logical systems that allow for non-trivial theories containing (apparent) contradictions. The motives for doing so, he said, were: to systematize theories that contain contradictions, particularly as they occur in dialectics; to study theories in which there are contradictions caused by vagueness; and to study empirical theories whose postulates or basic assumptions could be considered contradictory. He proposed the following problem:

> The task is to find a system of the sentential calculus which: 1) when applied to the contradictory systems would not always entail their [triviality], 2) would be rich enough to enable practical inference, 3) would have an intuitive justification.
>
> *Jaśkowski, 1948,* p.145

Jaśkowski himself devised a propositional calculus, which he called 'discursive', to satisfy these criteria.

Jaśkowski, 1948, appeared in Polish and was translated into English only in 1969. Independently, da Costa, beginning in *da Costa, 1963,* developed a sequence of logics $\mathbf{C}_n$ that allow for nontrivial theories based on (apparent) contradictions. His motives were similar to Jaśkowski's and are described in *da Costa, 1974,* where he summarizes the systems, their extensions to first-order logic, and his investigations of their use in resolving paradoxes of set theory. Those logics are presented entirely syntactically: no explanations of the connectives are given, though we can assume they are formal versions of 'not', 'if . . . then . . .', 'and', 'or'.

Due primarily to da Costa's influence much work has been done on these and other systems in which a nontrivial theory which may contain (apparent) contradictions, dubbed *paraconsistent logics* by F. Miró Quesada. *Arruda, 1980* and *1989,* surveys this work and the history of the subject, while *da Costa and Marconi, 1990,* also discuss the philosophical motivation.

In this chapter I want to study the paraconsistent logic $\mathbf{J}_3$, which was first proposed by *D'Ottaviano and da Costa, 1970,* as a solution to Jaśkowski's problem, and which was later developed by *D'Ottaviano, 1985 A, 1985 B,* and *1987.* It has also been studied extensively by Avron, especially in *Avron, 1991.*

B. The Semantics of $\mathbf{J}_3$

1. Motivation

How should we proceed when faced with apparently contradictory sentences both of which seem equally plausible (or equally implausible)?

It is raining	For $z = \{x: x \notin x\}$: $z \in z$
It is not raining	$z \notin z$

The classical logician cannot incorporate both into a theory: from a proposition and its negation one can deduce in classical logic any proposition of the semi-formal language. There is only one classically inconsistent formal theory, and that is the *trivial* one consisting of all wffs.

The classical logician resolves the matter by building separate theories based on first one and then the other proposition, comparing the consequences of each. Or he may say that the difficulty in the first pair is that the word 'raining' is vague or ambiguous, and he will strive to reach agreement on what that word means, making it sufficiently precise that one of the sentences is true, the other false. But for the second example, such methods won't work; the only course is to exclude the sentences as incoherent or place restrictions on what formulas define sets.

D'Ottaviano and da Costa, in introducing $\mathbf{J}_3$, said,

> In the preliminary phase of the formulation of a theory (mathematical, physical, etc.) contradictions can appear which, in the definitive formulation, are eliminated; 0, 1, $\frac{1}{2}$ are the truth-values, where 0 represents the "false", 1 the "truth" and $\frac{1}{2}$ the provisional value of a proposition A, such that A and ¬A are theses of the theory under consideration in its provisional formulation; in the definitive form of the theory, the value $\frac{1}{2}$ will be reduced, at least in principle, to 0 or 1. . . .
>
> The calculus $\mathbf{J_3}$ can also be used as a foundation for inconsistent and nontrivial systems . . . In this case, $\frac{1}{2}$ represents the logical value of a formula which is, really, true and false at the same time. . . .
>
> In the elaboration of a logic suitable to handle "exact concepts" and "inexact concepts" . . . $\mathbf{J_3}$ also constitutes a solution.
>
> *D'Ottaviano and da Costa, 1970,* p.1351

Thus, let us consider a 3-valued logic whose truth-values will be 0, $\frac{1}{2}$, 1. However, unlike the other many-valued logics we have studied in this volume, two of the truth-values will be designated: 1 and $\frac{1}{2}$.

In comparison, Łukasiewicz introduced the many valued logics $\mathbf{L_3}$, $\mathbf{L_4}$, . . . , $\mathbf{L_{\aleph}}$, but he required that only the value 1 represents truth. In fact, as D'Ottaviano says, he didn't open up the possibility of characterizing more truth, or degrees or levels of truth. He characterized only different degrees of *falsity.* The idea of absolute truth (the value 1) was maintained.

Instead, D'Ottaviano says, let us model not only absolute truth (value 1) and absolute falsity (value 0), but also different degrees, levels, or grades of truth and falsity. $\mathbf{J_3}$ has only 3 truth-values. But a further motivation is to generalize $\mathbf{J_3}$ to logics with *n* designated truth-values and *m* undesignated.

From a classical perspective we would build one theory based on a proposition which is possible, and another on its negation, comparing the consequences of each. But here it is not a matter of knowing whether the proposition is true or false, or which of the proposition and its negation is most fruitful to be taken as the basis of a theory. Rather, as with paradoxical sentences, the proposition is neither absolutely true nor absolutely false, and it and its negation are inseparable. The appropriate methodology is to base one theory on both the proposition and its negation.

A proposition and its negation may both be possible, and in this case we would assign them the value $\frac{1}{2}$. But unlike Łukasiewicz, we take propositions with value $\frac{1}{2}$ as suitable on which to build a theory.

2. The truth-tables

D'Ottaviano and da Costa, 1970, originally presented $\mathbf{J_3}$ in terms of three primitive connectives: negation, disjunction, possibility. In this section I will follow their presentation for the most part, though using conjunction rather than disjunction as primitive. In Section B.3 I will present $\mathbf{J_3}$ using different primitives that reflect a different view of paraconsistency. In that presentation the possibility operator will be derived as a metalogical abbreviation, avoiding the question of whether it involves a use-mention confusion if used as a connective.

We begin by introducing a new symbol, ~, for negation, the reasons for which I will discuss in Sections C, D, and G. The table for this *weak negation* is:

A	~A
1	0
$\frac{1}{2}$	$\frac{1}{2}$
0	1

This is the table for ¬ in $\mathbf{L_3}$. But Łukasiewicz, taking $\frac{1}{2}$ as undesignated, treats a proposition that is possible as provisionally false and its negation also as false. Here the import of the table is that both are treated as true.

The tables for conjunction and disjunction are:

A∧B		B: 1	$\frac{1}{2}$	0
A	1	1	$\frac{1}{2}$	0
	$\frac{1}{2}$	$\frac{1}{2}$	$\frac{1}{2}$	0
	0	0	0	0

A∨B		B: 1	$\frac{1}{2}$	0
A	1	1	1	1
	$\frac{1}{2}$	1	$\frac{1}{2}$	$\frac{1}{2}$
	0	1	$\frac{1}{2}$	0

These two tables have their usual classical meanings, in the sense that if the designated values (1 and $\frac{1}{2}$) are replaced by T, and the undesignated value (0) is replaced by F, then we have (with repetitions) the classical tables for ∧ and ∨. We may take either as primitive. I will take conjunction, and then:

$$A \vee B \equiv_{\mathrm{Def}} \sim(\sim A \wedge \sim B)$$

We also have that $\sim(\sim A \vee \sim B)$ has the same table as $A \wedge B$.

The table for the possibility operator is:

A	◇A
1	1
$\frac{1}{2}$	1
0	0

We define a $\mathbf{J_3}$-*evaluation* for the language $L(\sim, \wedge, \Diamond, p_0, p_1, \ldots)$ to be a map $\mathbf{e}: PV \to \{0, \frac{1}{2}, 1\}$ that is extended to all wffs by the tables for ~, ∧, ◇ above. The *designated values are* 1 *and* $\frac{1}{2}$, so that $\mathbf{e} \vDash A$ means $\mathbf{e}(A) = 1$ or $\frac{1}{2}$. And $\vDash A$ means that $\mathbf{e} \vDash A$ for all $\mathbf{J_3}$-evaluations $\mathbf{e}$. Finally, $\Gamma \vDash A$ means that for every $\mathbf{J_3}$-evaluation $\mathbf{e}$, if $\mathbf{e} \vDash B$ for all $B \in \Gamma$, then $\mathbf{e} \vDash A$. The collection of valid wffs we call $\mathbf{J_3}$, or the $\mathbf{J_3}$-*tautologies*. When these notions of validity and consequence are compared to others I'll write $\vDash_{\mathbf{J_3}}$.

Some other defined connectives are important in the study of $\mathbf{J_3}$. We define the necessity operator:

$$\Box A \equiv_{\text{Def}} \sim(\Diamond \sim A)$$

Its table is:

A	$\Box$A
1	1
$\frac{1}{2}$	0
0	0

Though the tables for $\wedge$, $\vee$, $\Diamond$, $\Box$, and $\sim$ (read as $\neg$) are the same as for $\mathbf{L_3}$, they function quite differently in the logic because both 1 and $\frac{1}{2}$ are designated. For example, in the case when A takes value $\frac{1}{2}$, $\sim(A \wedge \sim A)$ is true (designated) in $\mathbf{J_3}$ (and indeed the formula is valid), but $\neg(A \wedge \neg A)$ is false if A takes value $\frac{1}{2}$ in $\mathbf{L_3}$.

With two designated truth-values, the table for $\to$ of $\mathbf{L_3}$ is not appropriate. We take instead:

	$A \to B$	B 1	$\frac{1}{2}$	0
A	1	1	$\frac{1}{2}$	0
	$\frac{1}{2}$	1	$\frac{1}{2}$	0
	0	1	1	1

If the antecedent is false, then $A \to B$ is definitely true; if the antecedent is true, absolutely or provisionally, and the consequent is false, then $A \to B$ is false. Hence the table is again classical in the sense that if 1 and $\frac{1}{2}$ are replaced by T, and 0 is replaced by F we have the classical table. And *modus ponens* is a valid rule. The remaining cases are when the antecedent is true, either definitely or provisionally, and the consequent is provisionally true (possible). In that case it is correct to ascribe only provisional truth (possibility) to $A \to B$. I will let you show the *Semantic Deduction Theorem*:

$$\Gamma \cup \{A\} \vDash_{\mathbf{J_3}} B \quad \text{iff} \quad \Gamma \vDash_{\mathbf{J_3}} A \to B$$

It is not necessary to take $\to$ as primitive. We can define it:

$$A \to B \equiv_{\text{Def}} \sim(\Diamond A \wedge \sim B)$$

We also define:

$$A \leftrightarrow B \equiv_{\text{Def}} (A \to B) \wedge (B \to A)$$

Its table is:

$A \leftrightarrow B$		B: 1	$\frac{1}{2}$	0
A	1	1	$\frac{1}{2}$	0
	$\frac{1}{2}$	$\frac{1}{2}$	$\frac{1}{2}$	0
	0	0	0	1

If this seems puzzling, recall that a proposition that is provisionally true (possible) cannot be equivalent to a false one, while it can be provisionally equivalent to one that is true.

In most paraconsistent logics, and in particular $\mathbf{J_3}$ and the systems $\mathbf{C}_n$ of *da Costa, 1974,* two negations are distinguished. The first we have seen, $\sim$, which is called *weak negation.* The other (in this development) is a defined connective, $\neg$, and is called *strong* or *classical negation*:

$$\neg A \equiv_{\text{Def}} \sim\Diamond A$$

Its table is:

A	$\neg$A
1	0
$\frac{1}{2}$	0
0	1

This is classical in the sense that, unlike weak negation, the strong negation of a true proposition (one with designated value) is false, and of a false one is true. (This is not the first time we have seen two formalizations of a single English connective in one logic. The modal logics of Chapter VI used both '$\rightarrow$' and '$\supset$'.)

Finally, a metalogical abbreviation useful in axiomatizing $\mathbf{J_3}$ is:

$$©A \equiv_{\text{Def}} \neg(A \wedge \sim A)$$

It has table:

A	©A
1	1
$\frac{1}{2}$	0
0	1

With this we can assert that A has a classical (absolute) truth-value.

Note on notation: In *D'Ottaviano and da Costa, 1970,* and in D'Ottaviano's later work, $\Diamond$A and $\Box$A are written as ∇A and ΔA, and $\rightarrow$ is written as $\supset$. What is symbolized here as $\sim$ is written there as $\neg$, and what I write as $\neg$ they symbolize as $\neg^*$.

The classical rule of substitution fails in $\mathbf{J_3}$; letting C(B) be C(A) with B substituted for some but not necessarily all occurrences of A, the following is not valid:

$$\frac{A \leftrightarrow B}{C(A) \leftrightarrow C(B)}$$

We have $\vDash_{\mathbf{J_3}} (A \leftrightarrow A) \leftrightarrow (B \leftrightarrow B)$, whereas $\nvDash_{\mathbf{J_3}} \sim(A \leftrightarrow A) \leftrightarrow \sim(B \leftrightarrow B)$, for the latter may fail when $\mathbf{e}(A) = 1$ and $\mathbf{e}(B) = \frac{1}{2}$. *D'Ottaviano, 1985 A,* defines the stronger equivalence:

$$(A \equiv^* B) \equiv_{Def} (A \leftrightarrow B) \wedge (\sim A \leftrightarrow \sim B)$$

Then for every evaluation $\mathbf{e}$, $\mathbf{e} \vDash A \equiv^* B$ iff $\mathbf{e}(A) = \mathbf{e}(B)$. The appropriate rule of substitution for $\mathbf{J_3}$ is then:

$$\frac{A \equiv^* B}{C(A) \equiv^* C(B)}$$

Exercises for Sections B.2

1. a. Show that the table for $\vee$ is the same as for $\sim(\sim A \wedge \sim B)$.
   b. Show that, taking $\vee$ as primitive, $\sim(\sim A \vee \sim B)$ has the same table as $A \wedge B$.
   c. Show that the table in the text for $\Box A$ is the one for $\sim(\Diamond \sim A)$.
   d. Show that the table for $\rightarrow$ is the same as for $\sim(\Diamond A \wedge \sim B)$.
   e. Show that the table for $\leftrightarrow$ presented in the text is correct.
   f. Show that the table in the text for $\neg A$ is the one for $\sim \Diamond A$.
   g. Show that the table for ©A presented in the text is correct.
   h. Show that $\Box A \vee \Box \sim A$ has the same table as ©A.
   j. Show that $\sim(\Diamond A \wedge \Diamond \sim A)$ has the same table as ©A.
   k. Show that $\neg A$ has the same table as $\sim A \wedge$ ©A.
   l. Determine the table for $A \equiv^* B$.

2. Show that *modus ponens* is valid.

3. Prove the *Semantic Deduction Theorem.*

4. a. Show that $\sim(A \wedge \sim A)$ is valid.
   b. Show that $A \vee \sim A$ is valid.
   c. How can you reconcile (a) and (b) with the assumed non-bivalence of $\mathbf{J_3}$?

5. Determine which of the following are valid in $\mathbf{J_3}$:

   a. $\sim(\Box A \wedge \sim A)$
   b. $\Diamond A \rightarrow A$
   c. (©$A \wedge \sim A) \leftrightarrow \neg A$
   d. ©$\neg A$
   e. $A \rightarrow (B \rightarrow A)$
   f. $((A \rightarrow \sim A) \rightarrow A) \rightarrow A$
   g. $(\neg A \rightarrow A) \rightarrow \neg A$
   h. $(\sim A \rightarrow A) \rightarrow \sim A$
   j. $(\sim A \rightarrow \sim B) \rightarrow (B \rightarrow A)$
   k. $\sim A \rightarrow (A \rightarrow B)$

l. $\neg A \rightarrow (A \rightarrow B)$

m. $(A \rightarrow B) \rightarrow ((\sim A \rightarrow B) \rightarrow B))$

n. $(A \rightarrow B) \rightarrow ((\neg A \rightarrow B) \rightarrow B))$

6. a. Show that the usual rule of substitution of equivalents is not valid:

$$\frac{A \leftrightarrow B}{C(A) \leftrightarrow C(B)}$$

b. Show that the following rule of substitution of equivalents is valid:

$$\frac{A \equiv^* B}{C(A) \equiv^* C(B)}$$

7. Read $\neg$ in Wajsberg's axiomatization of $\mathbf{L_3}$ (Chapter VII.C.1.c) as $\sim$ here and determine which schema are valid in $\mathbf{J_3}$.

8. Show that $(\Diamond\sim A \vee B) \wedge (\Diamond B \vee \sim A)$ has the same table as $\rightarrow$ in $\mathbf{L_3}$.

3. Definability of connectives

The choice of primitives used in the development of $\mathbf{J_3}$ and how we symbolize those reflects the way in which we understand paraconsistency, the relation of $\mathbf{J_3}$ to classical logic, and the adequacy of the general form of semantics of Chapter IV. Before we can understand why this is so, we need to know what choice of primitives we can use.

By *definable* in the next theorem I mean the strong notion that for the connective in question there is a scheme built from the other connectives that has the same 3-valued table. For instance, we'll see that a unary connective that always takes the value $\frac{1}{2}$ cannot be defined. However, there is a scheme semantically equivalent to such a connective: $A \rightarrow A$, since both always take a designated value.

Theorem 1 ***a.*** $\vee, \rightarrow, \neg, ©$ are definable from $\sim, \wedge, \Diamond$

b. $\wedge, \rightarrow, \neg, ©$ are definable from $\sim, \vee, \Diamond$

c. $\vee, \Diamond, ©$ are definable from $\neg, \rightarrow, \wedge, \sim$

d. $\sim$ cannot be defined from $\wedge, \vee, \rightarrow, \Diamond$

e. $\sim$ cannot be defined from $\wedge, \vee, \rightarrow, \Diamond, \neg$

f. $\Diamond$ cannot be defined from $\sim, \wedge, \vee, \rightarrow$

g. © cannot be defined from $\sim, \wedge, \vee, \rightarrow$

h. $\neg$ cannot be defined from $\sim, \wedge, \vee, \rightarrow$

j. No scheme built from any of $\sim, \wedge, \Diamond, \vee, \rightarrow, ©, \neg$ takes only value $\frac{1}{2}$.

Proof: In this proof I will write '$\approx$' to mean 'has the same table as'.

a. $A \vee B \approx \sim(\sim A \wedge \sim B)$ $\quad$ $A \rightarrow B \approx \sim(\Diamond A \wedge \sim B)$

$\neg A \approx \sim\Diamond A$ $\quad$ $\Box A \approx \sim\Diamond\sim A$

$©A \approx \sim(\Diamond A \wedge \Diamond\sim A)$

b. $A \wedge B \approx {\sim}({\sim}A \vee {\sim}B)$

c. First note that for any B, $B \wedge \neg B$ takes value 0.

$\Diamond A \approx \neg(A \rightarrow (A \wedge \neg A))$

$©A \approx \neg(A \wedge {\sim}A)$

d. By induction on the length of a scheme S(A) built from $\wedge, \vee, \rightarrow, \Diamond$, we can show that if $\mathbf{e}(A) = 1$ or $\frac{1}{2}$, then $\mathbf{e}(S(A)) = 1$ or $\frac{1}{2}$. So $\sim$ cannot be defined.

e. By induction on the length of a scheme S(A) built from $\wedge, \vee, \rightarrow, \Diamond, \neg$, if $\mathbf{e}(A) = 1$ or $\frac{1}{2}$, then $\mathbf{e}(S(A)) = 1$ or $\frac{1}{2}$ *or* if $\mathbf{e}(A) = 1$ or $\frac{1}{2}$, then $\mathbf{e}(S(A)) = 0$. That is, we cannot separate the values 1 and $\frac{1}{2}$. So $\sim$ cannot be defined.

f. For any scheme S(A) built from $\sim, \wedge, \vee, \rightarrow$, if $\mathbf{e}(A) = \frac{1}{2}$, then $\mathbf{e}(S(A)) = \frac{1}{2}$.

g. Were © definable from $\sim, \wedge, \vee, \rightarrow$, then we could define $\Box A$ as ${\sim}(©A \rightarrow {\sim}A)$ and $\Diamond A$ as ${\sim}\Box{\sim}A$, contradicting part (f).

h. Were $\neg$ definable from $\sim, \wedge, \vee, \rightarrow$, then by (c), $\Diamond$ would be, too, contradicting part (f).

j. If **e** is any $\mathbf{J_3}$-evaluation such that $\mathbf{e}: PV \rightarrow \{0, 1\}$, then its extension satisfies $\mathbf{e}$: Wffs $\rightarrow \{0, 1\}$. So no connective taking only value $\frac{1}{2}$ can be defined. ■

There are two very different ways we may present $\mathbf{J_3}$. The first is to use $\sim, \wedge, \Diamond$ or $\sim, \vee, \Diamond$ as primitives. This is what we did above and is in accord with D'Ottaviano and da Costa's original motivation.

The alternative is to take $\neg, \rightarrow, \wedge, \sim$ as primitives. In what follows, *I will assume that the definitions of evaluation, validity, and semantic consequence for* $\mathbf{J_3}$ *are made with respect to either* $L(\sim, \wedge, \Diamond)$ or $L(\neg, \rightarrow, \wedge, \sim)$ as appropriate to the discussion at hand. Though we now have two different semantic consequence relations for two distinct languages, I will use the same symbol, $\models_{\mathbf{J_3}}$ or simply $\models$, for both. We can view them as "the same logic" formulated in two different languages (see the discussion in Chapter II.J.4 and Chapter X.B.6).

C. The Relationship Between $\mathbf{J_3}$ and Classical Logic

The way we see the relation between $\mathbf{J_3}$ and classical logic depends on how we understand the role of weak negation, $\sim$, in $\mathbf{J_3}$.

Theorem 2 *a.* $\mathbf{J_3}$ restricted to $L(\neg, \rightarrow, \wedge, \vee)$ is **PC**: $\Gamma \models_{\mathbf{J_3}} A$ iff $\Gamma \models_{\mathbf{PC}} A$

b. $\mathbf{J_3}$ restricted to $L(\sim, \rightarrow, \wedge, \vee)$ is contained in **PC** if we identify $\sim$ with $\neg$ in **PC**. That is, letting $A°$ be A with $\sim$ replaced by $\neg$ everywhere, if $\models_{\mathbf{J_3}} A$, then $\models_{\mathbf{PC}} A°$. However, $\{p, \neg p\} \models_{\mathbf{PC}} q$, but $\{p, {\sim}p\} \not\models_{\mathbf{J_3}} q$.

Proof: a. Suppose $\Gamma \vDash_{\mathbf{J_3}} A$. Let υ be any **PC**-valuation that validates Γ. Relabeling T as 1, F as 0, we have a $\mathbf{J_3}$-evaluation, **e**. Since the tables for $\neg$, $\rightarrow$, $\wedge$, $\vee$ are classical reading T for 1, F for 0, $\mathbf{e} \vDash \Gamma$. Hence $\mathbf{e} \vDash A$. Hence $\upsilon \vDash A$. So $\Gamma \vDash_{\mathbf{PC}} A$.

Suppose $\Gamma \vDash_{\mathbf{PC}} A$. Let **e** be any $\mathbf{J_3}$-evaluation that validates Γ. Consider $\upsilon: PV \rightarrow \{T, F\}$ defined by $\upsilon(p) = T$ iff $\mathbf{e}(p) = 1$ or $\frac{1}{2}$, $\upsilon(p) = F$ iff $\mathbf{e}(p) = 0$. Then for every B, $\upsilon \vDash B$ iff $\mathbf{e} \vDash B$. Hence $\upsilon \vDash \Gamma$, so $\upsilon \vDash A$, so $\mathbf{e} \vDash A$. Hence $\Gamma \vDash_{\mathbf{J_3}} A$.

The same methods establish $\Gamma \vDash_{\mathbf{J_3}} A$ iff $\Gamma \vDash_{\mathbf{PC}} A$.

b. Since ~ and $\neg$ agree on the values 0 and 1, we can use the same proof as in the first part of (a). To show $\{p, {\sim}p\} \nvDash_{\mathbf{J_3}} q$, let p take value $\frac{1}{2}$ and q value 0. ∎

If we identify ~ with negation in **PC**, then we have seen that the fragment of $\mathbf{J_3}$ in the language of ~, $\wedge$, $\vee$, $\rightarrow$ is contained in **PC**. However, this fragment does not equal **PC**, since $(A \wedge {\sim}A) \rightarrow B$ is not a $\mathbf{J_3}$-tautology. This accords with the design of $\mathbf{J_3}$ as a paraconsistent logic, ensuring that $\{A, {\sim}A\} \vDash_{\mathbf{J_3}} B$ is not a valid rule.

Here are some noteworthy classical tautologies that fail to be $\mathbf{J_3}$-tautologies when ~ is identified with negation in **PC**:

$${\sim}A \rightarrow (A \rightarrow B)$$

$$(A \rightarrow B) \rightarrow ((A \rightarrow {\sim}B) \rightarrow {\sim}A)$$

$$(A \rightarrow B) \rightarrow ({\sim}B \rightarrow {\sim}A)$$

$$((A \vee B) \wedge {\sim}A) \rightarrow B$$

However, the laws of excluded middle and bivalence translate into valid formulas:

$${\sim}(A \wedge {\sim}A) \quad \text{and} \quad (A \vee {\sim}A)$$

Viewed individually each proposition obeys these, but not so in terms of its consequences. Instead of $(A \wedge {\sim}A) \rightarrow B$ we have the tautology:

$$(A \wedge {\sim}A \wedge \copyright A) \rightarrow B$$

Instead of ${\sim}(A \vee {\sim}A) \rightarrow B$, we have the tautology:

$${\sim}(A \vee {\sim}A \vee \copyright A) \rightarrow B$$

So long as we restrict our attention to propositions that are classically true or false, we can reason classically in $\mathbf{J_3}$ in $L({\sim}, \wedge, \Diamond)$. Define a map * from the language $L(\neg, \rightarrow, \wedge, \vee)$ of **PC** to the language $L({\sim}, \wedge, \Diamond)$ of $\mathbf{J_3}$ by first taking:

$$A^\circ = A \text{ with } \neg \text{ replaced by } {\sim}$$

with the understanding that $\vee$ and $\rightarrow$ are the defined connectives of $\mathbf{J_3}$. Then set:

$$A^* = (\bigwedge_{\{p_i \text{ in } A\}} \copyright p_i) \rightarrow A^\circ$$

Here $\bigwedge$ means the conjunction of the indexed wffs, associating to the left.

Theorem 3 The map * is a validity preserving map from **PC** to $\mathbf{J_3}$: $\vDash_{\mathbf{PC}} A$ iff $\vDash_{\mathbf{J_3}} A^*$. However, the map does not preserve consequences.

Proof: First note that for any evaluation e, if $e(p) = \frac{1}{2}$ for some p in A, then the antecedent of A^* is evaluated to be 0. So we have:

$\vDash_{\mathbf{PC}} A$ iff for every **PC**-model v, $v(A) = \mathsf{T}$
iff for every $\mathbf{J_3}$-evaluation $e: PV \to \{0,1\}$, $e(A^\circ) = 1$
iff for every $\mathbf{J_3}$-evaluation e, $e(A^*) = 1$
iff $\vDash_{\mathbf{J_3}} A^*$

However, $\{p, \neg p\} \vDash_{\mathbf{PC}} q$, but $\{©p \to p,\ ©p \to {\sim}p\} \nvDash_{\mathbf{J_3}} ©q \to q$, for we may take $e(p) = \frac{1}{2}$ and $e(q) = 0$. ■

Alternatively, we can consider $\mathbf{J_3}$ *as an extension of* **PC**, as shown by Theorem 2.a. From this point of view $\mathbf{J_3}$ arises by adding a wholly intensional connective, ~, to classical logic.

Exercises for Section C

1. Given a $\mathbf{J_3}$-evaluation e, let $v: PV \to \{\mathsf{T}, \mathsf{F}\}$ be defined by: $v(p) = \mathsf{T}$ iff $e(p) = 1$ or $\frac{1}{2}$, $v(p) = \mathsf{F}$ iff $e(p) = 0$. Show for every B, $v \vDash B$ iff $e \vDash B$.
2. Why do the methods of Theorem 2 not work to prove the same for $\mathbf{L_3}$?
3. Show that ${\sim}({\sim}A \wedge {\sim}B)$ is semantically equivalent in $\mathbf{J_3}$ to $\neg(\neg A \wedge \neg B)$.
4. Prove that $(A \wedge {\sim}A \wedge ©A) \leftrightarrow (A \wedge \neg A)$ is valid.
5. What is the relation of $\mathbf{J_3}$ to $\mathbf{L_3}$?
   a. We can define a connective in $\mathbf{J_3}$ whose table is that of $\to$ in $\mathbf{L_3}$: $A \supset\!\!\to B \equiv_{\mathrm{Def}} (\Diamond{\sim}A \vee B) \wedge (\Diamond B \vee {\sim}A)$. Show that $\vee, \Diamond$ can be defined from ~ and $\supset\!\!\to$. (Hint: See Chapter VIII.C.1.a.)
   b. Show that we can formulate $\mathbf{J_3}$ in $L({\sim}, \supset\!\!\to)$. Is $\{A, A \supset\!\!\to B\} \vDash B$ valid?
   c. (Open) *D'Ottaviano, 1985 A,* has given a complete axiomatization of $\mathbf{J_3}$ in $L({\sim}, \supset\!\!\to)$. Determine whether it is strongly complete.
   d. (Open) Is there a translation of $\mathbf{L_3}$ into $\mathbf{J_3}$? Of $\mathbf{J_3}$ into $\mathbf{L_3}$?

D. Consistency vs. Paraconsistency

1. Definitions of completeness and consistency for $\mathbf{J_3}$ theories

A theory is consistent if it contains no contradiction. A proposition is a contradiction if it is false due to its form only, or semantically, if the corresponding wff is false in all models.

A theory is complete if it is as full a description as possible of "the way the world is" relative to the atomic propositions we've assumed and the semantics, informal or formal, that we employ. It might be inconsistent, but if not it will contain as many complex propositions as possible relative to the atomic ones while still being consistent.

So a complete and consistent theory corresponds to a possible description of the world, relative to our semantic intuitions and choice of atomic propositions. Therefore, if we have formal semantics, a complete and consistent theory should correspond to the collection of propositions true in a model.

In classical logic we take the standard form of a contradiction to be $A \wedge \neg A$. Then a theory (collection of sentences closed under deduction) is said to be consistent (with respect to classical logic) if for no A does it contain $A \wedge \neg A$; or, equivalently, for no A does it contain both A and $\neg A$. This reflects the classical semantic assumption that for no A can both A and $\neg A$ be true.

If we understand negation to be formalized by ~, then the classical notion of contradiction is not applicable to $\mathbf{J_3}$. And the associated criterion of consistency is inappropriate for $\mathbf{J_3}$, in which we specifically assumed that it is acceptable to build a theory on the basis of both A and ~A. By the semantic assumptions of $\mathbf{J_3}$, a theory that contains both A and ~A for some A is not necessarily contradictory, for it can reflect a possible way the world could be. To formulate an appropriate criterion of consistency for $\mathbf{J_3}$ I'll first let you prove the following.

Lemma 4 If **e** is a $\mathbf{J_3}$-evaluation, then

$$\mathsf{e} \vDash A \quad \text{iff} \quad \mathsf{e}(A) = 1 \text{ or } \tfrac{1}{2}$$
$$\mathsf{e} \vDash {\sim}A \quad \text{iff} \quad \mathsf{e}(A) = 0 \text{ or } \tfrac{1}{2}$$
$$\mathsf{e} \vDash \copyright A \quad \text{iff} \quad \mathsf{e}(A) = 0 \text{ or } 1$$

Thus, if all three of A, ~A, ©A are assumed by a theory, we can have no model of it. Such a theory is inconsistent.

In classical logic we took a theory to be complete if for every A it contains at least one of A, $\neg$A. That is, it must decide between these, and hence stipulate which is assumed to be true, or else embrace them both and be inconsistent. That criteria is inappropriate for $\mathbf{J_3}$ if we understand negation to be formalized by ~, for choosing one of A, ~A we have not stipulated the truth-value of A. To do that we need to choose two of A, ~A, ©A. So, relative to axiomatizations we will provide below, we define:

> Γ is *consistent relative to* $\mathbf{J_3}$ if for every A at *most* two of A, ~A, ©A are syntactic consequences of Γ
>
> Γ is *complete relative to* $\mathbf{J_3}$ if for every A at *least* two of A, ~A, ©A are in Γ

In Theorem 7 I'll demonstrate that these are the appropriate definitions for $\mathbf{J_3}$: Γ is complete and consistent relative to $\mathbf{J_3}$ iff there is a $\mathbf{J_3}$-evaluation that validates exactly Γ.

However, we may view $\mathbf{J_3}$ as an extension of classical logic formulated in the language $L(\neg, \rightarrow, \wedge, \sim)$. In that case the standard way to formalize 'not' is with $\neg$, and a theory is *classically consistent* means that at most one of $A, \neg A$ is a consequence of Γ; a theory is *classically complete* means that at least one of $A, \neg A$ is in Γ. We have:

$$\vDash_{\mathbf{J_3}} (A \wedge {\sim}A \wedge \copyright A) \leftrightarrow (A \wedge \neg A)$$

So in the next section we can demonstrate that *for* $\mathbf{J_3}$*-theories,* Γ *is complete and consistent relative to* $\mathbf{J_3}$ iff Γ *is classically complete and consistent.*

2. The status of negation in $\mathbf{J_3}$

Relative to the connectives $\{\wedge, \vee, \rightarrow\}$, both weak negation, $\sim$, and strong negation, $\neg$, are primitive (Theorem 1).

The approach favored by paraconsistent logicians is to view weak negation as primitive. Then possibility is formalized as a connective, with appropriate deference made to the use-mention controversy surrounding that decision. Strong negation, $\neg$, is taken as a defined connective. Only under favorable circumstances, where all the atomic propositions under discussion are classically (absolutely) true or false, can we view negation, $\sim$, as classical (Theorem 3). This is reflected by an alternative definition we can give of $\neg A$ as ${\sim}A \wedge \copyright A$.

From a classical point of view, though, we never suggested that all logically significant uses of 'not' can be properly modeled by classical negation. The simple propositions about a die 'Three faces are even numbered', 'Three faces are not even numbered', should convince us of that: both are true. The symbol $\neg$ should be reserved for formalizing 'not' in those cases where we agree that the proposition and the proposition with 'not' deleted cannot both be true (at the same time), and this is the analysis given in Chapter IV. We may choose to introduce a new connective, $\sim$, to formalize other uses of 'not' where the truth of the proposition and the truth of the proposition with 'not' deleted are (apparently) inseparable. Such propositions might be ones using vague terms, such as 'It is not raining', or paradoxical sentences such as 'This sentence is not true'. The division of logical uses of 'not' is thus the same as the paraconsistent logician's, but comes from a very different perspective.

Paraconsistent logics contain "inconsistent nontrivial theories" only from the application of classical criteria to an admittedly nonclassical connective, $\sim$. From their own semantic point of view such theories are consistent, corresponding to a possible description of the world. To call a theory inconsistent if it contains, for some A, both A and ${\sim}A$ is tantamount to understanding negation in its usual sense as assumed by all other logics in this volume: it cannot be that both A and ~A are true. That cannot be how we understand ~ in paraconsistent logics, for it would preclude building a theory based on both A and ~A. For $\mathbf{J_3}$ this strong view of negation is expressed by the table for ¬. Using that connective the classical criterion of consistency is apt, and we cannot have a nontrivial inconsistent theory: for all B, $(A \wedge \neg A) \vDash_{\mathbf{J_3}} B$.

I will return to this discussion of the status of negation in Sections F and G.

E. Axiomatizations of $\mathbf{J_3}$

In this section I will give two axiomatizations, intended to illustrate the discussion of the last section. In the first I take the viewpoint of the paraconsistent logician that ~ is the standard interpretation of 'not' and treat $\mathbf{J_3}$ as a modal logic. Strong negation, ¬, does not appear in the axiomatization, and the only notions of completeness and consistency used are relative to $\mathbf{J_3}$.

In the second axiomatization I view $\mathbf{J_3}$ as an extension of classical logic in the language of ¬, →, ∧, with a wholly intensional connective, ~. The notions of completeness and consistency will be the classical ones.

For both axiomatizations there are no inconsistent nontrivial theories.

1. As a modal logic

$\mathbf{J_3}$ *in* $L(\sim, \wedge, \Diamond)$

$A \rightarrow B \equiv_{\text{Def}} \sim(\Diamond A \wedge \sim B)$ $\quad\quad$ $©A \equiv_{\text{Def}} \sim(\Diamond A \wedge \Diamond\sim A)$

$A \leftrightarrow B \equiv_{\text{Def}} (A \rightarrow B) \wedge (B \rightarrow A)$

1. $B \rightarrow (A \rightarrow B)$
2. $(A \rightarrow (B \rightarrow C)) \rightarrow ((A \rightarrow B) \rightarrow (A \rightarrow C))$
3. $(B \rightarrow (A \rightarrow C)) \rightarrow ((A \wedge B) \rightarrow C)$
4. $A \rightarrow (B \rightarrow (A \wedge B))$
5. $(A \wedge \sim A \wedge ©A) \rightarrow B$
6. $((\sim A \wedge ©A) \rightarrow A) \rightarrow A$
7. $\sim\sim A \leftrightarrow A$
8. $©A \leftrightarrow ©\sim A$
9. $\sim\Diamond A \leftrightarrow (\sim A \wedge ©A)$
10. $©(\Diamond A)$
11. $[(A \wedge B) \wedge ©(A \wedge B)] \leftrightarrow [(A \wedge ©A) \wedge (B \wedge ©B)]$

rule $\dfrac{A,\ A \rightarrow B}{B}$

I will denote the consequence relation of this axiom system as $\vdash_{\mathbf{J_3},\Diamond}$. For the following proofs in Section E.1 I will write $\vdash$ for $\vdash_{\mathbf{J_3},\Diamond}$. I will leave to you to show that this system is sound for $\mathbf{J_3}$ (consult Theorem 2 for the first four schema).

Recall from the last section that Γ is *consistent relative to* $\mathbf{J_3}$ ($\mathbf{J_3}$*-consistent*) if for every A at most two of A, ~A, ©A are consequences of Γ. And Γ is *complete relative to* $\mathbf{J_3}$ ($\mathbf{J_3}$*-complete*) if for every A at least two of A, ~A, ©A are in Γ. As usual, Γ is a *theory* if Γ is closed under deduction.

Lemma 5 ***a.*** ***The Syntactic Deduction Theorem*** $\Gamma \cup \{A\} \vdash B$ iff $\Gamma \vdash A \to B$

b. $\vdash A \wedge B \to B$

c. $\vdash A \wedge B \to B \wedge A$

d. $\Gamma \cup \{A, B\} \vdash A \wedge B$

e. $\Gamma \cup \{A, B\} \vdash C$ iff $\Gamma \vdash (A \wedge B) \to C$

Proof: The proof of part (a) is as for **PC** (Theorem II.8, p. 71) using **axioms 1** and **2**. Part (b) follows from **axiom 3**, taking B for C, and **axiom 1**. Part (c) uses **axiom 3** taking $B \wedge A$ for C, and **axiom 4**. Part (d) uses **axiom 4**, and (e) uses **axioms 3** and **4**. ■

In the next lemma we show that $\mathbf{J_3}$-consistency is equivalent to Post-consistency, and $\mathbf{J_3}$-completeness is equivalent to Post-completeness.

Lemma 6 ***a.*** Γ is $\mathbf{J_3}$-consistent iff for some B, $\Gamma \nvdash B$.

b. If Γ is $\mathbf{J_3}$-complete and $A \notin \Gamma$, then for every B, $\Gamma \cup \{A\} \vdash B$.

c. If Γ is $\mathbf{J_3}$-complete and $\mathbf{J_3}$-consistent, then Γ is a theory.

d. If $\Gamma \nvdash A$, then $\Gamma \cup \{\sim A, ©A\}$ is consistent.

e. If Γ is $\mathbf{J_3}$-consistent, then one of $\Gamma \cup \{A, \sim A\}$, $\Gamma \cup \{A, ©A\}$, $\Gamma \cup \{\sim A, ©A\}$ is consistent.

Proof: a. From right to left is immediate. In the other direction, if Γ is $\mathbf{J_3}$-inconsistent, then for some A: $\Gamma \vdash A$, $\Gamma \vdash \sim A$, and $\Gamma \vdash ©A$. Hence by **axiom 5** and Lemma 5, for every B, $\Gamma \vdash B$.

b. Suppose Γ is $\mathbf{J_3}$-complete and $A \notin \Gamma$. Then ~A and ©A are in Γ, so $\Gamma \cup \{A\}$ is inconsistent, and the result follows by (a).

c. Suppose Γ is $\mathbf{J_3}$-complete and $\mathbf{J_3}$-consistent and $\Gamma \vdash A$. If $A \notin \Gamma$, then by the completeness of Γ, $\Gamma \cup \{A\}$ is $\mathbf{J_3}$-inconsistent. But $\text{Th}(\Gamma) = \text{Th}(\Gamma \cup \{A\})$, so Γ is $\mathbf{J_3}$-inconsistent, a contradiction. Hence $A \in \Gamma$.

d. Suppose $\Gamma \cup \{\sim A, ©A\}$ is $\mathbf{J_3}$-inconsistent. Then $\Gamma \cup \{\sim A, ©A\} \vdash A$. So by Lemma 5, $\Gamma \vdash (\sim A \wedge ©A) \to A$, and hence by **axiom 6**, $\Gamma \vdash A$.

e. Suppose $\Gamma \cup \{\sim A, ©A\}$ is $\mathbf{J_3}$-inconsistent. Then as in (d), $\Gamma \vdash A$. Suppose further $\Gamma \cup \{A, ©A\}$ is also $\mathbf{J_3}$-inconsistent. Then by **axiom 7** and **axiom 8**,

$\Gamma \cup \{ \sim\sim A, ©\sim A \}$ is $\mathbf{J_3}$-inconsistent. Hence $\Gamma \vdash (\sim\sim A \wedge ©\sim A) \rightarrow \sim A$, so using **axiom 6** again, $\Gamma \vdash \sim A$. Hence, $\Gamma \cup \{A, \sim A\}$ is $\mathbf{J_3}$-consistent. ■

Theorem 7 The following are equivalent:

a. Γ is $\mathbf{J_3}$-complete and $\mathbf{J_3}$-consistent

b. There is some $\mathbf{J_3}$-evaluation **e** such that $\Gamma = \{A : \mathbf{e} \vDash A\}$

c. There is some $\mathbf{J_3}$-evaluation **e** such that for all A,

$$\mathbf{e}(A) = 1 \text{ iff } A, ©A \in \Gamma$$

$$\mathbf{e}(A) = \tfrac{1}{2} \text{ iff } A, \sim A \in \Gamma$$

$$\mathbf{e}(A) = 0 \text{ iff } \sim A, ©A \in \Gamma$$

d. There is some $\mathbf{J_3}$-evaluation **e** such that for all A,

$$\mathbf{e}(A) = 1 \text{ iff } \sim A \notin \Gamma$$

$$\mathbf{e}(A) = \tfrac{1}{2} \text{ iff } ©A \notin \Gamma$$

$$\mathbf{e}(A) = 0 \text{ iff } A \notin \Gamma$$

Proof: The equivalence of (b), (c), and (d) comes from Lemma 4. I will show that Γ is $\mathbf{J_3}$-complete and $\mathbf{J_3}$-consistent iff (c).

First suppose there is an **e** as in (c). Then for every A exactly two of A, ~A, $©A \in \Gamma$. Hence Γ is $\mathbf{J_3}$-complete and, since Γ must be closed under deductions, Γ is consistent.

Now suppose Γ is $\mathbf{J_3}$-complete and $\mathbf{J_3}$-consistent. Then by Lemma 6.c, Γ is a theory. Define **e** as in (c). It remains to show that **e** is a $\mathbf{J_3}$-evaluation. We begin with the evaluation of weak negation:

$$\begin{aligned} \mathbf{e}(\sim A) = 1 \quad & \text{iff } \sim A, ©(\sim A) \in \Gamma \\ & \text{iff } \sim A, ©A \in \Gamma \qquad \text{by } \textbf{axiom 8} \\ & \text{iff } \mathbf{e}(A) = 0 \end{aligned}$$

$$\begin{aligned} \mathbf{e}(\sim A) = \tfrac{1}{2} \quad & \text{iff } \sim A, \sim\sim A \in \Gamma \\ & \text{iff } \sim A, A \in \Gamma \qquad \text{by } \textbf{axiom 7} \\ & \text{iff } \mathbf{e}(A) = \tfrac{1}{2} \end{aligned}$$

By process of elimination, $\mathbf{e}(\sim A) = 0$ iff $\mathbf{e}(A) = 1$. So **e** evaluates ~ correctly.

For the possibility operator, we have:

$$\begin{aligned} \mathbf{e}(\Diamond A) = 0 \quad & \text{iff } \sim\Diamond A, ©(\Diamond A) \in \Gamma \\ & \text{iff } \sim A, ©A \in \Gamma \qquad \text{by } \textbf{axiom 9} \text{ and } \textbf{10} \\ & \text{iff } \mathbf{e}(A) = 0 \end{aligned}$$

Since Γ is a theory, by **axiom 10**, $©(\Diamond A) \in \Gamma$. So $\mathbf{e}(\Diamond A) \neq \tfrac{1}{2}$. Thus $\Diamond$ is evaluated correctly by **e**.

For conjunction:

$\mathbf{e}(A \wedge B) = 0$ iff $A \wedge B \notin \Gamma$ — by consistency of Γ
iff $A \notin \Gamma$ or $B \notin \Gamma$ — by Lemma 5
iff $\sim A, \copyright A \in \Gamma$ or $\sim B, \copyright B \in \Gamma$
iff $\mathbf{e}(A) = 0$ or $\mathbf{e}(B) = 0$

$\mathbf{e}(A \wedge B) = 1$ iff $A \wedge B, \copyright(A \wedge B) \in \Gamma$
iff $A, \copyright A, B, \copyright B \in \Gamma$ — by **axiom 11**
iff $\mathbf{e}(A) = 1$ and $\mathbf{e}(B) = 1$

By process of elimination, **e** evaluates $\wedge$ correctly. ■

Lemma 8 If $\Gamma \nvdash A$, then there is some $\mathbf{J_3}$-complete and $\mathbf{J_3}$-consistent theory Σ such that $\Gamma \subseteq \Sigma$ and $A \notin \Sigma$.

Proof: Suppose $\Gamma \nvdash A$. Define $\Sigma_0 = \Gamma \cup \{\sim A, \copyright A\}$. This is $\mathbf{J_3}$-consistent by Lemma 6.d.

Let $B_0, B_1, \ldots$ be a listing of all wffs. Define:

$$\Sigma_{n+1} = \begin{cases} \Sigma_n \cup \{B_n, \sim B_n\} & \text{if that is } \mathbf{J_3}\text{-consistent; if not then} \\ \Sigma_n \cup \{B_n, \copyright B_n\} & \text{if that is } \mathbf{J_3}\text{-consistent; if not then} \\ \Sigma_n \cup \{\sim B_n, \copyright B_n\} & \end{cases}$$

$$\Sigma = \bigcup_n \Sigma_n$$

By Lemma 6.e, for every n, Σ_n is $\mathbf{J_3}$-consistent. Hence Σ is $\mathbf{J_3}$-consistent, and by construction it is $\mathbf{J_3}$-complete. As $\Sigma_0 \subseteq \Sigma$, $A \notin \Sigma$. ■

We can now prove strong completeness using Theorem 7 and Lemma 8.

Theorem 9 Strong Completeness of $\vdash_{\mathbf{J_3},\Diamond}$ $\Gamma \vdash_{\mathbf{J_3},\Diamond} A$ iff $\Gamma \vDash_{\mathbf{J_3}} A$

2. As an extension of classical logic

In Chapter II.N.6 I gave an axiomatization of **PC** in the language $L(\neg, \rightarrow, \wedge)$. If we allow any formula of the language $L(\neg, \rightarrow, \wedge, \sim)$ to be an instance of A, B, or C in those schema, then we have an *axiomatization of* **PC** *based on* $\neg, \rightarrow, \wedge$ *in the language of* $\mathbf{J_3}$.

$\mathbf{J_3}$ *in* $L(\neg, \rightarrow, \wedge, \sim)$

$\copyright A \equiv_{\text{Def}} \neg(A \wedge \sim A)$ $\quad\quad$ $A \leftrightarrow B \equiv_{\text{Def}} (A \rightarrow B) \wedge (B \rightarrow A)$

PC based on $\neg, \rightarrow, \wedge$

1. $\sim\sim A \leftrightarrow A$
2. $\sim\neg A \wedge \neg A \rightarrow \neg\sim A$

3. $\neg{\sim}(A \wedge B) \leftrightarrow \neg{\sim}A \wedge \neg{\sim}B$
4. $\neg A \to \neg{\sim}(A \to B)$
5. $(B \wedge \neg{\sim}B) \to \neg{\sim}(A \to B)$
6. $((A \wedge B) \wedge {\sim}B) \to {\sim}(A \to B)$

rule $\dfrac{A,\ A \to B}{B}$

I will denote the consequence relation of this axiom system as $\vdash_{\mathbf{J_3},\neg}$. For the following proofs in Section E.2 I will write $\vdash$ for $\vdash_{\mathbf{J_3},\neg}$.

Recall that a collection of wffs Γ is *classically consistent* if at most one of A, $\neg A$ are consequences of Γ. And Γ is *classically complete and consistent* iff for every A, exactly one of A, $\neg A$ is in Γ.

Throughout the following I will use results from **PC** as justified by this axiomatization. In particular, the *Syntactic Deduction Theorem* holds, and if Γ is classically complete and consistent, then Γ is a theory.

Theorem 10 If Γ is classically complete and consistent, then there is some $\mathbf{J_3}$-evaluation $\mathbf{e}$ such that for all A:

$$\mathbf{e}(A) = 1 \text{ iff } A, \copyright A \in \Gamma$$
$$\mathbf{e}(A) = \tfrac{1}{2} \text{ iff } A, {\sim}A \in \Gamma$$
$$\mathbf{e}(A) = 0 \text{ iff } {\sim}A, \copyright A \in \Gamma$$

Proof: Let Γ be classically complete and consistent. Given any A, suppose all three of $A, {\sim}A, \copyright A \in \Gamma$. Then by **PC** and the definition of $\copyright A$, $A \wedge {\sim}A \in \Gamma$, and $\neg(A \wedge {\sim}A) \in \Gamma$, a contradiction. So at most two of $A, {\sim}A, \copyright A \in \Gamma$. If $A \notin \Gamma$, then $\neg A \in \Gamma$, so ${\sim}A$ and $\copyright A \in \Gamma$. If $A \in \Gamma$ and $\copyright A \notin \Gamma$, then $\neg \copyright A \in \Gamma$. So by **PC**, $A \wedge {\sim}A \in \Gamma$, Hence, at least two of $A, {\sim}A, \copyright A \in \Gamma$. So we may define $\mathbf{e}$ as in the statement of the theorem. It remains to show that $\mathbf{e}$ evaluates the connectives correctly.

By **PC** and the definition of $\copyright A$ and **axiom 1**, $\vdash \copyright A \leftrightarrow \copyright{\sim}A$. So the proof that $\mathbf{e}$ evaluates ${\sim}$ correctly can proceed as in the proof of Theorem 7.

For the evaluation of $\neg$, if $\mathbf{e}(\neg A) = 1$, then $\neg A \in \Gamma$, so $A \notin \Gamma$, so ${\sim}A, \copyright A \in \Gamma$. Hence $\mathbf{e}(A) = 0$.

If $\mathbf{e}(A) = 0$, then ${\sim}A, \copyright A \in \Gamma$. By **PC** we have $\neg A \in \Gamma$. To show $\copyright\neg A \in \Gamma$, it is enough to show, by **PC**, $\neg{\sim}\neg A \in \Gamma$. But if not, then ${\sim}\neg A \in \Gamma$. So by **axiom 2**, $\neg{\sim}A \in \Gamma$, a contradiction on the consistency of Γ. So $\copyright\neg A \in \Gamma$. Hence $\mathbf{e}(\neg A) = 1$.

If $\mathbf{e}(A) = 1$ or $\frac{1}{2}$, then $A \in \Gamma$, so by completeness $\neg A \notin \Gamma$. Hence $\mathbf{e}(\neg A) = 0$.

If $\mathbf{e}(\neg A) = 0$, then $\neg A \notin \Gamma$. So by completeness, $A \in \Gamma$. Hence $\mathbf{e}(A) = 1$ or $\frac{1}{2}$. Hence $\mathbf{e}$ evaluates $\neg$ correctly.

For conjunction,

$$\begin{array}{llll}
\mathbf{e}(A\wedge B)=0 & \text{iff} & A\wedge B\notin\Gamma & \text{by the consistency of } \Gamma \\
 & \text{iff} & A\notin\Gamma \text{ or } B\notin\Gamma & \text{by } \mathbf{PC} \\
 & \text{iff} & \sim A, \copyright A\in\Gamma \text{ or } \sim B, \copyright B\in\Gamma & \\
 & \text{iff} & \mathbf{e}(A)=0 \text{ or } \mathbf{e}(B)=0 & \\
\mathbf{e}(A\wedge B)=1 & \text{iff} & A\wedge B, \copyright(A\wedge B)\in\Gamma & \\
 & \text{iff} & A, B, \neg\sim(A\wedge B)\in\Gamma & \text{by } \mathbf{PC} \\
 & \text{iff} & A, B, \neg\sim A, \neg\sim B\in\Gamma & \text{by } \mathbf{axiom\ 3} \\
 & \text{iff} & A, B, \copyright A, \copyright B\in\Gamma & \text{by } \mathbf{PC} \\
 & \text{iff} & \mathbf{e}(A)=1 \text{ and } \mathbf{e}(B)=1 &
\end{array}$$

So $\wedge$ is evaluated correctly, the other case following by process of elimination.

Finally, we consider the conditional. If $\mathbf{e}(A)\neq 0$ and $\mathbf{e}(B)=0$, then $A, \neg B\in\Gamma$ as in the proof for $\neg$ above. So $\neg(A\to B)\in\Gamma$. So, as in the proof for the evaluation of $\neg$, $\mathbf{e}(A\to B)=0$.

If $\mathbf{e}(A)=0$, then $\neg A\in\Gamma$. So by **PC** and **axiom 4**, $A\to B, \neg\sim(A\to B)\in\Gamma$. So by **PC**, $A\to B, \copyright(A\to B)\in\Gamma$. Hence $\mathbf{e}(A\to B)=1$.

If $\mathbf{e}(B)=1$, then $B, \copyright B\in\Gamma$. So by **PC**, $\neg\sim B\in\Gamma$. So by **axiom 5** and **PC**, $\neg\sim(A\to B), (A\to B)\in\Gamma$. And as above, $\mathbf{e}(A\to B)=1$.

If $\mathbf{e}(B)=\frac{1}{2}$, and $\mathbf{e}(A)\neq 0$, then $A, B, \sim B\in\Gamma$. Hence by **axiom 6** and **PC**, $\sim(A\to B), (A\to B)\in\Gamma$. So $\mathbf{e}(A\to B)=\frac{1}{2}$.

Hence, in all cases $\mathbf{e}$ evaluates $\to$ correctly. ∎

Lemma 11 If $\Gamma\nvdash A$, then there is a classically complete and consistent Σ such that $\Gamma\subseteq\Sigma$ and $A\notin\Sigma$.

The proof is as for **PC**.

Now it's routine to prove the following using Theorem 11 and Lemma 12.

Theorem 12 ***Strong Completeness of*** $\vdash_{\mathbf{J_3},\neg}$ $\Gamma\vdash_{\mathbf{J_3},\neg} A$ iff $\Gamma\vDash_{\mathbf{J_3}} A$

Exercises for Section E

1. Prove that the axiomatization $\vdash_{\mathbf{J_3},\Diamond}$ is sound.
2. Prove that the axiomatization $\vdash_{\mathbf{J_3},\neg}$ is sound.
3. (Open) Simplify the axiomatizations of this section to eliminate schema that are dependent
4. (Open) Give a strongly complete axiomatization of $\mathbf{J_3}$ in $L(\sim, \neg, \Diamond, \to, \wedge)$ such that if all axioms in which $\sim$ and $\Diamond$ appear are deleted, then we have a strongly complete axiomatization of $\mathbf{J_3}$ in $L(\neg, \to, \wedge)$; and similarly if we delete all axioms in which $\neg$ and $\to$ appear, we have a strongly complete axiomatization of $\mathbf{J_3}$ in $L(\neg, \to, \wedge)$. Use no defined connectives.

F. Set-Assignment Semantics for $\mathbf{J_3}$

In giving set-assignment semantics for $\mathbf{J_3}$ we must decide which negation, $\sim$ or $\neg$, should be modeled by the usual set-assignment table for negation. In this section I'll begin with $\neg$, which is most in keeping with the discussion in Section D.2 and Chapter IV.

Accordingly, we first take $L(\neg, \rightarrow, \wedge, \sim)$ as our language for $\mathbf{J_3}$. Then $<\mathsf{v},\mathsf{s},\mathsf{S}>$ is a $\mathbf{J_3}$-*model* for $L(\neg, \rightarrow, \wedge, \sim)$ if

$\neg, \rightarrow, \wedge$ are evaluated classically

$\mathsf{v}(\sim A) = \mathsf{T}$ iff $\mathsf{s}(A) \neq \mathsf{S}$

and s, S satisfy:

1. $\mathsf{S} \neq \varnothing$
2. $\mathsf{s}(p) \neq \varnothing$ iff $\mathsf{v}(p) = \mathsf{T}$
3. $\mathsf{s}(A) \subseteq \mathsf{s}(B)$ or $\mathsf{s}(B) \subseteq \mathsf{s}(A)$
4. $\mathsf{s}(A \wedge B) = \mathsf{s}(A) \cap \mathsf{s}(B)$
5. $\mathsf{s}(A \rightarrow B) = \begin{cases} \mathsf{s}(B) & \text{if } \varnothing \subset \mathsf{s}(A) \subset \mathsf{S} \\ \overline{\mathsf{s}(A)} \cup \mathsf{s}(B) & \text{otherwise} \end{cases}$
6. $\mathsf{s}(\neg A) = \begin{cases} \mathsf{S} & \text{if } \mathsf{s}(A) = \varnothing \\ \varnothing & \text{otherwise} \end{cases}$
7. $\mathsf{s}(\sim A) = \begin{cases} \mathsf{s}(A) & \text{if } \varnothing \subset \mathsf{s}(A) \subset \mathsf{S} \\ \overline{\mathsf{s}(A)} & \text{otherwise} \end{cases}$

From this point of view $\mathbf{J_3}$ *is classical logic with the addition of one wholly intensional connective,* $\sim$. To give a model we first assign each propositional variable (proposition) a truth-value (T or F), and then, choosing any collection of sets linearly ordered by inclusion, assign one set to each propositional variable in accord with condition (1), $\mathsf{s}(p) \neq \varnothing$ iff $\mathsf{v}(p) = \mathsf{T}$. Conditions (3)–(6) then allow an inductive definition of the set-assignment on Wffs, and hence of the valuation on all wffs.

Which are the false propositions? Those with no content. If A has some content, but not the full content S, then both A and its weak negation, $\sim A$, are true.

The proof of the following lemma is routine.

Lemma 13

a. $\mathsf{v}(A) = \mathsf{T}$ iff $\mathsf{s}(A) \neq \varnothing$

b. $\mathsf{v}(A \vee B) = \mathsf{T}$ iff $\mathsf{v}(A) = \mathsf{T}$ or $\mathsf{v}(B) = \mathsf{T}$

c. If there are only three possible content sets, $\varnothing \subset \mathsf{U} \subset \mathsf{S}$, then

$$\mathsf{s}(A \vee B) = \mathsf{s}(A) \cup \mathsf{s}(B)$$

Otherwise,

$$\mathsf{s}(A \vee B) = \begin{cases} \mathsf{s}(A) \cap \mathsf{s}(B) & \text{if both } \varnothing \subset \mathsf{s}(A) \subset \mathsf{S} \text{ and } \varnothing \subset \mathsf{s}(B) \subset \mathsf{S} \\ \mathsf{s}(A) \cup \mathsf{s}(B) & \text{otherwise} \end{cases}$$

d. $\upsilon(\Diamond A) = \mathsf{T}$ iff $\upsilon(A) = \mathsf{T}$

e.
$$\mathsf{s}(\Diamond A) = \begin{cases} \mathsf{S} & \text{if } \mathsf{s}(A) \neq \varnothing \\ \varnothing & \text{if } \mathsf{s}(A) = \varnothing \end{cases}$$

f.
$$\mathsf{s}(\Diamond A) = \begin{cases} \mathsf{S} & \text{if } \upsilon(A) = \mathsf{T} \\ \varnothing & \text{if } \upsilon(A) = \mathsf{F} \end{cases}$$

g. Given a $\mathbf{J_3}$-evaluation e: Wffs $\to \{0, \frac{1}{2}, 1\}$, define:

$\mathsf{s}(A) = \{x : x < \mathsf{e}(A) \text{ and } x \in [0,1]\}$

$\upsilon(p) = \mathsf{T}$ iff $\mathsf{e}(p) = 1$ or $\frac{1}{2}$

Then $<\upsilon, \mathsf{s}>$ is a $\mathbf{J_3}$-model and $<\upsilon, \mathsf{s}> \vDash A$ iff $\mathsf{e}(A) = 1$ or $\frac{1}{2}$.

h. Given a $\mathbf{J_3}$-model $<\upsilon, \mathsf{s}>$, define e: Wffs $\to \{0, \frac{1}{2}, 1\}$ by

$$\mathsf{e}(A) = \begin{cases} 1 & \text{if } \mathsf{s}(A) = \mathsf{S} \\ \frac{1}{2} & \text{if } \varnothing \subset \mathsf{s}(A) \subset \mathsf{S} \\ 0 & \text{if } \mathsf{s}(A) = \varnothing \end{cases}$$

Then e is a $\mathbf{J_3}$-evaluation and $\mathsf{e}(A) = 1$ or $\frac{1}{2}$ iff $<\upsilon, \mathsf{s}> \vDash A$.

Note that though $\Diamond A$ and A take the same truth-value in every model, they do not necessarily have the same content.

From parts (g) and (h), the consequence relation for these set-assignment semantics is the same as for the $\mathbf{J_3}$-matrix, which by Theorem 13 coincides with the syntactic consequence relation in this language.

Theorem 14 Strong Completeness of the Set-Assignment Semantics

$\Gamma \vDash_{\mathbf{J_3}} A$ iff every set-assignment $\mathbf{J_3}$-model that validates Γ also validates A

iff $\Gamma \vdash_{\mathbf{J_3}, \neg} A$

If we take $\mathbf{J_3}$ *to be formulated in the language* $L(\sim, \wedge, \Diamond)$, then we can define $<\upsilon, \mathsf{s}, \mathsf{S}>$ to be a $\mathbf{J_3}$-*model* if:

$\upsilon(A \wedge B) = \mathsf{T}$ iff $\upsilon(A) = \mathsf{T}$ and $\upsilon(B) = \mathsf{T}$

$\upsilon(\sim A) = \mathsf{T}$ iff $\mathsf{s}(A) \neq \mathsf{S}$

$\upsilon(\Diamond A) = \mathsf{T}$ iff $\upsilon(A) = \mathsf{T}$

and s, S satisfy:

1. $\mathsf{S} \neq \varnothing$
2. $\mathsf{s}(p) \neq \varnothing$ iff $\upsilon(p) = \mathsf{T}$

3. $s(A) \subseteq s(B)$ or $s(B) \subseteq s(A)$
4. $s(A \wedge B) = s(A) \cap s(B)$
5. $s({\sim}A) = \begin{cases} s(A) & \text{if } \varnothing \subset s(A) \subset S \\ \overline{s(A)} & \text{otherwise} \end{cases}$
6. $s(\Diamond A) = \begin{cases} S & \text{if } s(A) \neq \varnothing \\ \varnothing & \text{if } s(A) = \varnothing \end{cases}$

You can check that for these semantics we have:

$$v(\neg A) = T \quad \text{iff} \quad v(A) = F$$

$$v({\sim}A) = F \quad \text{iff} \quad v(A) = T \text{ and } s(A) = S$$

$$v(A \rightarrow B) = T \quad \text{iff} \quad v(A) = F \text{ or } v(B) = T$$

$$v(A \vee B) = T \quad \text{iff} \quad v(A) = T \text{ or } v(B) = T$$

And as above we can establish the following.

Theorem 15 $\Gamma \vDash_{J_3} A$ iff every set-assignment $\mathbf{J_3}$-model for $L({\sim}, \wedge, \Diamond)$ that validates Γ also validates A
iff $\Gamma \vdash_{J_3, \Diamond} A$

G. Truth-Default Semantics

Let us take the paraconsistent logician's point of view that the English 'not' is to be formalized as $\sim$, and $\neg$ is simply a defined connective that happens to correspond to classical negation. In that case it is not possible to give set-assignment semantics within the general framework of Chapter IV in such a way that we can translate $\mathbf{J_3}$-evaluations to set-assignment models and vice versa while satisfying:

$$e \vDash A \quad \text{iff} \quad <v,s> \vDash A$$

We do not have that if $e \vDash A$ then $e \nvDash {\sim}A$. The following table cannot be realized for any **N**:

A	N(A)	~ A
any value	fails	F
T	holds	F
F		T

The tables of the general framework of Chapter IV are based on the view that for a proposition to be true it must pass certain tests; if it fails any it is false. I have argued in Chapter IV and throughout this volume, as well as in *Epstein, 1992,* that this is correct: we analyze what it means for a proposition to be true, and every proposition that is not true is false.

The semantic intuitions behind $\mathbf{J_3}$ are a mirror image of this: we analyze what it means for a proposition to be false, and every proposition that is not false is true.

The general form of semantics of Chapter IV can be viewed as using falsity as the default truth-value. In $\mathbf{J_3}$ truth is taken as the default truth-value. Previously we have said: we cannot have both A and its negation true. For $\mathbf{J_3}$ we say: we cannot have both A and its negation false. It is precisely because these views are incompatible and yet are both represented in $\mathbf{J_3}$, albeit one of them derivatively, that I have used a different symbol, ~, for this negation. The appropriate form of the set-assignment table for (weak) negation is the following, where I have used the usual symbol for negation:

(1)

A	$\mathsf{N}(A)$	$\neg A$
any value	fails	T
T	holds	F
F		T

Because $\wedge$, $\vee$, and $\rightarrow$ are evaluated classically in $\mathbf{J_3}$ they can be presented by tables that take either truth or falsity as the default value, taking the relation governing the tables to be universal. Truth-default tables for these connectives have the following form:

(2)

A	B	$\mathsf{B}(A,B)$	$A \rightarrow B$
any	values	fails	T
T	T	holds	T
T	F		F
F	T		T
F	F		T

(3)

A	B	$\mathsf{C}(A,B)$	$A \wedge B$
any	values	fails	T
T	T	holds	T
T	F		F
F	T		F
F	F		F

(4)

A	B	$\mathsf{A}(A,B)$	$A \vee B$
any	values	fails	T
T	T	holds	T
T	F		T
F	T		T
F	F		F

Set-assignment semantics using tables (1)–(4) I call *truth-default set-assignment semantics.* I continue to call semantics of the general form of Chapter IV 'set-assignment semantics', but when there is a need to distinguish them from this alternate form, I will refer to them as *falsity-default set-assignment semantics.*

Similar definitions apply for relation based semantics. In Chapter IV.G.6 and 7, pp. 155–157, I discuss the general framework of truth-default semantics.

Consider now the set-assignment semantics for $\mathbf{J_3}$ in the language of $\sim, \wedge, \Diamond$ given in the last section. These are truth-default semantics, if we interpret $\sim$ as the standard formalization of 'not', because the evaluation of the connectives can be expressed equivalently as:

$$v(A \wedge B) = \mathsf{F} \quad \text{iff} \quad v(A) = \mathsf{F} \text{ or } v(B) = \mathsf{F}$$

$$v({\sim}A) = \mathsf{F} \quad \text{iff} \quad v(A) = \mathsf{T} \text{ and } s(A) \neq \mathsf{S}$$

$$v(\Diamond A) = \mathsf{F} \quad \text{iff} \quad v(A) = \mathsf{F}$$

and the defined connectives as:

$$v(A \rightarrow B) = \mathsf{F} \quad \text{iff} \quad v(A) = \mathsf{T} \text{ and } v(B) = \mathsf{F}$$

$$v(A \vee B) = \mathsf{F} \quad \text{iff} \quad v(A) = \mathsf{F} \text{ and } v(B) = \mathsf{F}$$

This points out that for all the set-assignment semantics for logics previously considered in this volume it is essential that the truth-value conditions "tag along" in the evaluation of the connectives for those logics to be considered falsity-default, even though the connective often could be evaluated as dependent only on the content of the constituent propositions.

We can also classify many-valued semantics as either falsity-default or truth-default. We say that a table for $\neg$, $\wedge$, $\vee$, or $\rightarrow$ in a many-valued matrix is *standard* if by renaming the designated values as T and the undesignated values as F, then the table is (with repetitions) the classical table for that connective. We say the table is *falsity-weighted* if, renaming as above, any row of the classical table that takes value F also takes value F in this table, whereas a row that takes value T in the classical table may take value F in the renamed many-valued table. A *truth-weighted* table is one in which any row of the classical table that takes value T also takes value T in the renamed many-valued table.

Every standard table is both falsity-weighted and truth-weighted. Every many-valued logic of Chapter VIII uses standard or falsity-weighted tables for each of $\neg$, $\rightarrow$, $\wedge$, $\vee$. Only the table for $\sim$ in $\mathbf{J_3}$ is truth-weighted.

A many-valued logic that uses truth-weighted tables for $\neg$, $\rightarrow$, $\wedge$, $\vee$, at least one of which is not standard, cannot be given falsity-default set-assignment semantics in such a way that we can translate many-valued models to set-assignment ones and vice versa while preserving validity in a model. It seems likely to me, though, that every truth-weighted many-valued matrix can be presented in terms of truth-default semantics, and every falsity-weighted many-valued matrix can be presented in terms of falsity-default semantics.

Exercises for Sections F and G

1. Prove Lemma 13.

2. For set-assignment semantics in $L(\sim, \wedge, \Diamond)$, show that:

   $v(\neg A) = T$ iff $v(A) = F$

   $v(\sim A) = F$ iff $v(A) = T$ and $s(A) = S$

   $v(A \rightarrow B) = T$ iff $v(A) = F$ or $v(B) = T$

   $v(A \vee B) = T$ iff $v(A) = T$ or $v(B) = T$

3. Prove Theorem 14.

4. Prove Theorem 15.

5. a. What is the difference between the usual set-assignment semantics and truth-default set-assignment semantics?
   b. Why do we call the usual set-assignment semantics 'falsity default'?
   c. Give examples from daily reasoning in which we seem to invoke falsity as the default truth-value.
   d. Give examples from daily reasoning in which we seem to invoke truth as the default truth-value.
   e. In writing computer programs, is it more common to choose truth or falsity as the default truth-value?
   f. Present the general tables for truth-default relation-based semantics.

6. What is meant by 'for logics previously considered in this volume it is essential that the truth-value conditions "tag along" in the evaluation of the connectives in order for those logics to be considered falsity-default'?

7. a. What is a standard table in a many-valued matrix? A falsity-weighted table? A truth-weighted table?
   b. Show that any many-valued matrix for $L(\neg, \rightarrow, \wedge, \vee)$ that is standard is strongly complete for **PC**.
   c. Prove that a many-valued logic that uses truth-weighted tables for $\neg$, $\rightarrow$, $\wedge$, $\vee$, at least one of which is not standard, cannot be given falsity-default set-assignment semantics in such a way that we can translate many-valued models to set-assignment ones and vice versa while preserving validity.
   d. (Open) In a uniform manner, for every falsity-weighted many-valued matrix give falsity-default set-assignment semantics.
   e. (Open) In a uniform manner, for every truth-weighted many-valued matrix give truth-default set-assignment semantics.

8. a. Why is it not appropriate to give truth-default set-assignment semantics to the minimal intuitionist logic **J**? (Hint: See the quotes by Johansson and Kolmogorov in Chapter VII.)
   b. Is there a translation of **J** into $\mathbf{J_3}$? Of $\mathbf{J_3}$ into **J**?

X Translations Between Logics

In the previous chapters we've seen many examples of interpretations of one logic in another. In Section A of this chapter Stanisław Krajewski and I formalize the notion of a translation, review the translations we've already encountered, present further examples, and raise some questions about them.

In Section B I attempt to give criteria for when a translation preserves meaning. I define what it means for a translation to be model preserving, and then define the stronger notion of a semantically faithful translation. I show that all the translations we have seen are semantically faithful, with the exception of the translations of classical logic into intuitionist logics. I conclude with a discussion of whether semantically faithful translations can be said to preserve meaning.

Wójcicki, 1988, §1.8.2, has a short history of other formalizations of the notion of translation.

A. Syntactic Translations

in collaboration with **Stanisław Krajewski**

1. A formal notion of translation

Throughout this volume I've referred to various interpretations of one logic into another as translations. In ordinary speech 'a translation' means a changing of some text from one language to another. Within the study of logic, the text to be changed is the logic itself. But what is 'a logic'?

If logics are presented to us as collections of theorems, then a translation should preserve theoremhood: theorems of one should be mapped into theorems of the other. But if this is all we mean by a translation, then the notion is trivial, for we can always enumerate the theorems of one logic and map them in order onto the theorems of another as we enumerate those. Similarly, the requirement that tautologies must be mapped to tautologies is trivial if we make the nonconstructive assumption that we can map any countable set into another. Nonetheless, particularly regular maps that preserve theoremhood or validity can be of interest, for example in establishing decidability or consistency, and those have sometimes been called 'translations' in the literature, so I will discuss them below. But preservation of validity or theoremhood alone seems too weak a criteria for a mapping to be a translation.

Generally we have considered logics to be either semantic or syntactic consequence relations. In that case what should be preserved by a translation is the consequence relation. This accords closely to the use of the term in the literature, and any such mapping preserves the essential syntactic aspect of the logic, for a consequence relation, whether presented semantically or syntactically, is a relation on collections of wffs.

For ease of exposition, throughout this chapter I will assume that *every logic is presented semantically.* By that I mean that either the logic is originally presented as a semantic consequence relation, or else there is a class of models that determines a semantic consequence relation that coincides with the syntactic consequence relation. All the definitions of Section A here apply equally to logics that are presented as syntactic consequence relations by replacing $\models$ by $\vdash$ throughout. I will use the boldface letters **L** and **M** to range over logics, and $L_{\mathbf{L}}$ and $L_{\mathbf{M}}$ for their respective languages. Given a mapping of languages, *, let us denote $\Gamma^* = \{A^*: A \in \Gamma\}$.

Definition 1 A *validity-preserving map* of a propositional logic **L** into a propositional logic **M** is a map * from $L_{\mathbf{L}}$ to $L_{\mathbf{M}}$ such that for every A,

$$\vDash_{\mathbf{L}} A \text{ iff } \vDash_{\mathbf{M}} A^*$$

A mapping is a *translation* if for every Γ and A,

$$\Gamma \vDash_{\mathbf{L}} A \text{ iff } \Gamma^* \vDash_{\mathbf{M}} A^*$$

We write $\mathbf{L} \hookrightarrow \mathbf{M}$ if there is a translation of **L** into **M**.

We do not require a translation to preserve the structure of the language being translated, but maps that are particularly regular are important to single out. The following definition generalizes to languages with different sets of connectives, fewer or more or different propositional variables, or propositional constants (for example, $\bot$), and I'll assume the appropriate generalizations below.

Definition 2 A map * from the language $L(\neg, \rightarrow, p_0, p_1, \ldots)$ to a formal language **L** is called *grammatical* if there are schema λ, φ, ψ of the language of **L** such that:

$$p^* = \lambda(p)$$
$$(\neg A)^* = \varphi(A^*)$$
$$(A \rightarrow B)^* = \psi(A^*, B^*)$$

Propositional constants may appear in these schema, but A* may contain no variables other than those appearing in $\lambda(p)$, where p appears in A. That is, the mapping may not depend on any *parameters*. Thus a grammatical map is a homomorphism between languages.

A map is *homophonic* for $\neg$ if $(\neg A)^* = \neg A^*$, homophonic for $\rightarrow$ if $(A \rightarrow B)^* = A^* \rightarrow B^*$, and homonphonic if it is homophonic for each connective.

A *grammatical translation* is a translation that is grammatical. We write $\mathbf{L} \twoheadrightarrow \mathbf{M}$ if there is a grammatical translation of **L** into **M**.

The proof of the following lemma is an exercise.

Lemma 3 ***a.*** The composition of translations is a translation.
b. The composition of grammatical maps is grammatical.

It might seem that if two logics **L** and **M** both have a deduction theorem and * is a grammatical map that preserves validity (theoremhood), then * must be a translation. The following theorem establishes that this is not so.

Theorem 4 There is a grammatical validity mapping between logics each of which has a deduction theorem that is not a translation.

Proof: The homophonic mapping of **PC** in $L(\neg, \wedge)$ to **Int** in $L(\neg, \rightarrow, \wedge, \vee)$ is grammatical but does not preserve consequences (Corollaries VII.17 and the remarks following Corollary VII.18, p. 292). ■

The definitions given here apply equally to logics that are presented as syntactic consequences relations if $\vDash$ is replaced by $\vdash$ and 'validity' by 'theorem'. In that case it is enough that the mapping preserves finite consequences, as syntactic consequence relations are compact. For logics that have both a syntactic and semantic presentation we have the following theorem, the proof of which I leave to you.

Theorem 5 If **L** and **M** are logics with strongly complete axiomatizations, then a mapping * from $L_{\mathbf{L}}$ to $L_{\mathbf{M}}$ is a translation iff for every *finite* Γ, $\Gamma \vDash_{\mathbf{L}} A$ iff $\Gamma^* \vDash_{\mathbf{M}} A^*$.

In some cases it is useful to be able to classify a map as preserving finite consequences, and I call such a map a *finite consequence translation.*

Exercises for Section A.1

1. a. What is a translation?
   b. What is a grammatical map?
   c. Why don't we classify a validity-preserving map as a translation?
   d. Give reasons why we should or should not consider a grammatical validity-preserving map to be a translation.
2. a. Prove that the composition of translations is a translation.
   b. Prove that the composition of grammatical maps is grammatical.
3. Prove Theorem 5.
4. Adapt the definition of grammatical map to a logic presented in $L(\rightarrow, \wedge, \vee, \bot)$.

2. Examples

Let's review the translations of the previous chapters in terms of our new terminology. In each case I will assume that the languages of both logics have the same stock of propositional variables $p_0, p_1, \ldots$. Unless noted otherwise, each propositional variable is translated to itself.

1. Classical logic to "itself"

$$\begin{array}{ccc} \mathbf{PC} & \twoheadrightarrow & \mathbf{PC} \\ L(\neg, \rightarrow, \wedge, \vee) & & L(\neg, \rightarrow) \end{array}$$

This was the first translation we saw (p. 55). The question arose there what it meant for "the same logic" to be formulated in different languages, which we'll return to in Section B.6.

2. *Classical logic to subject matter relatedness logic*

$$\begin{array}{ccc} \mathbf{PC} & \twoheadrightarrow & \mathbf{S} \\ L(\neg, \wedge) & & L(\neg, \rightarrow, \wedge) \end{array}$$

$$\begin{array}{ccc} \mathbf{PC} & \twoheadrightarrow & \mathbf{S} \\ L(\neg, \wedge) & & L(\neg, \rightarrow) \end{array}$$

When **S** is presented in $L(\neg, \rightarrow, \wedge)$ the homophonic map from $L(\neg, \wedge)$ is a translation (p. 112). If only $\neg$ and $\rightarrow$ are used as primitives for **S**, then $\neg$ is translated homophonically and $\wedge$ is translated to the defined version of $\wedge$ in **S** (p. 114):

$$(A \wedge B)^* = \neg(A^* \rightarrow (B^* \rightarrow \neg((A^* \rightarrow B^*) \rightarrow (A^* \rightarrow B^*))))$$

The same maps establish $\mathbf{PC} \twoheadrightarrow \mathbf{R}$ and $\mathbf{PC} \twoheadrightarrow \mathbf{Dual\ D}$. Similarly, $\mathbf{PC} \twoheadrightarrow \mathbf{D}$, except that the defined version of $\wedge$ within **D** is used (p. 177):

$$(A \wedge B)^* = \neg(((A^* \rightarrow B^*) \rightarrow (A^* \rightarrow B^*) \rightarrow A^*) \rightarrow \neg B^*)$$

By composing maps, presentations of classical logic in other languages can also be translated grammatically into these logics (Lemma 3).

3. *Dependence logic to dual dependence logic*
Dual dependence logic to dependence logic

$$\begin{array}{ccc} \mathbf{D} & \twoheadrightarrow & \mathbf{Dual\ D} \\ L(\neg, \rightarrow, \wedge) & & L(\neg, \rightarrow, \wedge) \end{array}$$

$$\begin{array}{ccc} \mathbf{Dual\ D} & \twoheadrightarrow & \mathbf{D} \\ L(\neg, \rightarrow, \wedge) & & L(\neg, \rightarrow, \wedge) \end{array}$$

The map translates $\neg$ and $\wedge$ homophonically and sets:

$$(A \rightarrow B)^* = \neg B^* \rightarrow \neg A^*$$

This is a translation *in both directions* (Theorem V.10, p. 185). Composing these we get translations of each logic to itself.

4. *The logic of equality of contents to dependence logic*

$$\begin{array}{ccc} \mathbf{Eq} & \twoheadrightarrow & \mathbf{D} \\ L(\neg, \rightarrow, \wedge) & & L(\neg, \rightarrow, \wedge) \end{array}$$

Both $\neg$ and $\wedge$ are translated homophonically, and

$$(A \rightarrow B)^* = (A^* \rightarrow B^*) \wedge (\neg B^* \rightarrow \neg A^*)$$

(Theorem V.13, p. 189)

The same translation establishes $\mathbf{Eq} \twoheadrightarrow \mathbf{Dual\ D}$.

5. *Any classical modal logic* **L** *to "itself"*

$$\underset{L(\neg,\wedge,\Box)}{\mathbf{L}} \twoheadrightarrow \underset{L(\neg,\rightarrow,\wedge)}{\mathbf{L}}$$

$$\underset{L(\neg,\rightarrow,\wedge)}{\mathbf{L}} \twoheadrightarrow \underset{L(\neg,\wedge,\Box)}{\mathbf{L}}$$

In each case both $\neg$ and $\wedge$ are translated homophonically. For the first map:

$$(\Box A)^* = \neg A^* \rightarrow A^*$$

For the second map:

$$(A \rightarrow B)^* = \Box(A^* \supset B^*)$$

(Theorem VI. 46, p. 262)

6. *Heyting's intuitionist logic to the classical modal logic* **S4**

$$\underset{L(\neg,\rightarrow,\wedge,\vee)}{\mathbf{Int}} \twoheadrightarrow \underset{L(\neg,\wedge,\Box)}{\mathbf{S4}}$$

Both $\wedge$ and $\vee$ are translated homophonically, and

$$p^* = \Box p$$
$$(A \rightarrow B)^* = \Box(A^* \supset B^*)$$
$$(\neg A)^* = \Box\neg(A^*)$$

(Theorem VII.12, p. 289)

The same map is a finite consequence translation of **Int** into **S4Grz**.

7. *The double negation translation of classical logic to Heyting's intuitionist logic*

$$\underset{L(\neg,\rightarrow,\wedge,\vee)}{\mathbf{PC}} \hookrightarrow \underset{L(\neg,\rightarrow,\wedge,\vee)}{\mathbf{Int}}$$

This translation (Theorem VII.16, p. 290) takes every A to $\neg\neg A$. It preserves the structure of the language being translated, but *the smallest unit of the language whose structure is preserved is an entire sentence.* Grammatical translations require that there be a specific structure corresponding to each connective.

8. *Gentzen's translation of classical logic to Heyting's intuitionist logic*

$$\underset{L(\neg,\rightarrow,\wedge,\vee)}{\mathbf{PC}} \twoheadrightarrow \underset{L(\neg,\rightarrow,\wedge,\vee)}{\mathbf{Int}}$$

In this translation (Theorem VII.20, pp. 292) each of $\neg$, $\rightarrow$, and $\wedge$ is translated homophonically and:

$$(p)^\circ = \neg\neg p$$
$$(A \vee B)^\circ = \neg(\neg A^\circ \wedge \neg B^\circ)$$

9. Classical logic to the minimal intuitionist calculus **J**

$$\mathbf{PC} \hookrightarrow \mathbf{J}$$
$$L(\neg, \rightarrow) \qquad L(\neg, \rightarrow)$$

Kolmogorov's translation (p. 309) preserves the structure of the language, though not grammatically, by preceding each subformula of a formula by $\neg\neg$.

10. Heyting's intuitionist logic to the minimal intuitionist calculus **J**

$$\mathbf{Int} \hookrightarrow \mathbf{J}$$
$$L(\neg, \wedge, \vee, \bot) \qquad L(\neg, \wedge, \vee, \bot)$$

This nongrammatical translation (Exercise 10, p. 310) translates each wff by replacing every one of its nonatomic subformulas B by $B \vee \bot$.

11. Classical logic to Łukasiewicz's 3-valued logic

$$\mathbf{PC} \twoheadrightarrow \mathbf{L_3}$$
$$L(\neg, \rightarrow) \qquad L(\neg, \rightarrow)$$

This translation (Corollary VIII.2, p. 322) sets:

$$(A \rightarrow B)^* = A^* \rightarrow (A^* \rightarrow B^*)$$
$$(\neg A)^* = A^* \rightarrow (A^* \rightarrow \neg(A^* \rightarrow A^*))$$

12. Classical logic to the paraconsistent logic $\mathbf{J_3}$

$$\mathbf{PC} \twoheadrightarrow \mathbf{J_3}$$
$$L(\neg, \rightarrow, \wedge) \qquad L(\neg, \rightarrow, \wedge, \sim)$$

$$\mathbf{PC} \twoheadrightarrow \mathbf{J_3}$$
$$L(\neg, \rightarrow, \wedge, \vee) \qquad L(\sim, \wedge, \Diamond)$$

When $\neg$ is taken as a primitive for $\mathbf{J_3}$ the homophonic map is a translation (Theorem IX.2, p. 357). That translation can be extended to the language $L(\neg, \rightarrow, \wedge, \vee)$ of **PC** by first translating **PC** to itself.

With primitives $\sim, \wedge, \Diamond$ for $\mathbf{J_3}$ the definitions of $\neg$ and $\rightarrow$ within $\mathbf{J_3}$ are used (Theorem IX.1, p. 356).

There is another mapping of classical logic into the paraconsistent logic $\mathbf{J_3}$ given by assuming that each proposition is "classical":

$$A^* = (\bigwedge_{\{p_i \text{ in } A\}} \copyright p_i) \rightarrow A'$$

Here A' is A with $\neg$ replaced by $\sim$. This nongrammatical mapping preserves validity but is not a translation (Theorem IX.3, p. 359).

Exercises for Section A.2

1. Give a grammatical translation of **PC** in $L(\neg, \vee)$ to **S** in $L(\neg, \rightarrow)$.
2. Give a grammatical translation of **PC** in $L(\neg, \vee)$ to **Dual D** in $L(\neg, \rightarrow)$.
3. In the translation of **D** to itself given by the composition maps in Example 3, what is the translation of the law of contraposition? of *Clavius' law*?
4. Prove that the map in Example 4 establishes **Eq** $\twoheadrightarrow$ **Dual D**.
5. Give a translation from $L(\neg, \rightarrow, \wedge)$ to $L(\neg, \wedge, \Diamond)$ of a modal logic to "itself". Is it grammatical?
6. Give a grammatical translation of **Int** in $L(\neg, \rightarrow, \wedge, \vee)$ to **S4** in $L(\neg, \rightarrow, \wedge)$.
7. In what sense is the structure of the language preserved in the validity mapping of classical logic into $\mathbf{J_3}$ given by assuming that each proposition is classical?

3. Logics that cannot be translated grammatically into classical logic

In the list above there is no translation of a logic into classical logic, other than classical logic itself. We'll see now that there can be no grammatical translation into **PC** of any logic we have studied. In doing so I'll outline a method due to Krajewski that can be applied to any logic that is presented in a language with $\neg$ and $\rightarrow$ among its primitives, though it easily generalizes to other languages.

Theorem 6 There is no grammatical translation into **PC** of any of the following logics, presented in languages with $\neg$ and $\rightarrow$ among their primitives:

R, **S**	(relatedness logics of Chapter III)
D, **Dual D**, **Eq**	(dependence logics of Chapter V)
S4, **S5**, **S4Grz**, **T**, **B**, **K**, **QT**, **MSI**, **ML**, **G**, **G***	(the classical modal logics of Chapter VI)
Int, **J**	(intuitionist logics of Chapter VII)
$\mathbf{L_3}$	(a many-valued logic from Chapter VIII)
$\mathbf{J_3}$	(a paraconsistent logic from Chapter IX)

Indeed, there is not even a validity-preserving grammatical map.

Proof: I'll present the method for an arbitrary logic **L** and then instantiate it for each of the logics above.

I will assume that **PC** is given in the language of $\neg$ and $\wedge$ and will use $\supset$ and $\equiv$ as defined symbols; the proof generalizes to any other presentation of **PC**. For the purposes of this proof only, I will write $\top$ for $\neg(p_1 \wedge \neg p_1)$ and $\perp$ for $(p_1 \wedge \neg p_1)$. I also adopt the convention of writing $\mathbf{L}\models$ for $\models_{\mathbf{L}}$ for a logic **L**.

To begin, note that for every wff A in the language of **PC** in which only one variable p appears, one of the following must hold:

$$\mathbf{PC} \vDash A \equiv p \qquad \mathbf{PC} \vDash A \equiv \neg p$$

$$\mathbf{PC} \vDash A \equiv \top \qquad \mathbf{PC} \vDash A \equiv \bot$$

Similarly, there are exactly sixteen truth-functions of 2 variables, and every B in the language of **PC** in which exactly two propositional variables appears must be evaluated semantically in **PC** by one of those. Thus, up to semantic equivalence, there are 256 possibilities for a triplet (λ, φ, ψ) (as in Definition 2 above) that establishes a grammatical map into **PC**.

Assume now that we are given a grammatical map * from the language of one of these logics, **L**, to the language of **PC** that is purported to preserve validity. We start by narrowing the possibilities for the map φ that translates negation, where I will leave to you to exhibit the necessary examples.

We cannot have $\mathbf{PC} \vDash \varphi(A^*) \equiv \bot$.
For each logic **L** we can find a wff A such that $\mathbf{L} \vDash \neg A$. If * were a translation, then we should have $\mathbf{PC} \vDash \varphi(A^*)$, but we have $\mathbf{PC} \nvDash \varphi(A^*)$.

We cannot have $\mathbf{PC} \vDash \varphi(A^*) \equiv \top$.
For each logic we can find a wff A such that $\mathbf{L} \nvDash \neg A$, yet we would have $\mathbf{PC} \vDash \varphi(A^*)$.

We cannot have $\mathbf{PC} \vDash \varphi(A^*) \equiv A^*$.
For each logic we can find a wff A such that $\mathbf{L} \vDash A$ and $\mathbf{L} \nvDash \neg A$, yet we cannot have both $\mathbf{PC} \vDash A^*$ and $\mathbf{PC} \nvDash A^*$.

Therefore, we must have $\mathbf{PC} \vDash \varphi(A^*) \equiv \neg A^*$. So without loss of generality we can assume $(\neg A)^* = \neg(A^*)$.

Though inessential to this proof it is worth investigating λ, the map that translates the variables.

We cannot have $\mathbf{PC} \vDash \lambda(p) \equiv \top$, since for each logic we have $\mathbf{L} \nvDash p$.

We cannot have $\mathbf{PC} \vDash \lambda(p) \equiv \bot$, for then, since $(\neg p)^* = \neg \lambda(p)$, we would have $\mathbf{PC} \vDash \neg \lambda(p)$, yet for each logic $\mathbf{L} \nvDash \neg p$.

Therefore, either $\mathbf{PC} \vDash \lambda(p) \equiv p$ or $\mathbf{PC} \vDash \lambda(p) \equiv \neg p$.

Consider now ψ, the map for the conditional. I will show a general form of counterexample to the various possibilities for ψ, and conclude the proof by instantiating the method for each of the logics. I will write, e.g., $\psi(\mathsf{F},\mathsf{T}) = \mathsf{F}$ to mean that the scheme of ψ is evaluated so in **PC**.

a. We cannot have $\psi(\mathsf{T},\mathsf{T}) = \mathsf{F}$.
For each logic we can find wffs A and B such that $\mathbf{L} \vDash A$, $\mathbf{L} \vDash B$, and $\mathbf{L} \vDash A \rightarrow B$. So, since ψ is a translation, we would have $\mathbf{PC} \vDash A^*$, $\mathbf{PC} \vDash B^*$, and $\mathbf{PC} \vDash \psi(A^*, B^*)$. Yet this evaluation of ψ should yield $\mathbf{PC} \nvDash \psi(A^*, B^*)$.

b. We cannot have $\psi(\mathsf{T},\mathsf{F}) = \mathsf{T}$.
For each logic we can find wffs A and B such that $\mathbf{L}\vDash A$, $\mathbf{L}\vDash \neg B$, and $\mathbf{L}\nvDash A \to B$. In that case, as $(\neg B)^* = \neg(B^*)$, we would have $\mathbf{PC}\vDash A^*$, $\mathbf{PC}\vDash \neg B^*$, and $\mathbf{PC}\nvDash \psi(A^*,B^*)$. But if $\mathbf{PC}\vDash \neg B^*$ then B^* is false in every model, so this evaluation of ψ should yield $\mathbf{PC}\vDash\psi(A^*,B^*)$.

c. We cannot have $\psi(\mathsf{F},\mathsf{T}) = \mathsf{F}$.
For each logic we can find wffs A and B such that $\mathbf{L}\vDash\neg A$, $\mathbf{L}\vDash B$, and $\mathbf{L}\vDash A\to B$. Then, as for (b), we would have $\mathbf{PC}\vDash\neg A^*$, $\mathbf{PC}\vDash B^*$, and $\mathbf{PC}\vDash\psi(A^*,B^*)$. And so A^* would have to be false in every **PC**-model and we should have $\mathbf{PC}\nvDash\psi(A^*,B^*)$.

d. We cannot have $\psi(\mathsf{F},\mathsf{F}) = \mathsf{F}$.
For each logic we can find wffs A and B such that $\mathbf{L}\vDash\neg A$, $\mathbf{L}\vDash\neg B$, yet $\mathbf{L}\vDash A\to B$. The argument then follows as in (c).

Hence we must have:

e. $\mathbf{PC}\vDash \psi(A,B) \equiv (A^*\supset B^*)$
For this possibility we have to find a counterexample peculiar to each logic.

I'll now exhibit the examples for (a)–(e) for each of the logics of the theorem, under the assumption that $(\neg A)^* = \neg(A^*)$.

If **L** is **R**, **S**, **D**, **Dual D**, **Eq**, any of the modal logics of Chapter VI, or **Int**, the following serve as examples for (a)–(d):

a. $\mathbf{L}\vDash(p_1\to p_1)$, and $\mathbf{L}\vDash(p_1\to p_1)\to(p_1\to p_1)$

b. $\mathbf{L}\vDash(p_1\to p_1)$, $\mathbf{L}\vDash\neg(p_1\wedge\neg p_1)$, and $\mathbf{L}\nvDash(p_1\to p_1)\to(p_1\wedge\neg p_1)$

c. $\mathbf{L}\vDash\neg(p_1\wedge\neg p_1)$, $\mathbf{L}\vDash(p_1\to p_1)$, and $\mathbf{L}\vDash(p_1\wedge\neg p_1)\to(p_1\to p_1)$

d. $\mathbf{L}\vDash\neg(p_1\wedge\neg p_1)$, and $\mathbf{L}\vDash(p_1\wedge\neg p_1)\to(p_1\wedge\neg p_1)$

And, except for **Int**, a counterexample for (e) is:

e. $\mathbf{L}\nvDash p_1\to(p_2\to p_1)$ but $\mathbf{PC}\vDash p_1{}^*\supset(p_2{}^*\supset p_1{}^*)$

For **Int**, to give a counterexample for (e) note that we would have $(\neg\neg p_1\to p_1)^* = \neg\neg p_1{}^*\supset p_1{}^*$, and

$\mathbf{Int}\nvDash\neg\neg p_1\to p_1$ but $\mathbf{PC}\vDash\neg\neg p_1{}^*\supset p_1{}^*$

For **J** the example for (a) above will do, and a counterexample for (e) is the same as for **Int**. For (b)–(d) the examples above work if the **J**-anti-tautology $\neg(p_1\to p_1)$ is substituted for $\neg(p_1\wedge\neg p_1)$.

If **L** is $\mathbf{L_3}$ then the example for (a) above will do. For parts (b)–(d) substitute the $\mathbf{L_3}$-anti-tautology $(p_1\wedge\neg p_1\wedge(p_1\leftrightarrow\neg p_1))$ for $(p_1\wedge\neg p_1)$ in the examples above. For part (e), let A be $(p_1\leftrightarrow\neg p_1)$. Then:

$\mathbf{L_3}\nvDash(\neg A\to A)\to A$ but $\mathbf{PC}\vDash(\neg A^*\supset A^*)\supset A^*$

If **L** is $\mathbf{J_3}$, then the the example for (a) above will do, and for parts (b)–(d) substitute the $\mathbf{J_3}$-anti-tautology $(p_1 \wedge {\sim}p_1 \wedge ©p_1)$ for $(p_1 \wedge \neg p_1)$. Here we are using $\neg$ as a primitive of the language for $\mathbf{J_3}$. We could instead use ~ , since the arguments concerning $\neg$ above apply equally to ~. So for (e) we have:
$\mathbf{PC} \models ((p_1 \wedge {\sim}p_1) \rightarrow p_2)^* \equiv (p_1{}^* \wedge \neg p_1{}^*) \supset p_2{}^*$. And then

$$\mathbf{J_3} \not\models (p_1 \wedge {\sim}p_1) \rightarrow p_2 \text{ but } \mathbf{PC} \models (p_1{}^* \wedge \neg p_1{}^*) \supset p_2{}^*.$$ ■

Exercises for Section A.3

1. Provide examples for each of the logics that show that a validity-preserving mapping of the logic into **PC** must translate $(\neg A)^* = \neg A^*$.
2. Prove that there is no grammatical translation of **D** into **Eq** (see p. 193).
3. Call a map *grammatical with parameters* if parameters are allowed in the schema of Definition 2.
   a. (*Krajewski, 1992*) Show that there is no grammatical translation with parameters of any of the logics of Theorem 6 into **PC**.
   b. (Open) Is there a "natural" example of two logics **L** and **M** such that there is a grammatical translation with parameters from **L** into **M**, but none without parameters?
4. (Open) Is there a nontrivial example of a logic that can be translated grammatically into **PC**? (By 'nontrivial' I mean to exclude *ad hoc* examples and fragments of **PC** or of other logics that wouldn't normally be considered well-motivated logics in their own right.)

4. Translations where there are no grammatical translations: R ↪ PC and S ↪ PC

There is no grammatical translation of **R** into **PC**. Nonetheless, despite the much richer semantic basis of **R**, I will show that there is a translation of **R** into **PC**. We can then modify that translation to produce one from **S** into **PC**. These maps preserve the structure of the language of **R** and **S**, though only in terms of sentences as wholes, and they seem to reflect the underlying semantic assumptions of **R** and **S**.

From the proof of Theorem 6 we see that there is no grammatical translation of **R** into **PC** because the only plausible candidate for $\psi(A, B)$ is $A^* \supset B^*$. But that map does not allow us to take into account whether the two formulas are related. If we consider the fact of whether p_i is related to p_j *as a new proposition* then we can produce a translation. Note that in the semantics for **R** every relation governing the table for $\rightarrow$ is reflexive (though not necessarily symmetric), so that p_i is related to p_i must be represented as a tautology.

In order to simplify the presentation of the translation I will translate from the language $L(\neg, \rightarrow, \wedge, \vee, p_0, p_1, \ldots)$ of **R** to a language for **PC** with variables $q_0, q_1, \ldots$ as well as $\{d_{ij}: i, j \text{ are natural numbers}\}$, and primitives $\neg$, $\wedge$, that is, $L(\neg, \wedge, q_0, q_1, \ldots, \{d_{ij}: i, j \text{ are natural numbers}\})$. I take $\supset$ and $\vee$ to be defined for **PC** in the usual way. It may seem that by using this language for **PC** we are introducing two sorts of variables and that, therefore, we are no longer working with the usual classical logic. But a variable is still just a variable, no matter how we designate it. We can achieve the same effect by partitioning the usual list of variables $p_0, p_1, \ldots$ into two classes, taking one class to be those with even indices and the other with odd, but that is harder to read (see Exercise 3 below).

I will use the notation $\bigvee\!\!\bigvee_{A \text{ in } \Gamma} A$ to mean the disjunction of the finite number of wffs in Γ, reading from left to right in the usual ordering of those wffs and associating to the left. I write $\{p_i \text{ in } A\}$ for $\{i: p_i \text{ appears in } A\}$.

The map * is given by:

$$(p_i)^* = q_i$$

$$(\neg A)^* = \neg(A^*)$$

$$(A \rightarrow B)^* =$$

$$(A^* \supset B^*) \wedge [\bigvee\!\!\bigvee_{\{p_i \text{ in } A,\, p_j \text{ in } B\}} d_{ij} \vee \bigvee\!\!\bigvee_{\{p_i \text{ in } A,\, p_i \text{ in } B\}} (d_{ii} \vee \neg d_{ii})]$$

That is, $A \rightarrow B$ is translated to the material implication conjoined with the disjunction of the propositions that are to be taken to assert that the variables appearing in A are related to those of B.

Theorem 7 The map * is a translation of **R** into **PC**.

Proof: For the defined symbols $\wedge$, R of the language of **R** we can prove:

$$\mathbf{PC} \vDash (A \wedge B)^* \equiv (A^* \wedge B^*)$$

$$\mathbf{PC} \vDash (R(A,B))^* \equiv$$

$$\bigvee\!\!\bigvee_{\{p_i \text{ in } A,\, p_j \text{ in } B\}} d_{ij} \vee \bigvee\!\!\bigvee_{\{p_i \text{ in } A,\, p_i \text{ in } B\}} (d_{ii} \vee \neg d_{ii})$$

In particular, for $i \neq j$, $\mathbf{PC} \vDash (R(p_i, p_j))^* \equiv d_{ij}$ and $\mathbf{PC} \vDash (R(p_i, p_i))^*$.

I will now reduce the problem to proving that * preserves validity. By Theorem 5 we need only show that for every finite Γ, $\Gamma \vDash_{\mathbf{R}} A$ iff $\Gamma^* \vDash_{\mathbf{PC}} A^*$. We have a *Semantic Deduction Theorem* for **R** in the form: $\{A_1, \ldots, A_n\} \vDash_{\mathbf{R}} A$ iff $\vDash_{\mathbf{R}} \neg((A_1 \wedge \cdots \wedge A_n) \wedge \neg A)$ (Exercise 9, p. 125). For **PC** we have the same (Exercise 8, p. 76). Because $\neg$ is translated homophonically and $A \wedge B$ is translated to a wff semantically equivalent to $A^* \wedge B^*$, it suffices to prove $\mathbf{R} \vDash A$ iff $\mathbf{PC} \vDash A^*$.

To show that if $\mathbf{R} \vDash A$ then $\mathbf{PC} \vDash A^*$ we can use the complete axiomatization of **R** given in Chapter III.K.3, p. 121. It is not difficult to check that if A is an axiom of **R** then $\mathbf{PC} \vDash A^*$, since $\mathbf{R} \vDash R(p,p)$ and

$$\mathbf{R} \vDash R(A,B) \text{ iff } \mathbf{R} \vDash \bigvee\!\!\bigvee_{\{p_i \text{ in } A,\, p_j \text{ in } B\}} R(p_i, p_j)$$

Moreover, if $\mathbf{PC} \vDash (A \to B)^*$ and $\mathbf{PC} \vDash A^*$, then $\mathbf{PC} \vDash B^*$, so by induction on the length of a proof of A, we have that if $\mathbf{R} \vdash A$, then $\mathbf{PC} \vDash A^*$.

Now suppose that $\mathbf{R} \nvDash A$. Then there is a model $\langle v, \mathsf{R} \rangle$ such that $\langle v, \mathsf{R} \rangle \nvDash A$. We want to define a **PC**-model w in which A^* is false; but I will do more. I will show that given any model $\langle v, \mathsf{R} \rangle$ of **R** there is a model w of **PC** such that for all B,

(1) $\qquad v(B) = \mathsf{T}$ iff $w(B^*) = \mathsf{T}$

That proof can then be read as establishing a 1-1 correspondence between **PC**-models and **R**-models, since the same definition converts any model w of **PC** into one for **R** satisfying (1).

We define the model by:

$w(q_i) = \mathsf{T}$ iff $v(p_i) = \mathsf{T}$

$w(d_{ij}) = \mathsf{T}$ iff $\mathsf{R}(p_i, p_j)$ holds

We can now prove that $\langle v, \mathsf{R} \rangle \vDash B$ iff $w \vDash B^*$ by induction on the length of the formula B. The only interesting step is if B is $C \to D$:

$v(C \to D) = \mathsf{T}$ iff $(v(C) = \mathsf{F}$ or $v(D) = \mathsf{T})$ and $\mathsf{R}(C, D)$

iff $(v(C) = \mathsf{F}$ or $v(D) = \mathsf{T})$ and (some p_i in C, some p_j in D, $\mathsf{R}(p_i, p_j)$ or some p_i appears in both C and D)

which by induction is

iff $(w(C^*) = \mathsf{F}$ or $w(D^*) = \mathsf{T})$ and (some p_i in C, some p_j in D, $w(d_{ij}) = \mathsf{T}$ or $(d_{ii} \vee \neg d_{ii})$ appears in $(C \to D)^*$)

iff $w((C^* \supset D^*) \wedge [\bigvee_{\{p_i \text{ in } C,\, p_j \text{ in } D\}} d_{ij} \vee \bigvee_{\{p_i \text{ in } C \text{ and } p_i \text{ in } D\}} (d_{ii} \vee \neg d_{ii})]) = \mathsf{T}$

So in particular, for the model that does not validate A we have $w(A) = \mathsf{F}$, and the proof is complete. ■

Theorem 8 $\quad \mathbf{S} \hookrightarrow \mathbf{PC}$

Proof: Recall (p. 121):

$\mathbf{S} = \mathrm{Th}(\mathbf{R} \cup \{\mathsf{R}(A,B) \to \mathsf{R}(B,A) : A, B \in \text{Wffs}\}$

Only a slight modification of the translation of **R** is needed: for $(A \to B)^*$ read $d_{ij} \wedge d_{ji}$ where d_{ij} appeared previously. You can check that **PC** validates the translation of $\mathsf{R}(A,B) \to \mathsf{R}(B,A)$, and the proof then is the same as for **R**. ■

In the proof of Theorem 7 we used that the logics of the translation have semantic deduction theorems to reduce the question of whether a map is a translation to whether it preserves validity. The method is often applicable, though not universally so (Theorem 4). In summary the method is:

(2) $\{A_1, \ldots, A_n\} \vDash_{\mathbf{L}} A$ iff $\vDash_{\mathbf{L}} \beta_n(A_1, \ldots, A_n, A)$ — *Deduction Theorem* for **L**

iff $\vDash_{\mathbf{M}} (\beta_n(A_1, \ldots, A_n, A))^*$ — as * preserves validity

iff $\vDash_{\mathbf{M}} \delta_n(A_1^*, \ldots, A_n^*, A^*)$ — because * is grammatical

iff $\vDash_{\mathbf{M}} \gamma_n(A_1^*, \ldots, A_n^*, A^*)$ — by some internal translation within **M**

iff $\{A_1^*, \ldots, A_n^*\} \vDash_{\mathbf{M}} A^*$ — *Deduction Theorem* for **M**

Though the mapping of **R** into **PC** is not grammatical, it does preserve the structure of the language of **R**. First reinterpret $A \rightarrow B$ in **R** using the defined connectives $\wedge$, R, and (truth-functional inclusive) $\vee$ as:

$$\gamma(A \rightarrow B) = \neg(A \wedge \neg B) \wedge$$
$$[\bigvee_{\{p_i \text{ in A},\, p_j \text{ in B}\}} R(p_i, p_j) \vee (\bigvee_{\{p_i \text{ in A},\, p_i \text{ in B}\}} (R(p_i, p_i) \vee \neg R(p_i, p_i))]$$

Then * preserves the structure of these sentences, mapping $R(p_i, p_j)$ to d_{ij} (compare the *Normal Form Theorem* for **S**, Corollary III.6, p. 115). The same holds for the translation of **S** into **PC**.

By the standards of subject matter relatedness logic many classically valid arguments are *enthymematic*. That is, one or more suppressed implicit premisses are needed to make them valid. For example, the following argument is valid in **PC**:

If $1+1=2$, then Mary has two children
If Mary has two children, then John loves Mary
Therefore:
If $1+1=2$, then John loves Mary

But it is enthymematic in **S**. We need the suppressed premisses:

'$1+1=2$' has something in common with 'Mary has two children'

'Mary has two children' has something in common with 'John loves Mary'.

'$1+1=2$' has something in common with 'John loves Mary'.

The translation of **S** into **PC** described in the previous paragraph is the formal counterpart of this view.

In what ways are grammatical translations preferable to nongrammatical ones? It can't be on the grounds of simplicity and regularity, as the double negation translation of **PC** into **Int** shows. But that is a global translation and does not reduce the connectives of the one logic to those of the other. Grammatical translations by preserving structure yield a *translation of the connectives.*

But then what about the translation of **R** into **PC**? The connectives are translated into specific structures, and once the semantic assumptions of these logics are understood, the translation is natural. However, this translation involves metalogical machinery that does not seem to be inherently propositional in nature: paying attention to the indices of variables or translating a compound proposition into an atomic one.

Grammatical translations are those that translate connectives and all the structure of propositions in a manner consonant with the propositional nature of the logics.

Exercises for Section A.4

1. For the translation of **R** into **PC** prove for the defined symbols $\wedge$ and R:

   PC$\models (A \wedge B)^* \equiv (A^* \wedge B^*)$

   PC$\models (R(A,B))^* \equiv \bigvee\!\!\bigvee_{\{p_i \text{ in } A,\ p_j \text{ in } B\}} d_{ij} \vee \bigvee\!\!\bigvee_{\{p_i \text{ in } A,\ p_i \text{ in } B\}} (d_{ii} \vee \neg d_{ii})$

2. Prove that given a decision procedure for **PC** we can produce a decision procedure for **R** based on it that takes no more than a simple polynomial function of the time of the original procedure.

3. Show that the following mapping is a translation of **R** into **PC** in $L(\neg, \rightarrow, p_0, p_1, \ldots)$:

   $(p_i)^* = p_{2i}$

   $(\neg A)^* = \neg(A^*)$

   $(A \rightarrow B)^* =$ as before reading $p_{3^i 5^j}$ for d_{ij}

4. a. (Open) Show that there is no map * that is a translation of **D** into **PC** for which $(p)^* = p$, $(\neg A)^* = \neg(A^*)$, and $(A \wedge B)^* = (A^* \wedge B^*)$.

   b. (Open) Is there any translation of **D** into **PC**?

   c. (Open) Is there a translation of **Int** into **PC**?

   d. (Open) Is there an example of two logics such that there is no translation from the first to the second?

B. Semantically Faithful Translations

What does it mean to say that a translation preserves meaning? The meanings of words and sentences in the one language should be reconstructed, understood, in terms of the meanings ascribed to words and sentences in the other language. To give a formal analysis of this idea requires some uniform conception of how meanings are ascribed to languages. This is the role of the general framework for semantics of Chapter IV.

I will propose a definition here that I believe captures necessary conditions for a translation between propositional logics to preserve meaning. Whether the definition expresses sufficient conditions cannot be resolved easily, for that depends in part on whether particular semantics for a logic can be said to give meanings to the language of the logic. However, for the translations we have seen in this volume I will show that the definition accords with our pre-formal intuitions. In particular, the best known example of a translation between nonclassical logics that is generally thought to recapture the meaning of the one within the other, **Int** $\twoheadrightarrow$ **S4**, will be classified as preserving meaning, while the problematic, unintuitive translations of **PC** into **Int** do not preserve meaning with respect to their intended models.

The definitions I propose will be for logics that have set-assignment semantics. I am not sure how to extend them to relation-based logics.

1. A formal notion of semantically faithful translation

How do we justify that a mapping is a translation from a logic **L** to a logic **M**? Given a mapping * from L_L to L_M we need to show:

(a) If $\Gamma \vDash_L A$, then $\Gamma^* \vDash_M A^*$.

(b) If $\Gamma^* \vDash_M A^*$, then $\Gamma \vDash_L A$.

Part (a) can sometimes be accomplished by using a strongly complete axiomatization of **L**: axioms of **L** are shown to translate into tautologies of **M**, and the rules of **L** correspond to derived rules on the set of translated wffs in L_M. Deduction theorems for the logics are then invoked to complete the proof, as in the proof of Theorem 7, (see (2) p. 387).

But even when available, that method does not work to prove (b), nor does it give any insight into how, or even whether the semantics of **L** can be reconstructed within those of **M**. In the text we have used semantic justifications, which proceed in the following way.

We prove (a) by showing that if $\Gamma^* \nvDash_M A^*$, then $\Gamma \nvDash_L A$. If $\Gamma^* \nvDash_M A^*$, then there is a model M of **M** such that $\mathsf{M} \vDash_M \Gamma^*$ and $\mathsf{M} \nvDash_M A^*$. So we find a model L of **L** such that for all B,

(3) $\quad \mathsf{L} \vDash_L B$ iff $\mathsf{M} \vDash_M B^*$

Then $\mathsf{L} \vDash_L \Gamma$ and $\mathsf{L} \nvDash_L A$. A correspondence of models is set up; if every model of **L** can be derived from some M satisfying (3), then we also have (b).

What can we say about this correspondence of models? Consider:

$$\text{Wffs}^* = \{A^* \in L_M : A \in L_L\}$$

The model M restricted to Wffs* determines the model L of **L** via (3). Or at least it determines what wffs L validates, since there may be other models of **L** that validate the same wffs.

The model L, on the other hand, need not determine M. For example, for the translation **PC** $\twoheadrightarrow$ $\mathbf{L_3}$ (Corollary VIII.2, p. 322) 2-valued models of **PC** are derived from 3-valued models of $\mathbf{L_3}$ via the condition $v(p) = \mathsf{T}$ iff $e(p) = 1$. The same

2-valued model arises whether the 3-valued evaluation assigns 0 or $\frac{1}{2}$ to variables to which it does not assign 1.

The correspondence, then, is from the class of models $\mathcal{M}$ of **M** to the class of models $\mathcal{L}$ of **L**. Denoting the model correspondence also by *, we have:

$$L_{\mathbf{L}} \xrightarrow{*} L_{\mathbf{M}}$$

$$\mathcal{L} \xleftarrow{*} \mathcal{M}$$

Where this method was used in the text, the model correspondences were established between the idiosyncratic models of each logic. How can we recast them within the general framework?

Given a model $<v,s,S>$ of **M**, how do we determine $<v^*,s^*,S^*>$? Condition (3) tells us what variables are accounted as true:

$$v^*(p) = v(p^*)$$

The analogous condition to impose on content is:

$$s^*(A) = s(A^*)$$

This may be too strong, for it requires that $S^* \subseteq S$. Sufficient would be a map $\theta : S \to S^*$ such that:

$$s^*(A) = \theta(s(A^*))$$

But such a θ would introduce serious complications into any discussion of whether a translation preserves meaning and appears unnecessary if appropriate content sets are used for the semantics. So I make the following definitions.

Definition 9 Let **L**, **M** be logics in languages $L_{\mathbf{L}}$, $L_{\mathbf{M}}$ respectively, and $\mathcal{L}$, $\mathcal{M}$ classes of set-assignment models that determine their semantics.

a. Two models L and L′ of $\mathcal{L}$ are *elementarily equivalent* if for all B,

$$L \vDash_{\mathbf{L}} B \text{ iff } L' \vDash_{\mathbf{L}} B$$

b. A mapping $* : L_{\mathbf{L}} \to L_{\mathbf{M}}$ is *model-preserving up to elementary equivalence* with respect to $\mathcal{L}$ and $\mathcal{M}$ if it induces a mapping $* : \mathcal{M} \to \mathcal{L}$, $M = <v,s,S> \mapsto <v^*,s^*,S^*> = M^*$, that satisfies:

$$S^* \subseteq S$$

$$s^*(A) = s(A^*) \quad \text{for every wff } A \in L_{\mathbf{L}}$$

$$v^*(A) = v(A^*) \text{ for every } A \in L_{\mathbf{L}}$$

And for every $L \in \mathcal{L}$ there is some $M \in \mathcal{M}$ such that M^* is elementary equivalent to L.

If the logic **L** is **PC** and $\mathcal{L}$ consists entirely of models that have no set-assignments, then all references to s and S in the definition are deleted.

c. A mapping $* : L_L \to L_M$ is *model-preserving* with respect to $\mathcal{L}$ and $\mathcal{M}$ if it is model-preserving up to elementary equivalence with respect to $\mathcal{L}$ and $\mathcal{M}$ and the mapping $* : \mathcal{M} \to \mathcal{L}$ is onto.

Theorem 10 If $*$ is model-preserving up to elementary equivalence, then $*$ is a translation.

Proof: Suppose that $*$ is model-preserving up to elementary equivalence with respect to models $\mathcal{L}$ and $\mathcal{M}$ that determine the logics **L** and **M**. Then:

$$\Gamma^* \nvDash_M A^* \text{ iff there is some } \mathsf{M} \in \mathcal{M} \text{ such that } \mathsf{M} \vDash_M \Gamma^* \text{ and } \mathsf{M} \nvDash_M A^*$$

$$\text{iff there is some } \mathsf{M} \in \mathcal{M} \text{ such that } \mathsf{M}^* \vDash_L \Gamma \text{ and } \mathsf{M}^* \nvDash_L A$$

So if $\Gamma^* \nvDash_M A^*$, then $\Gamma \nvDash_L A$. And if $\Gamma \nvDash_L A$, then there is some model $\mathsf{L} \in \mathcal{L}$ such that $\mathsf{L} \vDash_L \Gamma$ and $\mathsf{L} \nvDash_L A$. There is then some model $\mathsf{M} \in \mathcal{M}$ such that M^* is elementarily equivalent to L, so $\mathsf{M} \vDash_M \Gamma^*$ and $\mathsf{M} \nvDash_M A^*$. ■

Corollary 11 If $*$ is model-preserving, then $*$ is a translation.

Corollary 11 is easier to prove directly:

$$\Gamma \vDash_L A \text{ iff for every } \mathsf{L} \in \mathcal{L}, \text{ if } \mathsf{L} \vDash_L \Gamma \text{ then } \mathsf{L} \vDash_L A$$

$$\text{iff for every } \mathsf{M} \in \mathcal{M}, \text{ if } \mathsf{M}^* \vDash_L \Gamma \text{ then } \mathsf{M}^* \vDash_L A$$

$$\text{iff for every } \mathsf{M} \in \mathcal{M}, \text{ if } \mathsf{M} \vDash_M \Gamma^* \text{ then } \mathsf{M} \vDash_M A^*$$

$$\text{iff } \Gamma^* \vDash_M A^*$$

We can define a mapping to be *content variant model-preserving* (*up to elementary equivalence*) by requiring instead of $\mathsf{S}^* \subseteq \mathsf{S}$ in the definitions above that there be a map $\theta : \mathsf{S} \to \mathsf{S}^*$ such that $\mathsf{s}^*(A) = \theta(\mathsf{s}(A^*))$. I will not investigate such mappings here.

Earlier I said that for a translation to preserve meaning, the meanings of words and sentences in the one language should be reconstructed in terms of the meanings ascribed to words and sentences in the other language. Within the limited scope of propositional logic the only words that are of concern are those that are formalized by the connectives. To preserve the structure and meaning of those, we know from Section A the mapping should be grammatical.

Definition 12 A translation is *semantically faithful* if it is grammatical and is model preserving with respect to strongly complete semantics for the logics.

I'll leave the proof of the following theorem to you.

Theorem 13 ***a.*** The composition of maps that are model-preserving up to elementary equivalence is model-preserving up to elementary equivalence.

b. The composition of model-preserving maps is model-preserving.

c. The composition of semantically faithful maps is semantically faithful.

Why require semantically faithful translations to be model-preserving rather than just model preserving up to elementary equivalence? If we have a strongly complete class of models for a logic that contains elementary equivalent models, then either the distinctions between those models in terms of their content is significant, and hence should be respected by a mapping we wish to call semantically faithful. Or else the distinctions are inessential, and hence the class of models should be pared down to give a reduced class that is strongly complete for the logic and that still can be interpreted as giving meaning to the language.

With a semantically faithful translation we can completely reconstruct the semantics of the original logic within the semantics of the logic into which we are translating. In the correlation of models, the translation of a connective has the same table as the connective in the original logic. Suppose that * is a semantically faithful translation from **L** to **M**, and L is a model of **L**. Then for some model $<\mathsf{v},\mathsf{s}>$ of **M**, $\mathsf{L} = <\mathsf{v}^*,\mathsf{s}^*>$. And, for example,

(4) $\mathsf{v}^*(A \rightarrow B) = \mathsf{T}$ iff (not both $\mathsf{v}^*(A) = \mathsf{T}$ and $\mathsf{v}^*(B) = \mathsf{F}$) and $\mathsf{B}(\mathsf{s}^*(A), \mathsf{s}^*(B))$

But also, since * is grammatical,

$$\mathsf{v}^*(A \rightarrow B) = \mathsf{T} \text{ iff } \mathsf{v}((A \rightarrow B)^*) = \mathsf{T}$$
$$\text{iff } \mathsf{v}(\psi(A^*, B^*)) = \mathsf{T}$$

Since ψ is schematically constructed from the connectives of $L_{\mathbf{M}}$, it is truth-and-content functional in the semantics of **M**. That is,

$\mathsf{v}(\psi(A^*, B^*))$ depends only on $\mathsf{v}(A^*)$, $\mathsf{v}(B^*)$, $\mathsf{s}(A^*)$, and $\mathsf{s}(B^*)$

But then:

$\mathsf{v}(\psi(A^*, B^*))$ depends only on $\mathsf{v}^*(A)$, $\mathsf{v}^*(B)$, $\mathsf{s}^*(A)$, and $\mathsf{s}^*(B)$

And from (4) we know what that dependence must be:

$\mathsf{v}(\psi(A^*, B^*)) = \mathsf{T}$ iff (not both $\mathsf{v}^*(A) = \mathsf{T}$ and $\mathsf{v}^*(B) = \mathsf{F}$) and $\mathsf{B}(\mathsf{s}^*(A), \mathsf{s}^*(B))$

The translation reconstructs the original semantics for each connective as well as for sentences as wholes.

Exercises for Section B.1

1. a. When are two models elementarily equivalent?
   b. Give an example of two distinct models of a logic that are elementarily equivalent.

2. a. What is a model-preserving map?
   b. Distinguish between a model-preserving map and a map that is model-preserving up to elementary equivalence.
3. What is a semantically faithful translation? Give an example.
4. Prove:
   a. The composition of maps that are model-preserving up to elementary equivalence is model-preserving up to elementary equivalence.
   b. The composition of model-preserving maps is model-preserving.
   c. The composition of semantically faithful maps is semantically faithful.
5. Explain how a semantically faithful translation allows us to reconstruct the semantics of the logic being translated within the semantics of the logic into which it is translated.

2. Examples of semantically faithful translations

Let's review the translations we have seen in this volume, as summarized in Section A.2. The semantic justifications I quote are established in the text relative to models idiosyncratic to the particular logics. They can be converted into model correspondences relative to set-assignment semantics in a straightforward way, and I leave that to you.

a. Translations of a logic to "itself"
Every translation of a logic to itself that we have seen is semantically faithful. Each is designed to recreate models of the one language in terms of models of the other by defining the connectives in terms of others that are chosen to be primitive.

b. Translations of classical logic
Each grammatical translation of **PC** to another logic in the text is semantically faithful, as you can check. Except for the translation of **PC** into **Int**, which I discuss in Section B.4 below.

c. Translations of **D**, **Dual D**, *and* **Eq**
The translations **D** $\twoheadrightarrow$ **Dual D** and **Dual D** $\twoheadrightarrow$ **D** are the simplest examples I have of semantically faithful translations between nonclassical logics. Recall that $\neg$ is translated homophonically and $(A \rightarrow B)^* = \neg B^* \rightarrow \neg A^*$. Thus the relation governing the table for $\rightarrow$ in **D**, which is $\supseteq$, is converted to the relation containment governing the table for $\rightarrow$ for **Dual D**, $\subseteq$, and vice versa. This is the only example I have of different logics that can be translated semantically faithfully into each other. I will discuss in Section B.6 whether these translations should lead us to classify **D** and **Dual D** as being the "same" logic.

The translation **Eq** $\twoheadrightarrow$ **D** is semantically faithful, since the effect of translating

$(A \to B)^* = (A^* \to B^*) \wedge (\neg B^* \to A^*)$ is to recreate the relation of equality of contents governing the table for $\to$ for **Eq** from the containment $\supseteq$ governing the table for $\to$ for **D**. Similarly, **Eq** $\twoheadrightarrow$ **Dual D** is semantically faithful.

By varying slightly the semantics for **D**, the translation **Dual D** $\twoheadrightarrow$ **D** can provide an example of a grammatical translation that is *model-preserving up to elementary equivalence but not model-preserving* with respect to those semantics. Define the class of (formal) models for **D** as in the text with the proviso that the content set S = the real numbers; define the class of models for **Dual D** as in the text with the proviso that the content set S = the natural numbers. These classes are strongly complete for the logic, but the translation **Dual D** $\twoheadrightarrow$ **D** cannot be model-preserving with respect to them since the cardinality of the class for **D** is greater than that of the class for **Dual D**.

The remaining translations we have seen are those involving **Int** and those of relatedness logics to classical logic, which I discuss below.

3. The archetype of a semantically faithful translation: Int $\twoheadrightarrow$ S4

The best known translation between nonclassical logics that is generally thought to preserve meaning is that of **Int** to **S4**. I will show that it is semantically faithful.

Recall that the translation from the language $L(\neg, \to, \wedge, \vee)$ to $L(\neg, \wedge, \Box)$ is given by:

$$
\begin{aligned}
p^* &= \Box p \\
(A \wedge B)^* &= A^* \wedge B^* \\
(A \vee B)^* &= A^* \vee B^* \\
(\neg A)^* &= \Box\neg(A^*) \\
(A \to B)^* &= \Box(A^* \supset B^*)
\end{aligned}
$$

In this section I will use the symbol $*$ for this map only. It is grammatical.

We justified that $*$ is a translation by reconstructing possible-world models of **Int** within possible-world models of **S4** (Theorem VII.12, p. 289). The proof that $*$ is model-preserving will use the correlations of possible-world models to set-assignment models for each logic

For each logic we then need a strongly complete class of set-assignment models that can be put into 1-1 correspondence with the possible-worlds models. For **Int** we will use the class $\mathcal{I}$ of Theorem VII.25, p. 300. For **S4** we will use the class $\mathcal{K}$ described in Exercise 6, p. 256.

Theorem 14 The translation $*$ of **Int** to **S4** is semantically faithful with respect to $\mathcal{I}$ and the class of **S4** set-assignment models.

Proof: Consider the diagram:

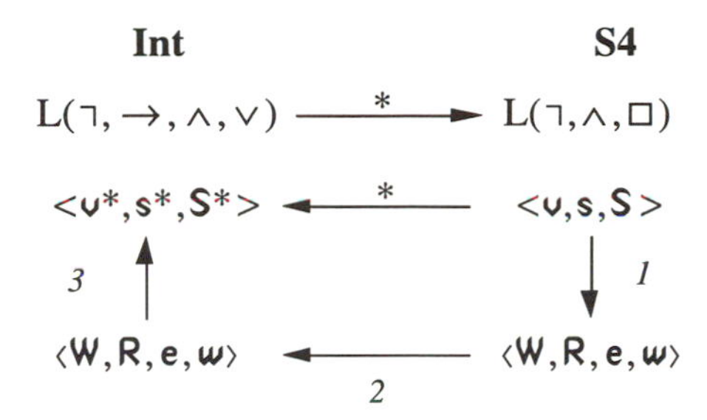

Map *1* is from the class K to the strongly complete class of reduced possible-world models ⟨W, R, e, w⟩ of **S4** where for all z, wRz. It is onto (Exercise 7, p. 256) and S = W − {w}.

Map *2* is from reduced possible-world models of **S4** to the strongly complete class of Kripke models for **Int** that have an initial point (p. 278) given by the re-interpretation of the connectives, Theorem VII.12 (p. 289). It is onto.

Map *3* is from Kripke models of **Int** with initial point to set-assignment models of **Int** that are in *I*, given in Theorem VII.25 and Lemma VII.23 (p. 297, p. 300). It is onto and S = W − {w}.

The composition of these maps applied to a model <v, s, S> of **S4** yields a model <v*, s*, S*> of **Int** given by:

$$S = S^*$$
$$v^*(p) = v(\Box p)$$
$$s^*(A) = s(A^*)$$

For that map $v^*(A) = v(A^*)$. Hence the composition of these maps, which is onto, is the induced map of set-assignment models, justifying that * is model-preserving. Since * is grammatical, it is a semantically faithful translation. ■

4. The translations of PC into Int

Intuitionists say that classical logic is incoherent, it makes no sense to them. The classical logician points to the translations of **PC** into **Int** and argues that classical logic can be perfectly well understood by the intuitionist. But do the translations justify that? I will show that, at least with respect to the standard models of **Int** and **PC**, the translations we have seen are not semantically faithful.

Lemma 15 Let *M* be a complete class of models for **Int** all of which use the intuitionist truth-conditions. Then for every propositional variable p there is some model <v, s> in *M* such that $v(\neg\neg p) = \mathsf{F}$ and $s(\neg\neg p) \neq \varnothing$.

Proof: Suppose to the contrary that for all <v, s> in *M*, $v(\neg\neg p) = \mathsf{T}$, or $v(\neg\neg p) = \mathsf{F}$ and $s(\neg\neg p) = \varnothing$. Then for each <v, s>, $v(\neg\neg p) = \mathsf{T}$ or $v(\neg\neg\neg p) = \mathsf{T}$, and hence

$v(\neg\neg p \lor \neg\neg\neg p) = T$. So by completeness we would have $\vdash_{\mathbf{Int}} \neg\neg p \lor \neg\neg\neg p$, which is a contradiction, for we can have a possible-world model of **Int** satisfying:

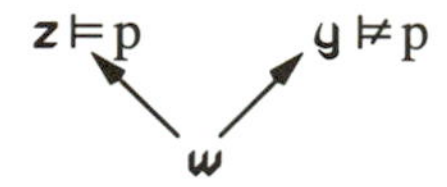

■

Recall that Gentzen's translation of **PC** into **Int** (p. 292) takes every variable to its double negation, translates $\neg$, $\rightarrow$, $\land$ homophonically, and takes $(A \lor B)^* = \neg(\neg A^* \land \neg B^*)$. The double negation translation of **PC** into **Int** simply takes every wff to its double negation (p. 290). So, for example, Gentzen's translation takes $p \lor \neg p$ to $\neg(\neg\neg\neg p \land \neg\neg\neg\neg p)$, while the double negation translation takes $p \lor \neg p$ to $\neg\neg(p \lor \neg p)$.

Theorem 16 Let $\mathcal{M}$ be a complete class of models for **Int** all of which use the intuitionist truth-conditions. Then neither Gentzen's translation of **PC** into **Int** nor the double negation translation of **PC** into **Int** is model-preserving with respect to $\mathcal{M}$ and the standard models for **PC**.

Proof: The proof is the same for both translations. Let $*$ be the translation, and suppose it is model-preserving. Let p be any propositional variable. By the previous lemma there is some $\langle v, s \rangle$ in $\mathcal{M}$ such that $v(\neg\neg p) = F$ and $s(\neg\neg p) \neq \varnothing$. So $v(\neg\neg\neg p) = F$, too. Let v^* be the **PC**-evaluation which is induced by $*$ from v. Then:

$$v^*(p) = v(p^*) = v(\neg\neg p) = F$$

$$v^*(\neg p) = v((\neg p)^*) = v(\neg\neg\neg p) = F$$

This contradicts that v^* is a **PC**-evaluation. Hence $*$ is not model-preserving. ■

This proof shows that even if $\mathcal{M}$ uses just the intuitionist truth-table for negation and the classical table for disjunction, then these translations cannot be model-preserving with respect to $\mathcal{M}$, even up to elementary equivalence.

Set-assignment semantics arising from any of the standard semantics for **Int** (possible-world, algebraic, topological, etc.) will, I believe, use the intuitionist truth-conditions, for in essence these semantics are all "isomorphic". Thus at least with respect to the semantics currently proposed for **Int**, neither the double negation translation nor Gentzen's translation is model-preserving.

Exercises for Section B.4

1. In twenty-five words or less, why are the translations of **PC** into **Int** not semantically faithful?
2. Is the translation of **PC** into **J** (p. 309) model-preserving?

3. Is the translation of **Int** into **J** given in Exercise 10, p. 310, model-preserving?

4. a. (Open) Is there some class of models for **Int** with respect to which Gentzen's translation is semantically faithful?

   b. (Open) Is there any semantically faithful translation of **PC** into **Int**? If there is one, then it would be grammatical, and hence we would have $(\neg A)^* = \varphi((A^*))$. So $v^*(\neg A) = \mathsf{T}$ iff $v(\varphi(A^*)) = \mathsf{T}$; and $v^*(A) = \mathsf{T}$ iff $v(A^*) = \mathsf{T}$. Thus $v(A^*) = \mathsf{T}$ iff $v(\varphi(A^*)) = \mathsf{F}$. That is, classical negation would be definable on the class of formulas Wffs* (cf. the query in Chapter VII.E.2.b, p. 298).

5. (Open) Define a class of set-assignment models for **S4Grz** such that the map * is a semantically faithful translation of **Int** into **S4Grz**.

5. The translation of S into PC

I remarked in Section A.4 that the translation of subject matter relatedness logic, **S**, into classical logic seems to reflect the underlying semantic assumptions. In this section I will show that it is model-preserving.

The translation cannot be model-preserving with respect to the standard models for **PC** because the set-assignments for the models for **S** must be derived from the set-assignments of the models for **PC**. So first define a new class of models for **PC** for the language $L(\neg, \wedge, q_0, q_1, \ldots, \{d_{ij}: i, j \text{ are natural numbers}\})$. Let $\boldsymbol{P}$ be the class of set-assignment models that use the classical tables for the connectives, S is countable, and for all i, j:

a. $s(q_i) \neq \varnothing$.

b. $v(d_{ij}) = \mathsf{T}$ iff $s(q_i) \cap s(q_i) \neq \varnothing$

c. $s(A) = \bigcup \{s(q_i): q_i \text{ appears in } A\}$

To see that $\boldsymbol{P}$ is strongly complete for **PC** it's enough to note that given any standard **PC**-evaluation v for this language, we can define:

$$\mathsf{S} = \{(i, j): i \geq 0 \text{ and } j \geq 0\}$$

$$s(d_{ij}) = \varnothing \text{ for all } i, j$$

$$s(q_i) = \{(i, i)\} \cup \textstyle\bigcup_j \{(i, j), (j, i): v(d_{ij}) = \mathsf{T}\}$$

This is a set-assignment satisfying (a), (b), and (c).

Theorem 17 The translation * of **S** into **PC** is model-preserving with respect to the standard models for **S** and the class of models $\boldsymbol{P}$ for **PC**.

Proof: Let $\langle w, t \rangle$ be a model of **S**. I will show that there is a model $\langle v, s \rangle$ in $\boldsymbol{P}$ such that for all A, $v(A^*) = w(A)$ and $s(A^*) = t(A)$.

Define for all i, j:

$\mathsf{v}(q_i) = \mathsf{T}$ iff $\mathsf{w}(p_i) = \mathsf{T}$

$\mathsf{v}(d_{ij}) = \mathsf{T}$ iff $\mathsf{t}(p_i) \cap \mathsf{t}(p_i) \neq \varnothing$

$\mathsf{s}(q_i) = \mathsf{t}(p_i)$

$\mathsf{s}(d_{ij}) = \varnothing$

$\mathsf{s}(B) = \bigcup\{\mathsf{s}(q_i) : q_i \text{ appears in } B\}$

Then $<\mathsf{v},\mathsf{s}>$ is in $\mathbf{P}$, and for all A, $\mathsf{s}(A^*) = \mathsf{t}(A)$. I will prove by induction that for all A, $\mathsf{v}(A^*) = \mathsf{w}(A)$.

The only interesting case is if A is $B \rightarrow C$. Then:

$\mathsf{w}(B \rightarrow C) = \mathsf{T}$ iff (not both $\mathsf{w}(B) = \mathsf{T}$ and $\mathsf{w}(C) = \mathsf{F}$) and $\mathsf{t}(B) \cap \mathsf{t}(C) \neq \varnothing$

iff (not both $\mathsf{v}(B^*) = \mathsf{T}$ and $\mathsf{v}(C^*) = \mathsf{F}$) and $\mathsf{s}(B^*) \cap \mathsf{s}(C^*) \neq \varnothing$

iff $\mathsf{v}(B^* \supset C^*) = \mathsf{T}$ and $\mathsf{s}(B^*) \cap \mathsf{s}(C^*) \neq \varnothing$

Now $\mathsf{s}(B^*) \cap \mathsf{s}(C^*) \neq \varnothing$ iff there is some i such that q_i appears in both B* and C*, or there are i, j such that q_i appears in B*, q_j appears in C*, and $\mathsf{s}(q_i) \cap \mathsf{s}(q_j) \neq \varnothing$ (Lemma III.1). Hence $\mathsf{s}(B^*) \cap \mathsf{s}(C^*) \neq \varnothing$ iff for some i, $(d_{ii} \vee \neg d_{ii})$ appears in $(B \rightarrow C)^*$, or there are i, j $\vdash$such that $\mathsf{v}(d_{ij}) = \mathsf{T}$ and $\mathsf{v}(d_{ij}) = \mathsf{T}$ and $(d_{ij} \wedge d_{ji})$ appears in $(B \rightarrow C)^*$. So $\mathsf{w}(B \rightarrow C) = \mathsf{T}$ iff $\mathsf{v}((B \rightarrow C)^*) = \mathsf{T}$. ■

6. Different presentations of the same logic and strong definability of connectives

What do we mean by saying that we have two different presentations of the same logic?

Model-preserving intertranslatability cannot be enough, as the example of **S** and **PC** shows. Perhaps, then, the criteria should be that there are semantically faithful translations in both directions. That would classify **D** and **Dual D** as the same. Are they?

The examples we have seen in this book where different presentations were called the same logic arose from choosing fewer or more connectives as primitive and defining others in terms of those. The notion of definability was, in each case, peculiar to the semantics under discussion, for example, the notion of definability with respect to the 3-valued semantics of $\mathbf{J_3}$ (Chapter IX.B.3, p. 356). Those various notions of definability can, I believe, be subsumed under one uniform notion, suggested by the discussion at (4), p. 392, that a semantically faithful translation recreates the table for each connective within the semantics of the logic into which it is translated.

Definition 18 Let **L** be a logic with language L with semantics given by a class of models $\mathcal{L}$ for L. Let α be a connective of L, and L/α be L with α deleted. Denote by $\mathcal{L}/\alpha$ the class of models of $\mathcal{L}$ viewed as models for the language $L_{\mathbf{L}}/\alpha$. Then α is *strongly definable* with respect to $\mathcal{L}$ if there is a semantically faithful translation from L to L/α with respect to $\mathcal{L}$ and $\mathcal{L}/\alpha$.

Does this notion of strong definability lead to a more stringent criteria for two presentations to be classified as the same logic? Or is semantically faithful inter-translatability all we want? I do not know, though I suspect the resolution of these questions depends in part on what we mean when we say that a translation preserves meaning.

Exercises for Section B.6

1. Show that $\Diamond$ is strongly definable in $L(\neg, \rightarrow, \wedge)$ for, e.g., **S5**.
2. (Open) Define a notion of functional completeness for a collection of connectives in terms of strong definability.

7. Do semantically faithful translations preserve meaning?

The translation $\mathbf{S} \hookrightarrow \mathbf{PC}$ is model-preserving but does not, I believe, preserve meaning. It is not grammatical, and, as discussed at the end of Section A.4, it relies on metalogical machinery that is not really propositional. Moreover, the proof that it is model-preserving suggests that any discussion of meanings being preserved should be reserved for translations that are semantically faithful with respect to classes of models that are agreed to ascribe meanings to the languages of the logics.

For **PC** the restriction would be clear: for a translation to preserve meaning it would have to be semantically faithful with respect to the standard 2-valued models of **PC**. For other logics it is not so evident how to restrict the choice: is there always a "standard" set of models? Indeed, even for **PC** the models $<v,s>$ where v evaluates the connectives classically and $\{s(A): A \in \text{Wffs}\}$ is a boolean algebra of sets ($s(A \wedge B) = s(A) \cap s(B)$, etc.) could plausibly be said to ascribe meanings for **PC**.

Semantically faithful translations preserve all that is formally significant about a logic. If, in addition, the practitioners of the logic agree that the classes of models are accurate formalizations of how they understand the formal language, what more could be required for a translation to preserve meaning?

Most intuitionists would object to the idea that the interpretation of the intuitionist connectives in **S4** is a faithful explication of the logical notions of intuitionism. You do not understand intuitionism if you only understand it via modal logic, they would say. Yet for those of us who are not native speakers of intuitionism and feel that we shall never be fluent in a way that reflects the same understanding of logic and mathematics as an intuitionist, the translation affords us projective knowledge of intuitionism. That is, we can communicate with intuitionists and be certain that, confined to the language of propositional logic, we will assert and deny the same propositions if we use the translation. That kind of projective knowledge is afforded by any translation. But here even the forms of how an intuitionist ascribes meaning to those sentences are recreated by us as we try to understand the language in the intuitionist's terms. Still, even those intuitionists who agree that the models of intuitionism accurately reflect their understanding of the logic continue to say,

> That is not it at all,
> That is not what I meant, at all.

Translations do not allow us to enter into another (logical) world view. But they can allow us projective knowledge of that world view. Whether we can ever do more, whether we can ever enter into another (logical) culture may be as problematic as whether we can even fully enter into another person's world view who shares our own culture.

XI The Semantic Foundations of Logic

Concluding Philosophical Remarks

It seems that we can understand all logics in the same way: We start with our everyday language and abstract away certain aspects of linguistic units and take into account certain others by making idealizations of them. The aspects we pay attention to determine our notion of truth.

There does not seem to be a difference between logical and pragmatic aspects of what we call propositions, at least any difference we can justify. What direct access have we to the world but our uncertain perceptions? And how can two of us share exactly the same perceptions or thoughts? Ambiguity seems essential to communication. So, it seems to me, to call a sentence true seems at best a hypothesis we hope to share with others. This sharing, which in a sense amounts to objectivity, is brought about by common understandings, which I call agreements. But this notion of truth is so basic to our experience and the fit of thought to the world that we can no longer allow ourselves to see that truth is in how we abstract, perceive, and agree, lest we have no language to talk.

The word 'agreement' is wrong, and 'convention' even more so. As *Searle, 1983,* Chapter 5§V, points out, we don't have good nonintentional words with which to describe our backgrounds. Almost all our conventions, agreements, assumptions are implicit, tacit. They needn't be either conscious or voluntary. Many of them may be due to physiological, psychological, or, perhaps, metaphysical reasons: for the most part we seem not to know. Agreements are manifested in lack of disagreement and the fact that people communicate. To be able to see that we have made (or been forced into, or simply have) a tacit agreement is to be challenged on it. If the assumption is sufficiently fundamental and widely held, we call the challenger 'mad'.

The explicit background is quite different from the implicit, though I use the same word 'agreement' for each. I think that's the best term, for it allows us to use the word 'disagreement' when our backgrounds clash. It points to the background as it affects our interactions with one another. And it is the drawing of the implicit background into explicitness by abstracting, idealizing, and simplifying that I am interested in, for that is the basis of formal logic and, as I see it, explicit objectivity.

It may be wrong to ascribe a uniformity to our backgrounds. As Peter Eggenberger has pointed out to me, from the comparatively uncontroversial 'For every act of communication there are some agreements on which it is founded', it does not follow that there are some agreements on which every act of communication is founded. But I have tried to show that this latter assumption is reasonable, at least with respect to logical discourse.

In a conversation John Searle tried to convince me that truth and referring can't be a matter of agreement. He argued that when he says 'The moon is risen' and I say 'Which moon?' then he has successfully referred even if I keep saying 'Which moon?'

But I believe it is a matter of agreement that we say the same object is in the sky each night, not 28 different ones, or one new one for each night of eternity. There may indeed be only one object "out there", but how can either of us ever know that with any certainty if that transcends our perceptual framework, our background? Even relative to our background, anything that is beyond direct immediate experience must be a matter of agreement for referring, while much that is of direct immediate experience is so theory-laden as to be called a matter of agreement for referring, too.

But, Peter Eggenberger argues, suppose we're playing a game of chess. The rules are completely explicit: a rook moves in this fashion, pawns in that. The game ends with either a checkmate, a stalemate, or an agreed draw, and so on. Now suppose you move and announce, 'Checkmate'. Or as he puts it, suppose I'm in a position that "says" I'm in checkmate. Then I can't get out of it by saying that I disagree.

But of course I can, though it's extremely unlikely (it seems to us) that I could do so in good faith, that is, really not recognize it as checkmate by your "objective" standards. If I do say 'I don't agree,' either through perversity or my actually not regarding such a position as checkmate, then pretty soon I'll find no one to play chess with me. Perhaps that's not such a great loss. But if I do not agree with the community's language agreements and assumptions, then I cannot get anyone to talk with me. That is a loss. Objectivity arises because we'd all go mad (or be mad) if we didn't act in conformity with some implicit agreements and rules. We are all built roughly the same, and we have to count on that in exchanging information about our experience. Given a particular shared background there will be plenty of room for experience and facts.

When we adopt a language we can't help but adopt the agreements on which it is based. For example, consider what happens when I teach real analysis, the theoretical foundations of calculus. I tend towards a constructivist view of mathematics and do not believe that there are any infinite entities, or at least none corresponding to what we call 'the real numbers'. But when I teach real analysis I have to adopt the language of the classical mathematician. I could preface each remark of mine with 'Of course, this is assuming that infinite totalities exist, which I don't really believe.' But I don't; I make that comment at the beginning of the course and then, slowly, forget it. I have to in order to be able to talk in the language of real analysis, a language that has grown out of human experience and is therefore accessible to me. I am only able to talk coherently in that language if I accept its background assumptions. The more I talk the language, the more likely I am to forget that my acceptance was hypothetical. I convince myself that I have not betrayed my beliefs by saying that the theory I am teaching is an idealization of experience. But it could be said that I have not betrayed my beliefs in the same way when I learned Polish and began to speak it in Poland, and in doing so adopted its assumptions and conventions. I have not forgotten the categories of the world of my language, I have just put them aside in order to communicate with people who do not have or use them.

But I mustn't be fooled into thinking that a logic or language can be justified by its utility. An example comes from *Wrigley, 1980,* and a discussion I had with Michael Wrigley.

We use Peano Arithmetic, **PA**, because we believe it's "right". Now suppose we encountered someone who used a deviant arithmetic, **DA**, which we see is inconsistent. Turing said that we could distinguish the two arithmetics because if we built bridges using **DA** they'd fall down (more often). Well, in a sense he's right and in a sense he's wrong.

We ought to suppose that the practitioners of **DA** have their own background assumptions, which surely must differ from ours, for with our background if we used **DA** to build bridges, they would fall down more often. But why shouldn't we assume that relative to their background assumptions the practitioners of **DA** would build perfectly fine bridges? After all, **DA** would reflect their background assumptions well or they wouldn't use it. There is a temptation here to say that we do not have to account for the deviant arithmetic because it's no arithmetic at all: they must not be talking about the natural numbers. But we were assuming that we had some good reason for calling it 'arithmetic' in the first place. Disagreements do not disappear by saying that the subject has been changed. Moreover, how do we know that even among ourselves we all understand **PA** in the same way, that there is a common subject? We are back again at the not unreasonable hypothesis that we have a shared background so that we can communicate. And only relative to a particular background do criteria of utility have force. Turing was trying to fill in the dotted line:

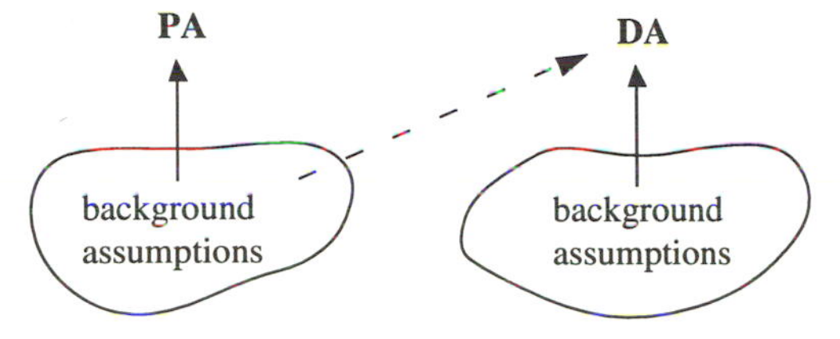

As Benson Mates pointed out to me, you don't believe it because it's useful; it's useful because you believe it. There's pragmatic value in believing that it's not a matter of pragmatics. Even to think that our fundamental background assumptions could be challenged is paralyzing to most users of language.

Well, then, why should we question them? Because we have disagreements. There *are* many logics. It isn't a "let's pretend" situation, as with the example of deviant arithmetic. The objectivity of our backgrounds really has been questioned.

Yet the classical logician, the intuitionist, the modal logician, the practitioner of relevance, or a many-valued logic all do communicate and work together. They write journal articles and read ones written by the others, they collaborate, however uneasily, they discuss. One possible explanation of this is that the usual daily languages of all these people share so many background assumptions that the practitioners of, say, intuitionism, cannot free themselves of the classical way of thinking any more than I can free myself of the classical real analysis way of thinking when I am helping my mathematics students. Perhaps intuitionism cannot be the challenge the intuitionists intended it to be because they only adopted a new way of mathematically talking (and thinking) while retaining the language and assumptions of the culture in which they live.

The relevance of this to doing logic and seeing the unity of logics is that because the practitioners of the various logics share the same background assumptions and agreements that come with the use of Western languages and culture (or perhaps that come with any human culture), those assumptions must be evident in their work. Those more fundamental assumptions are what I explicitly use in setting up the structural overview of propositional logic I've presented in Chapter IV: there is a common notion of a smallest linguistic unit, called 'a proposition', that can be called true, and common notions of the connectives as portrayed in the truth-tables I give. Those tables reflect that some notion of meaning or content is ascribed to propositions and a connection of meanings or contents must be made for a compound proposition to be accepted. If the connection is there, then only the truth-values of the constituents remain to be considered, and we evaluate the connectives according to the standard classical tables. Always present is the Yes–No, Accept–Reject dichotomy that we impose on (or is imposed on us by) experience.

The differences between the logics must be superficial relative to these assumptions, and the differences are, as I see it, in the choice of which, if any, less fundamental aspect (or possibly aspects) of propositions other than truth-values are to be taken into account in reasoning. Even then, some aspects are so close to fundamental, so near the heart of the background, that a challenge to them seems incomprehensible to most of us. Such aspects or notions are the ones we tend to label 'logical'. To me whether a notion is called logical is a measure of how fundamental it is to our reasoning, our communication, our view of reality, not how close it *is* to reality. The following diagram pictures this view.

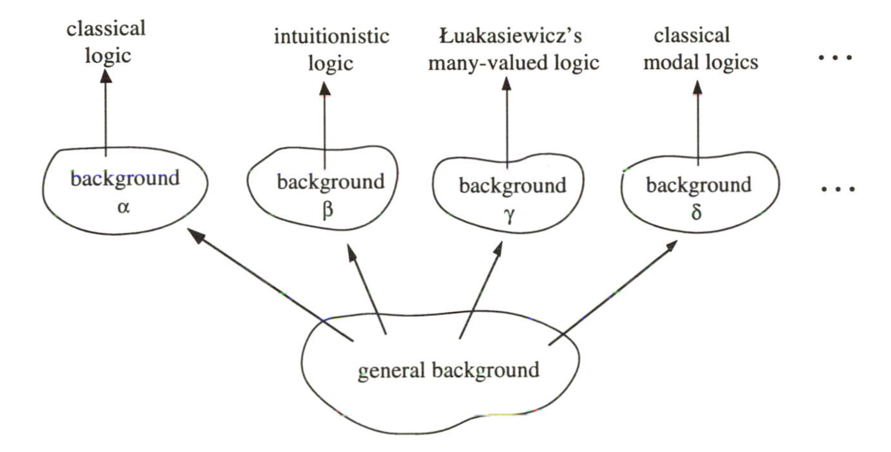

The backgrounds $\alpha, \beta, \gamma, \delta, \ldots$ are a spectrum of assumptions one can have relative to our more general background; some of these overlap, others are apparently incompatible.

I hope to have formalized the general background in the structural analysis I give. In the various chapters on modal, intuitionistic, many-valued, and paraconsistent logics I've tried to show that the fundamental assumptions on which I build that analysis are implicit in the reasoning of the practitioners of all those logics. I suspect that similar analyses could be made for other logics I have not included here. I may be wrong in my belief that I have uncovered an implicit background, yet if my view is compelling enough it may come to be seen as the general background.

Where is objectivity, where is 'must' in this picture? It used to be thought that it lay within one of the backgrounds $\alpha, \beta, \gamma, \delta, \ldots$; and such has been the basis of philosophical debates between practitioners of the various logics. I suspect it lies between the backgrounds $\alpha, \beta, \gamma, \delta, \ldots$ and their respective logics, as depicted in the following diagram:

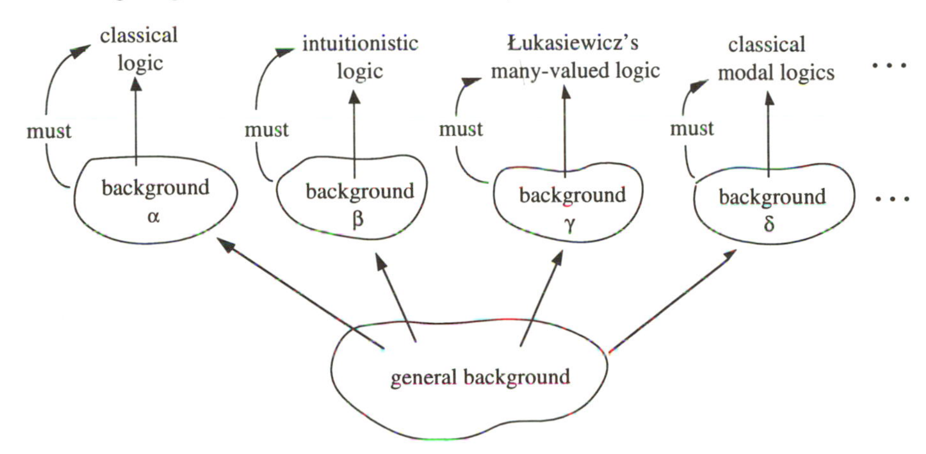

For someone with background α, classical logic seems objective: we must reason in accord with it, he argues. And that is because with that background it seems inconceivable to act otherwise. But it is not inconceivable to have another background, as I've argued, only extremely difficult to enter into one.

If there is a nonrelative sense of 'must' in this picture, it would seem to be in the general background. Nothing I have presented in this book depends on whether the general assumptions on which these logics are based are necessary truths. The metaphysical arguments may begin again, only relocated. We don't get rid of the background, for language must be anchored to the world. We only show that more or less of it as represented in language and logic can be considered universal.

Yet all our formalizations, all I've done in this book, are false if taken to be exact representations of our implicit backgrounds. What is thought remains thought, what is the external world remains the external world, and what we say is and can only be what we say, either about, because of, or through the world and thought. It seems to me false that what is said can be directly of these other realms, or a perfect representation of them. The explicit agreements we codify are abstractions, idealizations, simplifications, and hence a distortion of our real background. These explicit agreements may make it easier for us to communicate, and we may henceforth act in accord with them; that can sometimes feed back into our backgrounds becoming imperceptibly, in a new nonexplicit way, our unconscious, implicit background.

It seems hard for me to conceive of someone who reasons who doesn't have a smallest unit of language to which a Yes–No, Accept–Reject dichotomy is applied, and hence to conceive of someone who uses connectives that would not be expressible by the general tables of Chapter IV. So perhaps the story of propositional logics I give is universal. For it to be useful for us, for us to develop it, we will probably have to believe that.

Summary of Logics

- Unless noted otherwise, the axiomatization or semantics listed below is strongly complete for the logic: $\Gamma \vdash A$ iff $\Gamma \vDash A$.
- Axiomatizations are given in terms of schema even in those cases where the rule of substitution was used by the originators.
- A fully general abstraction is assumed for all the logics unless noted otherwise (see the index).
- For set-assignment semantics, unless noted otherwise **S** is countable.

A. Classical Logic, PC

A model for classical logic is a function $v: \mathrm{PV} \to \{\mathsf{T}, \mathsf{F}\}$ that is extended to all wffs of $L(\neg, \to, \wedge, \vee)$ by the following tables:

A	¬A
T	F
F	T

A	B	A→B
T	T	T
T	F	F
F	T	T
F	F	T

A	B	A∨B
T	T	T
T	F	T
F	T	T
F	F	F

A	B	A∧B
T	T	T
T	F	F
F	T	F
F	F	F

Axiomatizations

PC *in* $L(\neg, \rightarrow)$ (Chapter II.L.1)

1. $\neg A \rightarrow (A \rightarrow B)$
2. $B \rightarrow (A \rightarrow B)$
3. $(A \rightarrow B) \rightarrow ((\neg A \rightarrow B) \rightarrow B)$
4. $(A \rightarrow (B \rightarrow C)) \rightarrow ((A \rightarrow B) \rightarrow (A \rightarrow C))$

rule $\dfrac{A, A \rightarrow B}{B}$

Scheme 3 may be replaced by $(\neg A \rightarrow A) \rightarrow A$. (Exercise 6, p.76)

PC *in* $L(\neg, \rightarrow)$ (*Hilbert, 1922*)

1. $A \rightarrow (B \rightarrow A)$
2. $(A \rightarrow (A \rightarrow B)) \rightarrow (A \rightarrow B)$
3. $(A \rightarrow (B \rightarrow C)) \rightarrow (B \rightarrow (A \rightarrow C))$
4. $(B \rightarrow C) \rightarrow ((A \rightarrow B) \rightarrow (A \rightarrow C))$
5. $A \rightarrow (\neg A \rightarrow B)$
6. $(A \rightarrow B) \rightarrow ((\neg A \rightarrow B) \rightarrow B)$

rule $\dfrac{A, A \rightarrow B}{B}$

PC *in* $L(\neg, \rightarrow)$ (*Frege, 1879*)

1. $A \rightarrow (B \rightarrow A)$
2. $(A \rightarrow (B \rightarrow C)) \rightarrow ((A \rightarrow B) \rightarrow (A \rightarrow C))$
3. $(A \rightarrow (B \rightarrow C)) \rightarrow (B \rightarrow (A \rightarrow C))$
4. $(A \rightarrow B) \rightarrow (\neg B \rightarrow \neg A)$
5. $\neg\neg A \rightarrow A$
6. $A \rightarrow \neg\neg A$

rule $\dfrac{A, A \rightarrow B}{B}$

PC *in* $L(\neg, \rightarrow)$ (*Łukasiewicz and Tarski, 1930*)

$(p_1 \rightarrow p_2) \rightarrow ((p_2 \rightarrow p_3) \rightarrow (p_1 \rightarrow p_3))$

$(\neg p_1 \rightarrow p_1) \rightarrow p_1$

$p_1 \rightarrow (\neg p_1 \rightarrow p_2)$

rules $\dfrac{A, A \rightarrow B}{B}$ $\dfrac{\vdash A(p)}{\vdash A(B)}$ (substitution)

PC *in* $L(\neg, \rightarrow, \wedge, \vee)$ (Chapter II.M.6)

1. $\neg A \rightarrow (A \rightarrow B)$
2. $B \rightarrow (A \rightarrow B)$
3. $(A \rightarrow B) \rightarrow ((\neg A \rightarrow B) \rightarrow B)$
4. $(A \rightarrow (B \rightarrow C)) \rightarrow ((A \rightarrow B) \rightarrow (A \rightarrow C))$

5. $A \to (B \to (A \wedge B))$
6. $(A \wedge B) \to A$
7. $(A \wedge B) \to B$
8. $A \to (A \vee B)$
9. $B \to (A \vee B)$
10. $(A \to C) \to ((B \to C) \to ((A \vee B) \to C))$

rule $\dfrac{A,\ A \to B}{B}$

Scheme 10 may be replaced by either of: $((A \vee B) \wedge \neg A) \to B$ or $(A \vee B) \to (\neg A \to B)$. (Exercise 1, p. 84)

PC *in* $L(\neg, \to, \wedge)$ (Chapter II.M.6)

Axiom schema **1–7** of the axiomatization of **PC** in $L(\neg, \to, \wedge, \vee)$ with the rule of *modus ponens.*

PC *in* $L(\neg, \wedge)$ (Chapter II.M.7)

$$A \supset B \equiv_{\text{Def}} \neg(A \wedge \neg B)$$

1. $B \supset (A \supset B)$
2. $(A \supset (B \supset C)) \supset ((A \supset B) \supset (A \supset C))$
3. $(A \wedge B) \supset A$
4. $(A \wedge B) \supset B$
5. $A \supset (B \supset (A \wedge B))$
6. $\neg A \supset (A \supset B)$
7. $(A \supset B) \supset ((\neg A \supset B) \supset B)$

rule $\dfrac{A,\ A \supset B}{B}$ *material detachment*

Schema **6** and **7** may be replaced by: $(\neg A \supset A) \supset A$ and $(A \wedge \neg A) \supset B$. (Exercise 6, p. 84)

PC *in* $L(\neg, \wedge)$ (*Rosser, 1953*)

$$A \supset B \equiv_{\text{Def}} \neg(A \wedge \neg B)$$

$A \supset (A \wedge A)$

$(A \wedge B) \supset A$

$(A \supset B) \supset (\neg(B \wedge C) \supset \neg(C \wedge A))$

rule $\dfrac{A,\ A \supset B}{B}$

The fragment of **PC** *in* $L(\to)$ (Chapter II.M.8)

1. $B \to (A \to B)$
2. $(A \to (B \to C)) \to ((A \to B) \to (A \to C))$
3. $(B \to A) \to [(C \to A) \to (((B \to C) \to A) \to A)]$

rule $\dfrac{A,\ A \to B}{B}$

The fragment of **PC** *in* L($\rightarrow,\wedge,\vee$) (Chapter II.M.8)

1. $B \rightarrow (A \rightarrow B)$
2. $(A \rightarrow (B \rightarrow C)) \rightarrow ((A \rightarrow B) \rightarrow (A \rightarrow C))$
3. $(B \rightarrow A) \rightarrow [(C \rightarrow A) \rightarrow (((B \rightarrow C) \rightarrow A) \rightarrow A)]$
4. $A \rightarrow (B \rightarrow (A \wedge B))$
5. $(A \wedge B) \rightarrow A$
6. $(A \wedge B) \rightarrow B$
7. $A \rightarrow (A \vee B)$
8. $B \rightarrow (A \vee B)$
9. $(A \rightarrow C) \rightarrow ((B \rightarrow C) \rightarrow ((A \vee B) \rightarrow C))$

rule $\dfrac{A,\ A \rightarrow B}{B}$

The fragment of **PC** *in* L($\rightarrow,\wedge$) (Chapter II.M.8)

Axiom schema **1–6** of the axiomatization of **PC** in L($\neg, \rightarrow, \wedge, \vee$) with the rule of *modus ponens.*

The fragment of **PC** *in* L($\rightarrow,\vee$) (Chapter II.M.8)

Axiom schema **1–3**, **7–9** of the axiomatization of **PC** in L($\neg, \rightarrow, \wedge, \vee$) with the rule of *modus ponens.*

Axiomatizations of **PC** *in* L($\neg, \rightarrow, \wedge, \vee$) *relative to* **Int** (Theorem VII.21)

For axiomatizations of **Int** see pp. 424–425 below.

PC is the closure under *modus ponens* of:

Int plus any one of the following schema

$(A \rightarrow B) \rightarrow ((\neg A \rightarrow B) \rightarrow B)$

$\neg\neg A \rightarrow A$

$A \vee \neg A$

B. Relatedness and Dependence Logics S, R, D, Dual D, Eq, DPC

General Form of Semantics for **S**, **D**, **Dual D**, **Eq**

A model is a triplet $<v, s, S>$ where $v: PV \rightarrow \{T, F\}$, S is countable, and s: Wffs $\rightarrow$ Subsets of S. Then v is extended to all wffs by the classical tables for $\neg$ and $\wedge$, and for $\rightarrow$:

$$v(A \rightarrow B) = T \text{ iff } B(A,B) \text{ and (not both } v(A) = T \text{ and } v(B) = F)$$

The relation B in the respective logics is:

S	$s(A) \cap s(B) \neq \varnothing$, where for every A, $s(A) \neq \varnothing$
D	$s(A) \supseteq s(B)$
Dual D	$s(A) \subseteq s(B)$
Eq	$s(A) = s(B)$

Each of these logics uses *union set-assignments*:

$$s(A) = \bigcup\{s(p) : p \text{ appears in } A\}$$

These can be characterized as set-assignments satisfying:

U1. $s(\neg A) = s(A)$

U2. $s(A \wedge B) = s(A) \cup s(B)$

U3. $s(A \rightarrow B) = s(A) \cup s(B)$

For each of these logics alternate *relation-based semantics* are defined by a class of models $<v, B>$ where $v : PV \rightarrow \{T, F\}$ and $B \subseteq Wffs \times Wffs$; the class of relations for the logic is characterized below. The extension of v to all wffs uses the truth-conditions above.

S, Subject Matter Relatedness Logic

Set-assignment semantics (Chapter III.D, Chapter III.F.2)
The set-assignments are the union set-assignments that satisfy $s(A) \neq \varnothing$. The relation governing the truth-table for '$\rightarrow$' is $s(A) \cap s(B) \neq \varnothing$, and $\neg$ and $\wedge$ are classical.

$$R(A,B) \equiv_{Def} A \rightarrow (B \rightarrow B) \quad \text{and} \quad <v,s> \vDash R(A,B) \text{ iff } s(A) \cap s(B) \neq \varnothing$$

Relation-based semantics (Chapter III.C, Chapter III.F.1)
The class of *subject matter relatedness relations* governing the table for '$\rightarrow$' are those satisfying

R1. $R(A, A)$

R2. $R(A, B)$ iff $R(\neg A, B)$

R3. $R(A, B)$ iff $R(B, A)$

R4. $R(A, B \rightarrow C)$ iff $R(A, B)$ or $R(A, C)$

R5. $R(A, B \wedge C)$ iff $R(A, B \rightarrow C)$

Alternatively, R may be taken as any symmetric, reflexive relation on $PV \times PV$ extended to all wffs by R1–R5.

S *in* $L(\neg, \to)$ (Chapter III.K.1)

$$A \wedge B \equiv_{Def} \neg(A \to (B \to \neg((A \to B) \to (A \to B))))$$

$$A \supset B \equiv_{Def} \neg(A \wedge \neg B) \qquad R(A, B) \equiv_{Def} A \to (B \to B)$$

1. $R(A, B) \to R(\neg A, B)$
2. $R(\neg A, B) \to R(A, B)$
3. $R(A, B) \to R(B, A)$
4. $R(A, B) \to R(A, B \to C)$
5. $R(A, C) \to R(A, B \to C)$
6. $R(A, B \to C) \to (\neg R(A, B) \to R(A, C))$
7. $(A \to B) \to R(A, B)$
8. $(A \to B) \to (A \supset B)$
9. $A \to A$
10. $(A \supset B) \to (R(A, B) \to (A \to B))$
11. $\neg A \to (A \supset B)$
12. $B \to (A \supset B)$
13. $[A \supset (C \to D)] \to [(A \supset C) \to (A \supset D)]$
14. $(\neg A \supset \neg B) \to ((\neg A \supset B) \to A)$

rule $\dfrac{A,\ A \to B}{B}$

S *in* $L(\neg, \to, \wedge)$ (Chapter III.K.2)

$$R(A, B) \equiv_{Def} A \to (B \to B) \qquad A \supset B \equiv_{Def} \neg(A \wedge \neg B)$$

Schema **1–14** as for **S** in $L(\neg, \to)$, using these definitions, plus:

15. $R(A, B \wedge C) \to R(A, B \to C)$
16. $R(A, B \to C) \to R(A, B \wedge C)$
17. $A \wedge B \to A$
18. $A \wedge B \to B$
19. $A \to (B \to (A \wedge B))$

rule $\dfrac{A,\ A \to B}{B}$

S *in* $L(\neg, \to, \wedge)$ (Exercise 4, p. 124)

$$R(A, B) \equiv_{Def} A \to (B \to B) \qquad A \supset B \equiv_{Def} \neg(A \wedge \neg B)$$

1. $B \supset (A \supset B)$
2. $(A \supset (B \supset C)) \supset ((A \supset B) \supset (A \supset C))$
3. $(A \wedge B) \supset A$
4. $(A \wedge B) \supset B$
5. $A \supset (B \supset (A \wedge B))$
6. $(\neg A \supset A) \supset A$

7. $(A \to B) \to R(A, B)$
8. $(A \wedge \neg B) \supset \neg(A \to B)$
9. $\neg A \supset (R(A, B) \supset (A \to B))$
10. $B \supset (R(A, B) \supset (A \to B))$
11. $R(A, A)$
12. $R(A, B) \supset R(\neg A, B)$
13. $R(\neg A, B) \supset R(A, B)$
14. $R(A, B) \supset R(B, A)$
15. $R(A, B) \supset R(A, B \to C)$
16. $R(A, C) \supset R(A, B \to C)$
17. $R(A, B \to C) \supset (\neg R(A, B) \supset R(A, C))$
18. $R(A, B \to C) \supset R(A, B \wedge C)$
19. $R(A, B \wedge C) \supset R(A, B \to C)$

rule $\dfrac{A,\ A \supset B}{B}$

R, Non-Symmetric Relatedness Logic

Set-assignment semantics
No set-assignment semantics are known (except for the trivial ones, p. 140).

Relation-based semantics (Chapter III.F.3)
The conditions defining the relations are R1, R2, R4, and R5, as for **S**, p. 411, plus:

R6. $\mathsf{R}(B, A)$ iff $\mathsf{R}(B, \neg A)$

R7. $\mathsf{R}(B \wedge C, A)$ iff $\mathsf{R}(B \to C, A)$

R8. $\mathsf{R}(B \wedge C, A)$ iff $\mathsf{R}(B, A)$ or $\mathsf{R}(C, A)$

Alternatively, R may be taken to be any reflexive relation on PV×PV extended to all wffs by R1, R2, R4–R8.

$R(A, B) \equiv_{\text{Def}} A \to (B \to B)$ and $\langle v, \mathsf{R} \rangle \vDash R(A, B)$ iff $\mathsf{R}(A, B)$

R *in* $L(\neg, \to)$

As for **S** above except delete axiom scheme **3** and add:

$R(A, B) \to R(A, \neg B)$

$R(A, \neg B) \to R(A, B)$

$R(B, A) \to R(B \to C, A)$

$R(C, A) \to R(B \to C, A)$

$R(B \to C, A) \to (\neg R(B, A) \to R(C, A))$

rule $\dfrac{A,\ A \to B}{B}$

D, Dependence Logic

Set-assignment semantics (Chapter V.A.3)
Union set-assignments are used. The relation governing the truth-table for '$\rightarrow$' is $s(A) \supseteq s(B)$, and $\neg$ and $\wedge$ are classical.

$$D(A,B) \equiv_{Def} A\rightarrow(B\rightarrow B) \quad \text{and} \quad <v,s>\vDash D(A,B) \text{ iff } s(A)\supseteq s(B)$$

$$A\wedge B \text{ is equivalent to } \neg([(\{A\rightarrow B]\rightarrow[A\rightarrow B])\rightarrow A]\rightarrow\neg B)$$

Relation-based semantics (Chapter V.A.4)
The class of *dependence relations* relations governing the table for '$\rightarrow$' are those satisfying:

D is reflexive

D is transitive

$D(A,B)$ iff $D(A,p)$ for every p in B

Alternatively, the class of dependence relations can be characterized as those satisfying:

1. D is reflexive
2. D is transitive
3. $D(A,B\wedge C)$ iff $D(A,B)$ and $D(A,C)$
4. $D(A,B\rightarrow C)$ iff $D(A,B)$ and $D(A,C)$
5. $D(\neg A,A)$
6. $D(A,\neg A)$

D *in* $L(\neg,\rightarrow,\wedge)$ (Chapter V.A.9)

$$D(A,B) \equiv_{Def} A\rightarrow(B\rightarrow B) \qquad A\supset B \equiv_{Def} \neg(A\wedge\neg B)$$

1. $D(A, B)\wedge D(B, C) \rightarrow D(A, C)$
2. $D(A, B\wedge C) \rightarrow D(A, B)\wedge D(A, C)$
3. $D(A, B)\wedge D(A, C) \rightarrow D(A, B\wedge C)$
4. $D(A, B\wedge C) \rightarrow D(A, B\rightarrow C)$
   $D(A, B\rightarrow C) \rightarrow D(A, B\wedge C)$
5. $D(A,\neg A)$ $\quad D(\neg A, A)$
6. $(A\wedge B) \rightarrow A$ $\quad (A\wedge B) \rightarrow B$
7. $(A\rightarrow B) \rightarrow D(A, B)$
8. $(A\wedge\neg B) \rightarrow \neg(A\rightarrow B)$
9. $(D(A, B)\wedge\neg A) \rightarrow (A\rightarrow B)$
10. $(D(A, B)\wedge B) \rightarrow (A\rightarrow B)$
11. $(B\wedge D(A, A)) \rightarrow (A\supset B)$
12. $A\supset A$
13. $[(A\supset(C\rightarrow D))\wedge(A\supset C)] \rightarrow (A\supset D)$
14. $((\neg A\supset\neg B)\wedge(\neg A\supset B)) \rightarrow A$

rules $\quad \dfrac{A,\ A\rightarrow B}{B} \qquad \dfrac{A,\ B}{A\wedge B}$

Dual D, Dual Dependence Logic (Chapter V.C)

Set-Assignment Semantics

Union set-assignments are used. The relation governing the truth-table for '$\rightarrow$' is $s(A) \subseteq s(B)$, and $\neg$ and $\wedge$ are classical.

$$D(A,B) \equiv_{Def} A \rightarrow (B \rightarrow B) \quad \text{and} \quad \langle v,s \rangle \models D(A,B) \text{ iff } s(A) \subseteq s(B)$$

$$A \wedge B \text{ is equivalent to } \neg(A \rightarrow (B \rightarrow \neg[(A \rightarrow B) \rightarrow (A \rightarrow B)]))$$

Relation based semantics

Dual dependence relations are characterized as those satisfying the same laws as for dependence relations with the entries in reverse order, e.g., for (3) above, $\mathsf{D}(B \wedge C, A)$ iff $\mathsf{D}(B,A)$ and $\mathsf{D}(C,A)$.

Dual D *in* $L(\neg, \rightarrow, \wedge)$ (Chapter V.C)

$D(A,B) \equiv_{Def} A \rightarrow (B \rightarrow B)$ $\quad A \supset B \equiv_{Def} \neg(A \wedge \neg B)$

1. $(D(A, B) \wedge D(B, C)) \rightarrow (D(A, C) \wedge D(B, C))$
2. $D(A \wedge B, C) \rightarrow D(A, B) \wedge D(A, C)$
3. $D(A, C) \wedge D(B, C) \rightarrow D(A \wedge B, C)$
4. $D(A \wedge B, C) \rightarrow D(A \rightarrow B, C)$
   $D(A \rightarrow B, C) \rightarrow D(A \wedge B, C)$
5. $D(A, \neg A)$ $\qquad D(\neg A, A)$
6. $A \wedge B \rightarrow B \wedge A$ $\qquad A \rightarrow (B \rightarrow (A \wedge B))$
7. $(A \rightarrow B) \rightarrow D(A, B)$
8. $(A \wedge \neg B) \rightarrow \neg(A \rightarrow B)$
9. $(D(A, B) \wedge \neg A) \rightarrow (A \rightarrow B)$ $\qquad$ *rules* $\quad \dfrac{A,\ A \rightarrow B}{B} \quad \dfrac{A \wedge B}{B}$
10. $(D(A, B) \wedge B) \rightarrow (A \rightarrow B)$
11. $B \rightarrow (A \supset B)$
12. $A \supset A$
13. $[(A \supset (C \rightarrow D)) \wedge (A \supset C)] \rightarrow (A \supset D \wedge A \supset C)$
14. $(\neg A \supset \neg B) \wedge (\neg A \supset B) \rightarrow [A \wedge D(B, B)]$

Eq, a Logic of Equality of Contents (Chapter V.D)

Set-assignment Semantics

Union set-assignments are used. The relation governing the truth-table for $\rightarrow$ is $s(A) = s(B)$, and $\neg$ and $\wedge$ are classical.

$$E(A,B) \equiv_{Def} A \rightarrow (B \rightarrow B) \quad \text{and} \quad \langle v,s \rangle \models E(A,B) \text{ iff } s(A) = s(B).$$

Relation-based semantics
The *Eq-relations* are those satisfying:

1. E is an equivalence relation.
2. If $\mathsf{E}(A,B)$, then $\mathsf{E}(A \wedge C, B \wedge C)$
3. $\mathsf{E}(A, \neg A)$
4. $\mathsf{E}(A \wedge B, B \wedge A)$
5. $\mathsf{E}(A \rightarrow B, A \wedge B)$
6. $\mathsf{E}(A \wedge (B \wedge C), (A \wedge B) \wedge C)$
7. If $\mathsf{E}(A, B \wedge C)$ and $\mathsf{E}(B, A \wedge D)$, then $\mathsf{E}(A,B)$
8. $\mathsf{E}(A, A \wedge A)$

Eq *in* $L(\neg, \rightarrow, \wedge)$ (Chapter V.D.4)

$E(A, B) \equiv_{Def} A \rightarrow (B \rightarrow B) \qquad A \supset B \equiv_{Def} \neg(A \wedge \neg B)$

1. $E(A,A)$
2. $E(A,B) \rightarrow E(B,A)$
3. $(E(A,B) \wedge E(B,C)) \rightarrow (E(A,C) \wedge E(A,B))$
4. $(E(A,B) \wedge E(C,C)) \rightarrow E(A \wedge C, B \wedge C)$
5. $E(A, \neg A)$
6. $E(A \wedge B, B \wedge A)$
7. $E(A \rightarrow B, A \wedge B)$
8. $E(A \wedge (B \wedge C), (A \wedge B) \wedge C)$
9. $(E(A, B \wedge C) \wedge E(B, A \wedge D)) \rightarrow ((E(A,B) \wedge E(C,C)) \wedge E(D,D)$
10. $E(A, A \wedge A)$
11. $(A \wedge B) \rightarrow (B \wedge A)$
12. $(A \rightarrow B) \rightarrow E(A, B)$
13. $(A \wedge \neg B) \rightarrow \neg(A \rightarrow B)$
14. $(E(A, B) \wedge \neg A) \rightarrow (A \rightarrow B)$
15. $(E(A, B) \wedge B) \rightarrow (A \rightarrow B)$
16. $A \supset A$
17. $(E(A, A) \wedge B) \rightarrow (A \supset B)$
18. $[(A \supset (C \rightarrow D)) \wedge (A \supset C)] \rightarrow [(A \supset D) \wedge (A \supset C)]$
19. $(\neg A \supset \neg B) \wedge (\neg A \supset B) \rightarrow (A \wedge E(B, B))$

rules $\dfrac{A,\ A \rightarrow B}{B} \qquad \dfrac{A,\ B}{A \wedge B} \qquad \dfrac{A \wedge B}{B}$

DPC, Classically-Dependent Logic (Chapter V.F)

Set-Assignment Semantics
There is only one set-assignment, $\mathsf{s}(A) = \{B : A \vdash_{\mathbf{PC}} B\}$. The relation governing the truth-table for $\rightarrow$ is $\mathsf{s}(A) \supseteq \mathsf{s}(B)$, and $\neg$ and $\wedge$ are classical.

C. Classical Modal Logics

S4, S5, S4Grz, T, B, K, Q, T, MSI, ML, G, G*

Languages

These logics can be formulated in either $L(\neg, \wedge, \Box)$ or $L(\neg, \rightarrow, \wedge)$. For both languages:

$$A \supset B \equiv_{\text{Def}} \neg(A \wedge \neg B)$$

$$A \vee B \equiv_{\text{Def}} \neg(\neg A \wedge \neg B) \qquad A \rightarrow B \equiv_{\text{Def}} \Box(A \supset B)$$

$$A \equiv B \equiv_{\text{Def}} (A \supset B) \wedge (B \supset A) \qquad \Diamond A \equiv_{\text{Def}} \neg\Box\neg A$$

In $L(\neg, \wedge, \Box)$, $A \rightarrow B \equiv_{\text{Def}} \Box(A \supset B)$

In $L(\neg, \rightarrow, \wedge)$: $\Box A \equiv_{\text{Def}} \neg A \rightarrow A$

Possible-World Semantics (Chapter VI.B)

A *model* is $\langle \mathsf{W}, \mathsf{R}, \mathsf{e} \rangle$ where W is a nonempty set whose elements are called *possible worlds*, e is an *evaluation*, $\mathsf{e}: \mathsf{W} \rightarrow \text{Sub PV}$, and R is a binary relation on W called the *accessibility relation between possible worlds*. For $w \in \mathsf{W}$ an inductive definition of $w \vDash A$ is given, where $w \nvDash A$ means 'not $w \vDash A$':

$w \vDash p$ iff $p \in \mathsf{e}(w)$

$w \vDash A \wedge B$ iff $w \vDash A$ and $w \vDash B$

$w \vDash \neg A$ iff $w \nvDash A$

$w \vDash \Box A$ iff for all z such that $w\mathsf{R}z$, $z \vDash A$

The derived condition for $\rightarrow$ is:

* $w \vDash A \rightarrow B$ iff for all z such that $w\mathsf{R}z$, not both $z \vDash A$ and $z \nvDash B$

If '$\rightarrow$' is a primitive, then * is taken as definition, and the condition for $w \vDash \Box A$ is derived. In both languages:

$w \vDash \Diamond A$ iff for some z such that $w\mathsf{R}z$, $z \vDash A$

A *model with designated world* w is $\langle \mathsf{W}, \mathsf{R}, \mathsf{e}, w \rangle$.

A *frame* is $\langle \mathsf{W}, \mathsf{R} \rangle$. A frame is called *reflexive, symmetric,* etc., if the relation R is reflexive, symmetric, etc. It is finite if W is finite.

$\langle \mathsf{W}, \mathsf{R}, \mathsf{e}, w \rangle \vDash A$ iff $w \vDash A$

$\langle \mathsf{W}, \mathsf{R}, \mathsf{e} \rangle \vDash A$ iff for all $w \in \mathsf{W}$, $\langle \mathsf{W}, \mathsf{R}, \mathsf{e}, w \rangle \vDash A$

$\langle \mathsf{W}, \mathsf{R} \rangle \vDash A$ iff for all e, for all $w \in \mathsf{W}$, $\langle \mathsf{W}, \mathsf{R}, \mathsf{e}, w \rangle \vDash A$

The following logics were originally defined syntactically, but in the text are presented semantically as the collection of wffs valid in the following frames (Chapter VI.C, F, J.2,3) :

K	all frames
T	all reflexive frames
B	all reflexive and symmetric frames
S4	all reflexive and transitive frames
S5	all equivalence frames or all universal frames
S4Grz	all finite weak partial order frames (i.e., finite, reflexive, transitive, and anti-symmetric)
QT	all frames with designated world w such that wRw
G	all finite strict partial order frames (i.e., finite, anti-reflexive, transitive, and anti-symmetric)

For each logic, the subclass of finite frames yields the same valid formulas. (Corollary VI.16, p. 225, Theorem 31, p. 243)

For **G*** a different notion of validity is used. (Chapter VI.J.4)

There are two distinct consequence relations that can be associated to the semantics of the logics **K**, **T**, **B**, **S4**, and **S5** (Chapter VI.E.4):

$\Gamma \vDash_{\mathbf{L}} A$ iff for every $\langle W, R, e, w \rangle$ in the class listed, if $w \vDash \Gamma$, then $w \vDash A$

$\Gamma \vDash_{\mathbf{L}}^{\Box} A$ iff for every frame $\langle W, R \rangle$ in the class listed, for every evaluation e, if $\langle W, R, e \rangle \vDash \Gamma$, then $\langle W, R, e \rangle \vDash A$

Axiomatizations

If we allow any formula of $L(\neg, \wedge, \Box)$ (or of $L(\neg, \rightarrow, \wedge)$) to be an instance of A, B, or C in the schema of the axiomatization of **PC** in $L(\neg, \wedge)$ (p. 409 above) we have **PC** in the language of classical modal logic.

Complete axiomatizations in $L(\neg, \wedge, \Box)$ (Chapter VI.E.2)

logic	*axioms*	*rules*
K	**PC**	material detachment
	$\Box(A \supset B) \supset (\Box A \supset \Box B)$	necessitation
T	**PC**	material detachment
	$\Box(A \supset B) \supset (\Box A \supset \Box B)$	necessitation
	$\Box A \supset A$	
B	**PC**	material detachment
	$\Box(A \supset B) \supset (\Box A \supset \Box B)$	necessitation
	$\Box A \supset A$	
	$A \supset \Box\Diamond A$	

S4	**PC**	material detachment
	$\Box(A \supset B) \supset (\Box A \supset \Box B)$	necessitation
	$\Box A \supset A$	
	$\Box A \supset \Box\Box A$	
S5	**PC**	material detachment
	$\Box(A \supset B) \supset (\Box A \supset \Box B)$	necessitation
	$\Box A \supset A$	
	$\Box A \supset \Box\Box A$	
	$\Diamond A \supset \Box\Diamond A$	
ML	**PC** closed under necessitation	material detachment
	$\Box(A \supset B) \supset (\Box A \supset \Box B)$	
	$\Box A \supset A$	
	Presented only syntactically (Chapter VI.H.3)	
S4Grz	**PC**	material detachment
	$\Box(A \supset B) \supset (\Box A \supset \Box B)$	necessitation
	$\Box A \supset A$	
	$\Box A \supset \Box\Box A$	(Chapter VI.J.2)
	$(\Box(\Box(A \supset \Box A) \supset A)) \supset A$	
G	**PC**	material detachment
	$\Box(A \supset B) \supset (\Box A \supset \Box B)$	necessitation
	$\Box(\Box A \supset A) \supset \Box A$	(Chapter VI.J.3)
G*	**G**	material detachment
	$\Box A \supset A$	(Chapter VI.J.4)

A *normal* classical modal logic **L** is one that satisfies:

PC $\subseteq$ **L**

L is closed under the rule of material detachment, $\dfrac{A,\ A \supset B}{B}$

L contains all instance of the distribution scheme:

$$\Box(A \supset B) \supset (\Box A \supset \Box B)$$

L is closed under the rule of necessitation $\dfrac{A}{\Box A}$

The normal modal logics above can be described as containing the following axioms and closed under the rules of material detachment and necessitation:

K is **PC**	plus	$\Box(A \supset B) \supset (\Box A \supset \Box B)$
T is **K**	plus	$\Box A \supset A$
B is **T**	plus	$A \supset \Box\Diamond A$
S4 is **T**	plus	$\Box A \supset \Box\Box A$

S5 is **S4** plus $\Diamond A \supset \Box\Diamond A$

S4Grz is **S4** plus $(\Box(\Box(A \supset \Box A) \supset A)) \supset A$

G is **K** plus $\Box(\Box A \supset A) \supset \Box A$

A *quasi-normal* classical modal logic is one that contains **K**, all instances of the distribution schema, and is closed under material detachment. The quasi-normal modal logics can be described as those containing the following axioms and closed under material detachment:

QT is **K** plus $\Box A \supset A$

G* is **G** plus $\Box A \supset A$

Complete axiomatizations in L(¬, →, ∧) (Chapter VI.E.3)

logic	*axioms*	*rules*
K	**PC**	material detachment
	$(\neg(A \supset B) \to (A \supset B)) \equiv A \to B$	necessitation
	$(A \to B) \supset ((\neg A \to A) \supset (\neg B \to B))$	
T	**PC**	material detachment
	$(\neg(A \supset B) \to (A \supset B)) \equiv A \to B$	necessitation
	$(A \to B) \supset ((\neg A \to A) \supset (\neg B \to B))$	
	$(\neg A \to A) \supset A$	
B	**PC**	material detachment
	$(\neg(A \supset B) \to (A \supset B)) \equiv A \to B$	necessitation
	$(A \to B) \supset ((\neg A \to A) \supset (\neg B \to B))$	
	$(\neg A \to A) \supset A$	
	$A \supset ((A \to \neg A) \to \neg(A \to \neg A))$	
S4	**PC**	material detachment
	$(\neg(A \supset B) \to (A \supset B)) \equiv A \to B$	necessitation
	$(A \to B) \supset ((\neg A \to A) \supset (\neg B \to B))$	
	$(\neg A \to A) \supset A$	
	$(\neg A \to A) \supset (A \to (\neg A \to A))$	
S5	**PC**	material detachment
	$(\neg(A \supset B) \to (A \supset B)) \equiv A \to B$	necessitation
	$(A \to B) \supset ((\neg A \to A) \supset (\neg B \to B))$	
	$(\neg A \to A) \supset A$	
	$(\neg A \to A) \supset (A \to (\neg A \to A))$	
	$A \supset ((A \to \neg A) \to \neg(A \to \neg A))$	

Strongly complete axiomatizations (Chapter VI.E.4)
Two consequence relations are defined:

$\Gamma \vdash_{\mathbf{L}} A$ means there are wffs $B_1, \ldots, B_n = A$ such that each B_i is in $\mathbf{L}$, or is in Γ, or is a direct consequence of earlier B_i's by the rule of material detachment

$\Gamma \vdash_{\mathbf{L}}^{\Box} A$ means there are wffs $B_1, \ldots, B_n = A$ such that each B_i is in $\mathbf{L}$, or is in Γ, or is a direct consequence of earlier B_i's by the rule of material detachment or necessitation

For **L** any of **K**, **T**, **B**, **S4**, and **S5**:

$\Gamma \vdash_{\mathbf{L}} A$ iff $\Gamma \models_{\mathbf{L}} A$

$\Gamma \vdash_{\mathbf{L}}^{\Box} A$ iff $\Gamma \models_{\mathbf{L}}^{\Box} A$

Set-assignment semantics
A *modal semantics for (of) implication* is a set-assignment model or class of models $<\mathsf{v}, \mathsf{s}>$ for $L(\neg, \rightarrow, \wedge)$ satisfying:

M1. $\mathsf{s}(A \wedge B) = \mathsf{s}(A) \cap \mathsf{s}(B)$

M2. $\mathsf{s}(\neg A) = \overline{\mathsf{s}(A)}$

M3. $\mathsf{s}(A \rightarrow B) = \mathsf{s}(\Box(A \supset B))$

$\neg$ and $\wedge$ are evaluated classically, and '$\rightarrow$' is evaluated by the dual-dependence table for the conditional:

A	B	$\mathsf{s}(A) \subseteq \mathsf{s}(B)$	$A \rightarrow B$
any	values	fails	F
T	T		T
T	F	holds	F
F	T		T
F	F		T

Every modal semantics for implication satisfies:

† $\mathsf{v}(\Box A) = \mathsf{T}$ iff $\mathsf{v}(A) = \mathsf{T}$ and $\mathsf{s}(A) = \mathbf{S}$

$\mathsf{v}(\Diamond A) = \mathsf{T}$ iff $\mathsf{v}(A) = \mathsf{T}$ or $\mathsf{s}(A) \neq \varnothing$

Modal semantics for implication in $L(\neg, \wedge, \Box)$ replace the table for '$\rightarrow$' with the evaluation of $\Box$ by † and delete condition M3. The table for '$\rightarrow$' is then derivable.

A *weak modal semantics for (of) implication* is a set-assignment model or class of models $<\mathsf{v}, \mathsf{s}>$ for $L(\neg, \rightarrow, \wedge)$, where:

s satisfies M1–M3

$\neg$ and $\wedge$ are evaluated classically, and '$\rightarrow$' is evaluated by:

A	B	$s(A) \subseteq s(B)$	$A \to B$
any	values	fails	F
any	values	holds	T

Every weak modal semantics for implication satisfies:

$v(\Box A) = T$ iff $s(A) = S$

$v(\Diamond A) = T$ iff $s(A) \neq \varnothing$

Weak modal semantics for implication in $L(\neg, \wedge, \Box)$ replace the table for $\to$ with this evaluation of $\Box$, and delete condition M3. The table for $\to$ is then derivable.

Conditions on set-assignment models

M1 $s(A \wedge B) = s(A) \cap s(B)$

M2 $s(\neg A) = \overline{s(A)}$

M3 $s(A \to B) = s(\Box(A \supset B))$

M4 $s(\Box A) = S$ or $\varnothing$

M5 $s(\Box A) = S$ iff $s(A) = S$

M6 If $s(A) = S$, then $v(A) = T$

M7 If $s(A) = S$, then $s(\Box A) = S$

M8 $s(\Box(A \supset B)) \subseteq s(\Box A \supset \Box B)$

M9 $s(\Box A) \subseteq s(A)$

M10 $s(\Box A) \subseteq s(\Box\Box A)$

M11 $s(A) \subseteq s(\Box\Diamond A)$

M12 If $v(A) = T$ then $s(\Diamond A) = S$

M13 $s(\Box(\Box(A \supset \Box A) \supset A)) \subseteq s(A)$

M14 If $v(A) = F$, then $s(\Box(A \supset \Box A)) \nsubseteq s(A)$

M15 $s(\Box(\Box A \supset A)) \subseteq s(\Box A)$

M16 $s(\Box A) \subseteq s(A)$ iff $s(A) = S$

Further, s is a **K**-*set-assignment* if there are t, C, and $S \subseteq C$ such that t: Wffs $\to$ Sub C, t satisfies M1, M2, M7, M8, and for all A, $s(A) = t(A) \cap S$. S is called the *designated* subset.

A **T**-*set-assignment* is a **K**-set-assignment where t also satisfies M9.

A **B**-*set-assignment* is a **K**-set-assignment where t also satisfies M11.

Complete set-assignment semantics in $L(\neg, \wedge, \Box)$

The classes of *modal semantics for implication* $\langle v, s \rangle$ satisfying the following conditions are complete for the respective logics (Chapter VI.G):

S5 M1, M2, M4–6

S4 M1, M2, M7–10
Also may add:
If $\{y: \text{all } A, \text{ if } x \in s(\Box A) \text{ then } y \in s(A)\} \subseteq s(B)$, then $x \in s(\Box B)$.
(Exercises 6, 7 p. 256)

T s is a **T**-set-assignment

B s is a **B**-set-assignment and $\langle v, s \rangle$ satisfies M12

QT s is a **K**-set-assignment

S4Grz M1, M2, M7–10, M13, M14

G* M1, M2, M7, M8, M15, M16

MSI M1–M3

The classes of *weak* modal semantics for implication $\langle v, s \rangle$ satisfying the following conditions are complete for the respective logics:

K s is a **K**-set-assignment

G M1, M2, M7, M8, M15, M16

For all of these logics, except possibly **MSI**, we may restrict S to be finite.
For **L** any of **K**, **T**, **B**, **S4**, and **S5** the semantics are strongly complete:

$\Gamma \vdash_{\mathbf{L}} A$ iff $\Gamma \vDash A$

Complete set-assignment semantics in $L(\neg, \rightarrow, \wedge)$
Add condition M3 to each of the above.

Alternate simply presented complete set-assignment semantics can replace those above for **QT**, **T**, **B** and **K**. The models are $\langle v, s, T, S \rangle$ where $T \subseteq S$, and the truth-conditions are:

$\neg$ and $\wedge$ are evaluated classically

$v(\Box A) = T$ iff $v(A) = T$ and $T \subseteq s(A)$

In $L(\neg, \rightarrow, \wedge)$: $v(A \rightarrow B) = T$ iff
$(s(A) \cap T) \subseteq (s(B) \cap T)$ and (not both $v(A) = T$ and $v(B) = F$))

The conditions on the models for the various logics are:

QT s satisfies M1, M2, M7, M8

T s satisfies M1, M2, M7–M9

B s satisfies M1, M2, M7–10, M11, and: if $v(A) = T$ then $s(\Diamond A) \supseteq T$

For **K** the truth-conditions are different:

K s satisfies M1, M2, M7–M9, evaluates $\neg$ and $\wedge$ classically, and
$v(A \rightarrow B) = T$ iff $(s(A) \cap T) \subseteq (s(B) \cap T)$

D. Intuitionistic Logics, Int and J

Int, Heyting's Intuitionist Propositional Calculus

Axiomatizations

Int *in* L(¬, →, ∧, ∨) *Heyting, 1930* (Chapter VII.B.1)

I. $A \to (A \wedge A)$

II. $(A \wedge B) \to (B \wedge A)$

III. $(A \to B) \to ((A \wedge C) \to (B \wedge C))$

IV. $((A \to B) \wedge (B \to C)) \to (A \to C)$

V. $A \to (B \to A)$

VI. $(A \wedge (A \to B)) \to B$

VII. $A \to (A \vee B)$

VIII. $(A \vee B) \to (B \vee A)$

IX. $((A \to C) \wedge (B \to C)) \to ((A \vee B) \to C)$

X. $\neg A \to (A \to B)$

XI. $((A \to B) \wedge (A \to \neg B)) \to \neg A$

rules $\dfrac{A,\ A \to B}{B}$ $\dfrac{A,\ B}{A \wedge B}$

Heyting used the rule of substitution rather than schema.

Int *in* L(¬, →, ∧, ∨) *Dummett, 1977* (Chapter VII.C.3)

1. $A \to (B \to A)$
2. $A \to (B \to (A \wedge B))$
3. $(A \wedge B) \to A$
4. $(A \wedge B) \to B$
5. $A \to (A \vee B)$
6. $B \to (A \vee B)$
7. $(A \vee B) \to ((A \to C) \to ((B \to C) \to C))$
8. $(A \to B) \to ((A \to (B \to C)) \to (A \to C))$
9. $(A \to B) \to ((A \to \neg B) \to \neg A)$
10. $A \to (\neg A \to B)$

rule $\dfrac{A,\ A \to B}{B}$

Int *in* $L(\rightarrow, \wedge, \vee, \bot)$ *Segerberg, 1968*

$\neg A \equiv_{Def} A \rightarrow \bot$

1. $(A \wedge B) \rightarrow A$
2. $(A \wedge B) \rightarrow B$
3. $A \rightarrow (A \vee B)$
4. $B \rightarrow (A \vee B)$
5. $(A \rightarrow C) \rightarrow ((B \rightarrow C) \rightarrow ((A \vee B) \rightarrow C))$
6. $(A \rightarrow B) \rightarrow ((A \rightarrow C) \rightarrow (A \rightarrow (B \wedge C))$
7. $(A \rightarrow (B \rightarrow C)) \rightarrow ((A \rightarrow B) \rightarrow (A \rightarrow C))$
8. $A \rightarrow (B \rightarrow A)$
9. $\bot \rightarrow A$

rule $\dfrac{A,\ A \rightarrow B}{B}$

Kripke Semantics (Chapter VII.B.2)
In $L(\neg, \rightarrow, \wedge, \vee)$:
A *Kripke* model of **Int** is $\langle W, R, e \rangle$ where W is a nonempty set, R is a *reflexive*, *transitive* relation on W, and $e: PV \rightarrow \text{Sub } W$. e is called an *evaluation* and the model is *finite* if W is finite. The pair $\langle W, R \rangle$ is a *frame*. For $w \in W$ define $w \vDash A$ inductively, where $w \nvDash A$ means 'not $w \vDash A$':

1. $w \vDash p$ iff for all z such that wRz, $z \in e(p)$
2. $w \vDash A \wedge B$ iff $w \vDash A$ and $w \vDash B$
3. $w \vDash A \vee B$ iff $w \vDash A$ or $w \vDash B$
4. $w \vDash \neg A$ iff for all z such that wRz, $z \nvDash A$
5. $w \vDash A \rightarrow B$ iff for all z such that wRz, $z \nvDash A$ or $z \vDash B$

Then $\langle W, R, e \rangle \vDash A$ iff for all $w \in W$, $w \vDash A$.

In $L(\rightarrow, \wedge, \vee, \bot)$: Replace (4) above by: $w \nvDash \bot$.

A *Kripke tree* is an anti-symmetric Kripke model with designated first element w, i.e., w has no predecessors under R and is related by R to all other elements of W.

PV and Wffs are not necessarily assumed to be completed infinite totalities. Assignments or evaluations such as e can be understood as meaning that we have a method such that given any variable p_i we can produce a subset of W.

The class of all Kripke models and the class of all Kripke trees of cardinality less than or equal to that of the real numbers are both *strongly complete* for **Int**. (Chapter VII.C.2)

The class of all *finite* Kripke models and the class of all *finite* Kripke trees are both *finitely strongly complete* for **Int** and can be proved so by intuitionistically acceptable means. (Chapter VII.C.2, 3)

Set-Assignment Semantics (Chapter VII.E)
A set-assignment **Int**-model is a triplet $<v, s, S>$, where $v: PV \to \{T, F\}$, s: Wffs $\to$ Sub(S) satisfies conditions Int 1–Int 10 below, and v is extended to all wffs by the *intuitionist truth-conditions*:

$\wedge$ and $\vee$ are evaluated classically

$\to$ is evaluated by the dual dependence table (as for **S4**):

$v(A \to B) = T$ iff $s(A) \subseteq s(B)$ and (not both $v(A) = T$ and $v(B) = F$)

and $\neg$ is evaluated by the table for *intuitionist negation*:

A	$s(A) = \varnothing$	$\neg A$
any value	fails	F
T	holds	F
F		T

That is, $v(\neg A) = T$ iff $v(A) = F$ and $s(A) = \varnothing$.
A model is *finite* if S is finite.

Int 1. $s(A \wedge B) = s(A) \cap s(B)$

Int 2. $s(A \vee B) = s(A) \cup s(B)$

Int 3. $s(\neg A) \cup s(B) \subseteq s(A \to B)$

Int 4. $s(A) \cap s(A \to B) \subseteq s(B)$

Int 5. $s(A \to B) \subseteq s((A \wedge C) \to (B \wedge C))$

Int 6. $s(A \to B) \cap s(B \to C) \subseteq s(A \to C)$

Int 7. $s(A \to C) \cap s(B \to C) = s((A \vee B) \to C)$

Int 8. $s(A \to B) \cap s(A \to \neg B) = s(\neg A)$

Int 9. If $v(A) = T$, then $s(A) = S$

Int 10. $s(A \wedge \neg A) = \varnothing$

The class of all **Int**-models where S has cardinality less than or equal to that of the real numbers is *strongly complete* for **Int**. The class of finite **Int**-models is *finitely strongly complete* for **Int** and can be proved so by intuitionistically acceptable means.

I is the class of set-assignment models that use the intuitionist truth-conditions and satisfy Int 1, Int 2, Int 4, Int 8, Int 10, and:

Int K. If $\bigcap\{ s(C) : x \in s(C)\} \subseteq \overline{s(A)} \cup s(B)$, then $x \in s(A \to B)$

$\mathcal{I}$ is strongly complete for **Int**; the class of finite models in $\mathcal{I}$ is finitely strongly complete for **Int** and can be proved so by intuitionistically acceptable means (Theorem VII.25).

J, The Minimal Intuitionist Calculus

Axiomatizations (Chapter VII.F.1)

J *in* $L(\neg, \rightarrow, \wedge, \vee)$

Delete from Heyting's axiomatization of **Int** axiom scheme X:

$\neg A \rightarrow (A \rightarrow B)$

J *in* $L(\rightarrow, \wedge, \vee, \bot)$ *Segerberg, 1968*

$\neg A \equiv_{\text{Def}} A \rightarrow \bot$

Delete from Segerberg's axiomatization of **Int** axiom scheme **9**:

$\bot \rightarrow A$

Kol *in* $L(\neg, \rightarrow)$ *Kolmogorov, 1925*

$A \rightarrow (B \rightarrow A)$

$(A \rightarrow (A \rightarrow B)) \rightarrow (A \rightarrow B)$

$(A \rightarrow (B \rightarrow C)) \rightarrow (B \rightarrow (A \rightarrow C))$

$(B \rightarrow C) \rightarrow ((A \rightarrow B) \rightarrow (A \rightarrow C))$

$(A \rightarrow B) \rightarrow ((A \rightarrow \neg B) \rightarrow \neg A)$

rule $\dfrac{A, \ A \rightarrow B}{B}$

Kripke Semantics (Chapter VII.F.2)

A *Kripke model for* **J** in $L(\neg, \rightarrow, \wedge, \vee)$ is $\langle W, R, Q, e \rangle$ where:

$\langle W, R \rangle$ is transitive and reflexive

$Q \subseteq W$ is R-closed (i.e., if $w \in Q$ and wRz, then $z \in Q$)

$e : PV \rightarrow \text{Sub}\, W$

Q is to be thought of as those states of information that are inconsistent.

Validity in such a model is defined as in the Kripke semantics for **Int** with the exception of the evaluation of negations:

$w \vDash \neg A$ iff for all z such that wRz, $z \nvDash A$ or $z \in Q$

In the language $L(\rightarrow, \wedge, \vee, \bot)$: $w \vDash \bot$ iff $w \in Q$

Both the class of all Kripke models and all anti-symmetric Kripke models are strongly complete for **J**. The class of all finite anti-symmetric Kripke models is finitely strongly complete for **J**. (Theorem VII.29)

Set-Assignment Semantics (Chapter VII.F.5)
A set-assignment **J**-model in $L(\neg, \rightarrow, \wedge, \vee)$ is a triplet $<\upsilon, s, S>$, where $\upsilon: PV \rightarrow \{T, F\}$, s: Wffs $\rightarrow$ Subsets of S satisfies conditions Int 1, Int 2, Int 4–7, Int 9, and

Int 3′. $s(B) \subseteq s(A \rightarrow B)$ Int 8′. $s(A \rightarrow B) \cap s(A \rightarrow \neg B) \subseteq s(\neg A)$

And υ is extended to all wffs by the *minimal intuitionist truth-conditions*:

$\wedge$ and $\vee$ are evaluated classically

$\rightarrow$ is evaluated by the dual dependence table as for **Int**

and $\neg$ is evaluated by the *minimal intuitionist negation* table:

$\upsilon(\neg A) = T$ iff $s(A) \subseteq s(\neg A)$

E. Many-Valued Logics, L_3, L_n, $L_\aleph$, K_3, G_3, G_n, $G_\aleph$ Paraconsistent J_3

The general form of many-valued semantics can be found in Chapter VIII.B.

L_3, Łukasiewicz's 3-Valued Logic

Many-Valued Semantics

An L_3-*evaluation* is a map $e: PV \rightarrow \{0, \quad , 1\}$ that is extended to all wffs of $L(\neg, \rightarrow, \wedge, \vee)$ by the tables:

$A \wedge B$	B: 1	$\frac{1}{2}$	0
A 1	1	$\frac{1}{2}$	0
$\frac{1}{2}$	$\frac{1}{2}$	$\frac{1}{2}$	0
0	0	0	0

$A \vee B$	B: 1	$\frac{1}{2}$	0
A 1	1	1	1
$\frac{1}{2}$	1	$\frac{1}{2}$	$\frac{1}{2}$
0	1	$\frac{1}{2}$	0

A	$\neg A$
1	0
$\frac{1}{2}$	$\frac{1}{2}$
0	1

$A \rightarrow B$	B: 1	$\frac{1}{2}$	0
A 1	1	$\frac{1}{2}$	0
$\frac{1}{2}$	1	1	$\frac{1}{2}$
0	1	1	1

The only designated value is 1, so that $e \vDash A$ means $e(A) = 1$; $\vDash A$ means $e(A) = 1$ for all L_3-evaluations. And $\Gamma \vDash_{L_3} A$ means for every L_3-evaluation e, if $e(B) = 1$ for every B in Γ, then $e(A) = 1$.

The primitives may be reduced by defining:

$A \vee B \equiv_{Def} (A \to B) \to B \qquad A \wedge B \equiv_{Def} \neg(\neg A \vee \neg B)$

$A \leftrightarrow B \equiv_{Def} (A \to B) \wedge (B \to A)$

Other significant connectives (abbreviations) and their tables are:

$\Diamond A \equiv_{Def} \neg A \to A \qquad \Box A \equiv_{Def} \neg \Diamond \neg A \qquad IA \equiv_{Def} A \leftrightarrow \neg A$

A	◇A
1	1
$\frac{1}{2}$	1
0	0

A	□A
1	1
$\frac{1}{2}$	0
0	0

A	IA
1	0
$\frac{1}{2}$	1
0	0

$A \to_3 B \equiv_{Def} A \to (A \to B)$

$A \to_3 B$		B: 1	$\frac{1}{2}$	0
A	1	1	$\frac{1}{2}$	0
	$\frac{1}{2}$	1	1	1
	0	1	1	1

Axiomatizations

L₃ *in* L(¬, →) (*Wajsberg, 1931*)

$L_3 1.$ $A \to (B \to A)$

$L_3 2.$ $(A \to B) \to ((B \to C) \to (A \to C))$

$L_3 3.$ $(\neg A \to \neg B) \to (B \to A)$

$L_3 4.$ $((A \to \neg A) \to A) \to A)$

rule $\dfrac{A,\ A \to B}{B}$

L_3 *in* L(¬, →) (Chapter VIII.C.1.c)

Definitions of ∧, ∨, I, $\to_3$ are given above

1. $B \to_3 (A \to B)$
2. $(A \to_3 (B \to C)) \to ((A \to_3 B) \to (A \to_3 C))$
3. $\neg A \to (A \to B)$
4. $A \to_3 (IA \to B)$
5. $\neg A \to_3 (IA \to B)$
6. $(\neg A \to_3 A) \to ((IA \to_3 A) \to A)$
7. $\neg\neg A \to A$
8. $A \to \neg\neg A$
9. $A \to (\neg B \to \neg(A \to B))$
10. $A \to_3 (IB \to I(A \to B))$
11. $IA \to (\neg B \to I(A \to B))$
12. $IA \to (IB \to (A \to B))$

rule $\dfrac{A,\ A \to B}{B}$

Set-Assignment Semantics (Chapter VIII.C.1.d)
An $\mathbf{L_3}$-model for $L(\neg, \rightarrow)$ is a triplet $<\mathsf{v},\mathsf{s},\mathsf{S}>$, where $\mathsf{v}: PV \rightarrow \{\mathsf{T}, \mathsf{F}\}$, s: Wffs $\rightarrow$ Subsets of S, v is extended to all wffs by the intuitionist tables for $\neg$ and $\rightarrow$:

$\mathsf{v}(\neg A) = \mathsf{T}$ iff $\mathsf{s}(A) = \varnothing$ and $\mathsf{v}(A) = \mathsf{F}$

$\mathsf{v}(A \rightarrow B) = \mathsf{T}$ iff $\mathsf{s}(A) \subseteq \mathsf{s}(B)$ and (not both $\mathsf{v}(A) = \mathsf{T}$ and $\mathsf{v}(B) = \mathsf{F}$)

and s satisfies:

1. $$\mathsf{s}(\neg A) = \begin{cases} \overline{\mathsf{s}(A)} & \text{if } \mathsf{s}(A) = \varnothing \text{ or } \mathsf{s}(A) = \mathsf{S} \\ \mathsf{s}(A) & \text{otherwise} \end{cases}$$

2. $$\mathsf{s}(A \rightarrow B) = \begin{cases} \mathsf{S} & \text{if } \mathsf{s}(A) \subseteq \mathsf{s}(B) \\ \mathsf{s}(B) & \text{if } \mathsf{s}(B) \subset \mathsf{s}(A) \text{ and } \mathsf{s}(A) = \mathsf{S} \\ \mathsf{s}(A) & \text{if } \mathsf{s}(B) \subset \mathsf{s}(A) \text{ and } \mathsf{s}(A) \neq \mathsf{S} \end{cases}$$

3. $\mathsf{v}(p) = \mathsf{T}$ iff $\mathsf{s}(p) = \mathsf{S}$

4. If both $\varnothing \subset \mathsf{s}(A) \subset \mathsf{S}$ and $\varnothing \subset \mathsf{s}(B) \subset \mathsf{S}$, then $\mathsf{s}(A) = \mathsf{s}(B)$.

The defined connectives $\wedge, \vee$ are then evaluated classically, and

$\mathsf{s}(A \vee B) = \mathsf{s}(A) \cup \mathsf{s}(B) \qquad \mathsf{s}(A \wedge B) = \mathsf{s}(A) \cap \mathsf{s}(B)$

$\mathbf{L_3}$-models thus evaluate all four connectives $\{\neg, \rightarrow, \wedge, \vee\}$ by the intuitionist truth-conditions.

Alternate set-assignment semantics

A *rich* $\mathbf{L_3}$*-model* evaluates the connectives by:

$\mathsf{v}(\neg A) = \mathsf{T}$ iff $\mathsf{s}(A) = \varnothing$ and $\mathsf{v}(A) = \mathsf{F}$ [as before]

$\mathsf{v}(A \rightarrow B) = \mathsf{T}$ iff $(\mathsf{s}(A) \subseteq \mathsf{s}(B)$ or both $\varnothing \subset \mathsf{s}(A) \subset \mathsf{S}$ and $\varnothing \subset \mathsf{s}(B) \subset \mathsf{S})$
and (not both $\mathsf{v}(A) = \mathsf{T}$ and $\mathsf{v}(B) = \mathsf{F}$)

and satisfies conditions (1), (3) for $\mathbf{L_3}$-models, and condition (2) is replaced by:

2. $$\mathsf{s}(A \rightarrow B) = \begin{cases} \mathsf{S} & \text{if } \mathsf{s}(A) \subseteq \mathsf{s}(B) \text{ or (both } \varnothing \subset \mathsf{s}(A) \subset \mathsf{S} \text{ and } \varnothing \subset \mathsf{s}(B) \subset \mathsf{S}) \\ \mathsf{s}(B) & \text{if } \mathsf{s}(B) \subset \mathsf{s}(A) \text{ and } \mathsf{s}(A) = \mathsf{S} \\ \mathsf{s}(A) & \text{if } \mathsf{s}(B) \subset \mathsf{s}(A) \text{ and } \mathsf{s}(A) \neq \mathsf{S} \end{cases}$$

Every $\mathbf{L_3}$-model is a rich $\mathbf{L_3}$-model.

The Lukasiewicz Logics $\mathbf{L}_n$, $\mathbf{L}_{\aleph}$, $\mathbf{L}_{\aleph_0}$

Many-Valued Semantics (Chapter VIII.C.2.a)
An $\mathbf{L}$*-evaluation* is a map $\mathsf{e}: PV \rightarrow [0,1]$, the set of real numbers x such that $0 \leq x \leq 1$, that is extended to all wffs of $L(\neg, \rightarrow)$ by the following tables:

$\mathbf{e}(\neg A) = 1 - \mathbf{e}(A)$

$$\mathbf{e}(A \rightarrow B) = \begin{cases} 1 & \text{if } \mathbf{e}(A) \leq \mathbf{e}(B) \\ (1 - \mathbf{e}(A)) + \mathbf{e}(B) & \text{if } \mathbf{e}(B) < \mathbf{e}(A) \end{cases}$$

The connectives $\wedge$, $\vee$, and $\leftrightarrow$ are defined as for $\mathbf{L_3}$:

$A \vee B \equiv_{Def} (A \rightarrow B) \rightarrow B \qquad A \wedge B \equiv_{Def} \neg(\neg A \vee \neg B)$

$A \leftrightarrow B \equiv_{Def} (A \rightarrow B) \wedge (B \rightarrow A)$

These have the following tables:

$$\mathbf{e}(A \leftrightarrow B) = \begin{cases} 1 & \text{if } \mathbf{e}(A) = \mathbf{e}(B) \\ (1 - \mathbf{e}(A)) + \mathbf{e}(B) & \text{if } \mathbf{e}(A) > \mathbf{e}(B) \\ (1 - \mathbf{e}(B)) + \mathbf{e}(A) & \text{if } \mathbf{e}(B) > \mathbf{e}(A) \end{cases}$$

$\mathbf{e}(A \vee B) = \max(\mathbf{e}(A), \mathbf{e}(B))$

$\mathbf{e}(A \wedge B) = \min(\mathbf{e}(A), \mathbf{e}(B))$

For $n \geq 2$:

$\mathbf{L}_n = \{ A : \mathbf{e}(A) = 1$ for every $\mathbf{L}$-evaluation $\mathbf{e} : PV \rightarrow \{ \frac{m}{n-1} : 0 \leq m \leq n-1 \} \}$

$\mathbf{L}_{\aleph_0} = \{ A : \mathbf{e}(A) = 1$ for every $\mathbf{L}$-evaluation $\mathbf{e}$ that takes rational values in $[0, 1] \}$

$\mathbf{L}_{\aleph} = \{ A : \mathbf{e}(A) = 1$ for every $\mathbf{L}$-evaluation $\mathbf{e} \}$

When the values that $\mathbf{e}$ may take on PV are restricted in these definitions, then the extension of $\mathbf{e}$ to all wffs obeys the same restriction.

$\mathbf{L}_{\aleph_0} = \mathbf{L}_{\aleph}$ (Theorem VIII.13)

Axiomatization

$\mathbf{L}_{\aleph}$ *(and* $\mathbf{L}_{\aleph_0}$*) in* $L(\neg, \rightarrow)$ *(Turquette, 1959)*

In Wajsberg's axiomatization of $\mathbf{L_3}$ replace $L_3 4$ by:

$L_{\aleph}$. $(A \rightarrow B) \vee (B \rightarrow A)$

Set-Assignment Semantics for both $\mathbf{L}_{\aleph}$ *and* $\mathbf{L}_{\aleph_0}$

An $\mathbf{L}_{\aleph}$-model $\langle v, s \rangle$ for $L(\neg, \rightarrow)$ uses the intuitionist tables for $\neg$ and $\rightarrow$, as for $\mathbf{L_3}$-models, and satisfies:

L1. $s(A \rightarrow B) = S$ iff $s(A) \subseteq s(B)$

L2. $s(\neg A) = S$ iff $s(A) = \varnothing$

L3. $s(B) \subseteq s(A \rightarrow B)$

L4. $s(A \rightarrow B) \subseteq s((B \rightarrow C) \rightarrow (A \rightarrow C))$

L5. If $s(A) \subseteq s(B)$, then $s(B \rightarrow C) \subseteq s(A \rightarrow C)$.

L6. $s(\neg A \rightarrow \neg B) \subseteq s(B \rightarrow A)$

L7. $s(\neg B) \subseteq s(\neg A)$ iff $s(A) \subseteq s(B)$

L8. $s(A) \subseteq s(B)$ or $s(B) \subseteq s(A)$

L9. $\upsilon(p) = T$ iff $s(p) = S$

and for the defined connectives:

L10. $s(A \vee B) = s(A) \cup s(B)$

L11. $s(A \wedge B) = s(A) \cap s(B)$

The resulting tables for the defined connectives $\wedge$ and $\vee$ are then classical; thus all four connectives use the intuitionist truth-conditions.

K_3, Kleene's 3-Valued Logic

Many-Valued Semantics (Chapter VIII.D.1)
A K_3-*evaluation* is a map $e: PV \to \{T, F, U\}$ that is extended to all wffs of $L(\neg, \to, \wedge, \vee)$ by the tables for what Kleene calls the *strong connectives*:

$A \wedge B$	B: T	U	F
A T	T	U	F
U	U	U	F
F	F	F	F

$A \vee B$	B: T	U	F
A T	T	T	T
U	T	U	U
F	T	U	F

$A \to B$	B: T	U	F
A T	T	U	F
U	T	U	U
F	T	T	T

A	$\neg A$
T	F
U	U
F	T

The sole designated value is T. There are no tautologies. The logic is solely a consequence relation:

$\Gamma \vDash_{K_3} A$ iff for every K_3-evaluation e, if $e \vDash \Gamma$ then $e(A) = T$

Set-Assignment Semantics
A K_3-model for $L(\neg, \to, \wedge, \vee)$ is a triplet $<\upsilon, s, S>$, where $\upsilon: PV \to \{T, F\}$, s: Wffs $\to$ Subsets of S, υ is extended to all wffs by the classical tables for $\wedge$ and $\vee$, and:

$\upsilon(A \to B) = T$ iff $\upsilon(\neg A) = T$ or $\upsilon(B) = T$

$\upsilon(\neg A) = T$ iff $\upsilon(A) = F$ and $s(A) = \varnothing$

and s satisfies:

K1. $s(A) \subseteq s(B)$ or $s(B) \subseteq s(A)$

K2. $s(A \to B) = s(\neg A) \cup s(B)$

K3. $s(\neg A) = \begin{cases} s(A) & \text{if } \varnothing \subset s(A) \subset S \\ \overline{s(A)} & \text{otherwise} \end{cases}$

K4. $s(A \wedge B) = s(A) \cap s(B)$

K5. $s(A \vee B) = s(A) \cup s(B)$

K6. $v(p) = T$ iff $s(p) = S$

The entire difference between these semantics and the classical ones lies in the table and set-assignments for negation.

Conditions K1 and K2 may be replaced by:

K7. $s(A \to B) = \begin{cases} S & \text{if } s(\neg A) = S \text{ or } s(B) = S \\ \varnothing & \text{if } s(A) = S \text{ and } s(B) = \varnothing \\ s(B) & \text{otherwise} \end{cases}$

The Gödel Logics G_n and $G_\aleph$

Many-Valued Semantics (Chapter VIII.F)

A **G**-*evaluation* is a map $e: PV \to [0,1]$ that is extended to all wffs of $L(\neg, \to, \wedge, \vee)$ by the following tables:

$$e(\neg A) = \begin{cases} 1 & \text{if } e(A) = 0 \\ 0 & \text{if } e(A) \neq 0 \end{cases}$$

$$e(A \to B) = \begin{cases} 1 & \text{if } e(A) \leq e(B) \\ e(B) & \text{otherwise} \end{cases}$$

$$e(A \wedge B) = \min(e(A), e(B)) \qquad e(A \vee B) = \max(e(A), e(B))$$

The table for $A \leftrightarrow B \equiv_{\text{Def}} (A \to B) \wedge (B \to A)$ is:

$$e(A \leftrightarrow B) = \begin{cases} 1 & \text{if } e(A) = e(B) \\ \min(e(A), e(B)) & \text{otherwise} \end{cases}$$

For $n \geq 2$:

$\mathbf{G}_n = \{ A : e(A) = 1$ for every **G**-evaluation $e: PV \to \{ \frac{m}{n-1} : 0 \leq m \leq n-1 \} \}$

$\mathbf{G}_{\aleph_0} = \{ A : e(A) = 1$ for every **G**-evaluation e that takes rational values in $[0,1] \}$

$\mathbf{G}_\aleph = \{ A : e(A) = 1$ for every **G**-evaluation $e \}$

$\mathbf{G}_{\aleph_0} = \mathbf{G}_\aleph$ (Theorem VIII.23)

The tables for $\mathbf{G}_3$ are (*Heyting, 1930,* p. 56):

$A \wedge B$	B: 1	$\frac{1}{2}$	0
A: 1	1	$\frac{1}{2}$	0
$\frac{1}{2}$	$\frac{1}{2}$	$\frac{1}{2}$	0
0	0	0	0

$A \vee B$	B: 1	$\frac{1}{2}$	0
A: 1	1	1	1
$\frac{1}{2}$	1	$\frac{1}{2}$	$\frac{1}{2}$
0	1	$\frac{1}{2}$	0

A	$\neg A$
1	0
$\frac{1}{2}$	0
0	1

$A \rightarrow B$	B: 1	$\frac{1}{2}$	0
A: 1	1	$\frac{1}{2}$	0
$\frac{1}{2}$	1	1	0
0	1	1	1

Axiomatization

$\mathbf{G}_{\aleph}$ *(and* $\mathbf{G}_{\aleph_0}$*) in* $L(\neg, \rightarrow, \wedge, \vee)$ *(Dummett, 1959)*

Add to any of the axiomatizations of **Int** (pp. 424–425 above):

$(A \rightarrow B) \vee (B \rightarrow A)$

Set-Assignment Semantics (Chapter VIII.F)

For $\mathbf{G}_{\aleph}$ *and* $\mathbf{G}_{\aleph_0}$

A $\mathbf{G}_{\aleph}$-model for $L(\neg, \rightarrow, \wedge, \vee)$ is a triplet $\langle v, s, \mathsf{S} \rangle$, where $v: PV \rightarrow \{\mathsf{T}, \mathsf{F}\}$, s: Wffs $\rightarrow$ Subsets of S, v is extended to all wffs by the intuitionist truth-conditions:

$\wedge$ and $\vee$ are classical

$v(\neg A) = \mathsf{T}$ iff $v(A) = \mathsf{F}$ and $s(A) = \varnothing$ and

$v(A \rightarrow B) = \mathsf{T}$ iff $s(A) \subseteq s(B)$ and (not both $v(A) = \mathsf{T}$ and $v(B) = \mathsf{F}$)

and s satisfies:

G1. $s(\neg A) = \begin{cases} \mathsf{S} & \text{if } s(A) = \varnothing \\ \varnothing & \text{otherwise} \end{cases}$

G2. $s(A \rightarrow B) = \begin{cases} \mathsf{S} & \text{if } s(A) \subseteq s(B) \\ s(A) \cap s(B) & \text{otherwise} \end{cases}$

G3. $s(A \wedge B) = s(A) \cap s(B)$

G4. $s(A \vee B) = s(A) \cup s(B)$

G5. $v(p) = \mathsf{T}$ iff $s(p) = \mathsf{S}$

G6. $s(A) \subseteq s(B)$ or $s(B) \subseteq s(A)$

Alternatively, G6 can be added to the list of conditions for a set-assignment **Int**-model to obtain complete set-assignment semantics.

For $\mathbf{G_3}$
Replace G6 by: There is some U such that s: Wffs $\rightarrow \{\varnothing, \mathsf{U}, \mathsf{S}\}$.

Paraconsistent $\mathbf{J}_3$

Many-Valued Semantics (Chapter IX.B.2)
There are two distinct presentations of $\mathbf{J}_3$, in L(~, ∧, ◇) and in L(¬, →, ∧, ~).
A $\mathbf{J}_3$-*evaluation* is a map $\mathbf{e}$: PV $\rightarrow \{0, \frac{1}{2}, 1\}$ that is extended to all wffs of one of these languages by the appropriate tables below:

A	~A
1	0
$\frac{1}{2}$	$\frac{1}{2}$
0	1

A	¬A
1	0
$\frac{1}{2}$	0
0	1

A	◇A
1	1
$\frac{1}{2}$	1
0	0

A → B		B: 1	$\frac{1}{2}$	0
A	1	1	$\frac{1}{2}$	0
	$\frac{1}{2}$	1	$\frac{1}{2}$	0
	0	1	1	1

A ∧ B		B: 1	$\frac{1}{2}$	0
A	1	1	$\frac{1}{2}$	0
	$\frac{1}{2}$	$\frac{1}{2}$	$\frac{1}{2}$	0
	0	0	0	0

The *designated values are* 1 *and* $\frac{1}{2}$, so that $\mathbf{e} \vDash$ A means $\mathbf{e}$(A) = 1 or $\frac{1}{2}$. And $\vDash$ A means that $\mathbf{e} \vDash$ A for all $\mathbf{J}_3$-evaluations $\mathbf{e}$. And $\Gamma \vDash$ A means that for every $\mathbf{J}_3$-evaluation $\mathbf{e}$, if $\mathbf{e} \vDash$ B for all B $\in \Gamma$, then $\mathbf{e} \vDash$ A.

Four additional connectives (abbreviations) are important:

A ∨ B $\equiv_{\text{Def}}$ ~(~A ∧ ~B) A ↔ B $\equiv_{\text{Def}}$ (A → B) ∧ (B → A)

A ∨ B		B: 1	$\frac{1}{2}$	0
A	1	1	1	1
	$\frac{1}{2}$	1	$\frac{1}{2}$	$\frac{1}{2}$
	0	1	$\frac{1}{2}$	0

A ↔ B		B: 1	$\frac{1}{2}$	0
A	1	1	$\frac{1}{2}$	0
	$\frac{1}{2}$	$\frac{1}{2}$	$\frac{1}{2}$	0
	0	0	0	1

□A $\equiv_{\text{Def}}$ ~◇~A ©A $\equiv_{\text{Def}}$ ¬(◇A ∧ ◇~A)

A	□A
1	1
$\frac{1}{2}$	0
0	0

A	©A
1	1
$\frac{1}{2}$	0
0	1

Definitions are surveyed in Chapter IX.B.3.

Axiomatizations

J_3 *in* $L(\sim, \wedge, \Diamond)$ (Chapter IX.E.1)

$A \to B \equiv_{Def} \sim(\Diamond A \wedge \sim B)$ $\quad$ $©A \equiv_{Def} \sim(\Diamond A \wedge \Diamond \sim A)$

$A \leftrightarrow B \equiv_{Def} (A \to B) \wedge (B \to A)$

1. $B \to (A \to B)$
2. $(A \to (B \to C)) \to ((A \to B) \to (A \to C))$
3. $(B \to (A \to C)) \to ((A \wedge B) \to C)$
4. $A \to (B \to (A \wedge B))$
5. $(A \wedge \sim A \wedge ©A) \to B$
6. $((\sim A \wedge ©A) \to A) \to A$
7. $\sim\sim A \leftrightarrow A$
8. $©A \leftrightarrow ©\sim A$
9. $\sim\Diamond A \leftrightarrow (\sim A \wedge ©A)$
10. $©(\Diamond A)$
11. $[(A \wedge B) \wedge ©(A \wedge B)] \leftrightarrow [(A \wedge ©A) \wedge (B \wedge ©B)]$

rule $\dfrac{A, A \to B}{B}$

J_3 *in* $L(\neg, \to, \wedge, \sim)$ (Chapter IX.E.2)

$©A \equiv_{Def} \neg(A \wedge \sim A)$ $\quad$ $A \leftrightarrow B \equiv_{Def} (A \to B) \wedge (B \to A)$

PC based on $\neg, \to, \wedge$

1. $\sim\sim A \leftrightarrow A$
2. $\sim\neg A \wedge \neg A \to \neg\sim A$
3. $\neg\sim(A \wedge B) \leftrightarrow \neg\sim A \wedge \neg\sim B$
4. $\neg A \to \neg\sim(A \to B)$
5. $(B \wedge \neg\sim B) \to \neg\sim(A \to B)$
6. $((A \wedge B) \wedge \sim B) \to \sim(A \to B)$

rule $\dfrac{A, A \to B}{B}$

Set-Assignment Semantics (Chapter IX.F)

In $L(\sim, \wedge, \Diamond)$

A $\mathbf{J_3}$-model for $L(\sim, \wedge, \Diamond)$ is a triplet $<v, s, S>$, where $v: PV \to \{T, F\}$, s: Wffs $\to$ Subsets of S, v is extended to all wffs by:

$v(A \wedge B) = T$ iff $v(A) = T$ and $v(B) = T$

$v(\sim A) = T$ iff $s(A) \neq S$

$v(\Diamond A) = T$ iff $v(A) = T$

and s, S satisfy:

1. $S \neq \varnothing$
2. $s(p) \neq \varnothing$ iff $v(p) = T$
3. $s(A) \subseteq s(B)$ or $s(B) \subseteq s(A)$
4. $s(A \wedge B) = s(A) \cap s(B)$
5. $s(A \rightarrow B) = \begin{cases} s(B) & \text{if } \varnothing \subset s(A) \subset S \\ \overline{s(A)} \cup s(B) & \text{otherwise} \end{cases}$
6. $s(\neg A) = \begin{cases} S & \text{if } s(A) = \varnothing \\ \varnothing & \text{otherwise} \end{cases}$
7. $s({\sim}A) = \begin{cases} s(A) & \text{if } \varnothing \subset s(A) \subset S \\ \overline{s(A)} & \text{otherwise} \end{cases}$

The defined connectives $\neg$, $\rightarrow$, $\vee$ are then evaluated classically.

In $L(\neg, \rightarrow, \wedge, \sim)$

$\neg$, $\rightarrow$, $\wedge$ are evaluated classically

$v({\sim}A) = T$ iff $s(A) \neq S$

and s, S satisfy:

1. $S \neq \varnothing$
2. $s(p) \neq \varnothing$ iff $v(p) = T$
3. $s(A) \subseteq s(B)$ or $s(B) \subseteq s(A)$
4. $s(A \wedge B) = s(A) \cap s(B)$
5. $s(A \rightarrow B) = \begin{cases} s(B) & \text{if } \varnothing \subset s(A) \subset S \\ \overline{s(A)} \cup s(B) & \text{otherwise} \end{cases}$
6. $s(\neg A) = \begin{cases} S & \text{if } s(A) = \varnothing \\ \varnothing & \text{otherwise} \end{cases}$
7. $s({\sim}A) = \begin{cases} s(A) & \text{if } \varnothing \subset s(A) \subset S \\ \overline{s(A)} & \text{otherwise} \end{cases}$

Alternate truth-default set-assignment semantics for $\mathbf{J_3}$ are presented in Chapter IX.G.

Bibliography

- Only works cited in the text or elsewhere in the bibliography are listed.
- Page references are to the most recent English reference listed, unless otherwise noted.
- Quotation marks and logical notation in all quotations have been changed to conform with the conventions of this book (see pp. 6–7 for the use of quotation marks).

ÅQVIST, Lennart
1984 Deontic logic
Chapter II.11 of *Gabbay and Guenthner, 1984.*

ANDERSON, Alan R. and Nuel D. BELNAP, Jr.
1975 *Entailment*
Princeton University Press.

ARISTOTLE. *See* McKEON

ARRUDA, Ayda I.
1980 A survey of paraconsistent logic
Mathematical Logic in Latin America, ed. A. Arruda, R. Chuaqui, and N. C. A. da Costa, North-Holland.
1989 Aspects of the historical development of paraconsistent logic
Paraconsistent Logic, eds. G. Priest, R. Routley, J. Norman, Philosophia Verlag, pp. 99–130.

AUNE, Bruce
1976 Possibility
In *Edwards, 1967,* Vol. 6, pp. 419–424.

AVRON, Arnon,
1986 On an implication connective of RM
Notre Dame Journal of Formal Logic, vol. 27, pp. 201–209.
1991 Natural 3-valued logics–characterization and proof theory
Journal of Symbolic Logic, vol. 56, pp. 276–294.

BECKER, O.,
1930 Zur Logik der Modalitäten
Jahrbuch für Philosophie und Phänomenologische Forschung, vol. 11, pp. 497–548

BENNETT, Jonathan
1969 Entailment
Phil. Rev., vol. 78, pp. 197–235.

BERNAYS, Paul
1926 Axiomatische Untersuchung des Aussagen-Kalküls der *Principia mathematica*
Mathematische Zeitschrift, vol. 25, pp. 305–320.

BLOK, W. J. and KÖHLER, P.
1983 Algebraic semantics for quasi-classical modal logics
The Journal of Symbolic Logic, vol. 48, no. 4, pp. 941–963.

BLOK, W. J. and PIGOZZI, D.
1982 On the structure of varieties with equationally definable principal congruences I
Algebra Universalis, vol. 15, pp. 195–227.
198? The deduction theorem in algebraic logic
Typescript.
1989 *Algebraizable Logics*
Memoirs of the American Mathematical Society, no. 396.

BOCHENSKI, I. M.
1970 *A History of Formal Logic*
Chelsea. A revision and translation of the German *Formale Logik*, Verlag Karl Alber, Freiburg, 1956.

BOLZANO, Bernard
1837 *Wissenschaftslehre*
Sulzbach. The relevant passage is translated in *Bochenski, 1970,* pp. 220–282.

BOOLOS, George
1979 *The unprovability of consistency*
Cambridge University Press.
1980 A Provability, truth, and modal logic
J. Phil. Logic, vol. 9, no. 1, pp. 1–7.
1980 B On systems of modal logic with provability interpretations
Theoria, vol. 46, no. 1, pp. 7–18

BROUWER, L. E. J.
1907 Over de grondslagen der wiskunde
Dissertation, Amsterdam. Translated as 'On the foundations of mathematics' in *Brouwer, 1975*, pp. 11–101
1908 De onbetrouwbaarheid der logische principes
Tijdschrift voor wijsbegeerte, vol. 2, pp. 152–158.
Translated as 'The unreliability of the logical principles' in *Brouwer, 1975*, pp. 107–111.
1912 Intuitionisme en formalisme
Inaugural address, University of Amsterdam. Translated as 'Intuitionism and formalism' in *Bulletin of the American Math. Soc.*, vol. 20 (Nov. 1913), pp. 81–96, and reprinted in *Philosophy of Mathematics,* ed. P. Benacerraf and H. Putnam, Prentice-Hall, Englewood Cliffs N.J. 2nd edition, 1983, Cambridge University Press, pp. 77–89.
1928 Intuitionistische Betrachtungen über den Formalismus
Sitzungsberichte der Preussischen Akademie der Wissenschaften, Phys.-math. Kl., pp. 48–52. Translated as 'Intuitionistic reflections on formalism' in *van Heijenoort, 1967,* pp. 490–492.
1975 *The collected works of L. E. J. Brouwer*
ed. A. Heyting, North-Holland.

BULL, Robert and SEGERBERG, Krister
1984 Basic modal logic
Chapter II.1 of *Gabbay and Guenthner, 1984.*

CANTOR, Georg
1883 *Grundlagen einer allgemeinen Mannigfaltigkeitslehre*
Teubner, Leipzig.
The translation in the text comes from *Georg Cantor*, J. Dauben, Harvard University Press, 1979, pp. 128–129.

CARNAP, Rudolf
1954 *Einführung in die Symbolische Logik*
Springer.
Translated as *Introduction to Symbolic Logic,* Dover, 1958.

CARNIELLI, Walter A.
1987 A Methods of proof for relatedness and dependence logic
Reports on Mathematical Logic, vol. 21, pp. 35–46.
1987 B Systematization of finite many-valued logics through the method of tableaux
The Journal of Symbolic Logic, vol. 52, pp. 473–493.

CHELLAS, Brian
1980 *Modal Logic*
Cambridge University Press.

CHURCH, Alonzo
1956 *Introduction to Mathematical Logic*
Princeton University Press.

CHRISTENSEN, Niels Egmont
1973 Is there a "logic" or formal system based on the concept of a truth determinant?
Danish Yearbook of Philosophy, vol. 10, pp. 77–85.

CLEAVE, J. P.
1974 The notion of logical consequence in the logic of inexact predicates
Zeit. Math. Logik und Grundlagen, vol. 20, no. 4, pp. 307–324.

COPELAND, B. J.
1978 *Entailment, the Formalisation of Inference*
Doctor of Philosophy Thesis, Oxford.
1984 Horseshoe, hook, and relevance
Theoria, vol. L, pp. 148–164.

DA COSTA, Newton C.A.
1963 Calculs propositionnels pour les systèmes formels inconsistents
Comptes Rendus de l'Academie des Sciences de Paris, Série A, vol. 257, pp. 3790–3792.
1974 On the theory of inconsistent formal systems
Notre Dame Journal of Formal Logic, XV, no. 4, pp. 497–510.
See also D'OTTAVIANO and DA COSTA

DA COSTA, Newton C.A. and MARCONI , Diego
1990 An overview of paraconsistent logic in the 80s
The Journal of Non-Classical Logic, vol. 6, pp. 5–31.

DE MORGAN, Augustus
1847 *Formal Logic or the Calculus of Inferences Necessary and Probable*
London. Reprinted by Open Court, 1926.

DE SWART, H.

1976 Another intuitionistic completeness proof
Journal of Symbolic Logic, vol. 41, no. 3, pp. 644–662.

1977 An intuitionistically plausible interpretation of intuitionist logic
Journal of Symbolic Logic, vol. 42, no. 4, pp. 564–578.

D'OTTAVIANO, Itala M. L.

1985 A The completeness and compactness of a three-valued first-order logic
Revista Colombiana de Matemáticas, XIX, 1–2, pp. 31–42.

1985 B The model-extension theorems for $\mathbf{J}_3$-theories
Methods in Mathematical Logic, ed. C. A. Di Prisco, Lecture Notes in Mathematics, no. 1130, Springer-Verlag.

1987 Definability and quantifier elimination for $\mathbf{J}_3$-theories
Studia Logica, XLVI, 1, pp. 37–54.

D'OTTAVIANO, Itala M. L. and DA COSTA, Newton C. A.

1970 Sur un problème de Jaśkowski
C. R. Acad. Sc. Paris, 270, Série A, pp. 1349–1353.

DREBEN, Burton and VAN HEIJENOORT, Jean

1986 Note to Gödel's dissertation
In *Gödel, 1986,* pp. 44–59.

DUMMETT, Michael

1959 A propositional calculus with denumerable matrix
The Journal of Symbolic Logic, vol. 24, pp. 97–106.

1973 The philosophical basis of intuitionistic logic
In *Logic Colloquium '73*, ed. H. E. Rose and J. C. Shepherdson, North-Holland, Amsterdam. Reprinted in Dummett, *Truth and Other Enigmas*, Harvard University Press, 1978.

1977 *Elements of Intuitionism*
Clarendon Press, Oxford.

DUNN, J. Michael

1972 A modification of Parry's Analytic Implication
Notre Dame J. of Formal Logic, vol. 13, no. 2, pp. 195–205.

EDWARDS, Paul (ed.)

1967 *The Encyclopedia of Philosophy*
Macmillan and The Free Press.

EPSTEIN, Richard L.

1979 Relatedness and implication
Phil. Studies, vol. 36, no. 2, pp. 137–173.

1980 A (ed.) *Relatedness and Dependence in Propositional Logics*
Research Report of the Iowa State University Logic Group.

1980 B Relatedness and dependence in propositional logics
Abstract, *The Journal of Symbolic Logic*, vol. 46, no. 1, p. 202.

1987 The algebra of dependence logic
Reports on Mathematical Logic, vol. 21, pp. 19–34

1988 A general framework for semantics for propositional logics
In *Methods and Applications of Mathematical Logic*, Proceedings of the VII Latin American Symposium on Mathematical Logic, ed. W. Carnielli and L. P. de Alcantara, *Contemporary Mathematics*, American Math. Soc., no. 69.

1992 A theory of truth based on a medieval solution to the liar paradox

History and Philosophy of Logic, vol. 13, pp. 149–177.

EPSTEIN, Richard L. and CARNIELLI, Walter A.
1989 *Computability*
Wadsworth & Brooks/Cole.

EPSTEIN, Richard L. and MADDUX, Roger D.
1980 The algebraic nature of set assignments
In *Epstein, 1980 A.*

FEYS, R.
1937 Les logiques nouvelles des modalités
Revue Néoscholastic de Philosphie, vol. 40, pp. 517–553,
vol. 41, (1938), pp. 217–252.

FINE, Kit
1979 Analytic implication
Notre Dame J. of Formal Logic, vol. 27, no. 2, pp. 169–179.

FITTING, M. C.
1969 *Intuitionistic Logic, Model Theory and Forcing*
North-Holland.

FREGE, Gottlob
1879 *Begriffschrift*
L. Nebert, Halle. Translated as *Begriffschrift, a formula language, modeled upon that of arithmetic, for pure thought*, in *van Heijenoort, 1967,* pp. 1–82.
1892 Über Sinn und Bedeutung
Zeit. für Philosophie und philosophische Kritik, vol. 100, pp. 25–50. Translated as 'On sense and reference' in *Translations from the Philosophical Writings of Gottlob Frege,* ed. M. Black and P. Geach, Basil Blackwell, 1970, pp. 56–78.
1918 Der Gedanke: eine logische Untersuchung
Beträge zur Philosophie des deutschen Idealismus, pp. 58–77.
Translated by A. and M. Quinton as 'The thought: a logical inquiry' in *Mind,* (new series) vol. 65, pp. 289–311, and reprinted in *Philosophical Logic,* ed. P. F. Strawson, Oxford University Press, 1967, pp. 17–38.
1980 *Philosophical and Mathematical Correspondence*
University of Chicago Press.

GABBAY, D. and GUENTHNER, F.
1984 *Handbook of Philosophical Logic, Volumes I–IV*
D. Reidel.

GENTZEN, Gerhard
1936 Die Widerspruchsfreiheit der reinen Zahlentheorie
Mathematische Annalen, vol. 112, pp. 493–565.

GIRLE, Roderic A.
1989 'And/or' or 'Or but not both' or both
History and Philosophy of Logic, vol. 10, pp. 39–45.

GLIVENKO, V.
1929 Sur quelques points de la logique de M. Brouwer
Académie Royale de Belgique, Bulletins de la classe des sciences, ser. 5, vol. 15, pp. 183–188.

GÖDEL, Kurt
1932 Zum intuitionistischen Aussagenkalkül
Akademie der Wissenschaften in Wien, Math.-natur. Klasse, vol. 69, pp. 65–66.

Translated as 'On the intuitionistic propositional calculus' in *Gödel, 1986*, pp. 223–225.

1933 A Zur intuitionistischen Arithmetik und Zahlentheorie
Ergebnisse eines mathematischen Kolloquiums, vol. 4 (1931–32), pp. 34–38. Translated as 'On intutionistic arithmetic and number theory' in *The Undecidable*, ed. M. Davis, Raven Press, New York, 1965, pp. 75–81, and in *Gödel, 1986*, pp. 287–295.

1933 B Eine Interpretation des intuitionistischen Aussagenkalküls
Ergebnisse eines mathematischen Kolloquiums, vol. 4 (1931–32), pp. 39–40. Translated as 'An interpretation of the intuitionistic sentential logic', in *The Philosophy of Mathematics*, ed. J. Hintikka, Oxford University Press, 1969, pp. 128–129, and as 'An interpretation of the intuitionistic propositional calculus' in *Gödel, 1986*, pp. 301–303.

1933 C Über Unabhängigkeitsbeweise in Aussagenkalkül
Ergebnisse eines mathematischen Kolloquiums, vol. 4 (for 1931–32), pp.9–10. Translated as 'On independence proofs in the propositional calculus' in *Gödel, 1986*, pp. 269–271.

1986 *Collected Works, Volume 1*
ed. Feferman et al., Oxford University Press.

GOLDBLATT, Rob

1978 Arithmetical necessity, provability and intuitionistic logic
Theoria, vol. 44, pp. 38–46

1979 *Topoi, the categorial analysis of logic*
North-Holland.

GOLDFARB, Warren D.

1979 Logic in the twenties: the nature of the quantifier
The Journal of Symbolic Logic, vol. 44, pp. 351–369.

GRZEGORCZYK, Andrzej

1967 Some relational systems and the associated topological spaces
Fundamenta mathematicae, vol. 60, pp. 223–231.

HAACK, Susan

1974 *Deviant Logic*
Cambridge University Press.

HANSON, William H.

1980 First-degree entailments and information
Notre Dame J. of Formal Logic, vol. 21, no. 4, pp.659–671.

HENKIN, Leon

1954 Boolean representation through propositional calculus
Fundamenta Mathematicae, vol. 41, pp. 89–96.

HEYTING, Arend

1930 Die formalen Regeln der intuitionistischen Logik
Sitzungsberichte der Preussischen (Berlin) Akademie der Wissenschaften, Phys.-Math. Kl, pp. 42–56. The quotations in the text are from *Bochenski, 1970*, pp. 293–294.

HILBERT, David

1922 Die logischen Grundlagen der Mathematik
Mathematische Annalen, vol. 88 (1923), pp. 151–165.

HUGHES, G. E. and CRESSWELL, M. J.
1968 *An Introduction to Modal Logic*
Methuen. 2nd printing with corrections, 1971.
1984 *A Companion to Modal Logic*
Methuen and Co.
ISEMINGER, Gary
1986 Relatedness logic and entailment
The Journal of Non-classical Logic, vol.3, no.1, pp.5–23.
JAŚKOWSKI, S.
1948 Rachunek zdan dla systemów dedukcyjnych sprzecznych
Studia Societatis Scientiarum Torunensis, Section A, vol.1, no.5, pp.57–77. Translated as 'Propositional calculus for contradictory deductive systems', *Studia Logica*, XXIV, 1969, pp.143–157.
JOHANSSON, Ingebrigt
1936 Der minimalkalkül, ein reduzierter intutionistischer Formalismus
Compositio mathematica, vol.4, pp.119–136. The translation in the text is by D. Steiner.
KALMÁR, László
1935 Über die Axiomatisierbarkeit des Aussagenkalküls
Acta Scientiarum Mathematicarum, vol.7, pp.222–243.
KIELKOPF, Charles F.
1977 *Formal Sentential Entailment*
University Press of American, Washington D.C.
KLEENE, Stephen Cole
1952 *Introduction to Metamathematics*
North-Holland. Sixth reprint with corrections, 1971.
KNEALE, William and Martha
1962 *The Development of Logic*
Clarendon Press, Oxford.
KOLMOGOROV, A. N.
1925 Sur le principe de tertium non datur
Matématicéskij Sbornik, vol.32, pp.646–667. Translated as 'On the principle of excluded middle', in *van Heijenoort*, pp.416–437.
1932 Zur Deutung der intuitionistischen Logik
Mathematische Zeit., vol.35, pp.58–65
KRAJEWSKI, Stanisław
1986 Relatedness logic
Reports on Mathematical Logic, vol.20, pp.7–14.
1992 Note on grammatical translations of logical calculi
Archive of Mathematical Logic, vol. 31, pp. 259–262.
KRIPKE, Saul A.
1959 A completeness theorem in modal logic
The Journal of Symbolic Logic, vol.24, pp.1–14.
1965 Semantical analysis of intuitionistic logic, I
In *Formal Systems and Recursive Functions,* ed. J.N. Crossley and M. A.E. Dummett, North-Holland, Amsterdam, pp.92–130.
1972 Naming and necessity
In *Semantics of Natural Language,* ed. D.Davidson and G.Harman, pp. 253-355.

1975 Outline of a theory of truth
J. of Philosophy, vol. 72, pp. 690–716.

LEIVANT, Daniel
1985 Syntactic translations and provably recursive functions
Journal of Symbolic Logic, vol. 50, no. 3, pp. 682–688.

LEMMON, E. J.
1977 *An Introduction to Modal Logic*
In collaboration with Dana Scott, edited by K. Segerberg, *American Philosophical Quarterly,* Monograph 11, Basil Blackwell, Oxford.

LEWIS, C. I.
1912 Implication and the algebra of logic
Mind, vol. 21 (new series), pp. 522–531.
1918 *A Survey of Symbolic Logic*
University of California Press.

LEWIS, C. I. and LANGFORD, C. H.
1932 *Symbolic Logic*
The Century Company. 2nd edition with corrections, Dover, 1959.

LEWIS, David K.
1973 *Counterfactuals*
Harvard University Press.

ŁOS, Jerzy
1951 An algebraic proof of completeness for the two-valued propositional calculus
Colloquium Mathematicum, vol. 2, pp. 236–240.

ŁUKASIEWICZ, Jan
1920 O logice trójwartościowej
Ruch Filozoficzny, vol. 5, pp. 170–171. Translated as 'On three-valued logic' in *Łukasiewicz, 1970*, pp. 87–88, and in *McCall, 1967,* pp. 16–18.
1922 On determinism
Translation of the original Polish lecture, in *Łukasiewicz, 1970*, pp. 110–128, and in *McCall*, pp. 19–39.
1929 *Elementy logiki matematyczny*
Translated as *Elements of Mathematical Logic,* PWN–Polish Scientific Publishers, Warsaw, 1963.
1930 Philosophische Bemerkungen zu mehrwertigen Systemen des Aussagenkalküls
Comptes Rendus des Séances de la Société des Sciences et des Lettres de Varsovie, vol. 23, cl.iii, pp. 51–77. Translated as 'Philosophical remarks on many-valued systems of propositional logic' in *Łukasiewicz, 1970*, pp. 153–178, and in *McCall, 1967,* pp. 40–65.
1952 On the intuitionistic theory of deduction
Konikl. Nederl. Akademie van Wetenschappen, Proceedings, Series A, no. 3, pp. 202–212. Reprinted in *Łukasiewicz*, 1970, pp. 325–335.
1953 A system of modal logic
J. Computing Systems, vol. 1, pp. 111–149. Reprinted in *Łukasiewicz, 1970*, pp. 352–390.
1970 *Selected Works*
ed. L Borkowski, North-Holland.

ŁUKASIEWICZ, Jan and TARSKI, Alfred
1930 Untersuchungen über den Aussagenkalkül
Comptes Rendus des Séances de la Société des Sciences et des Lettres de Varsovie, vol. 23, cl. iii, pp. 39–50. Translated as 'Investigations into the sentential calculus' in *Łukasiewicz, 1970,* pp. 131–152, and in *Tarski, 1956*, pp. 38–59. References in the text are to the latter.

MARCISZEWSKI, Witold (ed.)
1981 *Dictionary of Logic*
Martinus Nijhoff.

MATES, Benson
1953 *Stoic Logic*
University of California Publications in Philosophy, Vol.26.
Reprinted by the University of California Press, 1961.
1965 *Elementary Logic*
Oxford University Press. 2nd edition, 1972.
1986 *The Philosophy of Leibniz*
Oxford University Press.

McCALL, Storrs (ed.)
1967 *Polish Logic*
Oxford University Press.

McCARTY, Charles
1983 Intuitionism: an introduction to a seminar
Journal of Philosophical Logic, vol. 12, pp. 105–149.

McKEON, Richard, ed.
1941 *The Basic Works of Aristotle*
Random House.

McKINSEY, J. C. C.
1939 Proof of the independence of the primitive symbols of Heyting's calculus of propositions
Journal of Symbolic Logic, vol. 4, pp.155–158.

MONTEIRO, A.
1967 Construction des algèbres de Łukasiewicz trivalentes dans les algèbres de Boole monadiques, I
Math. Japonicae, vol. 12, pp. 1–23.

PARRY, William Tuthill
1933 Ein Axiomensystem für eine neue Art von Implikation (analytische Implikation)
Ergebnisse eines mathematischen Kolloquiums, vol. 4, pp. 5–6.
1971 Comparison of entailment theories
Typescript of an address to the Association of Symbolic Logic, an abstract of which appears in *Journal of Symbolic Logic,* vol. 37 (1972), pp. 441–442.
197? Analytic implication: its history, justification, and varieties
Typescript of an address to the International Conference on Relevance Logics.

PERZANOWSKI, Jerzy
1973 The deduction theorem for the modal propositional calculi formalized after the manner of Lemmon, Part I
Reports on Mathematical Logic, vol. 1, pp. 1–12.

PIGOZZI, Donald. *See* BLOK and PIGOZZI

PORTE, Jean
1982 Fifty years of deduction theorems
In *Proceedings of the Herbrand Symposium Logic Colloquium '81,* ed. J. Stern, North-Holland, pp. 243–250.

POST, Emil L.
1921 Introduction to a general theory of elementary propositions
Amer. Journal of Math., vol. 43, pp. 163–185. Reprinted in *van Heijenoort, 1967,* pp. 264–283.

PRAWITZ, Dag and MALMNAS, P.-E.
1965 A survey of some connections between classical, intuitionistic and minimal logic
In *Contributions to Mathematical Logic,* eds. A. Schmidt, K. Schütte, and H. J. Thiele, North-Holland, pp. 215–229.

PRIOR, Arthur
1948 Facts, propositions and entailment
Mind, (new series) vol. 57, pp. 62–68
1955 *Formal Logic*
2nd edition with corrections, 1963, Clarendon Press, Oxford.
1960 The autonomy of ethics
Australasian J. of Phil., vol. 38, pp. 199–206. Reprinted in *Prior, 1976,* pp. 88–96.
1964 Conjunction and contonktion revisited
Analysis vol. 24, pp. 191–195. Reprinted in *Prior, 1976,* pp. 159–164.
1967 Many-valued logic
In *Edwards, 1967,* vol. 5, pp. 1–5.
1976 *Papers in Logic and Ethics*
ed. P. T. Geach and A. J. P. Kenny, University of Massachusetts Press.

QUINE, Willard Van Orman
1950 *Methods of Logic*
4th edition, Harvard University Press.
1970 *Philosophy of Logic*
Prentice-Hall.

RASIOWA, Helena
1974 *An Algebraic Approach to Non-classical Logics*
North-Holland.

RESCHER, Nicholas
1968 Many-valued logic
In *Topics in Philosophical Logic*, N. Rescher, D. Reidel.
1969 *Many-valued Logics*
McGraw-Hill.

RICHARDS, Thomas J.
1989 'Or' and/or 'And/or'
History and Philosophy of Logic, vol. 10, pp. 29–38.

ROSSER, J. Barkley
1953 *Logic for Mathematicians*
McGraw-Hill.

RUDIN, Walter
1964 *Principles of Real Analysis*
McGraw-Hill, 2nd edition.

RUSSELL, Bertrand . *See* WHITEHEAD and RUSSELL
SCOTT, Theodore Kermit
1966 *John Buridan: Sophisms on Meaning and Truth*
Appleton-Century-Crofts, New York.
SEARLE, John R.
1970 *Speech Acts*
Cambridge University Press.
1983 *Intentionality*
Cambridge University Press.
SEGERBERG, Krister
1968 Propositional logics related to Heyting's and Johansson's
Theoria, vol. 34, pp. 26–61.
1971 *An Essay in Classical Modal Logic*
Filosofiska Studier, no. 13, Uppsala University
SHAW-KWEI, MOH
1954 Logical paradoxes for many-valued logics
Journal of Symbolic Logic, vol. 19, no. 1, pp. 37–40.
SILVER, Charles
1980 A simple strong completeness proof for sentential logic
Notre Dame J. of Formal Logic, XXI, pp. 179–181.
SLUGA, Hans
1987 Semantic content and cognitive sense
In *Frege Synthesized*, eds. L Haaparanta and J. Hintikka, D. Reidel.
SMILEY, T. J.
1959 Entailment and deducibility
Proc. Aristotelian Soc., vol. 59, pp. 233–254.
1962 The independence of connectives
Journal of Symbolic Logic, vol. 27, no. 4, pp. 426–436.
1976 Comment on 'Does many-valued logic have any use?' by D. Scott,
in *Philosophy of Logic,* S. Körner, Univ. of California Press, 1976, pp. 74–88.
SMULLYAN, Raymond
1978 *What is the name of this book?*
Prentice-Hall.
SPECKER, Ernst
1960 Die Logik Nicht Gleichzeitig Entscheidbarer Aussagen
Dialectica, vol. 14, pp. 239–246. Translated as, 'The logic of propositions which are not simultaneously decidable', in *The Logico-Algebraic Approach to Quantum Mechanics*, vol. 1, ed. C. A. Hooker, D. Reidel, 1975, pp. 135–140.
STALNAKER, Robert C. and THOMASON, Richmond
1970 A semantic analysis of conditionals
Theoria, vol. 36, pp. 23–42.
SURMA, Stanisław J.
1972 A deduction theorem valid in certain fragments of Lewis' system S2 and the system T of Feys-von Wright
Studia Logica, vol. 31, pp. 127–138.
1973 A (ed.) *Studies in the History of Mathematical Logic*
Polish Academy of Sciences, Warsaw.

1973 B A history of the significant methods of proving Post's theorem about the completeness of the classical propositional calculus
In *Surma, 1973 A*, pp. 19–32.

TARSKI, Alfred

1930 Über einige fundamentale Begriffe der Metamathematik
Comptes Rendus des séances de la Société des Sciences et des Lettres de Varsovie, cl. iii, vol. 23, pp. 22–29. Translated as 'On some fundamental concepts of metamathematics' in *Tarski, 1956,* pp. 30–37.

1936 The concept of truth in formalized languages
Reprinted in *Tarski, 1956,* pp. 152–278, where a detailed publication history of it is given.

1956 *Logic, Semantics, Metamathematics*
2nd edition with corrections, 1983, edited by J. Corcoran, Hackett Publ., Indianapolis.

See also JÓNSSON and TARSKI, ŁUKASIEWICZ and TARSKI, McKINSEY and TARSKI

TROELSTRA, A. S. and VAN DALEN, Dirk

1988 *Constructivism in Mathematics*
North-Holland.

TURQUETTE, Atwell R.

1959 Review of papers by Rose and Rosser, Meredith, and Chang
Journal of Symbolic Logic, vol. 24, pp. 248–249.

VAN HEIJENOORT, Jean (ed.)

1967 *From Frege to Gödel: A source book in mathematical logic 1879–1931*
Harvard University Press.

VELDMAN, Wim

1976 An intuitionistic completeness theorem for intuitionistic predicate logic
Journal of Symbolic Logic, vol. 41, no. 1, pp. 159–166.

WAJSBERG, Mordechaj

1931 Aksjomatyzacja trójwartościowego rachunku zdan
Comptes Rendus des séances de la Société des Sciences et des Lettres de Varsovie, cl.iii, vol. 24, pp. 126–145. Translated as 'Axiomatization of the three-valued propositional calculus' in *McCall, 1967,* pp. 264–284.

WALTON, Douglas N.

1979 Relatedness in intensional action chains
Phil. Studies, vol. 36, no. 2, pp. 175–225.

1982 *Topical Relevance in Argumentation*
John Benjamins, Philadelphia.

1985 *Arguer's Position*
Greenwood Press, London.

WHITEHEAD, Alfred North and RUSSELL, Bertrand

1910–13 *Principia Mathematica*
Cambridge University Press.

WILLIAMSON, Colwyn

1968 Propositions and abstract propositions
In *Studies in Logical Theory,* ed. N. Rescher, *American Philosophical Quarterly,* Monograph no. 2, Basil Blackwell, Oxford.

WÓJCICKI, Ryszard
1988 *Theory of Logical Calculi*
D. Reidel.
WOJTYLAK, Piotr
1984 An example of a finite though finitely non-axiomatizable matrix
Reports on Mathematical Logic, vol. 17, pp. 39–46.
WOLF, Robert G.
1977 A survey of many-valued logic
In *Modern Uses of Multiple-Valued Logic,* eds. J. M. Dunn and G. Epstein, D. Reidel, 1977.
WRIGLEY, Michael
1980 Wittgenstein on inconsistency
Philosophy, vol. 55, pp. 471–484.
YABLO, Stephen
1985 Truth and reflection
J. Phil. Logic, vol. 14, pp. 297–349.

Glossary of Notation

General

iff	'if and only if'
■	end of proof

Formal Languages

$L(\neg, \rightarrow, \wedge, \vee, p_0, p_1, \ldots)$	a formal language, 13, 47, 150
$p_0, p_1, \ldots$	propositional variables of the formal language
$p, q, q_0, q_1, \ldots$	metavariables ranging over $\{p_0, p_1, \ldots\}$
PV	the collection of all propositional variables $p_0, p_1, \ldots$, 43, 150
$A, B, C, A_0, A_1, \ldots$	metavariables ranging over wffs of the formal language or propositions, 13
real (A)	the realization of A in the semi-formal language, 24
Wffs	the collection of all wffs of the formal language, 47, 150
$\Gamma, \Sigma, \Delta, \ldots$	collections of propositions or wffs
L, **M**	logics
$L_{\mathbf{L}}$	the formal language for logic **L**
L/α	language L with connective α deleted, 399
γ	a formal connective, 150
$\equiv_{\text{Def}}$	equivalent by definition, used for introducing defined connectives or abbreviations
$\rightarrow$-wff	a wff with principal connective '$\rightarrow$', 49
C(A)	A is a subformula of C, 50
A(q)	q is a variable appearing in A, 50
PV(A)	the collection of all propositional variables appearing in A, 50
$[\![A]\!]$	the Gödel number of A, 264

Connectives and Abbreviations in the Formal Language

$\neg$	formalization of 'not', 11
$\wedge$	formalization of 'and', 11
$\vee$	formalization of 'or', 11
$\rightarrow$	formalization of 'if . . . then . . .', 'implies', etc., 11
$\leftrightarrow$	formalization of 'if and only if', defined as $(A \rightarrow B) \wedge (B \rightarrow A)$, 34, 76
$\supset$	the material conditional, dependent on truth-values only; defined as $\neg(A \wedge \neg B)$, 83
$\equiv$	material equivalence, defined as $(A \supset B) \wedge (B \supset A)$
$\Box$	necessity operator in classical modal logics, primitive, or $\neg A \rightarrow A$, or $A \rightarrow (A \rightarrow A)$, 208–209 in $\mathbf{L_3}$, $\neg \Diamond \neg A$, 321 in $\mathbf{J_3}$, $\sim \Diamond \sim A$ or $\neg(A \rightarrow (A \wedge \neg A))$, 353, 357
$\Diamond$	possibility operator in classical modal logics, $\neg \Box \neg A$ or $\neg(A \rightarrow \neg A)$, in $\mathbf{L_3}$, $\neg A \rightarrow A$, 320–321 in $\mathbf{J_3}$, primitive or $\neg(A \rightarrow (A \wedge \neg A))$, 357
$\prec$ (strict implication hook)	formalization of 'implies' in classical modal logics; in this text '$\rightarrow$' is used, 200
$\sim$	weak negation in $\mathbf{J_3}$, 352
$\perp$	falsity, a propositional constant, 151, 305
$\mid$	Sheffer stroke, 'nand', 55
$\downarrow$	formalization of 'neither . . . nor . . .', 55
$\bigwedge$	conjunction of all the wffs in the collection, associating to the left
$\bigvee$	disjunction of all the wffs in the collection, associating to the left
R(A, B)	abbreviation of $A \rightarrow (B \rightarrow B)$, true in a model of **R** or **S** iff R(A, B), 114
D(A, B)	abbreviation of $A \rightarrow (B \rightarrow B)$ true in a model of **D** iff $s(A) \supseteq s(B)$, true in a model of **Dual D** iff $s(A) \subseteq s(B)$, 186
E(A, B)	abbreviation of $A \rightarrow (B \rightarrow B)$ true in a model of **Eq** iff $s(A) = s(B)$, 177
$\neg^n$	$\neg$ repeated n times, 282
$A \rightarrow_3 B$	abbreviation of $A \rightarrow (A \rightarrow B)$ in $\mathbf{L_3}$, 322
IA	abbreviation of $A \leftrightarrow \neg A$ in $\mathbf{L_3}$, indeterminacy, 320
©A	abbreviation of $\neg(A \wedge \sim A)$ in $\mathbf{J_3}$, indicating that the wff has a classical (absolute) truth-value, 354
D_n	wff used to determine whether a logic has n-valued semantics, 333, 339

Syntactic Consequence Relations

$\vdash A$	A is a theorem, 64
$\Sigma \vdash A$	A is a syntactic consequence of Σ, 65
$A \vdash B$	B is a syntactic consequence of $\{A\}$, 65
$\Sigma, A \vdash B$	abbreviation for $\Sigma \cup \{A\} \vdash B$
$\mathrm{Th}(\Sigma)$	the theory of Σ (the collection of all syntactic consequences of Σ), 66
$\vdash_{\mathbf{L}}^{\Box}$	a modal logic consequence relation using the rule of necessitation, 242
$\vdash_{\mathbf{J_3}, \Diamond}$	the consequence relation for $\mathbf{J_3}$ where $\Diamond$ is primitive, 363
$\vdash_{\mathbf{J_3}, \neg}$	the consequence relation for $\mathbf{J_3}$ where $\neg$ is primitive, 366

Semantic Terms

T	true, 24
F	false, 24
U	unknown, undefined, 336
M	a model
v	a valuation, 24, 130
	a model for classical logic (**PC**), 24
$\in$	abbreviation for 'is an element of', 43
(α, β)	an ordered pair, 43
$U \times V$	collection of ordered pairs, first element from U, second from V, 105
$\langle v, s \rangle$	a set-assignment model
s	a set-assignment, 99, 130
$\mathsf{S}, \mathsf{U}, \mathsf{T}, \mathsf{C}$	content sets, 130
Sub S	the collection of subsets of S, 135
$\varnothing$	the empty set, 43
$s(A)$	the content set assigned to A
$\overline{s(A)}$	the complement of $s(A)$
$\mathsf{B}, \mathsf{C}, \mathsf{A}, \mathsf{N}$	relations governing the tables for $\rightarrow, \wedge, \vee, \neg$ respectively, 132–133
$\langle v, R \rangle$	a model for **S**, 104; a model for **R**, 106
$\langle v, D \rangle$	a model for **D**, 170
W	a collection of possible worlds, 205
	a collection of states of information, 278
R	an accessibility relation, 205–206
e	an evaluation for: a classical modal logic, 205–206 the intuitionistic logic **Int**, 277 Johansson's minimal calculus **J**, 305 a many-valued logic, 317
w	a possible world, 205–206

$\langle W, R, e, w \rangle$	a possible-worlds model with designated world for a classical modal logic, 206
	a Kripke model for the intuitionistic logic **Int**, or a Kripke tree, 277, 278
$\langle W, R, e \rangle$	a possible-worlds model for a classical modal logic, 206
$\langle W, R \rangle$	a frame for a classical modal logic, 206
$\langle W, R, Q, e \rangle$	a model for the minimal intuitionistic calculus **J**, 305
Q	a collection of inconsistent states of information, 305
F	a class of frames, 206

Semantic Consequence Relations

$\vDash A$	A is valid, A is a tautology, 51, 138
$\Gamma \vDash B$	B is a semantic consequence of Γ (every model that validates every wff in Γ also validates B), 29, 51, 138
$A \vDash B$	B is a semantic consequence of A, 51, 138
$v \vDash A$	A is true in the classical model v, i.e., $v(A) = T$, 51
$<v,s> \vDash A$	A is true in model $<v,s>$, i.e., $v(A) = T$
$<v,B> \vDash A$	A is true in model $<v,B>$, i.e., $v(A) = T$
$M \vDash A$	A is true in model M, 151
$w \vDash A$	A is true in (at) world w, 206
	the state of information w verifies A, 277
$\langle W, R, e \rangle \vDash A$	A is true in model $\langle W, R, e \rangle$, 206
$\langle W, R \rangle \vDash A$	frame $\langle W, R \rangle$ validates A, 206
$\vDash^{(j)}$, $\vDash^{*}$	definitions of validity for **G***, 268
M	a formal semantic structure, 151
$\mathbf{M}$	a class of models, 390
SA	the class of all formal set-assignment semantic structures for all formal languages, 151
RB	the class of all formal relation based semantic structures for all formal languages, 152
C(SA)	the collection of all consequence relations for SA, 152
C(RB)	the collection of all consequence relations for RB, 152

Translations

$\mathbf{L} \hookrightarrow \mathbf{M}$	there is a translation from logic **L** to logic **M**, 376
$\mathbf{L} \twoheadrightarrow \mathbf{M}$	there is a grammatical translation from logic **L** to logic **M**, 376
A^*	the translation of A by *
Γ^*	the collection of translations by * of all wffs in Γ, 375
M^*	the model correlated to M by translation *, 390

Index of Examples

Italicized page numbers indicate an example presented in the full example format, described on p. 34. Examples presented only in the exercises are not listed.

Index

- *Italicized* page numbers indicate a definition, theorem, or quotation.
- Numbers in parentheses indicate the pages of the cited section.
- Only exercises in which new material is introduced are indexed; they are marked by an x .
- The Summary of Logics is not indexed.